LEONHARDI EULERI

OPERA OMNIA

LEONHARDI EULERI
OPERA OMNIA

SUB AUSPICIIS
ACADEMIAE SCIENTIARUM NATURALIUM
HELVETICAE

EDENDA CURAVERUNT

HANS-CHRISTOPH IM HOF
WALTER HABICHT
THOMAS STEINER
GUSTAV ANDREAS TAMMANN

SERIES SECUNDA
OPERA MECHANICA ET ASTRONOMICA
VOLUMEN TRICESIMUM PRIMUM

AUCTORITATE ET IMPENSIS
ACADEMIAE SCIENTIARUM NATURALIUM HELVETICAE

BASILEAE MCMXCVI

VENDITIONI EXPONUNT
BIRKHAUSER BASILEAE

LEONHARDI EULERI

COMMENTATIONES MECHANICAE ET ASTRONOMICAE AD PHYSICAM COSMICAM PERTINENTES

EDIDIT

ERIC J. AITON †

APPENDICEM ADDIDIT

ANDREAS KLEINERT

AUCTORITATE ET IMPENSIS
ACADEMIAE SCIENTIARUM NATURALIUM HELVETICAE

BASILEAE MCMXCVI

VENDITIONI EXPONUNT
BIRKHAUSER BASILEAE

Supported in part by the Swiss National Science Foundation

The CIP catalogue record for this book is No. 14–18992
at the Library of Congress, Washington, D.C., USA

Die Deutsche Bibliothek – CIP-Einheitsaufnahme

Euler, Leonhard:
[Opera omnia]
Leonhardi Euleri opera omnia / sub auspiciis Academiae
Scientiarum Naturalium Helveticae. – Basileae : Birkhäuser.
Ser. 2, Opera mechanica et astronomica / ed. curaverunt
Hans-Christoph Im Hof ...
NE: Im Hof, Hans-Christoph [Hrsg.]; Euler, Leonhard: [Sammlung]
Vol. 31. Commentationes mechanicae et astronomicae ad
physicam cosmicam pertinentes / ed. Eric J. Aiton. – 1996
ISBN 3-7643-1459-1
NE: Aiton, Eric J. [Hrsg.]

Printed on acid-free paper produced from chlorine-free pulp TCF ∞
Typography: Albert Gomm SGD, Basel
Printed in Switzerland
ISBN 3-7643-1459-1
ISBN 0-8176-1459-1

9 8 7 6 5 4 3 2 1

VORWORT

Der vorliegende Band beschließt die zweite Serie der Opera omnia Leonhardi Euleri. Die redaktionellen Arbeiten haben sich so verzögert, daß es dem wissenschaftlichen Herausgeber, ERIC J. AITON, nicht vergönnt war, das Erscheinen des Bandes zu erleben.

Kurz vor der Drucklegung ist eine anonym erschienene Schrift über die Schwerkraft als eine Arbeit EULERS erkannt worden. Diese Schrift, die im ENESTRÖMschen Verzeichnis nicht aufgeführt ist, wird hiermit, zusammen mit einem Kommentar von ANDREAS KLEINERT, in die Opera omnia aufgenommen.

Basel, im Dezember 1994

Die Redaktion

INTRODUCTION

On several occasions during the period of his scientific career, EULER contributed papers on themes relating to cosmic physics. These are assembled in the present volume. The first was written soon after his arrival in St. Petersburg in 1727 and contained an analysis of the constitution of the atmosphere based on the ideas of JOHANN BERNOULLI. Over fifty years later, he presented a new paper on the same subject to the St. Petersburg Academy of Sciences. Another paper dating from the time of his first stay in St. Petersburg was an essay on the tides, which shared the prize awarded by the Paris Academy of Sciences in 1740. This was supplemented by a paper presented to the St. Petersburg Academy in 1775, in which he employed more recently discovered techniques to simplify the mathematical demonstrations. Meanwhile in 1748 he began to compose an essay on the detection of ocean currents with the intention of entering it for the prize of the Paris Academy of Sciences in 1749, but this remained incomplete and unpublished during his lifetime. In 1746 he wrote a review of D'ALEMBERT's prize essay on the winds, which also remained unpublished.

In 1744, three years after his move to Berlin, EULER published his treatise *Theoria motuum planetarum et cometarum*[1]), which contained determinations of the orbits of the comets of 1680 and 1681 besides that which had appeared at the beginning of 1744. In the same year, he published anonymously two popular tracts on comets in German. These took the form of a response to questions that had been raised following the spectacular appearances of the most recent comet. Two years later he presented a memoir to the Berlin Academy of Sciences on the constitution of the tails of comets, in which he claimed that these had a common origin with the northern lights and the zodiacal light.

Also in 1746 EULER published a paper on the resistance of the aether to the motion of the planets and comets. This led to correspondence with KASPAR WETTSTEIN, chaplain to the Prince of Wales, which was published in the *Philosophical Transactions*, and also with Bishop ERIK PONTOPPIDAN, the Scandinavian theologian, who published the letters in one of his books. In 1762 EULER's son JOHANN ALBRECHT received honourable mention by the Paris Academy of Sciences for an essay based on his father's paper concerning the resistance of the aether.

An annular eclipse of the sun observed in Berlin on 25 July 1748 provided the data on which EULER based his argument for the existence of a lunar atmosphere, which he presented to the Berlin Academy of Sciences at the end of the same year.

Although at the time of his first stay in St. Petersburg, EULER had hoped for a collaboration with MAUPERTUIS on the resolution of problems concerning the figure of the

---

1) E 66, *Opera omnia* II 28, pp. 105–251.

earth, this did not in fact take place and he seems to have lost interest in the subject. During his second stay in St. Petersburg, however, EULER returned to the problem and in 1775 he presented a paper to the St. Petersburg Academy of Sciences, in which he attempted to explain discrepancies between theory and observation.

In the commentary that follows, a brief introduction to each topic preceeds detailed expositions of EULER's contributions to that topic. For the convenience of the reader, many of EULER's diagrams have been repeated here to illustrate the explanations and in order to prevent any confusion, they have been numbered as in the original texts.

EULER's prize essay on the tides merits special attention on two counts. First, it is the most substantial contribution included in the present volume and secondly, it contains innovative elements which have retained a significant role in the course of later developments in theories of the tides up to the present day. For these reasons, a more general account of the historical context of EULER's theory of the tides has been included at the end of this commentary.

CONSTITUTION OF THE ATMOSPHERE (E 7 and E 527)

In his letter to JOHANN BERNOULLI (1667–1748) of 5 November 1727[2]), EULER remarked that he had first explained the constitution of the atmosphere, on the basis of the theory of elasticity of his former teacher, in an essay entitled *Dissertatio physica de sono*[3]). He had published this work in Basel earlier in the year in the hope that it would secure his appointment as Professor of Physics in his native town. Having arrived in St. Petersburg, he elaborated his theory of the atmosphere in a paper entitled *Tentamen explicationis phaenomenorum aeris* (E 7).

Following JOHANN BERNOULLI[4]), EULER attributed the elasticity of the air to the centrifugal force of a subtle matter rotating inside the small spherical particles of which he believed the air to be composed. This subtle matter, he supposed, was confined within the globules of air by means of an aqueous crust, whose thickness varied according to the humidity. The expansion of the globules was restricted by the weight of those pressing down on them, so that in equilibrium, their elastic force balanced the force exerted on them by

2) R 190, *Bibliotheca mathematica*, 3rd Series, 4 (1903), 346–348. Here and in the following, a number preceded by R refers to the corresponding entry in *Opera omnia* IV A 1.

3) E 2, *Opera omnia* III 1, pp. 181–196.

4) EULER cited JOHANN BERNOULLI, "Discours sur les loix de la communication du mouvement." This essay received honourable mention in the competition for the prize of the Paris Academy of Sciences in 1724 and 1726. It was first published in 1727. *Recueil des pièces qui ont remporté les prix de l'Académie Royale des Sciences*, vol. 1 (Paris, 1752), pièce VII. See especially the "Addition", pp. 81–108.

the weight of the atmosphere above. Evidently the subtle matter[5]) was envisaged by EULER as the proper matter of the air. For in the state of greatest compression, the subtle matter entirely filled the interior but as the globule expanded, a virtual vacuum was created at the centre[6]).

When he returned to the problem over fifty years later, in his paper[7]) *Coniectura circa naturam aeris pro explicandis phaenomenis in atmosphaera observatis* (E 527), EULER perceived his major innovation to be a clear explanation of the decrease in pressure which is observed as the humidity increases. At the time of the original memoir, he noted, the theory of fluids had not been sufficiently refined for him to be in a position to give an explanation. As in the earlier memoir, EULER supposed the globule of air to consist of three parts, though these are described rather differently. Beneath the outer crust of aqueous matter is a middle shell, consisting of the proper matter of the air[8]), and an inner core of aether[9]), destitute of gravity and exciting the proper matter of the air to rapid motion, more or less according to the temperature. The centrifugal force engendered by the circulation of the proper matter of the air produced the elasticity or pressure of the air.

E 7

Tentamen explicationis phaenomenorum aeris

After giving an outline of his theory, according to which the globules of air consist of an aqueous crust enclosing a middle section of rotating subtle matter with a virtual vacuum at the centre (§§1–9), EULER first considers a globule in the state of greatest compression (§11). Let the radii of the outer and inner surfaces of the aqueous crust (fig. 1)

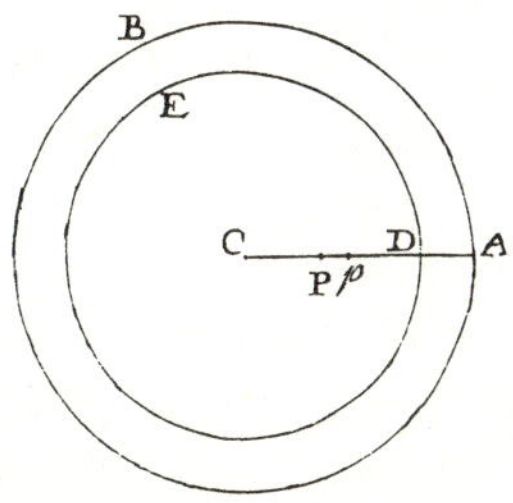

Fig. 1

5) The concept of small elastic vortices was first introduced by NICOLAS MALEBRANCHE (1638–1715) to describe the particles of the elastic medium for the transmission of light; in other words, the aether. For his definitive account, see his *Recherche de la vérité*, 6th edition (Paris, 1712), vol. 4, pp. 435–544.

6) A short review of EULER's theory, in the form of a summary, was published in the *Acta Eruditorum*, July 1731, pp. 304–305.

7) Presented to the St. Petersburg Academy on 14 August 1780.

8) This corresponded to the subtle matter of the earlier account.

9) EULER used the term *aether*. In the earlier account the core was described as a virtual vacuum.

be g and h respectively. Then, taking the specific gravity of the crust to be m and that of the subtle matter to be n, the mass of the subtle matter[10] $= \frac{4}{3}\pi nh^3$ and the mass of the crust $= \frac{4}{3}\pi m(g^3 - h^3)$. The centrifugal force of the subtle matter of the globule $= 4\pi nkh^2$, where the speed of the subtle matter is that acquired by a heavy body falling freely from a height k (§12).

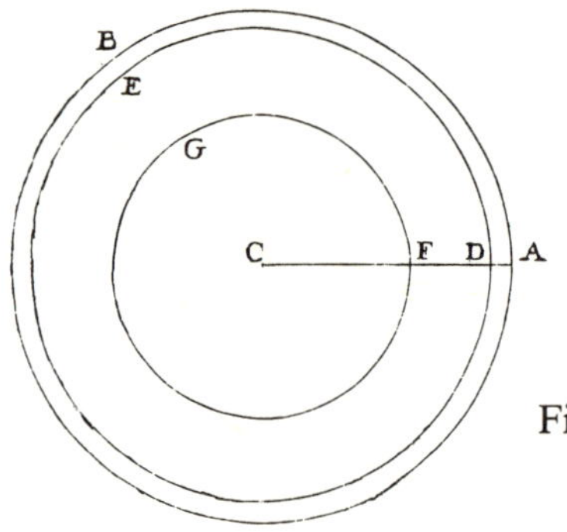

Fig. 2

When the globule expands (fig. 2), it consists of three parts, the aqueous crust, the middle shell of rotating subtle matter and the inner core of virtually empty space (§13). Taking the outer radii of the three parts as a, b and c respectively, the weights[11] of the crust and middle part are $\frac{4}{3}\pi m(a^3 - b^3)$ and $\frac{4}{3}\pi n(b^3 - c^3)$. If i is taken to be the specific gravity of the air, then the weight of the whole globule $= \frac{4}{3}\pi ia^3$, from which it follows that $ia^3 = m(a^3 - b^3) + n(b^3 - c^3)$. Also, from the condition that the quantities of matter in the crust and interior are the same whether the globule is expanded or compressed, it follows that (§14)

$$b^3 = \frac{(i - m + n)\,a^3 - ng^3}{n - m} \quad \text{and} \quad c^3 = a^3 - g^3.$$

Substituting these values in the last equation gives $h^3 = (ia^3 - mg^3)/(n - m)$. In this way the dimensions of the internal parts of the globule are eliminated from the calculation.

Since the centrifugal force of the subtle matter in the expanded globule is $4\pi nk(b^2 - c^2)$, on substituting for b and c, and putting $h^3 = pg^3$, the centrifugal force is found to be equivalent to a weight (§15)

$$\frac{4\pi nka^2}{(m - pm + pn)^{2/3}}\left((m - i + pi - pm + pn)^{2/3} - (m - pm + pn - i)^{2/3}\right).$$

As this force acts over the surface of the globule, whose area is $4\pi a^2$, the force on unit area, or in other words the pressure, is (§§16–17)

---

10) In this commentary π represents the ratio of the circumference of a circle to its diameter. At this time, however, EULER used the symbol to denote the ratio of the circumference to the radius.

11) EULER uses the term *weight*, although in the previous paragraph he had used the term *mass*.

$$\frac{nk}{(m-pm+pn)^{2/3}}\left((m-i+pi-pm+pn)^{2/3}-(m-pm+pn-i)^{2/3}\right).$$

EULER applies this result first to the case of dry air, when the aqueous crust is vanishingly thin, so that g may be taken equal to h and consequently p equal to 1. In this case the pressure is found to be proportional to $n^{2/3}-(n-i)^{2/3}$, supposing k to be a constant (§18). If the air is also nearly in its state of greatest compression, $n-i$ will be almost zero, so that the pressure will be almost constant. On the other hand, if the air is extremely rarefied, in which case i is small compared with n, the pressure will be proportional to the density i (§19). Thus EULER's theory agrees with the law established by ROBERT BOYLE (1627–1691) in the case of very rarefied air.

If the air is not quite so rarefied, the pressure may be taken to be proportional to $6ni+i^2$. Then using the experiments of BOYLE, it should be possible to determine the ratio of n to i and in this way the greatest and least densities to which the air can be compressed (§20). Using the first entry in the table of BOYLE's experiments which he transcribes (§21), EULER finds that $n=(8v-233i^2)/(1398i-48v)$, where v is the pressure. Substituting the value of v for the last entry, when $i=4$, gives $n=54{,}64$. For $i=3$, the value of n becomes $58\frac{1}{12}$, but for $i=2$ it is only 17 (§24). Instead of discarding his theory of the constitution of the air in view of this huge discrepancy, EULER concludes that the experiments were not sufficiently accurate, and he suggests that the tube may have been very narrow so that the mercury stuck to the side (§25)[12]).

Following a description of some geometrical constructions for his equations (§§26–31), EULER draws further inferences from his theory. Equating the weight of the column of mercury to the total elastic force of the hemisphere of the globule on which it presses, he arrives at the equation

$$f=\frac{nk}{r(m-pm+pn)^{2/3}}\left((m-i+pi-pm+pn)^{2/3}-(m-i-pm+pn)^{2/3}\right),$$

where f is the height of the column of mercury[13]) and r the specific gravity of mercury (§§32–33). Again taking i small in comparison with n, which EULER believes to correspond with the natural state of the air (§34), the expression reduces to

$$f=\frac{2pnik}{3r(m-pm+pn)}.$$

For dry air, p may be taken equal to 1, so that $f=2ik/(3r)$. In general, putting $p=1-q$,

12) In fact the product of pressure and volume varies only between 349 and 354.

13) The height of the mercury barometer should be included in f. That this was EULER's intention seems implicit in his statements in other passages.

$$f=\frac{2ik}{3r}\left(1-\frac{qm}{qm+n(1-q)}\right).$$

As q increases, the thickness of the aqueous crust, and hence the humidity, increases. For air in its natural state, f represents the height of the barometer (§35). EULER deduces that the barometer ascends when the humidity decreases and falls when the humidity increases, which corresponds to experience (§§36–37). If it sometimes happens that low pressure is not accompanied by great humidity, the reason is that the height of the barometer depends on the state of the whole atmosphere, whereas the hygrometer only measures the local humidity (§38).

Finally, EULER observes that, when the air is dry, the value of k, and hence the speed of the subtle matter, can be derived from the height of the barometer (§39). He finds this speed to be $1518\frac{1}{2}$ Rhenish feet in one second (§40).

E 527

Coniectura circa naturam aeris, pro explicandis phaenomenis in atmosphaera observatis

Having described the general structure of the globules of air, consisting of an outer aqueous crust, middle section of proper matter of the air and inner core of aether destitute of gravity (§§1–3), EULER notes that observation of water full of foam and of soap-bubbles indicates the reality of the aqueous crust. From these observations he infers that, if vapours ascend in the atmosphere, they surround the elements of air like skins, so that the more vapour there is, the thicker the skins become. It may happen, however, that vapour becomes confusedly dispersed inside the globules of air. In this case it disturbs the refraction and passage of rays of light, giving rise to fog in the lower atmosphere and to clouds above (§4).

Two reasons are advanced by EULER in support of the circular motion of the proper matter of the air. First, there is no other way of explaining the elasticity. Secondly, since heat consists of motion, the matter must retain some motion in the globules and, in the confined space, must therefore circulate (§5). To explain the equal pressure in all directions, the proper matter of the air must move in great circles. Yet such motion is incompatible with the principles of mechanics (§§6–7). EULER shows, however, that although there are collisions between particles moving in great circles, the centrifugal force, directed away from the centre of the globule, is the same as if the individual particles moved unhindered in great circles (§§9–11). Consequently the centrifugal force of a particle of unit mass at a distance x from the centre $=c^2/(2gx)$, where c is the speed and g the vertical distance fallen by a heavy body in 1 second (§8)[14].

Let the radius of the globule (fig. 2) be r, the inner radius of the aqueous crust be s and the radius of the core of aether be t. Suppose now that the proper matter of the air

14) EULER's g is half the acceleration due to gravity at the earth's surface.

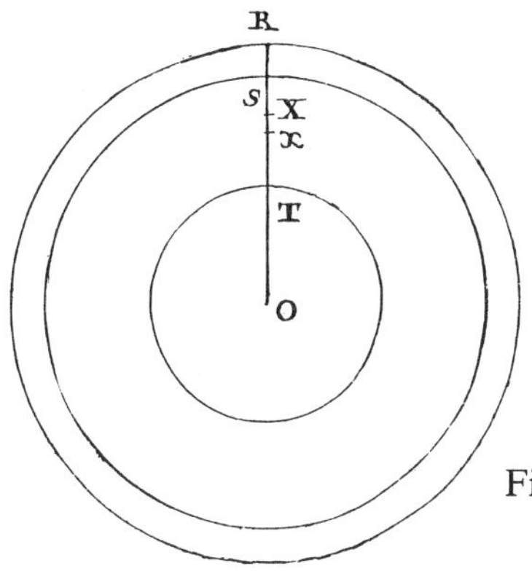

Fig. 2

is completely compressed and that in this state, it has the same density as water (§ 14). Let this density be 1. Then the mass of the globule $= \frac{4}{3}\pi(r^3 - t^3)$. Let the pressure in X be p and that in x be $p + dp$. Then EULER supposes that dp balances the centrifugal force by which the element Xx is repelled, so that $dp/dx = c^2/(2gx)$, which on integration gives[15])

$$p = \frac{c^2}{2g} \log \frac{x}{t},$$

taking $p = 0$ when $x = t$. It follows that, at the inner surface of the aqueous crust, the pressure

$$p = \frac{c^2}{2g} \log \frac{s}{t}.$$

This is the pressure by which the aqueous crust and the whole globule endeavours to expand, and in fact expands unless compressed by a similar force (§ 16).

EULER now supposes that the quantity of matter in a globule is constant and equal to that of a water globule of radius a (§ 17). It follows that, if the mass of water in the aqueous crust is $\frac{4}{3}\pi a^3 \lambda$, the mass of proper matter of air is $\frac{4}{3}\pi a^3(1-\lambda)$. Hence $r^3 - s^3 = \lambda a^3$ and $s^3 - t^3 = (1-\lambda)a^3$, so that $s^3 = r^3 - \lambda a^3$ and $t^3 = r^3 - a^3$. Substituting these values in the formula for the pressure, this becomes (§ 18)[16])

$$p = \frac{c^2}{6g} \log \frac{r^3 - \lambda a^3}{r^3 - a^3}.$$

15) In the earlier memoir, instead of basing his calculation on the dynamics of the rotating fluid, he simply calculated the total centrifugal force. As he remarked at the beginning, the theory of fluids was not sufficiently refined at the time he composed the original memoir.

16) Thus the humidity λ enters directly into the formula for p, which facilitates the promised explanation of the decrease in pressure as the humidity increases.

In this way, the internal dimensions of the globule are eliminated. Moreover, since the mass of a globule $=\frac{4}{3}\pi a^3$ and its volume $=\frac{4}{3}\pi r^3$, the density of air $q = a^3/r^3$, so that the formula for the pressure becomes (§ 19)[17])

$$p = \frac{c^2}{6g} \log \frac{1-\lambda q}{1-q}.$$

Since q is very small, p may be taken as

$$\frac{c^2}{6g}\left((1-\lambda)q + \tfrac{1}{2}(1-\lambda^2)q^2 + \tfrac{1}{3}(1-\lambda^3)q^3\right)$$

and for most purposes the first term is sufficient (§§ 20–21). For example, when the humidity $\lambda = 0$ and the atmospheric pressure is a maximum, say 34 feet in the water barometer, $q = 1/1000$ and the first term gives $c = 1790$ feet per second, assuming the value of $6g$ to be $93\frac{3}{4}$ Rhenish feet. Since these are the conditions when the temperature of the air is a maximum, EULER takes this speed of circulation of the proper matter of the air inside the globule to correspond with 100° on the DELISLE scale of temperature (§ 24)[18]). Similarly, assuming the conditions for the coldest air temperature (200° on the DELISLE scale), namely $p = 31$ feet, $q = 1/700$ and again taking $\lambda = 0$, c is found to be 1430 feet per second (§ 25). Assuming a linear relationship, EULER is able to assign a speed to any temperature. In particular, the speed corresponding to the boiling point of water (0° on the DELISLE scale) is found to be 2150 feet per second (§ 30). EULER's formula agrees with experience. For, according to the formula, the speed increases when the pressure increases, while the boiling point is observed to increase when the pressure increases. Although the speed is large, EULER notes that incomparably greater speeds exist in nature, such as the speed of rays of light (§ 31).

It seemed to EULER that a combination of theory and experiment would serve not only to confirm the theory but also to improve the precision of the measuring instruments. For example, experiments were needed to assign speeds more accurately to the particular degrees of the DELISLE scale of temperature (§ 25), while the effect of the variable density of mercury on the height of the barometer needed to be investigated and allowed for (§ 26).

17) As $c^2/(6g)$ will be shown to be a measure of temperature, this equation relates the pressure, temperature, density and humidity.

18) DELISLE's scale of temperature is based on one fixed point (the boiling point of water) and the contraction of the mercury relative to the containing vessel. It runs from 0° at the top (the boiling point of water), through 150° (the melting point of ice) to about 200°, the coldest air temperature. The scale is described in JOSEPH NICOLAS DE L'ISLE (1688–1768), *Mémoires pour servir à l'histoire et au progrès de l'astronomie, de la géographie et de la physique* (St. Petersburg, 1738), pp. 267–284. See also W. E. KNOWLES MIDDLETON, *A history of the thermometer* (Baltimore, 1966), pp. 87–89 and p. 96.

Again, the degree of accuracy of existing hygrometers was still very low (§28). Finally, the manometer designed by PIERRE VARIGNON[19]) (1654–1722) also needed much development. This instrument, measuring the density of the air and seen by EULER as important for the exploration of his theory, was at that time regarded as almost obsolete (§27).

Although the formula can be used to determine any one of the four elements (pressure, temperature, density and humidity) given the values of the other three (§§32–35), EULER considers that it is especially suited to the resolution of the most important problem of air measurement, namely the determination of the force needed to compress a given quantity of air into a given volume (§36).

Let the initial density of the air be equal to b and the initial pressure equal to a. Then

$$a = h \log \frac{1-\lambda b}{1-b},$$

or $a = (1-\lambda)hb$ approximately, where $h = c^2/(6g)$ (§37). Suppose now that the air is compressed into a volume n times less. Then the density $q = nb$, so that

$$p = h \log \frac{1-\lambda nb}{1-nb}.$$

If n is not very large, p may be taken equal to $h(1-\lambda)nb$ and it follows that $p = na$. In this case, therefore, the prediction is in agreement with BOYLE's law (§38). For very large compressions, however, the logarithmic formula

$$p = \frac{1}{(1-\lambda)b} \log \frac{1-\lambda nb}{1-nb}$$

must be used, so that p is generally greater than na, and indeed when $nb = 1$, an infinite force is needed (§40). Taking $a = 1$ atmosphere and $b = 1/800$, corresponding to the natural state of the air, and also assuming the humidity $\lambda = 0$, EULER calculates a table of values of p and n up to the limit $n = 800$ (§42)[20]). The effect of the humidity, he notes, is to increase the values of p still further (§43).

Finally, EULER considers the variation in the state of the atmosphere with altitude (§44). Let the heights of the water barometer at A and Z (fig. 3) be a and p respectively, the corresponding densities b and q, the humidities λ and φ, and the speeds c and u.

Fig. 3

19) P. VARIGNON, "Manometre ou machine pour trouver le raport des raretés ou raréfactions de l'air naturel," *Mémoires de l'Académie Royale des Sciences* (Paris), (1705), 1706, pp. 300–331. See also CHRISTIAN WOLFF (1679–1754), *Aerometricae elementa* (Leipzig, 1709), pp. 282–285.

20) Apart from the first few entries, the table had little practical relevance.

Also let $c^2/(6g) = h$ and $u^2/(6g) = v$. The values of c, of course, are related to the temperature and EULER includes a table of corresponding values of c and h (§45). Since q is small, the first term of the series for p suffices, so that $p = v(1-\varphi)q$. Now let the altitude AZ be z. Then, if the pressure at z is $p + dp$, then $dp = -qdz$, so that

$$-\frac{dp}{p} = \frac{dz}{v(1-\varphi)}.$$

On using the condition that $p = 0$ when $z = 0$, the integral becomes (§46)

$$\int \frac{dz}{(1-\varphi)v} = \log\frac{a}{p}.$$

EULER notes that, as only the ratio of a to p is involved on the right-hand side, the height of the mercury barometer may be used in place of that of the water barometer. To facilitate the use of the formula, he gives a table of hyperbolic logarithms for the numbers 36 to 6.5, corresponding to the relevant heights in inches of the mercury barometer (§47).

Since the temperature decreases with altitude, EULER adopts the working hypothesis that $v = fh/(f+z)$, so that in altitude f, the speed is reduced to half the speed at sea level. Substituting this value of f and also assuming the humidity to be constant, so that $\varphi = \lambda$, the equation on integration becomes (§§49–50)

$$\frac{1}{h(1-\lambda)}\left(z + \frac{z^2}{2f}\right) = \log\frac{a}{p}.$$

Then substituting $vz/(h-v)$ in place of f, after rearrangement this becomes (§51)

$$z = \frac{2(1-\lambda)hv}{h+v}\log\frac{a}{p},$$

where v is derived from the observed temperature at the altitude z.

By means of this formula, therefore, the altitude of the place can be derived from readings of the barometer and thermometer (§52). Because of the uncertainty in the values of c and u, however, EULER recommends that the formula should first be used in cases where the altitude is already known, in order to obtain more accurate values of c and u, and thereby improve the precision of the thermometer (§53).

At the beginning of his paper EULER remarked that, even if his hypothesis of the constitution of air should have to be regarded as fictitious, in the same sense as the hypothesis of universal attraction, it was nevertheless useful on account of its capacity to explain the phenomena and in particular, the relationship between pressure and humidity, which, he claimed, no one else had clearly explained. A different hypothesis had been proposed by DANIEL BERNOULLI in his *Hydrodynamica*[21]) and defended in a letter to EULER of 20 March 1745[22]). Although BERNOULLI took no account of the humidity, he introduced for the first time a kinetic theory of the constitution of the air, which enabled him to explain the basic gas laws and to anticipate, though in an incomplete form, the VAN DER WAALS equation of state.

THE THEORY OF THE TIDES (E 57 and E 546)

Having dispatched his essay, *Inquisitio physica in causam fluxus ac refluxus maris* (E 57), to the Paris Academy of Sciences on 14 July 1739, EULER became concerned when he had not received any acknowledgement by the last date for submission of entries for the prize[23]). On 2 October 1739, he therefore wrote to BERNARD LE BOVIER DE FONTENELLE (1657–1757), the permanent secretary, offering to send a copy if the original had not arrived[24]). The original manuscript in fact arrived in January 1740 and was only allowed to compete for the prize after the intervention of the President of the St. Petersburg Academy of Sciences through the Russian ambassador in Paris[25]).

Writing to EULER on 14 April 1740, PIERRE-LOUIS MOREAU DE MAUPERTUIS (1698–1759)[26]) informed him that his essay was one of four to share the prize. The authors of the other prize essays were COLIN MACLAURIN (1698–1746), DANIEL BERNOULLI (1700–1782) and an 'unknown' Cartesian ANTOINE CAVALLERI (1698–1765)[27]). Like MACLAURIN and BERNOULLI, EULER based his explanation of the tides on the Newtonian principle of universal gravitation. In a letter of 30 April 1740, BERNOULLI told EULER that his essay had been especially praised for the determination of how the figure of the earth is changed by the action of the moon and also for the way in which the inertia of the water is skilfully drawn into consideration[28]).

21) DANIEL BERNOULLI, *Hydrodynamica* (Strasbourg, 1738), Sectio decima, pp. 200–203.

22) R 158, PAUL HEINRICH FUSS, *Correspondance mathématique et physique de quelques célèbres géomètres du XVIIIeme siècle* (St. Petersburg, 1843), vol. 2, pp. 573–575.

23) The last date for the receipt of entries for the prize was 1 September 1739.

24) R 604, *Leonard Ejler, Pis'ma k učenym*, ed. T. N. KLADO et al. (Moscow and Leningrad, 1963), pp. 271–274.

25) EULER, *Opera omnia* IV A 5, p. 80, note 3.

26) R 1498, EULER, *Opera omnia* IV A 6, pp. 44–45.

27) ANTOINE CAVALLERI was a Jesuit professor of mathematics and theology who had received prizes from the Bordeaux Academy in 1738 and 1739 for essays on transparency and opacity of bodies and on the cause of heat.

28) R 136, P. H. FUSS, *Correspondance*, vol. 2, pp. 458–460.

In the autumn of 1740, ALEXIS-CLAUDE CLAIRAUT (1713–1765) initiated a correspondence with EULER, who responded[29] by praising the papers on the figure of the earth which CLAIRAUT had published in London[30] and referring to similar studies of his own[31]. By means of CLAIRAUT's theory and a knowledge of the true ratio of the axis of the earth to the diameter of the equator, he remarked, it was possible to draw conclusions concerning the internal structure of the earth[32]. A particular merit of EULER's essay, CLAIRAUT observed, was its illustration of the art of approximations in the kind of studies on which they were both engaged[33].

The first serious criticism of EULER's theory was made at the beginning of 1742 by DANIEL BERNOULLI[34]. Although he was displeased with EULER at the time, in consequence of a clash of opinions on the scientific merits of CHRISTIAN GOLDBACH (1690–1764), his complaint concerning the way in which EULER had treated NEWTON's theory was justified. According to EULER, NEWTON had overestimated the height of the tides as a result of erecting a superstructure on a very weak foundation. As BERNOULLI pointed out, however, NEWTON had taken account of the attraction of the heaped-up water, whereas EULER had not. By neglecting this factor, he added, EULER had made the problem easy. On the other hand, both he and MACLAURIN had estimated the effect of this attraction and arrived at results in agreement with those of NEWTON.

On 13 June 1744, DANIEL BERNOULLI[35] informed EULER of the publication in Geneva of THOMAS LE SEUR and FRANÇOIS JACQUIER's edition of NEWTON's *Principia*, with the three prize essays on the tides[36] as an appendix. A year later he sought the help of EULER in demonstrating from the phenomena of the tides that his calculation of the position of the centre of gravity of the earth and the moon was more accurate than that of NEWTON[37].

29) R 387, EULER, *Opera omnia* IV A 5, pp. 71–78.

30) A. C. CLAIRAUT, "Investigationes aliquot, ex quibus probetur terrae figuram secundum leges attractionis in ratione inversa quadrati distantiarum maxime ad ellipsin accedere debere," *Philosophical Transactions* 40 (1737–38), 19–25; "An enquiry concerning the figure of such planets as revolve about an axis, supposing the density continually to vary from the centre towards the surface," *ibid.*, pp. 277–306.

31) Evidently his theory of the tides.

32) EULER explains to DANIEL BERNOULLI, however, that CLAIRAUT's inference on the hypothesis that the density varies from the centre of the earth to the surface has no bearing on his theory of the tides, as he is only concerned with the matter at the surface. Letter of EULER to D. BERNOULLI, 15 September 1740; R 138, *Bibliotheca mathematica*, 3rd Series, 7 (1906–1907), 145–153.

33) Letter of CLAIRAUT to EULER, 26 December 1740; R 388, EULER, *Opera omnia* IV A 5, pp. 78–81.

34) Letter of D. BERNOULLI to EULER, 20 January 1742; R 143, P. H. FUSS, *Correspondance*, vol. 2, pp. 479–483.

35) R 155, P. H. FUSS, *Correspondance*, vol. 2, pp. 555–560.

36) That is, the essays of BERNOULLI, EULER and MACLAURIN.

37) Letter of D. BERNOULLI to EULER, 7 July 1745; R 160, P. H. FUSS, *Correspondance*, vol. 2, pp. 576–578.

In his memoir, *De statu equilibrii maris a viribus solis et lunae sollicitati* (E 546), presented to the St. Petersburg Academy of Sciences in 1775, EULER expounded the theory of the tides[38]) on the basis of the principle that, in the state of equilibrium, the potential (which he described as the action of the forces) is constant over the surface of the water.

E 57

Inquisitio physica in causam fluxus ac refluxus maris

Chapter 1

While he acknowledges that ISAAC NEWTON (1642–1727) had assigned the true cause of the tides in the *Principia*, EULER claims that NEWTON had only sketched the details, whereas his own analysis demonstrates the perfect agreement of the observed phenomena with theory (§15). EULER accepts that the gravitational forces of the sun and the moon had been established from the phenomena (§§7–9) but he rejects the concept of an attraction at a distance. However, in following the Cartesians by attributing the gravitational forces to the actions of a solar and a lunar vortex, he remarks that, although these actions are still to some extent obscure, there seems to be a general agreement that a new investigation of the cause of gravity would not be profitable (§§12–14). From the properties of the solar vortex[39]), he deduces that the solar force at a distance x from the centre of the sun is S/x^2, where $S = 227\,512$ times the natural gravity at the surface of the earth. For the corresponding lunar force L/x^2, he adopts NEWTON's[40]) value of $l = 1/40$, deduced from a comparison of the heights of the solar and lunar tides (§14).

Before expounding his own theory, EULER reviews the various explanations of the tides that had already been proposed. First, he claims to refute GALILEO's[41]) theory,

38) This theory, he claimed, had been completely established in the three essays awarded a prize by the Paris Academy of Sciences in 1740. He made no mention of the prize essay of CAVALLERI.

39) The first mathematical analysis of vortex motion was made by CHRISTIAAN HUYGENS (1629–1695) between the presentation of his vortex theory of gravity to the Paris Academy of Sciences in 1669 and the publication of NEWTON's *Principia* in 1687. See CHRISTIAAN HUYGENS, *Discours sur la cause de la pesanteur*, in *Oeuvres complètes* (La Haye, 1888–1950), vol. 21. Cf. E. J. AITON, "The Cartesian theory of gravity", *Annals of Science* 15 (1959), 27–49. Attempts to reconcile vortex motion with an inverse-square law of force were made by PHILIPPE VILLEMOT (1651–1713), *Nouveau système ou nouvelle explication du mouvement des planètes* (Lyon, 1707) and JOSEPH PRIVAT DE MOLIÈRES (1677–1742), "Loix générales du mouvement dans le tourbillon sphérique," *Mémoires de l'Académie Royale des Sciences* (Paris), (1728), 1730, pp. 245–267. Cf. E. J. AITON, *The vortex theory of planetary motions* (London and New York, 1972).

40) ISAAC NEWTON, *Philosophiae Naturalis Principia Mathematica* (London, 1687), Book III, proposition 37, corollary 4.

41) GALILEO GALILEI (1564–1642), *Dialogo sopra i due massimi sistemi del mondo* (Florence, 1632). The Fourth Day.

according to which the tides are caused by the combination of the diurnal and annual motions of the earth. For if the earth had only its diurnal motion, he explains, this would not affect the sea in any way, except to elevate it under the equator, from which cause the earth in fact receives the figure of an oblate spheroid. He adds that, if the earth is then given a uniform motion in a straight line, in addition to the rotation, the figure will remain the same, because that "uniform motion in a straight line will be impressed equally on all parts of a system of bodies, so that it can cause absolutely no change in the motion and situation of parts relative to one another" (§3). Finally, he states without further justification, that even in the case when the earth moves in a circle or an ellipse, no motion of reciprocation of the sea can be produced[42]). However, when EULER adds that the combination of the two motions is the cause of the supposed universal east-west current of the ocean (§4), which had been generally assumed from antiquity[43]), his rejection of the principle underlying GALILEO's theory becomes somewhat equivocal[44]). It is not the motion of the earth in itself that could give rise to this effect, any more than to the tides, but rather the forces of gravitation. In fact EULER deduces the effect from the imaginary hypothesis that the particles of water are, as it were, freely orbiting the sun.

The theory of DESCARTES is dismissed by EULER on the grounds that the pressure of the moon on the ocean (transmitted through the terrestrial vortex) is introduced solely for the purpose of explaining the tides, and that the theory predicts low water at a time when high water is generally observed. Moreover, he objects that DESCARTES fails to explain how the moon hidden under the earth produces the same effect as if it moved over the horizon[45]).

42) This conclusion is demonstrated in effect by DANIEL BERNOULLI in his prize essay, though he does not in fact mention GALILEO's theory. For if the earth were supposed to move in an ellipse (without rotation on its axis) about the centre of mass of the earth and the moon, each point of the earth and the ocean would move in an equal ellipse. From the equality of the ellipses, he infers that each part has the same centrifugal force and that the directions of the centrifugal forces are parallel to one another. It follows that the centrifugal forces, which are the effect of the motion in the ellipse, acting equally on every part of the ocean (and of the earth), cannot give rise to any relative motion of the parts. The same argument applies to the annual motion of the earth about the sun. D. BERNOULLI, "Traité sur le flux et reflux de la mer", *Recueil des pièces qui ont remporté les prix de l'Académie Royale des Sciences*, vol. 4 (Paris, 1752), pp. 79–80. Cf. *Die Werke von Daniel Bernoulli*, ed. D. SPEISER et al., vol. 3, (Basel, 1987), pp. 348–349.

43) For the history of observations and explanations of this supposed phenomenon, see GIAMBATTISTA RICCIOLI (1598–1671), *Almagestum novum* (Bologna, 1651), vol. 1, part 1, pp. 72–74 and part 2, pp. 363–365. Cf. BERNHARDUS VARENIUS (1622–1650), *Geographia generalis* (Amsterdam, 1650), Book II, chapter 14, proposition 7, pp. 178–179.

44) The literature is bedevilled by false assertions of a *Galileo effect* of the earth's double motion, despite the clear and decisive demonstration of DANIEL BERNOULLI that this double motion cannot give rise to any relative motion of the earth and the ocean.

45) This last criticism is not well-founded. DESCARTES explains the simultaneous tides on opposite sides of the earth by a displacement of the earth from the centre of the terrestrial vortex. RENÉ DESCARTES (1596–1650), *Principia philosophiae* (Amsterdam, 1644), part 4, §49. *Oeuvres de Descartes*, ed. CHARLES ADAM and PAUL TANNERY, vol. 8 (Paris, 1905).

Chapter 2

In this chapter EULER calculates the magnitudes of the tide-generating forces. First, he explains that, although the absolute value of the gravitational attraction of the moon is much smaller than that of the sun, nevertheless the lunar tide is greater than the solar tide, because the tide-generating forces are the differences between the attraction at the different points on the earth's surface and that at the earth's centre. These differences are greater in the case of the moon than in the case of the sun (§17). Next he establishes that a sphere, in which the density depends only on the distance from the centre, is attracted as if the whole mass were concentrated at the centre (§§19–21). Although the earth is not a perfect sphere, he considers the difference from this to be of no consequence in the calculation of the tide-generating forces. To investigate the relative motion, EULER in fact reduces the earth to rest by applying to the earth and the ocean a force equal and opposite to the force of attraction at the centre of the earth (§§22–23).

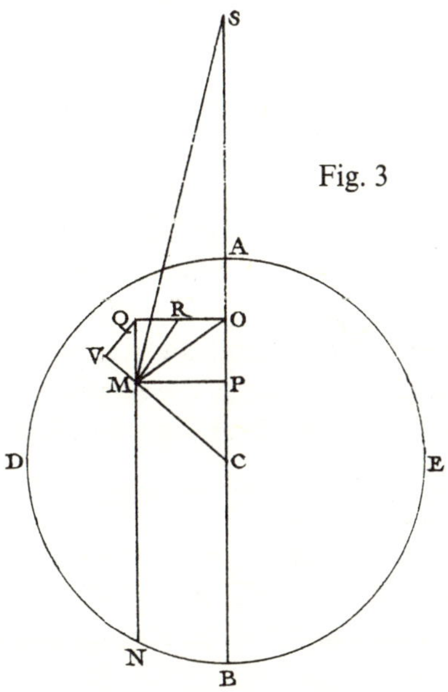

Fig. 3

Let $CS = a$, $CM = r$ and angle $MCS = \varphi$ (fig. 3)[46]. Then the attraction of the sun at the point M (§24) is $S/(a^2 - 2ar\cos\varphi + r^2)$. Resolving into components in the directions MQ and MC respectively, and subtracting S/a^2, the attraction at the centre of the earth, from the former, the disturbing forces in the directions MQ and MC are found to be

46) EULER does not use the abbreviations for the trigonometrical ratios in these calculations, though he introduces them later in the essay. For greater clarity in the commentary, EULER's formulae are transformed into the trigonometrical notation.

$$\frac{Sa}{(a^2 - 2ar\cos\varphi + r^2)^{3/2}} - \frac{S}{a^2} \quad \text{and} \quad \frac{Sr}{(a^2 - 2ar\cos\varphi + r^2)^{3/2}}.$$

EULER now resolves these forces in the directions MC and MR, thus obtaining the following formulae for the vertical and horizontal components of the tide-generating forces (§26):

Vertical component (downwards) $= \dfrac{S\cos\varphi}{a^2} - \dfrac{S(a\cos\varphi - r)}{(a^2 - 2ar\cos\varphi + r^2)^{3/2}}$

Horizontal component (in direction of φ decreasing) $= \dfrac{Sa\sin\varphi}{(a^2 - 2ar\cos\varphi + r^2)^{3/2}} - \dfrac{S\sin\varphi}{a^2}$.

Expanding in terms of $1/a$ and neglecting powers beyond the fourth, these become (§27):

Vertical component $= \dfrac{Sr(1 - 3\cos^2\varphi)}{a^3} + \dfrac{3Sr^2\cos\varphi(3 - 5\cos^2\varphi)}{2a^4}$

Horizontal component $= \dfrac{3Sr\sin 2\varphi}{2a^3} + \dfrac{3Sr^2\sin 3\varphi}{2a^4}$.

Considering just the first term in each case (§29), it may be inferred that (at the surface of the earth, where $r = 1$), the vertical force has its greatest upward value, $2S/a^3$, when the sun is in the zenith or the nadir, and its greatest downward value, S/a^3, when the sun is on the horizon. The horizontal component, on the other hand, has its maximum value when the zenith distance of the sun is 45°. Since the earth's gravity at the surface of the earth can be taken as 1, it follows that the plumb-line should be inclined at an angle to the vertical whose tangent is $\dfrac{3S\sin 2\varphi}{2a^3}$. Taking the most favourable case of a conjunction, when the disturbing forces of the sun and moon reinforce each other, EULER shows by calculation that the predicted variation in the period of a pendulum and the inclination of a plumb-line to the vertical are many orders of magnitude too small to be observed (§30)[47]).

Chapter 3

Having calculated the disturbing forces, EULER next determines the equilibrium figure of the ocean that these forces are tending at any time to produce. Assuming that whatever changes the disturbing forces would induce in the surface of the water surrounding a spherical earth would also happen to the spheroidal form actually acquired as a result of

47) Failure to detect these effects could not therefore be used as an argument against the theory proposed.

the rotation, he is able to suppose the earth to be a perfect sphere for the purpose of determining the elevation or depression from the natural state (§33). In the equilibrium position, the resultant force is normal to the surface of the water, and as the vertical component of the disturbing force is negligible in comparison with the natural gravity, that is 1, it follows that the inclination of the surface of the water to the level depends only on the horizontal component of the disturbing force[48]). Solving the differential equation

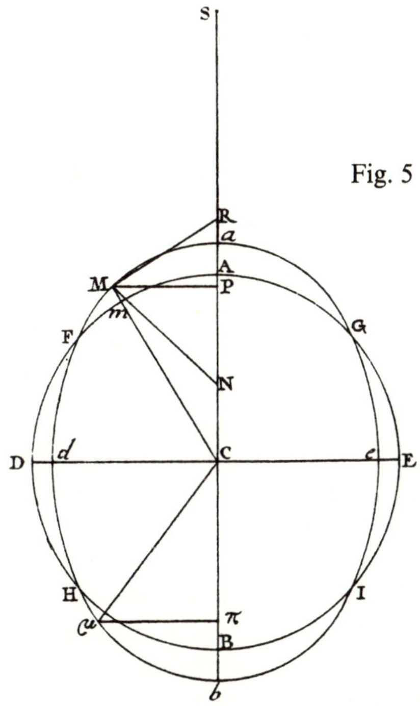

Fig. 5

obtained by equating tan CMN (fig. 5) to the horizontal component of the disturbing force, Euler obtains the equation of the free surface *adb* in the form (§35)

$$z = c + \frac{3Sc^2 \cos^2 \varphi}{2a^3} + \frac{Sc^3 \cos \varphi \, (5\cos^2 \varphi - 3)}{2a^4}$$

where $z = \mathrm{CM}$ and $\varphi = $ angle ACM. The constant c, determined by the condition that the total volume of the ocean must be the same as it would be in the undisturbed state, is found to be $1 - \frac{1}{2}S/a^3$, so that the equation of the free surface becomes (§36)

$$z = 1 + \frac{S(3\cos^2 \varphi - 1)}{2a^3} + \frac{S\cos \varphi \, (5\cos^2 \varphi - 3)}{2a^4}.$$

48) Euler's recognition that the effective tide-generating force is the horizontal component of the disturbing force was his principal contribution to the theory of the tides.

The elevation above the natural level is given by the last two terms.

In the case of the solar tide, EULER deduces (§ 38) that the maximum elevation, which occurs when the sun is in the zenith or nadir, is $S/a^3 = 0{,}5072$ or approximately half a Paris foot, and the maximum depression, which occurs when the sun is on the horizon, is about a quarter of a foot. These are the values when the sun is at its mean distance. In the apogee or perigee the effect is diminished or increased respectively in the proportion of the inverse cube of the distance.

EULER criticizes NEWTON's calculation of the height of the tide, which gives larger values. By comparing the sun's disturbing force with the centrifugal force at the equator arising from the rotation and supposing that the former elevates the sea in the same proportion as the latter elevates the earth under the equator, he claims, NEWTON erects a superstructure on a very weak foundation (§ 39)[49]). If any doubt concerning his own values remains, on account of the use of approximations in the integration, this will be removed, EULER promises, when he obtains the same solution by a different method[50]).

At this point EULER explains that the value of S/a^3 depends only on the sidereal year and by a new calculation he confirms the value already supposed (§40)[51]). For the calculation of the lunar tide (§41), he adopts NEWTON's value of $L = 1/40$.

Although the method of calculation described so far works easily for the sun or moon separately, it would present insuperable difficulties, EULER explains, if the sun and moon were supposed to act together. He therefore proposes a new method, based on the principle that, if columns of water are imagined leading from different points on the surface of the earth to the centre, they will be in equilibrium among themselves (§42). Let AC (fig. 6) be a column of water reaching from the point A on the surface of the earth down to the centre C. Also let $AC = h$, $CM = z$ and $Mm = dz$. Then supposing the force of gravity inside the earth[52]) to be proportional to z^n, so that the force of gravity on the element Mm is $z^n dz$, it follows that the downward force arising from the gravity of the whole column $AC = n^{n+1}/(n+1)$. Only the vertical component of the disturbing force contributes to the

---

49) Cf. NEWTON, *Principia*, Book III, proposition 36.

50) NEWTON's calculation, though more intuitive than analytical, does allow for the effect of the attraction of the heaped-up water, whereas EULER's does not.

51) EULER's calculation implies a value of the acceleration due to gravity $g = 30{,}32$ Paris feet per second^2. The centripetal force (acting on unit mass) needed to retain the earth in orbit is

$$\left(\frac{2\pi}{t}\right)^2 \times \frac{\text{distance of earth from sun}}{g} = \left(\frac{2\pi}{t}\right)^2 \times \frac{19\,695\,539\,a}{g},$$

taking the natural gravity at the surface of the earth $= 1$. But this force $= S/a^2$. Hence

$$\frac{S}{a^3} = \left(\frac{2\pi}{t}\right)^2 \times \frac{19\,695\,539}{g} \quad \text{or} \quad t = \frac{5064\tfrac{1}{2} a\sqrt{a}}{\sqrt{S}}.$$

Then taking the length of the year $t = 31\,558\,164$ seconds, $S/a^3 = 0{,}50723$ feet.

52) EULER adopts this arbitrary hypothesis because he knows that the the gravity inside the earth is not constant. It will appear in the course of the calculation that the conclusion is independent of the value of n chosen.

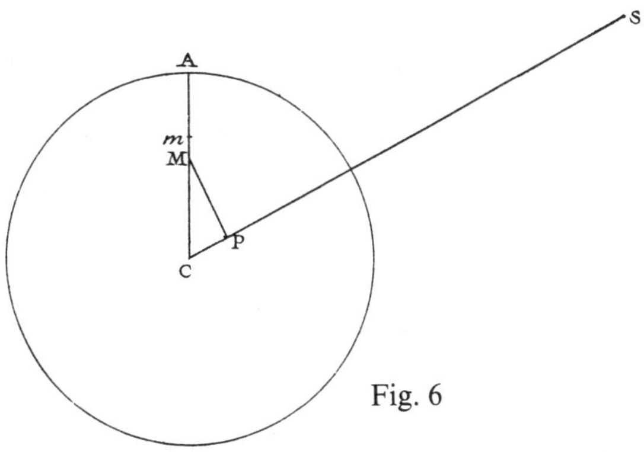

Fig. 6

pressure at the base of the column, so the horizontal component, which played an essential role in the previous method of calculation, is here neglected (§43). Now the vertical component of the disturbing force acting on the element Mm is (§27)

$$\frac{Sz(1-3\cos^2\varphi)\,dz}{a^3}+\frac{3Sz^2\cos\varphi\,(3-5\cos^2\varphi)\,dz}{2a^4}$$

where $\varphi =$ angle ACS. Hence, by integration, the total force at the base of the column, arising from the disturbing force of the body S, is

$$\frac{Sh^2(1-3\cos^2\varphi)}{2a^3}+\frac{Sh^3\cos\varphi\,(3-5\cos^2\varphi)}{2a^4}.$$

Consequently, the total force (including gravity) at the base of the column is

$$\frac{h^{n+1}}{n+1}+\frac{Sh^2(1-3\cos^2\varphi)}{2a^3}+\frac{Sh^3\cos\varphi\,(3-5\cos^2\varphi)}{2a^4}.$$

But the force, arising from gravity alone, at the base of a column equal to the earth's radius 1 in its natural state $=1/(n+1)$. Since the forces at the bases of the two columns must be equal, it follows that

$$h^{n+1}=1+(n+1)\left\{\frac{Sh^2(3\cos^2\varphi-1)}{2a^3}+\frac{Sh^3\cos\varphi\,(5\cos^2\varphi-3)}{2a^4}\right\}.$$

Now let $h=1+X$, so that X is the elevation above the natural level. Then $h^{n+1}=1+(n+1)X$ to a sufficient approximation, and since h can be taken equal to 1 in the terms on the right,

$$X=\frac{S(3\cos^2\varphi-1)}{2a^3}+\frac{S\cos\varphi\,(5\cos^2\varphi-3)}{2a^4}.$$

This is the same as the value already determined by the first method.

Taking account of the contributions to the force at the base of the column AC of the disturbing forces of both the sun and the moon (§44), it follows easily that the total elevation is simply the sum of the elevations that would be produced by the sun and moon separately. By a similar calculation (§45), Euler demonstrates that the elevation of the water above the spheroidal surface of the earth, when the rotation is taken into account, is the same as it would be above the spherical surface that the earth would have if it were assumed to be without rotation.

Chapter 4

Turning now to the determination of the tides at any given point on the earth's surface, Euler first considers the effect of the moon, since the lunar tide is the principal phenomenon and the effect of the sun can be regarded as a modification of this. He was mistaken in supposing that the delay in the tide after the transit of the disturbing body across the meridian was the effect of the inertia of the water (§46). Pierre Simon Laplace (1749–1827) demonstrated that, in the absence of friction, the transit of the disturbing body across the meridian would coincide with either high water or low water, depending on the latitude of the place[53]). Euler was thus mistaken in supposing that his results depended on the hypothesis that the water lacked inertia. In this chapter he in fact establishes the predictions of the equilibrium theory in the absence of friction.

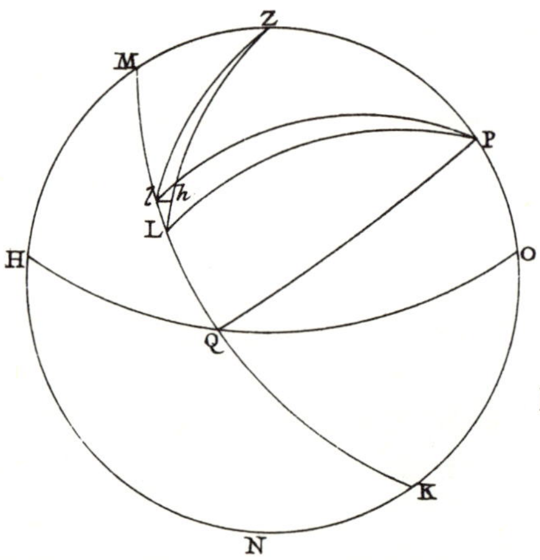

Fig. 7

Let Z (fig. 7) be the zenith of a given point on the earth, HQO the horizon, P the north pole, PZHNO the meridian and MLK the parallel of latitude in which the moon in L circulates with the diurnal motion. Also let the moon's hour angle MPL be h, the

53) P. S. Laplace, *Oeuvres complètes* (Paris, 1878–1912), vol. 2, p. 212. Laplace's prediction of an inversion of the tide at the 18th parallel of latitude is in accordance with observation. The inversion is not predicted by the equilibrium theory.

moon's declination δ and the latitude of the place λ. Then, if ψ is the zenith distance of the moon, by spherical trigonometry (§47),

$$\cos\psi = \cos h \cos\lambda \cos\delta + \sin\lambda \sin\delta\,.$$

But the elevation of the water above the natural level

$$= \frac{L(3\cos^2\psi - 1)}{2b^3} + \frac{L\cos\psi\,(5\cos^2\psi - 3)}{2b^4},$$

where $b =$ the mean distance of the moon from the earth[54]). Since the second term is very small, the elevation of the surface of the water is approximately

$$\frac{L(3\cos^2\psi - 1)}{2b^3}.$$

Consequently, the elevation is zero when $\cos\psi = 1/\sqrt{3}$; that is, when the altitude of the moon is 35° 16′ (§49). If the moon has an altitude greater than 35° 16′ above or below the horizon, the water is elevated, but if less than 35° 16′, it is depressed. The greatest depression occurs when the moon is on the horizon. After remarking that there is no tide at the poles, EULER tabulates the height of the tide for selected regions; namely places on the equator and in latitudes 30° and 60° (§50). In each case he gives values for the moon on the equator and also when the moon has declination 20° north and south. Further, he includes the angle MPQ, which defines the time of rising or setting of the moon when the greatest depression below the natural level occurs.

The effects of the sun and of both luminaries together are easily derived from the same formulae. Thus EULER explains the spring tides in the syzygies and the neap tides in the quadratures (§§52–54). From the spring and neap tides at the time of the equinoxes (§§ 55), he uses a comparison of theory and observation to determine the ratio of the absolute forces of the sun and the moon, a method NEWTON had used but, in EULER's view, not with enough precision. According to theory, the ranges of the tidal oscillations in the syzygies and quadratures at the time of the equinoxes are respectively

$$3\cos^2\lambda\left(\frac{S}{2a^3} + \frac{L}{2b^3}\right) \quad \text{and} \quad \frac{3L(\cos^2\lambda\cos^2\delta + \sin^2\lambda\sin^2\delta)}{2b^3} - \frac{3S\cos^2\lambda}{2a^3}.$$

54) The quantities L, ψ and b for the moon correspond to the quantities S, φ and a for the sun, in terms of which the formula for the elevation of the water has been previously calculated.

Observations at Le Havre-de-Grâce (Portus Gratiae)[55]) revealed the ratio to be 17:11, so that

$$17\left(\cos^2\delta + \frac{\sin^2\lambda\sin^2\delta}{\cos^2\lambda} - 11\right):28 = \frac{S}{a^3}:\frac{L}{b^3}.$$

Taking $\lambda = 50°$ and $\delta = 23°\,29'$, the result is $S/a^3 : L/b^3 = 7{,}1356:28$, so that the force of the moon is confirmed to be about four times the force of the sun, as NEWTON had concluded from the observations available to him.

Chapter 5

Turning to the consideration of the times of the tides, EULER first explains a rule found by JACQUES CASSINI (1677–1756) from observations made at Dunkirk and Le Havre in 1701 and 1702. CASSINI found that, if the new moon or full moon precedes or follows the transit of the sun across the meridian by n hours, then the high water respectively follows or precedes the transit of the moon across the meridian by $2n$ minutes[56]). In the quadratures, however, the sun being on the horizon when the moon crosses the meridian, the time of high water coincides exactly with the transit (§61).

In order to determine the time of high water more precisely, EULER calculates the rate of change of the disturbing forces as the luminaries move towards or away from the horizon.

Let the hour angle of the moon, angle MPL (fig. 7), be h and angle LPl $= d\theta$, which represents the element of time. Also let $v = \cos\psi = \cos h\cos\lambda\cos\delta + \sin\lambda\sin\delta$. The force of the moon in L

$$= \frac{L(3v^2-1)}{2b^3} \quad \text{and the increment of this force} \quad = \frac{3Lv\,dv}{b^3}.$$

Since $dv = \sin h\cos\lambda\cos\delta\,d\theta$, the increment of the force becomes (§62)

$$\frac{3L(\cos h\cos\lambda\cos\delta + \sin\lambda\sin\delta)\sin h\cos\lambda\cos\delta\,d\theta}{b^3}.$$

There is a similar force for the sun. Although the same $d\theta$ does not represent equal times for the sun and moon, the difference is so small that they can be taken for equal. Suppose now the sun to be in L, the moon having already crossed the meridian and the water at

55) Systematic observations had been instituted at Le Havre by the Paris Academy of Sciences in 1701 and 1702.

56) *Mémoires de l'Académie Royale des Sciences* (Paris), (1710), 1712, pp. 318–341; cf. *Histoire de l'Académie Royale des Sciences* (Paris), (1710), 1712, pp. 4–10.

its greatest elevation (§ 63). Then the decrement of the force of the moon in an infinitesimal element of time is equal to the increment of the force of the sun. Equating these changes, taking $\cos h = 1$, since the moon is close to the meridian, and using the approximation $L/b^3 = 4S/a^3$, the equation becomes

$$\sin h = \frac{\sin H \cos \Delta (\cos H \cos \lambda \cos \Delta + \sin \lambda \sin \Delta)}{4 \cos \delta (\cos \lambda \cos \delta + \sin \lambda \sin \delta)},$$

where H is the hour angle of the sun and Δ is the declination of the sun. This gives the time of high water after the transit of the moon across the meridian.

Under the equator, where $\lambda = 0$,

$$\sin h = \frac{\sin 2H \cos^2 \Delta}{8 \cos^2 \delta}.$$

If the declinations are assumed equal, $\sin h$ becomes $\frac{1}{8} \sin 2H$. This is a maximum if $H = 45°$, in which case $h = 7° 11'$, which indicates that the high water occurs about 30 minutes after the transit of the moon across the meridian. Thus, at the equator, if the moon approaches the meridian at 9 o'clock in the morning or evening, high water occurs 30 minutes later. If, on the contrary, the moon approaches the meridian at 3 o'clock, high water occurs 30 minutes earlier. In other regions, though the effect is somewhat irregular, it can be estimated near enough from the formula.

The declinations in fact depend on the longitudes of the luminaries, so that $\sin \Delta = \sin L \sin \varepsilon$ and $\sin \delta = \sin l \sin \varepsilon$, where L and l are the longitudes of the sun and moon respectively and ε is the obliquity of the ecliptic. Moreover, EULER modifies his formula (§ 64) to give the moon's hour angle at the time of high water in terms of the angular separation of the sun and moon. Let angle SpL (fig. 8) be θ. Then

$$\theta = \tan^{-1}(\cos \varepsilon \tan L) - \tan^{-1}(\cos \varepsilon \tan l)$$

and

$$\sin h = \frac{\sin \theta \cos \Delta (\cos \theta \cos \lambda \cos \Delta + \sin \lambda \sin \Delta)}{4 \cos \delta (\cos \lambda \cos \delta + \sin \lambda \sin \delta) + \cos 2\theta \cos \lambda \cos^2 \Delta + \cos \theta \cos \Delta \sin \Delta \sin \lambda}.$$

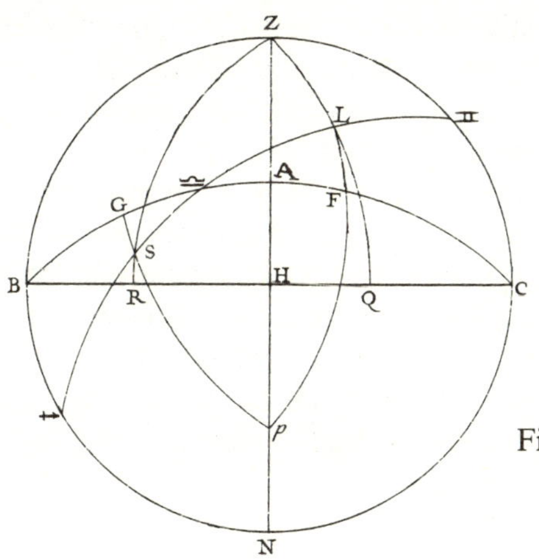

Fig. 8

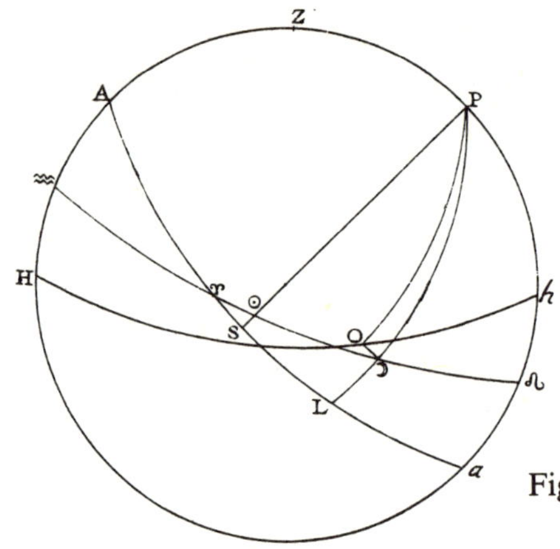

Fig. 9

As an illustration of the application of his formula, EULER deduces that at the time of the equinoxes, when the moon is in the quadratures, high water coincides with the transit of the moon across the meridian (§65). In another illustration, he shows that in latitude 45°, when the sun is in the summer solstice, high water occurs in the first quadrature 27 minutes before the transit of the moon across the meridian, and in the last quadrature 27 minutes after the transit. The opposite happens if the moon crosses the meridian under the horizon or if the sun is in the winter solstice (§66).

In a similar way, EULER establishes a formula (fig. 9)

$$\sin \mathrm{OP}☾ = \frac{\sin H \cos \lambda \cos \Delta (\cos H \cos \lambda \cos \Delta + \sin \lambda \sin \Delta)}{4(\cos^2 \lambda - \cos^2 \delta)}$$

giving the interval of time before the rising or after the setting of the moon at which low water occurs (§§67–69).

Chapter 6

The magnitude of the tides observed in ports is generally many times greater and the time of high water consistently later than the prediction of the equilibrium theory described so far. Although recognizing that a large part of these discrepancies should be attributed to the particular configurations of coasts and continents, EULER was also aware of the limitations of the equilibrium theory itself (§71). However, he mistakenly believed that the principal defect of the equilibrium theory was its supposed failure to take account of the inertia of the water, to which he chiefly (and erroneously) attributed the retardation. Yet the absence of a sufficiently general theory of fluid motion led him to despair of finding a dynamical solution. As an improvement on the equilibrium theory, he therefore offered one based on what he described as geometrical and analytical considerations (§75). In fact he supposed the vertical motion of the water to be a forced harmonic oscillation whose amplitude and phase were influenced by the inertia of the water (§76).

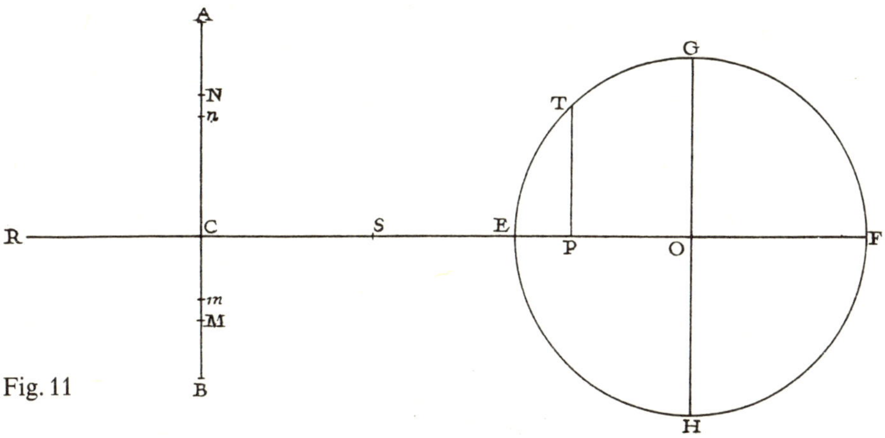

Fig. 11

Let s be the distance CM (fig. 11) of the water below the natural level. Then the natural gravity tends to raise it with a force s/g, where g is a constant to be determined (§75)[57]). For simplicity, suppose also that the place of observation is on the equator and the declination of the moon zero, so that the moon appears to revolve in the great circle EGFH through the zenith of the place crossing the equator. Now suppose that the moon is in T, where $PT = y$ is the sine of the moon's altitude, so that the force of the moon to raise the water is (§76)

$$\frac{L(3y^2-1)}{2b^3} = \frac{3y^2-1}{h}, \quad \text{where } h \text{ denotes } \frac{2b^3}{L}.$$

Then taking $y = \sin z$, where z is the arc ET representing the time elapsed since the rising of the moon in E, the differential equation for the vertical oscillation of the water at M becomes (§77)

$$2\frac{d^2s}{dz^2} + \frac{s}{g} + \frac{(1-3\cos 2z)}{2h} = 0.$$

First EULER solves

$$2\frac{d^2s}{dz^2} + \frac{s}{g} = 0,$$

obtaining

$$s = c \sin \frac{z}{\sqrt{(2g)}},$$

where c is an arbitrary constant. For a solution of the original equation, he then puts

$$s = u \sin \frac{z}{\sqrt{(2g)}},$$

57) EULER's constant g should not be confused with the acceleration due to gravity. EULER uses a system of units in which the natural gravity at the surface of the earth is 1.

leading to the equation

$$2\frac{d^2u}{dz^2}\sin\frac{z}{\sqrt{(2g)}}+\frac{4}{\sqrt{(2g)}}\frac{du}{dz}\cos\frac{z}{\sqrt{(2g)}}+\frac{1-3\cos 2z}{2h}=0\,.$$

Now putting $du/dz = p$ gives a first-order linear equation from which an expression for p is found (§78). Integrating this gives u and finally (§79)

$$s = C\cos\frac{z}{\sqrt{(2g)}}+D\sin\frac{z}{\sqrt{(2g)}}-\frac{g}{2h}+\frac{3g\cos 2z}{2h(1-8g)}\,.$$

Assuming the motion to be reduced to uniformity, so that the water returns to the same state when the moon returns to the same position, it is necessary that $C = D = 0$, so that only the forced oscillation remains[58]).

In the case of the equilibrium theory, where the water at every instant is in equilibrium with the forces of the moon and its own gravity, these forces must themselves balance, so that (§81) $s = g(1-3y^2)/h$. Consequently, the greatest depression, CM $= g/h$, occurs when the moon is on the horizon, and the greatest elevation, CN $= 2g/h$, when the moon crosses the meridian. Thus the range MN $= 3g/h$. This has previously been found to be 3,372 feet. According to the new theory, however, the range

$$\mathrm{AB} = \frac{3g}{h(1-8g)}\,.$$

Now the period of the free oscillation $= 2\pi\sqrt{(2g)}$, where π, the semicircle with radius 1, corresponds to 12 lunar hours (§82). If the time of restoration to the normal level under the action of gravity alone (that is, one-quarter of a period) is taken to be $12/n$ hours, then $g = 2/n^2$, from which it follows that the shorter the period of the free oscillation, the smaller is the excess of AB over MN. EULER supposes, as a reasonable hypothesis, that AB is almost twice MN, so that he estimates n to be 6. In this case, the time of restoration would be 2 hours and AB = 6,025 feet. Moreover, in order to avoid resonance, n needs to be greater than 4, from which EULER concludes that it must be 5 or 6.

Now generalising the problem to take account of the latitude λ of the place of observation and the declination δ of the moon, EULER establishes the equation (§83)

$$2\frac{d^2s}{dz^2}+\frac{s}{g}+\frac{3\cos^2\lambda\cos^2\delta+6\sin^2\lambda\sin^2\delta-2}{2h}+\frac{3\sin 2\lambda\sin 2\delta\cos z}{2h}$$
$$+\frac{3\cos^2\lambda\cos^2\delta\cos 2z}{2h}=0\,,$$

58) At this point (§80), EULER notes that he has taken no account of the horizontal motion of the water.

where z is the hour angle of the moon, which is assumed to have already crossed the meridian. Following the steps in the solution of the previous equation, he finds, in the case where the motion has been reduced to regularity so that only the forced oscillation remains (§ 84),

$$s = -\frac{g(3\cos^2\lambda\cos^2\delta + 6\sin^2\lambda\sin^2\delta - 2)}{2h} - \frac{3g\sin 2\lambda\sin 2\delta\cos z}{2h(1-2g)} - \frac{3g\cos^2\lambda\cos^2\delta\cos 2z}{2h(1-8g)}.$$

In this case

$$\mathrm{AB} = \frac{3g}{h(1-8g)}\left\{\cos\lambda\cos\delta \pm \frac{(1-8g)\sin\lambda\sin\delta}{1-2g}\right\}^2,$$

where the plus sign corresponds to the transit of the moon across the meridian over the horizon and the minus sign to the transit under the horizon (§ 86). EULER estimates the value $g = 2/25$ to be most in agreement with observation. Assuming this value, the greatest tide, for a given latitude, occurs when $\tan\delta = \frac{3}{7}\tan\lambda$ (§ 87). He gives a table for the different latitudes, showing the declination of the moon corresponding to the highest tide. Since the observations indicate that the latitude at which the greatest tides occur when the moon has its maximum declination is 60° rather than that predicted in the table, he further revises his estimate of the value of g to be $1/10$.

Assuming the depth of the ocean to be constant, the speed of ascent of the water is proportional to the decrement of the speed of the horizontal motion. Since the speed of ascent is $-ds/dz$, it follows that the horizontal speed is proportional to $-s$, that is (§ 95) to

$$\frac{g(3\cos^2\lambda\cos^2\delta + 6\sin^2\lambda\sin^2\delta - 2)}{2h} + \frac{3g\sin 2\lambda\sin 2\delta\cos z}{2h(1-2g)} + \frac{3g\cos^2\lambda\cos^2\delta\cos 2z}{2h(1-8g)}.$$

At the equator, when the declination of the moon is also zero, the horizontal speed is therefore proportional to

$$\frac{g}{2h} + \frac{3g\cos 2z}{2h(1-8g)} \quad \text{or} \quad 15\cos 2z + 1,$$

taking $g = 1/10$. The speed of ascent is thus proportional to $-\sin 2z$. EULER gives a table for the horizontal and vertical speeds at intervals of an hour (§ 96). For other regions, he adds, the speeds can easily be derived from the formulae.

EULER concludes the chapter with a brief description of the retardation of the tides, which he attributes to the effect of the inertia of the water.

Chapter 7

In this chapter EULER brings together the explanations of the principal phenomena of the tides in the open oceans and in the vicinity of islands, which are scattered in previous chapters. He arranges them systematically and compares them with observations. Above the Arctic circle, the theory predicts one daily tide but this prediction, he notes, has not yet been confirmed by observations [59]).

Chapter 8

To complete his account of the tides, EULER now considers the effects of the earth and coasts. The horizontal motion is clearly impeded by coasts, so that, for example, the retardation increases in passing from the open Atlantic through the English channel (§114). In seas like the Mediterranean and the Baltic, separated from the vast ocean by narrow straits, the tide can be evanescent, unless they are of such amplitude that the forces of the sun and moon are able to produce in them their own tides (§116). The condition for tides to be generated as in the open ocean is evidently that the longitude of the tract of water should be more than 90°, so that the water may be elevated in one place and depressed in another (§118). If the tract is so small that the extremities are attracted or depressed by almost equal forces, no tides can arise. Examples are the Baltic, Caspian, Black Sea and other smaller lakes.

Consider a tract of ocean (fig. 15) bounded by the lines of longitude PM and PN and by the parallels of latitude MN and *mn*. For simplicity, EULER takes account only of the action of the moon, which he supposes to be in L. Let the declination of the moon be δ and the latitudes of M, *m* and X be λ_M, λ_m and λ respectively. Also let angle LPX be h, angle LPM be l and angle MPN be θ. Then using the condition that the total volume of water in the tract must remain constant, the elevation of the surface of the water at X above the natural level is found to be (§124)

$$\frac{3L}{2b^3}(\cos h\cos\delta\cos\lambda+\sin\delta\sin\lambda)^2+\frac{L(1-3\sin^2\delta)(\sin^2\lambda_m+\sin\lambda_m\sin\lambda_M+\sin^2\lambda_M)}{4b^3}$$
$$-\frac{3L\cos^2\delta}{4b^3}+\frac{L}{4b^3\theta(\sin\lambda_m-\sin\lambda_M)}\Big\{\cos^2\delta\sin\theta\cos(2l+\theta)(\sin\lambda_m-\sin\lambda_M)(\sin^2\lambda_m$$
$$+\sin\lambda_m\sin\lambda_M+\sin^2\lambda_M-3)+2\sin2\delta(\cos^3\lambda_M-\cos^3\lambda_m)(\sin l-\sin(l+\theta))\Big\}.$$

59) In a letter of 15 September 1740 to DANIEL BERNOULLI, EULER explains that observations received from the Arctic ocean since the essay was written were still not sufficient to conclude anything definite. R 138, *Bibliotheca mathematica*, 3rd Series, 7 (1906–1907), 145–153. In his reply, BERNOULLI remarked that the observations appeared to conform not at all to their theory, though he had no doubts about the theory. R 139, *ibid.*, pp. 153–156.

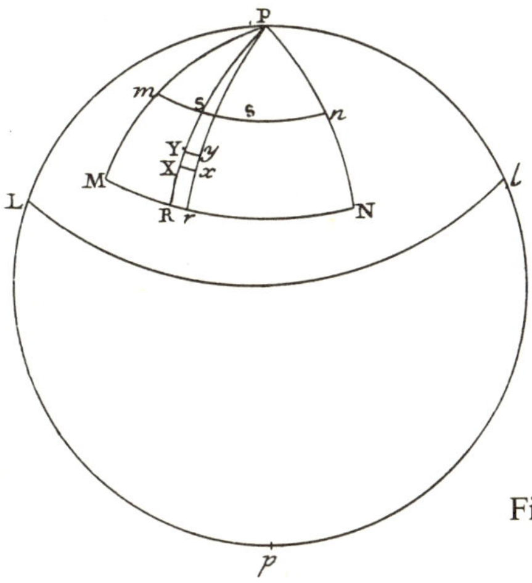

Fig. 15

In the case when the ocean covers the whole earth, the last term is zero and the first three terms reduce to

$$\frac{L(3v^2-1)}{2b^3},$$

where $v = \cos h \cos \delta \cos \lambda + \sin \delta \sin \lambda$, thus confirming the previous result.

As examples of the application of the general formula, EULER considers a narrow longitudinal tract of length 90° (§§126–127) and then a narrow latitudinal tract (§129). From the properties of the former, he derives an explanation of the tides in the Mediterranean, which may be thought of as a longitudinal tract of length 60° (§128). In the case of the latter, he finds that the tide vanishes if the moon is on the equator, but that if the moon is north or south of the equator, there is a single tide each day. These are the phenomena observed in the Gulf of Tonkin, and EULER claims that his theory offers a better explanation than that of NEWTON (§130). For he supposed that the ocean in this region could be regarded as a narrow tract, extending beyond the equator to about 45° of latitude.

E 546

De statu aequilibrii maris a viribus solis et lunae sollicitati

In this memoir, presented to the St. Petersburg Academy of Sciences on 2 November 1775, EULER employs a new principle of the equilibrium of fluids (§4), which he had stated about twenty years earlier in a memoir presented to the Berlin Academy of Sciences[60]). This

60) E 225, "Principes généraux de l'état d'équilibre des fluides", *Mémoires de l'Académie Royale des Sciences de Berlin* [11] (1755), 1757, 217–273 = *Opera omnia* II 12, pp. 2–53. See especially p. 15.

principle, in effect, employs the concept of potential, which EULER describes as the action of the forces [61]). He also refutes the view that the height of the tide depends on the density of the water (§§1–3).

For simplicity, EULER considers the terraqueous globe to be completely surrounded by water. As in the prize essay, he supposes further, that whatever changes the external forces would induce in the surface of the water, assuming the earth to be without rotation, would also happen to the spheroidal form acquired as a result of the rotation (§5), so that in the determination of the elevation or depression from the natural state, he is able to suppose the earth to be a sphere.

EULER takes the mean distance of the sun from the centre of the earth to be $a = 24000$ earth-radii and that of the moon to be $b = 60\frac{1}{4}$ earth-radii (§6). Taking the force at a distance α from the centre of the sun equal to 1, the force of gravity at the earth's surface, the force of the sun at any distance z is α^2/z^2. Similarly, the force of the moon at a distance z is β^2/z^2, where β is the distance from the moon at which the force is 1. From the latest determinations of the solar parallax, α is taken to be 596, while EULER prefers DANIEL BERNOULLI's value of $\beta = 1/8$ to the somewhat larger value of NEWTON [62]).

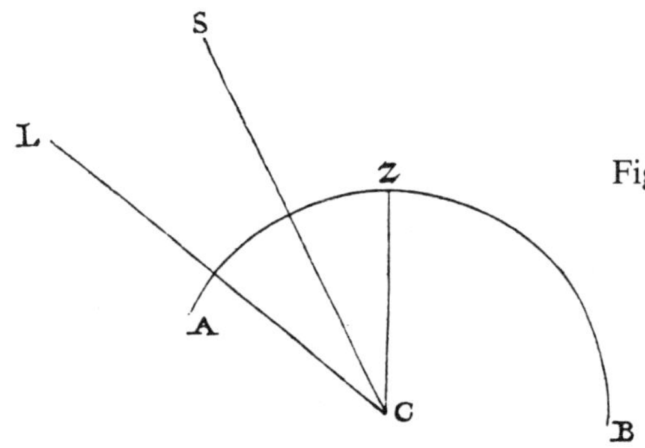

Fig. 4

Let Z (fig. 4) be a point on the surface of the ocean at a distance r from the centre of the earth, S the sun with zenith distance φ and L the moon with zenith distance ψ; then the quantity of action (that is, the potential) of the disturbing forces and the earth's gravity at the point Z is (§§7–9) [63])

61) Despite EULER's reference to MAUPERTUIS, his principle has nothing to do with the principle of least action. More appropriate would have been an acknowledgement of his debt to CLAIRAUT who, in his *Théorie de la figure de la terre* (Paris, 1743), pp. 94–105, demonstrated that level surfaces are what would now be called equipotential surfaces.

62) On the basis of a comparison of the heights of the solar and lunar tides, NEWTON takes β^2 to be 1/40 approximately. *Principia,* Book III, proposition 37, cor. 4. DANIEL BERNOULLI's calculation, based on a comparison of the times of the solar and lunar tides, implies a value of $\beta^2 = 1/64$. "Traité sur le flux et reflux de la mer", *Recueil des pièces qui ont remporté les prix de l'Académie Royale des Sciences*, vol. 4 (Paris, 1752), pp. 113–117 = *Die Werke von Daniel Bernoulli*, vol. 3 (Basel, 1987), pp. 379–382.

$$r - \frac{\alpha^2}{\sqrt{(a^2 - 2ar\cos\varphi + r^2)}} + \frac{\alpha^2 r\cos\varphi}{a^2} - \frac{\beta^2}{\sqrt{(b^2 - 2br\cos\psi + r^2)}} + \frac{\beta^2 r\cos\psi}{b^2}.$$

In the equilibrium state, this quantity is constant over the surface of the ocean. Expanding by the Binomial theorem and putting $r = k + v$, where $k = 19\,601\,352$ Paris feet is the radius of the earth (§ 11), the equation for v becomes (§ 12)

$$\begin{aligned} v\Big(1 &+ \frac{\alpha^2 k}{a^3}(1 - 3\cos^2\varphi) + \frac{3\alpha^2 k^2}{2a^4}\cos\varphi\,(3 - 5\cos^2\varphi) \\ &+ \frac{\beta^2 k}{b^3}(1 - 3\cos^2\psi) + \frac{3\beta^2 k^2}{2b^4}\cos\psi\,(3 - 5\cos^2\psi)\Big) \\ &= c - \frac{\alpha^2 k^2}{2a^3}(1 - 3\cos^2\varphi) - \frac{\alpha^2 k^3}{2a^4}\cos\varphi\,(3 - 5\cos^2\varphi) \\ &- \frac{\beta^2 k^2}{2b^3}(1 - 3\cos^2\psi) - \frac{\beta^2 k^3}{2b^4}\cos\psi\,(3 - 5\cos^2\psi), \end{aligned}$$

where c is a constant.

For a numerical solution, EULER first takes the sun and moon to be in their mean distances. Then the coefficient of v may be taken equal to 1, so that the equation becomes (§ 13)

$$\begin{aligned} v &= c - 0{,}25183\,(1 - 3\cos^2\varphi) - 0{,}70896\,(1 - 3\cos^2\psi) \\ &- 0{,}00001\cos\varphi\,(3 - 5\cos^2\varphi) - 0{,}01182\cos\psi\,(3 - 5\cos^2\psi). \end{aligned}$$

In general

$$v = c + m\,(3\cos^2\varphi - 1) + n\,(3\cos^2\psi - 1) + \nu\cos\psi\,(5\cos^2\psi - 3).$$

Taking the eccentricity of the earth's orbit $= 1/60$ and that of the moon's orbit $= 1/18$, EULER calculates the values of m, n and ν when the luminaries are in apogee and perigee, obtaining the following values (§ 14)[64]:

---

63) The first term is the potential of the earth's gravity at the point Z, the second term that of the sun's attraction at the point Z and the third term (with sign reversed) the potential at the point Z of the sun's attraction on the earth as a whole. The potential of the disturbing force of the sun is the difference represented by the second and third terms. Similarly, the potential of the disturbing force of the moon is the difference represented by the fourth and fifth terms.

64) EULER calculates the value m for the apogee as $0{,}25183\,(1 + 1/60)^{-3} = 0{,}25183\,(1 - 1/20)$ and for the perigee as $0{,}25183\,(1 - 1/60)^{-3} = 0{,}25183\,(1 + 1/20)$. Similarly, the value of n for the apogee is $0{,}70896\,(1 - 1/6)$ and for the perigee $0{,}70896\,(1 + 1/6)$.

	minimum	mean	maximum
m	0,23924	0,25183	0,26442
n	0,59080	0,70896	0,82712
v	0,00931	0,01197[65])	0,01463

Turning to applications of the formula for v, the elevation of the water above the natural level, EULER first supposes that the ocean *abcd* (fig. 5) extends over the whole

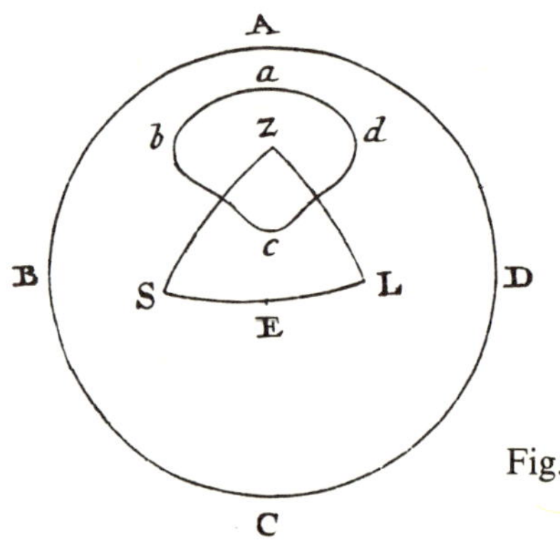

Fig. 5

surface of the earth (§§17–18). Then the maximum elevation occurs at some point E between S and L. Taking the point Z to be in E, arc SE $= \varphi$, arc LE $= \psi = \zeta - \varphi$, where ζ is the angular separation of the luminaries (assumed constant) and consequently $d\psi = -d\varphi$. Equating dv to zero and neglecting v in comparison with m and n, the points of maximum elevation and depression are determined by the condition

$$\tan 2\varphi = \frac{n \sin 2\zeta}{m + n\cos 2\zeta}.$$

The solutions are of the form $\varphi = \frac{1}{2}\theta$, $\varphi = \frac{1}{2}\theta + 90°$, $\varphi = \frac{1}{2}\theta + 180°$, $\varphi = \frac{1}{2}\theta + 270°$. The first gives a position of greatest elevation and the second (distant a quadrant from the first) gives a position of greatest depression. The third and fourth (diametrically opposite the first and second) give the other points of greatest elevation and depression respectively. At the point S, where the sun is in the zenith, the elevation becomes

65) This should be 0,01182. The minimum and maximum values of v reflect the error in the mean value, cf. note 8, p. 339.

$$v = c + 2m + n(3\cos^2\zeta - 1) + \nu\cos\zeta\,(5\cos^2\zeta - 3)$$

and at L

$$v = c + m(3\cos^2\zeta - 1) + 2n + 2\nu\,.$$

At the point Z a quadrant distant from both luminaries, however, $v = c - m - n$, so that the surface of the water will always have its greatest depression.

The value of c is determined by the condition that the volume of the ocean remains constant (§16). For an ocean completely surrounding the earth, EULER shows that $c = 0$ (§§20–24). In other cases, it is generally not necessary to compute c, since only the difference of elevation in different places is required (§29).

Finally, EULER solves the general problem: to determine, in the equilibrium hypothesis, the elevation or depression at any point for given positions of the sun and moon (§25). Taking the longitudes of the sun and moon (fig. 7) to be ζ and η, the complements of their

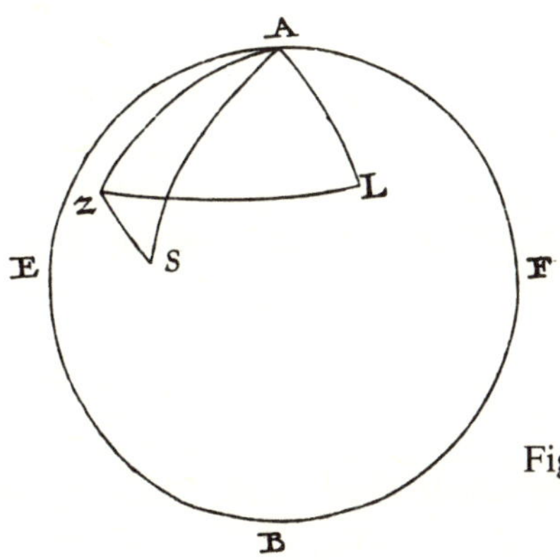

Fig. 7

latitudes to be f and g, and the corresponding quantities for the point Z to be ω and z, then by spherical trigonometry

$$\cos\varphi = \cos f \cos z + \sin f \sin z \cos(\zeta - \omega)$$

and

$$\cos\psi = \cos g \cos z + \sin g \sin z \cos(\eta - \omega)\,.$$

Substituting these values in the general formula

$$v = m(3\cos^2\varphi - 1) + n(3\cos^2\psi - 1) + \nu\cos\psi\,(5\cos^2\psi - 3)\,,$$

the problem is solved. As corollaries (§§26–28), EULER derives the elevation at the poles and equator, and also at the times when the luminaries are in conjunction and opposition.

Ocean Currents and Winds
(E 850 and E 851)

Among Euler's posthumous papers is an unfinished essay on ocean currents, which he had intended to enter for the prize of the Paris Academy of Sciences in 1749[66]), and also a review of Jean le Rond D'Alembert's (1717–1783) essay on the winds, which was awarded the prize of the Berlin Academy of Sciences in 1746[67]).

E 850
Recherches sur la découverte des courants de la mer

In a general introduction (§§1–6), Euler outlines the methods by which ocean currents can be investigated. The first method is provided by astronomy, for a comparison of changes in latitude and longitude obtained at sea from astronomical observations with those estimated by dead-reckoning will reveal the direction and strength of the ocean current. Observation near land provides another method. Since currents are usually most rapid at the surface, methods of observing the differences at the surface and at various depths are needed.

The essay was planned to consist of four parts, but only the beginning of the first part (§§7–11), describing the astronomical method, was actually written.

E 851
Recensio dissertationis de ventis

Euler praised D'Alembert's prize essay, *Réflexions sur la cause générale des vents*, especially for the mathematical analysis, in which there is extensive application of partial differential equations. The aspects of the problem most amenable to mathematical analysis are the effects of the attraction of the sun and moon, so that the essay, and also the review, seems to be more concerned with atmospheric tides than the proper subject of the prize[68]).

---

66) See *Commentationes arithmeticae collectae*, ed. Paul Heinrich Fuss and Nicolaus Fuss (St. Petersburg, 1849), vol. 1, p. xvi. The prize was withheld in 1749 and a double prize awarded to Daniel Bernoulli in 1751.

67) In a letter of 28 May 1746, Euler informed Goldbach that D'Alembert had been awarded the prize. R 819, *Leonhard Euler und Christian Goldbach, Briefwechsel 1729–1764*, ed. Adolf Pavlovič Juškevič and Eduard Winter (Berlin, 1965), p. 250.

68) Daniel Bernoulli was less impressed, remarking to Euler in his letter of 26 January 1750, that the essay really contained nothing of significance on the winds. R 180, P. H. Fuss, *Correspondance*, vol. 2, pp. 648–650. For a modern assessment, see Clifford Ambrose Truesdell in: Euler, *Opera omnia* II 12, p. lvii.

D'ALEMBERT treats this briefly at the end of the essay, subjecting the effects of the heat of the sun, which EULER regards as the primary cause of the winds, to approximate calculation.

COMETS
(E 67, E 68 and E 103)

EULER's authorship of the anonymous tract, *Beantwortung verschiedener Fragen über die Beschaffenheit, Bewegung und Würckung der Cometen*, published in Berlin in 1744, is confirmed by a letter of CHRISTIAN GOLDBACH written in St. Petersburg on 12 March 1746[69]). The tract was evidently composed in haste, as EULER notes in his answer to GOLDBACH[70]), in reply to some *Fragen* of which GOLDBACH possessed a copy. Although it seems likely that EULER's work was a response to a published text, such a tract has not been identified among the several works of this kind that were published in 1744, following the spectacular appearance of the comet at the beginning of that year.

Later in the same year, EULER published a continuation of his tract, also anonymously, with the title, *Fortgesetzte Beantwortung der Fragen über die Beschaffenheit, Bewegung und Würckung der Cometen.*

In 1746 EULER presented to the Berlin Academy of Sciences a paper on the constitution of the tails of comets, in which he claimed to find a common explanation of these phenomena and also of the northern lights and zodiacal light in the effect of the sun's rays on the atmosphere of the appropriate body. This paper was published in 1748 in the second volume of the *Mémoires* of the Berlin Academy of Sciences[71]). Reviews appeared in the *Nouvelle Bibliothèque Germanique*[72]), the *Nova Acta Eruditorum*[73]) and the *Berlinische Bibliothek*[74]).

E 67
Beantwortung verschiedener Fragen über die Beschaffenheit, Bewegung und Würckung der Cometen

In this popular account, published anonymously at the time of the appearance of the comet of 1744, EULER describes all the aspects of comets that were of contemporary interest. First he distinguishes comets from fixed stars on the basis of their appearances in the telescope. Although the comets are opaque bodies and are seen only by reflected light, they

69) R 815, *Leonhard Euler und Christian Goldbach, Briefwechsel*, p. 243.

70) R 816, *ibid.*, p. 244.

71) EULER advised GABRIEL CRAMER (1704–1752) of the imminent publication in a letter of 17 March 1747; R 471, EULER, *Opera omnia* IV A 7.

72) *Nouvelle Bibliothèque Germanique* 5 (1748), 1749, 14–16.

73) *Nova Acta Eruditorum*, 1750, pp. 600–601.

74) *Berlinische Bibliothek* 2 (1748), 377–383.

differ from planets in a number of ways. For example, they are surrounded by a thick atmosphere, extending several diameters from the surface. The resulting meteorological conditions, together with the extremes of temperature as the comets approach near to the sun and then recede to great distances, render them unsuitable as abodes of life. EULER had no doubt that the planets of the solar system and also those surrounding the fixed stars – invisible on account of their small angular separation from the central body – were inhabited. According to the system of WILLIAM WHISTON (1667–1752)[75]), he remarks, the earth must have been similar to a comet at the time of creation, so that these bodies may perhaps be intended for human habitation in the future. Again, he adds, their purpose may be to bring about certain changes in the planets, as WHISTON supposed when he attributed NOAH's Flood to the effect of a comet. However, EULER believes that our understanding is too weak to decide such questions.

Returning to the description of the physical nature of comets, EULER explains that observation is unable to reveal whether the solid core rotates. Almost the whole surface of the comet is illuminated on account of the refraction of the sun's rays in passing through the very high atmosphere, though the extent of this atmosphere is also the reason why the appearance of the comet is so faint. Rejecting NEWTON's view that the tail is a very fine vapour emitted from the core on account of the heat, EULER holds that the tail is nothing else than a great collection of dust particles driven from the core by the sun's rays and gathered together in the neighbourhood behind the comet, where they reflect the sunlight. This explanation is supported by the observation that the beard on the side facing the sun appears to be somewhat compressed, which would be a necessary consequence of the force of the sun's rays. Since the bright tail of a comet has some similarity with the northern lights, EULER speculates that these phenomena have the same cause.

Like the planets, the comets move around the sun in ellipses, but these are very elongated, so that the visible parts of their paths almost take the form of parabolas. Moreover, the orbits of comets are not restricted, as those of planets, to a narrow band near the ecliptic. In order to determine the orbit of a comet, and so predict its appearances, six elements are needed; namely the perihelion and aphelion distances, the nodes and inclination of the orbit to the ecliptic, the longitude of perihelion and the periodic time[76]). While there is no obstacle in theory[77]), the short time for which the comet is visible presents practical difficulties to the determination of these elements.

Although the distances of the aphelia are not known accurately, there can be no doubt, EULER explains, that they are less than the distances of the fixed stars, so that the comets must remain in the solar system. Again there are many reasons why the exact

75) WILLIAM WHISTON, *The cause of the Deluge demonstrated*, appended to *A new theory of the earth*, 2nd edition (London, 1708). WHISTON takes the comet of 1680, with a period of 575 years, to be the cause of NOAH's Flood.

76) EULER acknowledges the contribution of EDMOND HALLEY (1656–1743), *Astronomiae cometicae synopsis* (Oxford, 1705); *Philosophical Transactions* 24 (1704–1705), 1882–1899.

77) EULER determined the orbit of the comet of 1744 in E 66, *Theoria motuum planetarum et cometarum* (Berlin, 1744) = *Opera omnia* II 28, pp. 105–251, in particular pp. 180–210.

number of comets is not known. For example, some small comets may be masked by the sun when they are close enough to be visible, while others may be visible in the southern hemisphere, where observations have not been made. Yet EULER has no doubt that the number is very large.

Concerning the question whether comets have influence on the earth, EULER recommends extreme caution in attributing their appearances to the wrath of God and he believes that astrology has for the most part been abandoned. Since the comets, like other bodies, have gravitational attraction, they are in principle capable of causing perturbations in the earth's orbit, but in EULER's view none has ever approached close enough to do so. A close encounter could cause another catastrophic Flood or bring about the destruction of the earth by robbing it of its atmosphere. However, EULER allays the readers' fears by referring to Biblical texts which exclude the possibilities of such events.

E 68

Fortgesetzte Beantwortung der Fragen über die Beschaffenheit, Bewegung und Würckung der Cometen

As the title indicates, this is a continuation of the previous work. Having explained why the tail sometimes appears straight, sometimes curved and sometimes branched, EULER gives credit to JOHANNES KEPLER (1571–1630) for the basic idea of the theory that he adopts. According to KEPLER[78]), the sun's rays pass through the body of the comet and instantaneously take something of its matter with them on their path away from the sun.

Rejecting the surmise of JOHANNES HEVELIUS (1611–1687) that the comets describe spirals beyond the sphere of Saturn[79]), EULER explains that the apparent zig-zag path of some comets towards the end of their visibility should be ascribed to the motion of the earth and not to an irregularity in the real orbit of the comet. Other questions resolved by EULER relate to the density of the atmospheres of comets, the rotations of comets on their axes and the reason why the image of a comet or planet is brighter in a small telescope than in a large one. For the resolution of many questions concerning the orbits, such as the reason, for example, why the small force of the sun's gravity at aphelion is able to bring the comet back, EULER recommends a study of the mathematics. In particular, he explains that a collision of a comet with the earth is extremely unlikely.

For the reader interested in astronomy, EULER includes a demonstration of the true path of the comet of 1744 based on the observations of the astronomers in Lausanne, Paris and Berlin. Finally, he adds a detailed daily record of observations of the comet of 1744

78) J. KEPLER, *De cometis libelli tres* (1619) = *Gesammelte Werke,* vol. 8 (Munich, 1963), ed. FRANZ HAMMER, p. 226.

79) J. HEVELIUS, *Cometographia* (Danzig, 1668), p. 670. The frontispiece includes a diagram of the spiral path in the region beyond Saturn.

made in Berlin by an accomplished lady[80]). These observations comprise first, an account of the motion of the comet among the stars from its first appearance on the evening of 3 January to 25 February, after which it was hidden by clouds for several days, and secondly, a description of the tail, including the appearance of the remarkable multiple tail seen before sunrise in the early part of March, following the perihelion[81]).

E 103

Recherches physiques sur la cause de la queüe des cometes, de la lumiere boreale, et de la lumiere zodiacale

While a comparison of the tails of comets with the northern lights indicates a common origin for these phenomena, EULER rejects the view of JEAN JACQUES D'ORTOUS DE MAIRAN (1678–1771)[82]) that this cause is to be found in the zodiacal light (§1). It seems more probable to EULER that the zodiacal light proceeds from a cause similar to that of the other phenomena (§2). Since neither the tails of comets nor the northern lights are optical

80) The accomplished lady was without doubt MARGARETHE KIRCH, daughter of GOTTFRIED (1639–1710) and MARIA MARGARETHE KIRCH (1670–1720). The observations are contained in a manuscript diary entitled, "Gewitter Observationes 1744. M.K.", preserved in the Crawford Library of the Royal Observatory, Edinburgh (Shelf Mark 2.67). In the entry for 13 April, M.K. identifies herself as the observer, while on 7 March she notes that she had been to the Observatory with C.K., evidently her better known sister CHRISTINE (1696–1782). In the weather diary for 1743 (Shelf Mark 2.66), attributed to CHRISTINE, it is frequently mentioned that the comet of 1743 was observed by M.K. In the *Astronomische Nachrichten* 2 (1824), 377–382, H.W.M. OLBERS published observations of the comet of 1743 contained in a manuscript entitled, "Von dem Cometen 1743, wie er von Margarethe Kirchin observirt worden", which was then in his possession.

Both MARGARETHE KIRCH and EULER's accomplished lady employed a 2-foot and a 6-foot telescope for their observations of the comet of 1744. The occurrence of many identical phrases in the reports leaves no doubt that they were written by the same person.

EULER's accomplished lady was first identified by J. L. E. DREYER in his paper "On the multiple tail of the great comet of 1744", *Copernicus* 3 (1884), 104–110.

81) The multiple tail was also observed at Lausanne by JEAN PHILIPPE LOYS DE CHÉSEAUX (1718–1751) and described in his *Traité de la comete qui a paru en Decembre 1743 et en Janvier, Fevrier et Mars 1744* (Lausanne and Geneva, 1744), pp. 158–160. J. N. DE L'ISLE observed the multiple tail at St. Petersburg but did not publish his observations.

82) J.J. D'ORTOUS DE MAIRAN, *Traité physique et historique de l'aurore boréale* (Paris, 1733). The second edition (Paris, 1754) contains an extensive critique of EULER's theory, pp. 307–378. This critique was first published in 1752 in the *Mémoires de l'Académie Royale des Sciences* (Paris) for 1747, pp. 363–435. Although DE MAIRAN agrees with EULER in supposing that the solar rays have a role in the production of the tails of comets, he regards this as only a conjecture given at the end of his *Traité*. Concerning the other phenomena, DE MAIRAN holds the zodiacal light to be identical with the solar atmosphere and he explains the northern lights in terms of this, without the intervention of the solar rays.

illusions – though he agrees that some aspects of the latter are produced by reflection and refraction – EULER concludes that there exist luminous or illuminated particles which give rise to the appearances. In the case of the zodiacal light, the particles emanate from the sun, and in the cases of the tails of comets and the northern lights, they emanate from the bodies of the comets and the earth respectively. In each case, the particles are detached from the atmosphere of the appropriate body by the force of the sun's rays and while they are naturally opaque, in passing from a dense atmosphere into a thinner air they can acquire the vibration needed to form luminous rays (§3). As the stars are generally visible through these phenomena, the particles must be widely separated. It is for this reason, EULER notes, that NEWTON had supposed the matter of the tail of a comet capable of compression into a very small space[83]).

If light consisted of corpuscles, as NEWTON believed, it would be easy to believe that it could expel the most subtle particles floating in the atmosphere. But if light was propagated in waves, as EULER had demonstrated in his theory of light and colours[84]), the explanation seemed more difficult (§4). However, the analogy with the vibratory motion of sound, which excites a real motion in the very light dust floating in the air, suggests a similar effect of the vibratory motion of light. This effect seems to be confirmed by the motion excited in very small particles placed at the focus of a burning mirror. It can be assumed that each ray has a similar effect, though much smaller, so that for particles to be driven a large distance they must remain exposed to the rays of the sun for a long time.

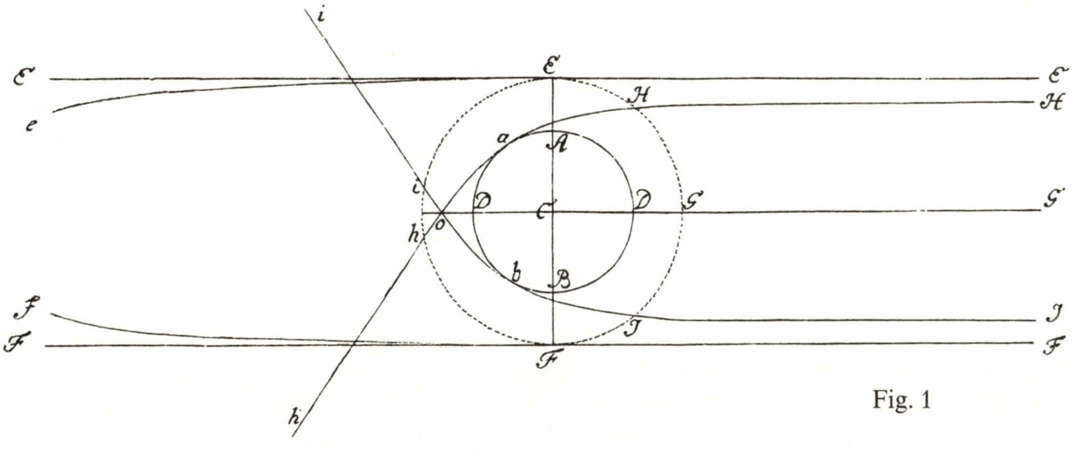

Fig. 1

Let the body of the comet (fig. 1) be the sphere *aADBb* and the atmosphere *iEHGIFh*. Suppose that the sun is on the right. Then the rays *EE* and *FF*, grazing the atmosphere, drive the small particles in the directions *EE* and *FF*. On account of their gravity, however, they move in the curves *Ee* and *Ff*. Rays such as *HHa*, which are refracted in passing through the atmosphere, drive the particles in directions such as *hh*. It

83) NEWTON, *Principia*, Book III, proposition 41.

84) E 88, "Nova theoria lucis et colorum", *Opera omnia* III 5, pp. 1–45.

follows that, after a time, these particles fill the space *EiefhF* behind the comet, and because of the rays *ahh* and *bii*, this space will be a diverging cone. The divergence is considerably restricted, however, by the rays external to the atmosphere, so that the tails are formed as we see them (§ 5). The length of the tail depends on the proximity to the sun and the time for which the rays act, while the luminosity is proportional to the extent of the atmosphere and the number of particles which have been expelled from it. Because of the refraction, more than half of the surface of the comet is illuminated by the sun, and indeed the atmosphere is so large that even the part not directly illuminated receives light from the atmosphere, so that the comet appears to be surrounded by shining matter (§ 6).

EULER explains the diversity in the appearance of the tails as a consequence of the motions of the comets. First he supposes that the solar rays are less effective in detaching particles from the atmosphere when the particles are moving swiftly across them. Consequently the length and brightness of the tail can be reduced on account of the rotation of the comet on its axis, depending on the orientation of this axis in relation to the sun (§ 7). Secondly, the orbital motion of the comet results in a small curvature of the tail. At perihelion, it can happen that the force of the rays engenders a new tail before the first can follow the motion, so that the comet develops two or more tails. This was observed in the comet of 1744 (§ 8). Again, subsidiary tails can be produced as a result of the spheroidal shape of a comet. The same thing can happen even when the body of the comet is spherical. For, if the obliquity of the rays *hh* and *ii* is so great that the divergence of the expelled particles cannot be corrected by the remaining rays, the tail, near the body of the comet, may appear separated into three (§ 9).

Turning to the problem of why the planets do not have tails, EULER finds the reason in the small extent of the atmospheres, which he considers to be negligible in comparison with those of comets (§ 10). On this account, the particles, when they have been expelled, are still so near the earth that their gravity is strong enough to pull them back. The greatest obstacle to the generation of tails, however, he holds to be the motion of rotation, which prevents the same particles from remaining exposed to the action of the solar rays for long. Nevertheless, the solar rays cannot be without effect on the particles floating in the atmosphere (§ 11). At sunrise these particles are driven towards the west, where they become visible to inhabitants of the earth for whom the sun has not yet risen. In the same way, at sunset the particles are driven towards the east and become visible to inhabitants for whom the sun has already set. EULER does not hesitate to affirm that this is the true cause of twilight, which is not explicable in terms of atmospheric refraction. The effect depends on the length of time the sun remains near the horizon and is therefore less marked near the equator.

Near the poles, especially at the time of the equinoxes when the sun remains for several days close to the equator, the effect is very great, so that the particles are driven a considerable distance from the earth and appear as the northern lights (§ 12). This phenomenon is very variable owing to meteorological conditions, and on account of the earth's rotation, precedes a little the place opposite the sun, in agreement with the observation that the appearance declines from the true north (§ 13). Similar phenomena are not seen about the poles of other planets, probably because of the great distances combined with the feebleness of the light. In the case of the moon, the almost complete absence of an atmosphere explains why the phenomenon does not exist there (§14).

EULER infers the existence of a solar atmosphere from the appearance of sun-spots, which he believes are observed at a considerable distance from the surface of the sun (§15). Because of the large gravity, the expulsion of the particles would be entirely prevented if the sun did not rotate. But at the equator, the centrifugal force arising from this motion is quite large and suffices to dilate considerably the figure of the solar atmosphere, in which both DE MAIRAN[85]) and GIAN DOMENICO CASSINI (1625–1712)[86]), who discovered the phenomenon, locate the zodiacal light. Thus it seems probable to EULER that the zodiacal light is no other than the view of the solar atmosphere extended about the equator. Finally, EULER calculates the form of the solar atmosphere and believes it possible to infer from the cubic equation he obtains that the atmosphere must assume the form of a ring, similar to that surrounding Saturn (§16)[87]). However, he notes that the observations do not permit us to decide whether the zodiacal light is contiguous to the sun or placed at some distance from it in the form of a ring.

THE RESISTANCE OF THE AETHER (E89, E183, E184 and E218)

Writing to DELISLE[88]) on 15 February 1746, EULER announced the forthcoming publication of a volume of his dissertations by AMBROSIUS HAUDE[89]). In particular, EULER mentioned his new theory of light and colours, different from that of NEWTON, but more important, he added, was his discovery that the length of the year was not constant. All the observations of the eighteenth century, he explained, were in agreement in showing the year to be shorter than that given in the tables of CASSINI or revealed by the observations of the three previous centuries. Although the observations of PTOLEMY appeared to show the opposite, EULER defended both PTOLEMY's observations and his own conclusion. For he located the apparent contradiction in an error of a day made by the chronologists in the reduction of PTOLEMY's times to the Julian calendar. This decrease in the length of the year,

85) J.J. D'ORTOUS DE MAIRAN, loc. cit., p. 3.

86) G.D. CASSINI, "Découverte de la lumière celeste qui paroist dans le zodiaque", *Mémoires de l'Académie Royale des Sciences depuis 1666 jusqu'à 1699* (Paris, 1730), vol. 8, pp. 121–209. Cf. *Histoire de l'Académie Royale des Sciences depuis 1666 jusqu'à 1686* (Paris, 1733), vol. 1, pp. 378–381. CASSINI first observed the phenomenon on 18 March 1683.

87) In a letter to CLAIRAUT of 26 October 1751, EULER acknowledged that he had erred in believing to be able to deduce the ring formation from his cubic equation; R438a, EULER, *Opera omnia* IVA5, pp. 217–218.

88) R517, *Leonard Ejler, Pis'ma k učenym*, ed. T.N. KLADO et al. (Moscow and Leningrad, 1963), pp. 113–120.

89) *Euleri opuscula varii argumenti* (Berlin, 1746), vol. 1.

EULER explained to DELISLE, he attributed to the resistance of the aether. For in his dissertation, *De relaxatione motus planetarum* (E 89)[90]), he had demonstrated that, if the aether has a resistance, the periodic times of the planets and comets, and also the eccentricities of their orbits, must diminish. In the case of the comets, the effect should be considerable and this had been confirmed by the observations of HALLEY's comet in 1682 and in its two previous appearances.

There was an essential connection between the two dissertations that EULER had mentioned in his letter to DELISLE. For his wave theory implied the need for some kind of medium or aether for the transmission of light[91]).

One of the conclusions reached by EULER was that the resistance had no effect on the motion of the apogee. Consequently, the Cartesian explanations of this motion as the effect of an impulsion of the vortex, equivalent to a negative resistance, could not be true. The problem of the motion of the moon's apogee was the subject of correspondence between EULER and CLAIRAUT in the autumn and winter of 1747. In order to solve this problem, CLAIRAUT had suggested an amendment of the inverse-square law of attraction by a small term proportional to the inverse-fourth power of thc distance. EULER agreed[92]) that the forces acting on the moon did not follow NEWTON's law exactly. This circumstance, he added, rendered very probable the hypothesis that the forces of gravitation were caused by vortices, since it was then easy to conceive that these forces were modified when they were transmitted by some other vortex, as for example, when the force of the sun on the moon at an opposition is transmitted through the earth's vortex[93]). CLAIRAUT, in his reply[94]), was critical of EULER's appeal to vortices. In particular, he remarked that EULER's demonstration of the lack of effect of the resistance on the motion of the apogee diminished the probability that the problem could be solved by the action of vortices[95]). EULER then emphasized that, in his lunar tables[96]), he had modified his formula for the moon's motion

90) On the title page of the volume, this essay is entitled, *De perturbatione motus planetarum a resistentia aetheris orta.*

91) In his essay on the resistance of the aether EULER does not mention the problem of the cause of gravitation.

92) Letter of EULER to CLAIRAUT, 30 September 1747; R 420, EULER, *Opera omnia* IV A 5, pp. 175–177. Cf. the letter of EULER to MAUPERTUIS, 2 December 1747; R 1525, EULER, *Opera omnia* IV A 6, pp. 94–95. See also EULER, E 112, "Recherches sur le mouvement des corps célestes en général", *Opera omnia* II 25, pp. 1–44.

93) This explanation was entirely consistent with the conclusion that the impulsion of the matter of the vortex could have no effect. For it involved the centrifugal forces of the circulating matter in terms of which the Cartesians explained gravitation.

94) Letter of CLAIRAUT to EULER, 7 December 1747; R 421, EULER, *Opera omnia* IV A 5, pp. 177–179.

95) In the context of Cartesian theory, the resistance and the centrifugal force giving rise to gravity were quite independent.

96) E 87, "Tabulae astronomicae solis et lunae", *Opuscula varii argumenti* (Berlin, 1746), vol. 1, pp. 137–168 = *Opera omnia* II 23, pp. 1–10.

in accordance with observations received indirectly from JAMES BRADLEY (1693–1762). Although he agreed that the motion was not in accordance with NEWTON's law, he had not on that account thought to correct the theory by some modification of the expression of the forces[97]). Towards the end of 1748 CLAIRAUT succeeded in reconciling the motion of the moon's apogee with NEWTON's inverse-square law. He presented his new findings to the Paris Academy of Sciences on 17 May 1749 and informed EULER in a letter of 19 June 1749[98]).

The theologians JOHANN KASPAR WETTSTEIN (1695–1760), chaplain to the Prince of Wales, and ERIK PONTOPPIDAN (1698–1764), Bishop of Bergen, were interested in EULER's conclusions concerning the resistance of the aether, because they provided scientific evidence for the beginning and end of the world. WETTSTEIN published extracts of two letters[99]) he received from EULER on the subject in the *Philosophical Transactions* (E 183 and E 184). After seeing a German translation[100]) of the first of these, which appeared in Berlin, PONTOPPIDAN wrote to EULER on 26 March 1754, asking six questions of a technical nature. This letter (R 2020), together with EULER's reply of 11 May 1754 (R 2021, E 218), was included by PONTOPPIDAN in an essay which he presented to the Danish Academy of Sciences in 1755 and published in Copenhagen in 1757 with the title *Afhandling om Verdens Nyehed*[101]).

E 89
De relaxatione motus planetarum

Although the Newtonian principle of universal gravitation was generally accepted as an empirically established fact, EULER was of the opinion that this force could not operate in a vacuum. Moreover, the penetration of light, whose rays no one doubted were joined

97) Letter of EULER to CLAIRAUT, 6 January 1748; R 422, EULER, *Opera omnia* IV A 5, pp. 180–182. The implication is that his suggestion of a modification of forces in the course of transmission by vortices was a pure speculation having no role in the development of his mathematical theories.

98) R 425, EULER, *Opera omnia* IV A 5, pp. 186–188.

99) R 2763, R 2765, cf. note 113.

100) *Physikalische Belustigungen* 1 (1751), 313–315. This translation was based on an extract published in *The London Magazine* 20 (1751), 121.

101) ENESTRÖM locates the first publication in E. PONTOPPIDAN, *Essays sur la nouveauté du monde* (Copenhagen, 1755). This title is not included, however, in the bibliography of PONTOPPIDAN given in R. NYERUP og J. E. KRAFT, *Litteraturlexicon for Danmark, Norge, og Island* (Copenhagen, 1820). Moreover, the German translation, *Abhandlung von der Neuigkeit der Welt* (Copenhagen and Leipzig, 1758) is made from the Danish edition of 1757 not mentioned by ENESTRÖM. It seems that ENESTRÖM misinterpreted PONTOPPIDAN's remark, in the letter, that he was writing 'un essay sur la nouveauté du monde', supposing this to be the actual title rather than a description of the theme. The letters were published again by B. HANSTED in *Bulletin des sciences mathématiques et astronomiques*, deuxième série, 3 (1879), 26–32.

with subtle matter, showed that the whole of space was filled with this matter, henceforth called aether (§1). The question therefore arises whether the planets and comets in their motions show any evidence of the effect of the resistance of the aether. Since the tails of comets are not deflected by the aether, having a constant direction away from the sun, it seems likely that the aether is at rest in the heavens and therefore does not circulate with the planets (§2). Consequently the planets and comets should be expected to show the effects of the resistance in a decrease of their periodic times and the eccentricities of their orbits (§3). No changes of this kind had apparently been detected in the case of the planets, taking account of over 4000 years of astronomical observations, but the period of HALLEY's comet had evidently decreased by over a year between its last two appearances.

Using NEWTON's formula[102]) $v = \sqrt{(e/d)}$ for the velocity of waves in an elastic fluid, where e is the elasticity and d the density, and comparing the propagation of sound in air with that of light in the aether, EULER estimates mn to be about 4×10^{11}, where m is the ratio of the density of air to that of the aether and n is the ratio of the elasticity of the aether to that of air (§4)[103]). Supposing n to be 1000 – he thinks it can hardly be less – EULER infers that m is about 4×10^8. As the resistance depends on the density, it follows that the resistance of the aether should be very small. Before investigating the effect of this resistance on the motion of the planets, EULER first considers the effect on a sphere freely advancing through the aether.

Let the radius of the sphere be a and its density k times that of the air, or in other words km times that of the aether. Also suppose that the sphere has initial speed corresponding to fall from a height b, and that, when it has moved a distance x, its speed V corresponds to a fall from a height v. Then using NEWTON's proposition[104]) that the sphere meets with resistance, which is to the force by which the whole motion may be taken away in the time required to describe $\frac{8}{3}$ of a diameter moving uniformly as the density of the medium to the density of the sphere, EULER derives the equation (§5)[105])

$$dv = -\frac{3v\,dx}{8kma}.$$

102) NEWTON, *Principia* (1726), 3rd edition, Book II, proposition 48. Cf. C. A. TRUESDELL's introduction in EULER, *Opera omnia* II 13, pp. xxi–xxiv.

103) Cf. E 88, "Nova theoria lucis et colorum", *Opera omnia* III 5, pp. 16–18.

104) NEWTON, *Principia* (1726), 3rd edition, Book II, proposition 35, Scholium.

105) Let the sphere have unit mass and the time taken to describe 8/3 of a diameter moving uniformly with speed V be T. Then $T = 16a/(3V)$ and the force needed to stop the sphere in this time $= V/T = 3V^2/(16a) = 3gv/(8a)$, since $V^2 = 2gv$, where g is the acceleration due to gravity. Hence the resistance $= 3vg/(8kma)$. But the deceleration caused by the resistance $= -dV/dt = -\frac{1}{2}d(V^2)/dx = -g\,dv/dx$, where t is the time at which the speed is V. Hence $g\,dv/dx = -3vg/(8kma)$ and consequently $dv = -3v\,dx/(8kma)$.

Cf. J. BERNOULLI, "Discours sur les loix de communication des mouvements", (1727), *Recueil des pièces qui ont remporté les prix de l'Académie Royale des Sciences*, vol. 1 (Paris, 1752), pp. 71–76.

Integrating, this becomes

$$\log \frac{b}{v} = \frac{3x}{8kma}.$$

As an example he finds that a sphere of iron would have to travel 149×10^8 radii in order to lose a thousandth part of its speed. Since the earth is porous, the resistance would be further reduced, perhaps by a factor of 10, so EULER assumes the resistance to the earth to be $v/(27km)$ and estimates $27km$, henceforth denoted by μ, to be about 10^{14} (§6).

EULER explains that, in the case of the planets and comets, as the speed is diminished by the resistance, the force of the sun draws the body nearer and increases the speed again, so that the distance from the sun and the periodic time should both decrease (§7).

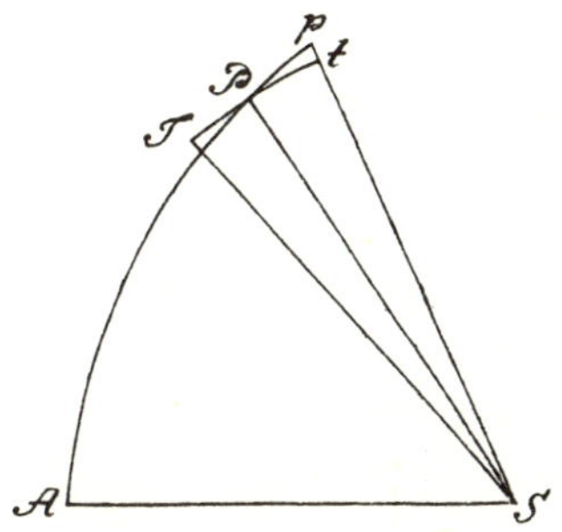

Fig. 1

Let S (fig. 1) be the centre of the sun, A the major axis of the earth's orbit, T the periodic time of the earth's motion about the sun and f the distance from the sun at which the centripetal force is equal to the force of gravity at the earth's surface. Then EULER finds that[106])

$$T = \frac{\sqrt{(1000)}}{250} \frac{\pi A^{3/2}}{f},$$

where A is measured in Rhenish feet. The value of f he finds to be $320\,708a$, where a is the radius of the earth. For any comet or planet, the force of attraction is f^2/y^2, where y is the distance from the sun and the gravity at the earth's surface is taken as the unit of force.

Now suppose that when the planet is in the perihelion A, the speed is that corresponding to a fall b. Also let $AS = h$ and ST be the perpendicular to the tangent PT.

106) The attraction at distance x from the sun is f^2g/x^2, where g is the acceleration due to gravity. At the mean distance of the earth,

$$\frac{f^2g}{x^2} = \left(\frac{2\pi}{T}\right)^2 x, \quad \text{or} \quad T = \frac{\pi}{\sqrt{(2g)}} \frac{A^{3/2}}{f},$$

where the major axis $A = 2x$. Thus EULER takes $\sqrt{(2g)} = 250/\sqrt{(1000)}$ approximately.

Suppose again that SP $= y$, ST $= p$, arc AP $= s$ and the speed of the planet at P to be that corresponding to a fall v. Take the radius of the planet to be α and let $\mu\alpha = c$. Then already in his *Mechanica* (1736)[107]), EULER had set up the equations (§8)

$$v = \frac{bh^2}{p^2} e^{-s/c} \quad \text{and} \quad \frac{f^2}{y^2} = \frac{2bh^2\,dp}{p^3\,dy} e^{-s/c}.$$

First he considers the case in which the resistance is zero, so that c is infinite (§9). Then denoting $2bh^2/f^2$ by g, the second equation becomes[108])

$$\frac{dy}{y^2} = \frac{g\,dp}{p^3},$$

representing a conic section. The solution is

$$p^2 = \frac{gh^2 y}{2h^2 + (g-2h)\,y}.$$

Now taking angle ASP $= t$, so that

$$dt = \frac{p\,dy}{y\sqrt{(y^2 - p^2)}},$$

it follows that

$$y = \frac{gh}{h + (g-h)\cos t} \quad \text{and} \quad p = \frac{gh}{\sqrt{(h^2 + 2h(g-h)\cos t + (g-h)^2)}}.$$

When the planet is in aphelion, $t = 1$, so that the aphelion distance $= gh/(2h-g)$. Adding this to the perihelion distance h gives the major axis $= 2h^2/(2h-g)$. The difference of the two distances divided by the major axis gives the eccentricity $= (g-h)/h$.

In the case of resistance, the arc length s is involved in the calculation. From the relation $ds = \sqrt{(dy^2 + y^2\,dt^2)}$ and the formula for y, it follows that

$$ds = \frac{g\,dt}{(1+\zeta\cos t)^2}\sqrt{(1 + 2\zeta\cos t + \zeta^2)},$$

where the eccentricity $(g-h)/h$ is represented by ζ. In particular, for a nearly circular ellipse[109]), in which ζ is small (§10),

107) E15, *Opera omnia* II1, p. 364.

108) EULER's constant g must not be confused with the acceleration due to gravity.

109) EULER also finds s (and the periodic time) in the case of an ellipse which is almost a parabola, but he makes no further application of these results (§11).

$$ds = g\,dt\,(1 - \zeta\cos t + \tfrac{3}{4}\zeta^2 + \tfrac{1}{4}\zeta^2\cos 2t)$$

and

$$s = g\,(t - \zeta\sin t + \tfrac{3}{4}\zeta^2 t + \tfrac{1}{8}\zeta^2\sin 2t)\,.$$

In order to take account of the resistance, EULER modifies the equation for y already found for motion in a vacuum to become

$$\frac{1}{y} = \frac{1}{g} + \frac{\zeta}{g}\cos t + P,$$

where P and its derivatives $dP/dt = Q$ and $dQ/dt = R$ are functions of t. From

$$dt = \frac{p\,dy}{y\sqrt{(y^2 - p^2)}},$$

it follows that

$$\frac{1}{p^2} = \frac{1}{y^2} + \frac{dy^2}{y^4\,dt^2}\,.$$

Substituting the value for y, differentiating and neglecting squares of P and Q, it follows that

$$g\frac{dp}{p^3} = \zeta\sin t\,dt - Q\,dt + \zeta P\sin t\,dt + \zeta R\sin t\,dt\,.$$

Now substituting the values of dy/y^2 and dp/p^3 in the equation

$$\frac{e^{s/c}\,dy}{y^2} = \frac{g\,dp}{p^3},$$

and taking $e^{s/c} = 1 + s/c$, since s/c is very small, leads to the equation

$$\frac{s\,dt^2}{cg} = d^2P + P\,dt^2,$$

which gives

$$P = \frac{\sin t}{cg}\int s\cos t\,dt - \frac{\cos t}{cg}\int s\sin t\,dt\,.$$

The formulae for Q and R follow by differentiation and the initial conditions require that the integrals should be zero when $t = 0$ (§12).

When the planet is in aphelion or perihelion, $dy = 0$ and consequently $(\zeta\sin t)/g = Q$. Having found the value of t from this equation, the perihelion and aphelion distances can then be calculated from the equation (§13)

$$\frac{1}{y} = \frac{1}{g} + \frac{\zeta \cos t}{g} + P.$$

Unable to find a general solution (§ 14), EULER investigates the particular case of an ellipse with small eccentricity (§ 15). Using the value of s already found, he derives the values

$$P = \frac{1}{c}(t - \sin t - \tfrac{1}{2}\zeta \sin t + \tfrac{1}{2}\zeta t \cos t + \tfrac{3}{4}\zeta^2 t - \tfrac{2}{3}\zeta^2 \sin t - \tfrac{1}{24}\zeta^2 \sin 2t)$$

and

$$Q = \frac{1}{c}(1 - \cos t - \tfrac{1}{2}\zeta t \sin t + \tfrac{3}{4}\zeta^2 - \tfrac{2}{3}\zeta^2 \cos t - \tfrac{1}{12}\zeta^2 \cos 2t).$$

It follows, therefore, that the planet is in aphelion or perihelion when (§ 16)

$$\frac{\zeta c}{g} \sin t = 1 - \cos t - \tfrac{1}{2}\zeta t \sin t + \tfrac{3}{4}\zeta^2 - \tfrac{2}{3}\zeta^2 \cos t - \tfrac{1}{12}\zeta^2 \cos 2t.$$

As this equation is satisfied when $t = 0, 2\pi, 4\pi, \ldots$, the place of perihelion remains fixed. Consequently, EULER remarks, any observed motion of the apsides cannot be explained as an effect of the resistance, nor can such motion be explained (as by the Cartesians) by the circulation of the vortex, whose effect would be equivalent to a negative resistance.

Evidently the equation is not satisfied when $t = \pi, 3\pi, 5\pi, \ldots$, but when $t = \pi - 2g/\zeta c, \ldots$, so that the aphelion is not exactly diametrically opposite the perihelion but very slightly in advance of this point (§ 17).

Substituting the appropriate values of P in the equation

$$\frac{1}{y} = \frac{1}{g} + \frac{\zeta}{g} \cos t + P,$$

EULER finds the perihelion and aphelion distances after i revolutions to be respectively (§§ 18–20)

$$\frac{g}{1+\zeta} - \frac{2i\pi(1+\frac{1}{2}\zeta)g^2}{c(1+\zeta)^2}$$

and

$$\frac{g}{1-\zeta} - \frac{(2i+1)\pi(1-\frac{1}{2}\zeta)g^2}{c(1-\zeta)^2}.$$

It follows that, after i revolutions, the major axis

$$= \frac{2g}{1-\zeta^2} - \frac{4i\pi g^2}{c(1-\zeta^2)^2},$$

the distance of the foci

$$= \frac{2\zeta g}{1-\zeta^2} - \frac{2i\pi\zeta g^2(3-\zeta^2)}{c(1-\zeta^2)^2}$$

and the eccentricity

$$= \zeta - \frac{i\pi\zeta g}{c}.$$

The periodic time after i revolutions, derived by KEPLER's third law from the ratio of the initial and final values of the major axis, is found to be

$$T\left(1 - \frac{3i\pi g}{c(1-\zeta^2)}\right),$$

where T is the initial periodic time (§21). For planets of similar density, the decrease is evidently inversely proportional to the diameter (since $c = \mu\alpha$) and is greater when the eccentricity is large.

In order to apply these results to the earth, EULER first supposes that the determinations of the maximum equation by TYCHO BRAHE (1546–1601) and NEWTON can be regarded as reliable (§22). From these, he infers that the maximum equation, which may be taken proportional to the eccentricity, had decreased by 7′ in 100 years. Starting with TYCHO's value of 2° 3′ 15″ and assuming a decrease of 7′ in 100 years, it follows that

$$2^\circ\, 3'\, 15'' = \left(1 - \frac{100\pi g}{c}\right) 1^\circ\, 56'\, 15'',$$

so that $100\pi g/c = 1/17$ approximately. This means that the decrease in the periodic time, namely $300\pi gT/c$, should be about $\frac{1}{6}$ of the initial value T. As this is absurd, EULER concludes that, even if the decrease in the maximum equation really exists, it must have another cause[110]).

On the basis of the historical record of determinations of the length of the year, EULER believes that it is reasonable to suppose a decrease in the period of the earth's orbital revolution of about 5″ in a century (§23). Then equating

$$\frac{3i\pi gT}{c(1-\zeta^2)}$$

to 5″, and taking $i = 100$ and $\zeta = 1/60$, gives $c/g = 5\,950\,231\,000$. Since $g = 15866a$ and $c = \mu a$, it follows that $c/g = \mu/15866$, so that $\mu = 94\,406\,365 \times 10^6$, or approximately 10^{14},

110) There is in fact a very small secular decrease in the eccentricity but here it is engulfed in the errors of the older determinations, which arise from the use of faulty values of refraction and solar parallax. See C. A. WILSON, "Perturbations and Solar Tables from LACAILLE to DELAMBRE: the rapprochement of observation and theory. Part 1", *Archive for history of exact sciences* 22 (1980), 53–188, p. 100.

the value previously conjectured (§24). As the corresponding decrease in the maximum equation, arising from the resistance of the aether, is only 1″ in 2720 centuries, it follows that the decrease in the eccentricity is quite imperceptible (§25). Also the predicted advance of the position of aphelion in front of the point diametrically opposite the perihelion is only 1/244th part of 1″ and therefore negligible. As far as the planets are concerned, EULER has thus shown that the effects to be expected from the resistance of the aether are entirely consistent with the observations.

In the case of the comets, EULER explains that the formula for s is too complicated to permit him to derive any precise results (§26). Since, however, the decrease in the periodic time of the planets has been shown to be greater according to the size of the eccentricity, he believes it to be not surprising that HALLEY's comet should return to perihelion after 76 years and 75 years in successive revolutions[111]).

E 183

Part of a letter from EULER to WETTSTEIN, concerning the gradual approach of the earth to the sun

Between 1746 and 1749 EULER corresponded with WETTSTEIN on scientific matters, especially those associated with the affairs of the Academies[112]). This translation of an extract from EULER's letter of 28 June 1749[113]) was read to the Royal Society on 2 November 1749 and published in the *Philosophical Transactions* for that year, which appeared in 1752.

Having informed WETTSTEIN that the astronomer PIERRE-CHARLES LE MONNIER (1715–1799) intends to publish a translation of an Arabic manuscript which he has found in Leiden, EULER remarks that he is impatient to see such a work, containing observations not so ancient as those of PTOLEMY[114]) but older than those of BERNHARD WALTHER[115]),

111) HALLEY had himself attributed the variations in period to the perturbations of Jupiter and Saturn. Some years after EULER composed the paper, CLAIRAUT investigated the perturbations. See, for example, A. C. CLAIRAUT, "Mémoire sur la comète de 1759, dans lequel on donne les périodes qu'il est le plus à propos d'employer, en faisant usage des observations faites sur cette comète dans les quatre dernières apparitions", *Mémoires de l'Académie Royale des Sciences* (Paris), (1759), 1765, pp. 115–120.

112) R 2748–R 2804. The correspondence has been published in *Die Berliner und die Petersburger Akademie der Wissenschaften im Briefwechsel Leonhard Eulers*, ed. A. P. JUŠKEVIČ und E. WINTER, Teil 3 (Berlin, 1976), pp. 256–366. Most of the original manuscripts are in the library of the Wellcome Institute for the History of Medicine (London), Western Manuscripts 5152. While 56 of EULER's letters are extant, only one of WETTSTEIN's letters is known.

113) R 2763. The French original of this letter is not known.

114) Although EULER does not mention the fact in the letter, the observations of PTOLEMY were of less use to him in establishing the shortening of the year, as he had explained to DELISLE, because he believed there was an uncertainty of at least one day in the reduction of PTOLEMY's times to the Julian calendar.

on which he has based his conclusion concerning the shortening of the year. The reason for the shortening, he goes on to explain, is the resistance of the aether, and this entails the gradual approach of the earth to the sun. It follows that the system of planets as we know it had a beginning and must also have an end.

E 184

Part of a letter from EULER to WETTSTEIN, concerning the contraction of the orbits of the planets

This translation of an extract from EULER's letter of 20 December 1749 (R 2765) was read to the Royal Society on 1 March 1750 and published in the same volume of the *Philosophical Transactions* as the previous letter.

While confirming his belief that the orbits of the planets are continually contracted and their periodic times consequently reduced, EULER remarks that, in order to remove any doubt, good ancient observations are needed. He then explains the problem with the observations of PTOLEMY; namely, that the times of PTOLEMY's observations were probably one or two days earlier than the chronologers had calculated. The Arabic observations, he adds, should not have been liable to this inconvenience[116]).

The remainder of the published extract contains a number of speculations unrelated to the effect of the resistance of the aether. Following a reference to HALLEY's remark that the revolutions of the moon are quicker than in the time of the ancient Chaldeans, EULER raises the question whether the length of the day has always been the same. Even if the ancient Chaldeans, like the people of EULER's time, had been able to measure the length of the day by pendulum oscillations, no conclusion could be drawn unless gravity were supposed to be constant, and a variation in gravity, he adds, would be undetectable. Yet EULER believes that the force of Jupiter so accelerates the earth in its orbit that the consequent diminution in the length of the year would be too great to be in agreement with observation, unless the diurnal motion had been accelerated in the same proportion.

115) BERNHARD WALTHER (1430–1504) was a student of REGIOMONTANUS. There is a facsimile of his solar observations in *Joannis Regiomontani opera collectanea*, ed. FELIX SCHMEIDLER (Osnabrück, 1972), pp. 627–641.

116) In a letter of 27 September 1749 (R 2764), EULER had informed WETTSTEIN that he was depending on the Arabic manuscript discovered by LE MONNIER, though he added that the Arabic star-catalogues were useless because they only gave the conclusions and not the observations on which these were based. *Die Berliner und die Petersburger Akademie der Wissenschaften im Briefwechsel Leonhard Eulers*, Teil 3, pp. 283–285.

E 218

Letter of Euler to Pontoppidan

On reading in the *Physikalische Belustigungen* Euler's letter to Wettstein concerning the gradual approach of the earth to the sun, Pontoppidan sought clarifications that would help him in the composition of an essay on the newness of the world. Of the six questions posed by Pontoppidan, the first three concerned the aether. If this was simply the matter of light, it seemed to Pontoppidan that it should drive the earth away from the sun rather than towards it. Moreover, the continuous emanation of luminous matter from the sun should cause the sun to decrease in volume and the solar vortex to become brighter. Pontoppidan next asked whether the inhabitants would have been able to accommodate themselves to the climate, if the earth had at one time been at the distance of Saturn. To this question he offered his own answer, suggesting that the sun was then larger and diminished in proportion to the approach of the earth and the other planets. Lastly, he asked whether the shortening of the year had really been demonstrated.

In his reply, written on 11 May 1754 (R 2021), Euler confirms his belief that the gradual approach of the planets towards the sun provides a convincing demonstration that the system of the world, as we know it today, has had a beginning and will have an end. For the purpose of proving this *a priori*, Euler first demonstrates the existence of an aether. Although Newton had supposed space to be void, his theory of light, Euler remarks, was inconsistent with this hypothesis. For if luminous matter emanates from the sun with the tremendous speed of light revealed by experiment, the whole of space must be filled with this matter. If on the other hand, light is a phenomenon similar to sound, as Euler maintains, then an elastic fluid filling space is needed for its transmission. Such a fluid cannot fail to resist the motions of the planets and the approach of the planets to the sun follows by the laws of mechanics.

Answering Pontoppidan's first question concerning the aether, Euler agrees that the luminous matter does in fact tend to drive the earth away from the sun, but this effect is overwhelmed by the gravity, which Pontoppidan had neglected. The other questions involving the aether, he explains, do not require an answer, for they are based on the false hypothesis that light is an actual emanation of matter from the sun. Concerning the question whether the inhabitants of the earth could have accommodated themselves to the climate at the distance of Saturn, Euler answers in the negative[117]), adding that, even if the earth had existed from eternity (which of course would be inconsistent with the contraction of the orbits), a special creation of inhabitants would have been needed at a certain time.

The remaining question concerns the observational confirmation of the shortening of the year. Euler explains that he had hoped to demonstrate this by means of the observations of Ptolemy and the length of the year calculated by Ptolemy[118]). But having found

117) Pontoppidan's own suggestion is rendered invalid by his false view of the nature of light.

118) Ptolemy confirms the length of the year in his time to be the same as that found by Hipparchus. *Almagest*, Book III, chapter 1. See, for example, *Ptolemy's Almagest*, translated and annotated by Gerald James Toomer (London, 1984), p. 139.

that the length of the year was no longer in the time of PTOLEMY than it is today[119]), he falls back on the explanation given in the second of his letters to WETTSTEIN on the subject; namely, that although the years have become shorter, so have the days, in such a way that the years have continued to contain the same number of days.

JOHANN ALBRECHT EULER ON THE RESISTANCE OF THE AETHER (A 8 and A 8^2)

In June 1761 JOHANN ALBRECHT EULER read to the Berlin Academy of Sciences an essay entitled, *Recherches sur la résistance du milieu dans lequel les planètes se meuvent* (A 8), in which he developed the researches introduced by his father in the memoir, *De relaxatione motus planetarum*[120]). This was published in 1762 and a copy was presented to the Academy on 7 December. A more polished version of the same essay, under the title, *Mémoire dans lequel on examine si les planètes se meuvent dans un milieu dont la résistance produise quelque effet sensible sur leur mouvement?* (A 8^2), received honourable mention in the competition for the prize of the Paris Academy of Sciences in 1762[121]). When this was published in 1771 in the *Recueil des pièces qui ont remporté les prix*, it received a brief notice in the *Journal encyclopédique*[122]) and a more extended review in the *Journal des Sçavans*[123]).

---

119) JOHANN TOBIAS MAYER (1723–1762) had published his conclusion that the length of the year was constant (assuming the length of the day to be constant) in *Commentarii Societatis Regiae Scientiarum Gottingensis* 1 (1751), 1752, 373.

In a letter of 22 August 1753 (R 1645), MAYER had explained to EULER that, from a consideration of PTOLEMY's observations of equinoxes and lunar eclipses, the intercalation of a day could be ruled out. *The Euler-Mayer Correspondence (1751–1755)*, edited by E. G. FORBES (London and New York, 1971), pp. 73–76. EULER evidently accepted this conclusion. In a further letter of 25 November 1753 (R 1646), MAYER admitted that a very weak inequality in the Earth's diurnal motion, which he had not been able to detect, could possibly be in existence, and he declared that he was satisfied with the reasons that EULER had used to support the resistance of the aether (*ibid.*, pp. 76–78).

120) E 89, pp. 195–220 of this volume.

121) The prize was awarded to the Abbé CHARLES BOSSUT (1730–1814) for his essay "Recherches sur les altérations que la résistance de l'Ether peut produire dans le mouvement moyen des Planetes", *Recueil des pièces qui ont remporté les prix de l'Académie Royale des Sciences*, vol. 8 (Paris, 1771).

122) *Journal encyclopédique* 6, Part I (1771), 17–18. There is also a brief notice in JOHANN III BERNOULLI's *Recueil pour les astronomes* 2 (1772), 250–251. At the time, however, the writer had not actually seen the essay.

123) *Journal des Sçavans*, March 1772, pp. 77–79.

A 8[2]

Mémoire dans lequel on examine si les planètes se meuvent dans un milieu dont la résistance produise quelque effet sensible sur leur mouvement?

In the first part of the essay, J. A. EULER considers the nature of the medium in which the planets move. The diffusion of light shows that the whole of space is filled with a subtle matter, called aether, while experiments in which sound waves are shown to pass through the same point without mutual interference suggest that light is propagated in a similar way (§§1 – 3). Then comparing the velocities of light and sound, it follows that, if the elasticity of the aether is at least 1000 times greater than that of air (which it would need to be in order to explain the cohesion of terrestrial bodies), its density must be at least $3{,}6 \times 10^8$ times less than that of air (§§4–5). The question arises whether it is possible for a matter to be so subtle and yet fill so much space.

J. A. EULER distinguishes between the true volume or geometric solidity of a body, which he defines as the space it would occupy if deprived of its pores, and the quantity of matter of the same body. Also he supposes that the aether, which is not subject to gravitation, is different in kind from gross matter, whose gravitation is an effect of the aether (§§8–10). This enables him to envisage properties of the aether which are different from those of gross bodies. For example, although the matter of all gross bodies is homogeneous, in the sense that the inertia or quantity of matter is proportional to the true volume, it is possible to suppose that a given true volume of aether has much less inertia than an equal true volume of gross matter (§§11–12). Again it is possible to suppose that the aether is endowed with a force of extending further into its own substance without thereby creating pores or void spaces (§§13–14). Although seemingly miraculous, such a property is needed to explain the elasticity of the aether.

In the second part of the essay, concerning the resistance of the aether, J. A. EULER first rejects the suggestion of some mathematicians that an elastic aether should offer no resistance (§15). Their arguments, he asserts, are destroyed by experiments in air, which indicate a resistance of this elastic fluid even greater than that predicted by the ordinary rules of mechanics (§16)[124]). According to the ordinary rules, the resistance experienced by a sphere of radius a and mass A moving with the speed acquired by a heavy body in falling freely through a distance v is $3Av/(8na)$, where $1:n$ is the ratio of the density of the aether to that of the sphere (§19). Because the aether passes freely through the pores of the sphere, EULER supposes the true resistance to be $3Av/(8\lambda na)$, where λ may well

124) He cites experiments of JOHAN LULOFS (1711–1768) with windmills: "Wiskundige en Werktuigkundige Beschouwinge der Wind-Molens", *Verhandelingen uitgegeeven door de Hollandse Maatschappy der Weetenschappen te Haarlem*, vol. 2, 1755, pp. 525–621. LULOFS writes: "Wy zagen ... dat de snelheid, die men in der daad bespeurt in de Windmolens, veel grooter is, dan die volgens de gewoone *Theorie* met de snelheid van den wind zoude overeenkomen", p. 605. Cf. "Men ziet ... dat de Molens veel meer doen, dan de Wisen Werktuigkundige Bespiegeling beloofde", p. 600.

surpass 10. For a sphere having the density of water, he estimates that $3/(8\lambda n)$, henceforth denoted by μ, is about 10^{-13}, or taking the density of the aether ten times less, as little as 10^{-14} (§20).

In order to obtain some idea of the effect to be expected, J. A. EULER applies this result to the case of a planet moving as it were in a circular canal which prevents it from leaving a circular orbit (§§21–25). For the earth, the effect would be an increase in the length of the year of 1 second in 6 years. In fact, when the planet is allowed to leave the circular orbit, the decrease in speed caused by the resistance leads to a decrease in the distance from the sun and a consequent decrease in the periodic time. This result is established in the more profound research undertaken in the third part of the essay.

Fig. 1

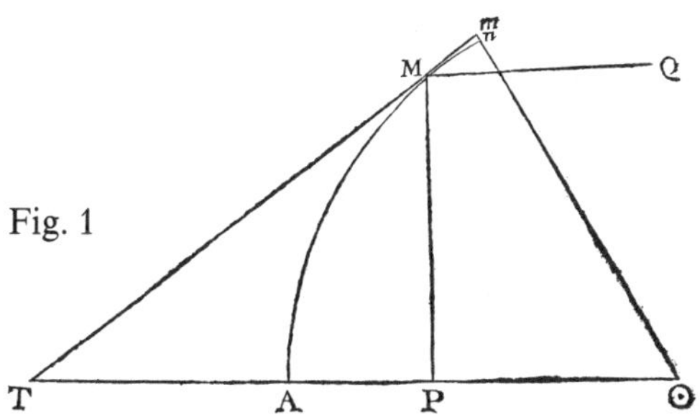

J. A. EULER first develops a general theory. Suppose the planet (fig. 1) of radius a describes the arc AM in time t. Let angle $A \odot M = \varphi$ and the distance of the planet from the sun $\odot M = z$. Also let the mean distance of the planet from the sun be c and the periodic time Θ. Then the accelerative force of the sun, acting in the direction $M\odot$, is f^2/z^2, where $f^2 = 2\pi^2 c^3/(g\Theta^2)$, g being the distance through which a heavy body at the surface of the earth falls in 1 second[125]). The components of this force in the directions MP and MQ are $(f^2/z^2)\sin\varphi$ and $(f^2/z^2)\cos\varphi$ respectively (§26).

Now let $\odot P = x$ and $PM = y$. Suppose $Mm = ds$ is traversed in time dt. Since $Mn = z\,d\varphi$ and $mn = dz$, it follows that $ds = \sqrt{(dz^2 + z^2\,d\varphi^2)}$. Also $ds = 2\sqrt{(gv)}\,dt$, so that $v = ds^2/(4g\,dt^2)$. Hence the accelerative force of the resistance, acting along MT, is $\mu v/a = \mu\,ds^2/(4ga\,dt^2)$ and the components of this force in the directions PT and MP are $-\mu\,dx\,ds/(4ga\,dt^2)$ and $\mu\,dy\,ds/(4ga\,dt^2)$ respectively (§27). It follows that the equations of motion are

$$ddx = -\frac{2gf^2\,dt^2\cos\varphi}{z^2} - \frac{\mu\,dx\,ds}{2a}$$

and

$$ddy = -\frac{2gf^2\,dt^2\sin\varphi}{z^2} - \frac{\mu\,dy\,ds}{2a}.$$

125) This has already been established in the second section (§§21–22). Taking the speed of the planet to be that acquired by a heavy body falling freely through a distance v, it follows that $2\pi c/\Theta = 2\sqrt{(gv)}$. Equating the accelerative force of the sun to the centrifugal force of the planet at the mean distance gives $f^2/c^2 = 2v/c$. Hence $f^2 = 2\pi^2 c^3/(g\Theta^2)$.

On eliminating x and y, these equations become (§28)

$$-2\,dz\,d\varphi - z\,dd\varphi = \frac{\mu z\,d\varphi\,ds}{2a} \tag{1}$$

and

$$ddz - z\,d\varphi^2 = -\frac{2gf^2\,dt^2}{z^2} - \frac{\mu\,dz\,ds}{2a}. \tag{2}$$

On division by $z\,d\varphi$, equation (1) becomes (§29)

$$\frac{2dz}{z} + \frac{dd\varphi}{d\varphi} + \mu\frac{ds}{2a} = 0,$$

of which the integral is

$$z^2\,d\varphi\,e^{\mu s/2a} = G\,dt, \tag{3}$$

where G is a constant. Also multiplying (1) by $-z\,d\varphi$ and (2) by dz and then adding gives

$$ds\,dds = -\frac{2gf^2\,dt^2\,dz}{z^2} - \frac{\mu\,ds^3}{2a}. \tag{4}$$

Dividing (4) by $\frac{1}{2}dt^2$ and integrating gives[126])

$$\frac{ds^2}{dt^2} = 4gf^2\left(\frac{1}{z}+\frac{1}{b}\right) - \frac{4\mu g f^2}{a}\int ds\left(\frac{1}{z}+\frac{1}{b}\right), \tag{5}$$

where b is a constant. Also using $z^2\,d\varphi^2 = ds^2 - dz^2$ to eliminate $d\varphi$ from (3) gives

$$e^{\mu s/a}z^2(ds^2 - dz^2) = G^2\,dt^2. \tag{6}$$

If at first the terms affected by μ are neglected, then eliminating dt from (5) and (6) gives (§30)

$$ds = \frac{dz}{\sqrt{\left(1-\dfrac{bh}{z(z+b)}\right)}}$$

where $4gf^2/G^2 = 8\pi^2c^3/(G^2\Theta^2) = 1/h$. It suffices to use this value of ds in the small terms of the equations affected by μ. Thus if

126) Since μ is small, ds^2/dt^2 in the second term can be taken equal to $4gf^2\left(\frac{1}{z}+\frac{1}{b}\right)$.

$$\frac{dz}{\sqrt{\left(1-\dfrac{bh}{z(z+b)}\right)}}$$

is called $d\sigma$, the equations become (§31)

$$e^{\mu\sigma/a}z^2(ds^2-dz^2)=G^2dt^2=\frac{8\pi^2c^3h}{\Theta^2}dt^2, \tag{7}$$

$$\frac{ds^2}{dt^2}=\frac{8\pi^2c^3}{\Theta^2}\left\{\frac{1}{z}+\frac{1}{b}-\frac{\mu}{a}\int d\sigma\left(\frac{1}{z}+\frac{1}{b}\right)\right\} \tag{8}$$

and

$$e^{\mu\sigma/2a}z^2d\varphi=\frac{2\pi c}{\Theta}\sqrt{(2ch)}\,dt. \tag{9}$$

From these equations, expressions for ds, dt and $d\varphi$ are obtained in terms of z.

In order to apply these results to the planets, J. A. EULER puts $h=\frac{1}{2}k$ and $b=-2k/(1-n^2)$, where k is the semi latus-rectum and n the eccentricity of the orbital ellipse. Then in the case where there is no resistance,

$$z=\frac{k}{1+n\cos\omega},$$

where $d\varphi=d\omega$ and ω is the true anomaly calculated from perihelion (§32). Using this value of z, he finds that

$$\sigma=\int\frac{k\,d\omega\sqrt{(1+2n\cos\omega+n^2)}}{(1+n\cos\omega)^2}.$$

Then supposing that, when the resistance is taken into account,

$$z=\frac{k}{1+n\cos\omega+\mu u}, \tag{10}$$

he finds that (§34)

$$u=\frac{1}{a}\left(\sin\omega\int\sigma\,d\omega\cos\omega-\cos\omega\int\sigma\,d\omega\sin\omega\right).$$

Provided that $n<1$, which will always be the case for an ellipse (§35), the arc σ can be reduced to an expression such as

$$\sigma=k(A\omega+Bn\sin\omega+Cn^2\sin 2\omega+Dn^3\sin 3\omega+En^4\sin 4\omega+\ldots),$$

from which it follows that

$$u = \frac{k}{a}\left(A\omega - \tfrac{1}{2}Bn\omega\cos\omega - \tfrac{1}{3}Cn^2\sin 2\omega - \tfrac{1}{8}Dn^3\sin 3\omega - \tfrac{1}{15}En^4\sin 4\omega - \ldots\right).$$

Substituting this value of u in equation (10), the value of z is found for a given true anomaly, and it follows from equation (9), taking

$$e^{\mu\sigma/2a} = 1 + \frac{\mu\sigma}{2a},$$

that (§ 36)

$$\frac{2\pi c\sqrt{c}}{\Theta k\sqrt{k}}\,t = \int\frac{d\omega}{(1+n\cos\omega)^2} - \frac{2\mu k}{a}\int\frac{d\omega\,(A\omega - \tfrac{1}{2}Bn\omega\cos\omega - \tfrac{1}{3}Cn^2\sin 2\omega - \ldots)}{(1+n\cos\omega)^3}$$

$$+\frac{\mu k}{2a}\int\frac{d\omega\,(A\omega + Bn\sin\omega + Cn^2\sin 2\omega + \ldots)}{(1+n\cos\omega)^2}.$$

J. A. Euler applies these results to three cases; first he considers a circular orbit, then an orbit of small eccentricity and finally an orbit of medium eccentricity. For a circular orbit (§ 38), $n = 0$ and $A = 1$, so that $\sigma = k\omega$,

$$z = \frac{k}{1+\dfrac{\mu k\omega}{a}}$$

and

$$\frac{2\pi c\sqrt{c}}{\Theta k\sqrt{k}}\,t = \omega - \frac{3\mu k}{4a}\,\omega^2.$$

After m revolutions, $\omega = 2\pi m$, so that

$$z = \frac{k}{1+\dfrac{2\mu\pi m k}{a}}$$

and

$$t = \frac{k\sqrt{k}}{c\sqrt{c}}\,\Theta m\left(1 - \frac{3\mu k m\pi}{2a}\right).$$

Similarly the time t' for $m+1$ revolutions will be

$$\frac{k\sqrt{k}}{c\sqrt{c}}\,\Theta(m+1)\left(1 - \frac{3\mu k(m+1)\pi}{2a}\right),$$

so that the periodic time after m revolutions will be

$$t' - t = \frac{k\sqrt{k}}{c\sqrt{c}}\,\Theta\left(1 - \frac{3\mu k\pi}{2a}(2m+1)\right),$$

where $\dfrac{k\sqrt{k}}{c\sqrt{c}}\,\Theta$ is the periodic time initially.

The resistance therefore reduces the distance of the planet from the sun and also the periodic time, though in any one revolution the orbit remains circular. Taking $\mu = 10^{-14}$, J. A. Euler found that, in the case of the earth, the decrease in the length of the year should be about 5″ per century, which seemed to be consistent with the observations (§ 39).

In the case of an orbit with small eccentricity, where n^2 and higher powers can be neglected, $\sigma = k(\omega - n\sin\omega)$ so that $A = 1$ and $B = -1$. The decrease in the periodic time is found to be exactly the same as in the case of a circular orbit (§§ 40–43). After m revolutions, the mean distance is found to be

$$k\left(1 - \frac{2\pi\mu mk}{a}\right)$$

and the eccentricity

$$n\left(1 - \frac{\pi\mu mk}{a}\right).$$

In the case of the earth, these changes are shown to be entirely insignificant even after 100 000 years.

The third case considered by J. A. Euler is that in which the powers of the eccentricity up to the fourth are retained (§§ 44–47). As a result of the involved calculation needed in this case, he finds that only the decrease in the periodic time is significant and that this is exactly the same as in the previous cases.

In the last part of the essay (§§ 48–60), J. A. Euler compares his theory with the observations of J. Cassini and the Abbé Nicolas Louis de Lacaille (1713–1762). As a result, he finds that, for agreement with the observations, the value of μ should be even smaller than he had supposed. In the absence of a precise value for μ, he estimates the relative effect of the resistance on the motions of the six planets.

J. A. Euler believed he had shown that the planets do experience some resistance from the aether, though he acknowledged that it was impossible to determine the true quantity by the theory alone. Finally, he commented critically on the arguments of Johann Tobias Mayer (1723–1762), who maintained that the length of the year was constant. If this were true, then without his father's hypothesis of a decrease in the length of the day, which J. A. Euler does not mention, there could be no question of the planets encountering any resistance from the aether.

Lunar Atmosphere
(E 142 and E 839)

From his observations of the annular eclipse of the sun visible in Berlin on 25 July 1748, Euler concluded that the moon possessed a thin atmosphere extending a little way from the surface. On 5 December of the same year, he presented his findings to the Berlin Academy of Sciences in a paper written in Latin[127]). A French version of this paper was published in 1750 in the *Mémoires* of the Academy for the year 1748 (E 142). Reviews, in the form of summaries, appeared in the *Nouvelle Bibliothèque Germanique*[128]), the *Bibliothèque impartiale*[129]) and the *Nova Acta Eruditorum*[130]).

E 142
Sur l'atmosphere de la Lune prouvée par la derniere eclipse annulaire du Soleil

Having already given an account of the more accurate determination of the parallax and mean motion of the moon[131]), which was the principal object of his observations of the annular eclipse of 25 July 1748, Euler seeks to demonstrate in this memoir that certain phenomena arising from the eclipse serve to establish the existence of a thin atmosphere surrounding the moon (§ 1).

Although cloud prevented him from seeing the beginning of the eclipse, Euler was able to observe the whole of the annular phase (§ 3)[132]). The telescope was set up in a darkened room so that the image of the sun fell on a white screen perpendicular to the axis and at such a distance that the image exactly filled a circle drawn on the screen in advance (§ 2). As the moon covered the disc of the sun, so that the visible part became a crescent (fig. 1), the horns were observed to protrude outside the circle drawn on the screen, though the limb of the sun elsewhere remained within the circle. When the horns came together and the sun appeared as an annulus, the whole of its disc formed on the screen an image much greater than the circle within which it had been previously confined (§§ 4–5).

127) E 839, "De atmosphaera lunae ex eclipsi solis annulari evicta". This was published in 1862 in the second volume of Euler's *Opera postuma*.

128) *Nouvelle Bibliothèque Germanique* 7 (1750), 34–36.

129) *Bibliothèque impartiale* 1 (1750), 329–331.

130) *Nova Acta Eruditorum*, 1751, pp. 67–68.

131) E 117, "Réflexions sur la dernière éclipse du Soleil du 25 juillet a. 1748", *Mémoires de l'Académie Royale des Sciences de Berlin* [3] (1747), 1749, 250–273 = *Opera omnia* II 30, pp. 51–72.

132) The times had been worked out by Johann Kies (1713–1781), "Observation de l'éclipse annulaire du soleil le 25 Juillet 1748 selon le nouveau stile à Berlin", *Mémoires de l'Académie Royale des Sciences de Berlin* [4] (1748), 1750, 99–102.

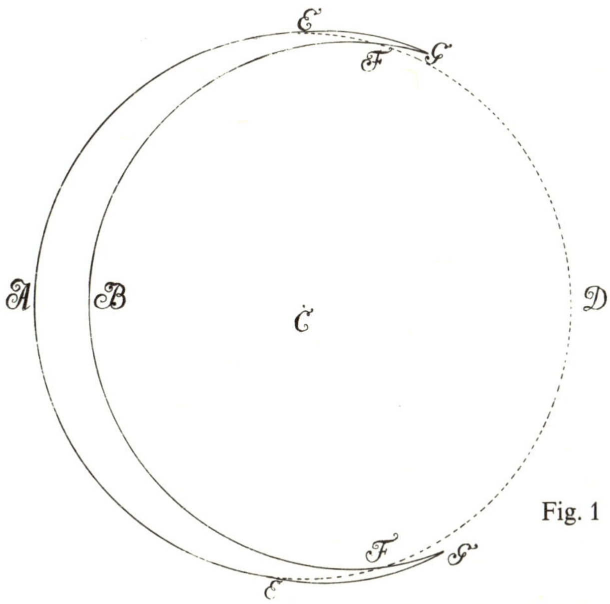

Fig. 1

More information on the observations is given by EULER in a letter to Marshal SAMUEL VON SCHMETTAU[133]). He used a 9 foot telescope and the diameter of the circle filled by the image of the sun before and after the eclipse was about half a foot. EULER was evidently assisted in his observations by CHRISTINE and MARGARETHE KIRCH. EULER simply remarks to VON SCHMETTAU that they carefully marked all the phases of the sun during the eclipse. In her weather diary, however, CHRISTINE[134]) records that she arrived with her sister at EULER's house before 9 o'clock, at which time the sun was shining though there were clouds in the sky. Although her notes provide no further information on the eclipse itself, they contain a detailed record of the weather during the time of the eclipse.

Suppose that, towards the middle of the eclipse, *AZBN* (fig. 2) represents the disc of the sun, where *Z* is the highest point and *AB* is horizontal, *A* being on the east side and *B* on the west. Suppose also that *azbn* represents the disc of the moon and *EF* the line through the centres. Then the angle between *EF* and *ZN* was found to be about 40°, while the minimum width *Ee* of the annulus EULER estimated to be about a quarter of the maximum width *Ff* (§5).

Observations before the formation of the annulus confirmed the theoretical values for the radius of the sun $=952''$ and of the moon $=898''$. The minimum distance between the centres, *Cc*, was found to be 53″. Whence, if it is supposed that the sun had no dilatation in the region of maximum width *Ff* (since the dilatation occurs when the limb of the moon

133) R 2092, *Russko-francuzskie naučnye svjazi.* A.T. GRIGORJAN and A.P. JUŠKEVIČ, Leningrad 1968, pp. 274–276.

134) CHRISTINE KIRCH, "Gewitter Observationes 1748", Entry for 25 July. MS (Shelf Mark 2.71), Crawford Library of the Royal Observatory, Edinburgh.

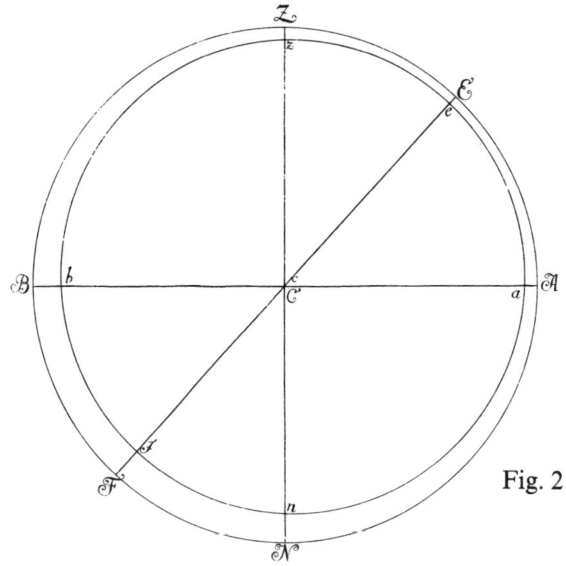

Fig. 2

touches that of the sun), it follows that $Ff = 107''$ (§6). Then the minimum thickness Ee should have been $1''$, whereas the observations indicated a value of $26''$ (§7). Observations of the eclipse at Frankfurt an der Oder gave a similar result[135]).

It had been suggested by observers of other annular eclipses that the diameter of the moon appeared smaller in the eclipse than outside the sun (§10)[136]). EULER's measurements, however, had shown that this was not the case, so that he concluded the dilatation of the annulus and the protrusion of the horns to be real. These phenomena, he claimed, demonstrated the existence of a lunar atmosphere which refracted the sun's rays passing through it (§11).

Now let the distance MN (fig. 5) from the surface of the moon be observed at the earth T as an angle x. Because both the height of the whole atmosphere MK and the refraction itself are small, the apparent position s of the star S may also be taken to be at a distance x from the surface of the moon. The true position of the star is nearer the centre of the moon by the distance $sS = 2$ angle $SP\sigma$, where angle $SP\sigma$ is the horizontal refraction at the moon (§17). From the observed dilatation of the annulus at the point of minimum thickness from $1''$ to $26''$ in the eclipse, it follows that, when the star appears to

135) The dilatation, seen independently by two observers in different locations, may possibly be ascribed to the sudden appearance of the 'blue' image, which had previously been masked by the brightness of the sun's image focussed in the middle of the spectrum. EULER himself comments that his estimate of the magnitude of the dilatation could be in error by several seconds because, first, it had not been possible to measure the size of the annulus exactly, and secondly, the colours in which the limbs of the sun and moon were constantly clothed, the limb of the sun in violet and that of the moon in red, prevented the extremities of these limbs from being distinguished.

136) See, for example, the report of COLIN MACLAURIN on an observation of the annular eclipse of 18 February 1737, *Philosophical Transactions* 40 (1738), 177–195, p. 183.

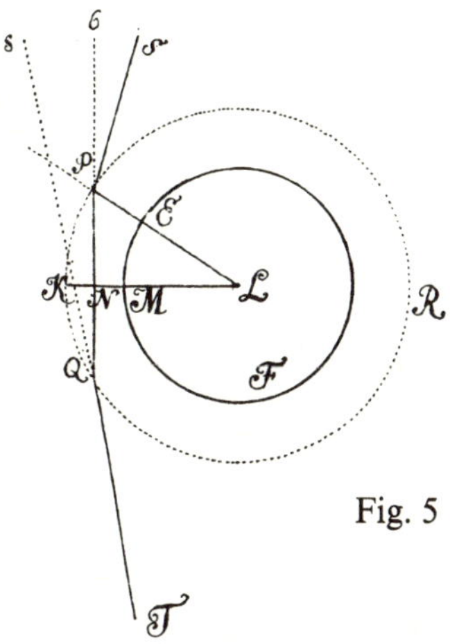
Fig. 5

be 25″ distant from the moon[137]), its true distance is zero. Again a wealth of observations shows that, when the distance becomes only 1″, the effect of refraction becomes entirely imperceptible (§ 18). Since it is impracticable to apply the theory of refraction, EULER proceeds by setting up a formula in agreement with the observations (§ 20). If z'' is the correction for refraction required at distance x'' from the moon, then he regards

$$z = \frac{40}{1 + x^2/400}$$

as a suitable formula. For when $x = 20$, the correction $z = 20$ and when $x = 60$, $z = 4$. Using this formula, EULER constructs a table of corresponding values of x and z (§ 21).

At the limb of the moon, where $x = 0$, the refraction is evidently 40″ as seen from the earth, or in other words, the horizontal refraction at the moon is 20″ (§ 22). It follows, therefore, that as soon as the limb of the sun is hidden behind the moon by less than 40″, it should become visible. One consequence of this is that an eclipse which is not in fact annular can appear annular (§ 23). Moreover, EULER remarks, the annulus when very narrow is invisible, like stars masked by the sun's rays, so that it appears suddenly, when it has already gained a considerable size (§ 24). For a similar reason, the horns appear blunted on the screen (§ 26).

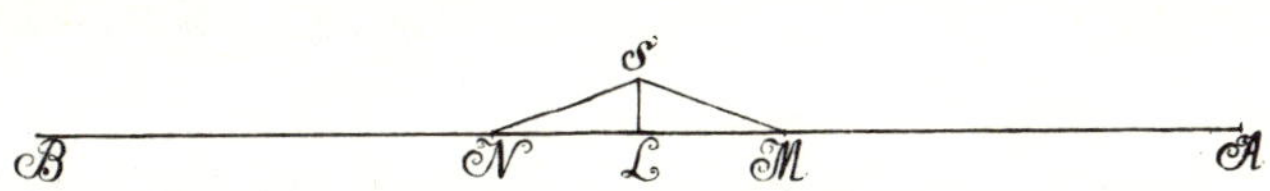
Fig. 6

EULER now attempts to define an annular eclipse by calculation (§ 28). Let AB (fig. 6) be the apparent path of the centre of the moon in relation to the centre of the sun, imagined to be stationary in S. The minimum distance of the centres $SL = c$. As soon as the centre

137) Because of the uncertainty of the data, this quantity, EULER remarks, could be 20″ or even 15″.

of the moon arrives at A the eclipse begins, and it ends when this centre arrives at B. Taking a and b as the radii of the sun and moon respectively, $SA = SB = a + b$, and since SL is very small, the line AB can be regarded as equal to $SA + SB = 2a + 2b$. Now let the difference of the radii of the sun and moon be d. Then the annulus begins to appear when the centre of the moon arrives at M, where $SM = d$ increased first by 40″ and then decreased by the size the annulus must have before it becomes visible (§29). Thus the annulus begins to appear when $SM = d + 40'' - 20'' = d + 20''$, and it ends when $SN = d + 20''$. The eclipse will therefore be annular if $d + 20''$ is greater than c.

Now suppose the duration of the eclipse to be t and the period of visibility of the annulus to be θ, so that $\theta = (MN/AB)\,t$. Then (§30)

$$c = \sqrt{\left((d + 20'')^2 - (a + b)^2 \frac{\theta^2}{t^2}\right)}.$$

Substituting the data from the observed eclipse gives $c = 72\frac{3}{4}''$, though the theoretical value is only 51″. Believing that he had overestimated the horizontal refraction at the moon, EULER tries a smaller value, 15″, which makes $c = 62\frac{1}{2}''$, exceeding the calculated value by just $11\frac{1}{2}''$ (§§31–32). It seems probable, he remarks, that the elements of the eclipse need this small correction. However, he goes on to say that the horizontal refraction at the moon hardly exceeds 10″, which would in fact make the correction unnecessary (§33).

Advocating further investigation of the lunar atmosphere, EULER suggests that the circular micrometer designed by GOTTFRIED KIRCH (1639–1710)[138]) could be employed to measure small changes in the distance of two close stars as one of them is occulted by the moon (§39). The stars would have to be not more than 15′ apart, so that both could be seen simultaneously in the telescope (§35). Let the true distance of the stars, measured before the

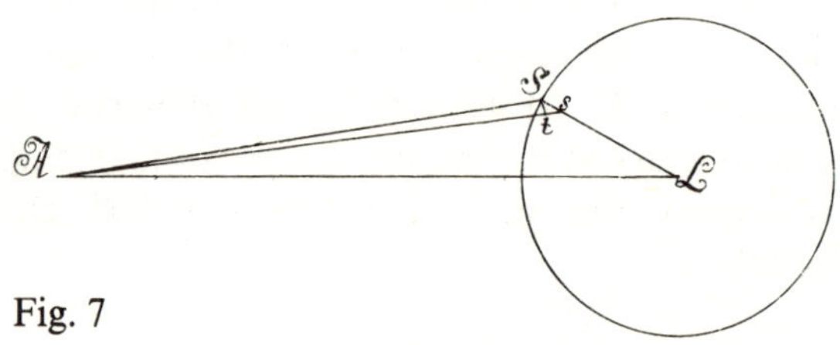

Fig. 7

occultation, be m''. Suppose the first star is at A (fig. 7) and the distance AS, measured at the time of occultation $= m'' - \mu''$. Now let angle $ASL = \varphi$ and the true position of the star S be s. Then $As = m''$, and taking $At = AS$, $st = \mu''$. Since angle $Sst = 180° - \varphi$ approximately, $Ss = -\mu''/\cos\varphi$, and $\frac{1}{2}Ss$ is the horizontal refraction at the surface of the moon (§36). Occultations at both the dark limb and the bright limb should be observed,

138) GOTTFRIED KIRCH, "De simplicissima quadam specie micrometri", *Miscellanea Berolinensia* 1, 1710, 202–205.

EULER suggests, in order to discover whether there is any difference in the refraction owing to the sun's heat (§ 37).

EULER finally remarks (§ 40) that, while the existence of a lunar atmosphere is beyond doubt, the small horizontal refraction of less than 20″ shows that it is almost 200 times less dense than the earth's atmosphere[139]).

THE FIGURE OF THE EARTH (E 619)

On receiving a copy of MAUPERTUIS' *Figure de la terre*, EULER remarked in a letter to the author dated 23 November 1738[140]), that the flattening of the earth at the poles revealed by the geodetic measurements was nearly twice as great as NEWTON had estimated it. EULER had, in fact, greatly exaggerated the discrepancy between theory and observation. Although the collaboration with MAUPERTUIS on this topic for which he had hoped did not take place, EULER returned to the problem in a paper presented to the St. Petersburg Academy of Sciences on 2 November 1775. This was published in 1788 in the *Nova Acta* for 1784.

E 619
Enodatio difficultatis super figura Terrae a vi centrifuga oriunda

According to observation, EULER declares, the ratio of the diameter of the equator to the axis of the earth is nearly 201 : 200, whereas estimates based on the force of gravity in combination with the centrifugal force arising from the rotation do not exceed 578 : 577 (§ 1). Now the ratio 578 : 577 is the result obtained by HUYGENS[141]), whereas NEWTON, applying his theory of the attraction of a spheroid, found the ratio 230 : 229, which is only

139) JOHANN TOBIAS MAYER, who claimed to have proved that the moon has no atmosphere, raised the question with EULER in a letter of 15 November 1751; R 1631, *The Euler-Mayer Correspondence (1751–1755)*, edited by E. G. FORBES (London and New York, 1971), p. 43. In his reply of 25 December 1751, EULER explained that the bending of light from the eclipsed sun that his observations had revealed could not be in doubt, but whether the cause of this was the effect of a thin lunar atmosphere or some other cause, such as the moon's attraction, could be considered as yet undecided (R 1632, *ibid.*, p. 45). In a further letter of 18 March 1752, EULER pointed out that he had suggested observations of occultations of stars by the moon to decide the matter and he now recommended to MAYER that he should make these observations himself since an excellent micrometer was available for the purpose (R 1634, *ibid.*, p. 55).

140) R 1496, EULER, *Opera omnia* IV A 6, pp. 39–42. Cf. NEWTON, *Principia*, 3rd edition (1726), Book III, proposition 19.

141) CHRISTIAAN HUYGENS, *Discours de la cause de la pesanteur* (Leiden, 1690), pp. 155–156. *Oeuvres complètes* (La Haye, 1888–1950), vol. 21, pp. 468–469.

about 10% less than the value found from geodetic measurements. Since NEWTON had used approximations, a rigorous calculation taking into account the attraction of a spheroid would seem to offer the best prospect of a solution. It is curious that EULER, in his paper, neither quotes NEWTON's result[142]) nor undertakes a critical examination of the principles on which it is based.

First, EULER demonstrates that, in the figure of the earth determined by the gravity towards the centre and the centrifugal force arising from the rotation, the ratio of the diameter of the equator to the axis is 577:576, whatever the law of gravity inside the earth of the form z^{n-1}, where z is the distance from the centre (§2). The discrepancy between this result and that derived from the geodetic measurements, he believes, shows that in addition to gravity to the centre of the earth and the centrifugal force, the parts of the earth must also be subject to a lateral force (§3). Although the lateral force evidently arises from universal gravitation[143]), EULER does not attempt to derive it from consideration of the attraction of the spheroid, but instead to define it in terms of the required mathematical properties (§4)[144]).

Fig. 1

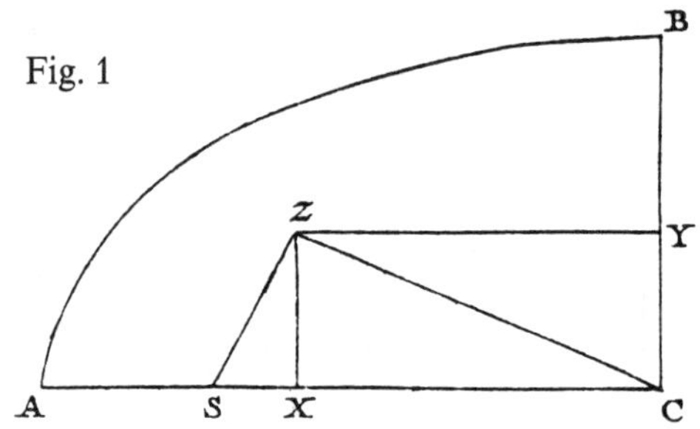

Let C (fig. 1) be the centre of the earth (assumed fluid), B the north pole and A a point on the equator. Also let Z be a point distant z from C, p the pressure and q the density of the fluid at Z. Then if Z represents the force of gravity at the point Z directed

142) He had quoted NEWTON's result in the letter to MAUPERTUIS (R1496).

143) In a spheroid, the gravitation is in general directed towards points other than the centre.

144) Important contributions based on analysis of the attraction of a spheroid were made by C. MACLAURIN in his prize essay on the tides, "De causa physica fluxus et refluxus" (1741), *Recueil des pièces qui ont remporté les prix de l'Académie Royale des Sciences*, vol. 4 (Paris, 1752), and A.C. CLAIRAUT in his *Théorie de la figure de la terre* (Paris, 1743). MACLAURIN's work was better presented in his *Treatise of fluxions* (Edinburgh, 1742), where he demonstrates, for example, that in a fluid spheroid of uniform density, similar concentric surfaces would be level surfaces at all depths, and the forces with which equal particles at those surfaces tend towards the spheroid are measured by normals to the surfaces terminated either by the plane of the equator or by the axis of the spheroid. *Treatise of fluxions*, vol. 2, pp. 530–531. CLAIRAUT extended the theory to a spheroid consisting of layers of different density. He showed that the ellipses separating the layers become more flat towards the poles as the distance from the centre increases. CLAIRAUT, *Théorie de la figure de la terre*, pp. 155–156, 198–232 and 262–296.

towards C, S the lateral force in the direction ZS perpendicular to CZ and $1/f$ is the centrifugal force at the equator, the total forces acting at Z in the directions CX and CY are respectively

$$P = \frac{x}{f} - \frac{Zx}{z} + \frac{Sy}{z}$$

and

$$Q = -\frac{Zy}{z} - \frac{Sx}{z},$$

where $x = \mathrm{CX}$ and $y = \mathrm{CY}$.

For the equilibrium of the fluid at Z, $dp/q = P\,dx + Q\,dy$, which may be written (§ 5)

$$\frac{dp}{q} = \frac{x\,dx}{f} - Z\,dz + \frac{S}{z}(y\,dx - x\,dy).$$

Taking angle ACZ $= \varphi$, so that $x = z\cos\varphi$ and $y = z\sin\varphi$, the equation becomes

$$\frac{dp}{q} = \frac{x\,dx}{f} - Z\,dz - Sz\,d\varphi.$$

Assuming the earth to be in equilibrium, it follows that this equation must be integrable (§ 6). Since $Sz\,d\varphi$ is integrable only if Sz is a function of φ, EULER takes this function to be $\alpha \sin\varphi \cos\varphi$, which is the simplest function satisfying the necessary conditions that it should be zero when $\varphi = 0$ and when $\varphi = 90°$ (§§ 7–9). For dp/q to be integrable, either q must be assumed constant, or through the particular strata, whenever q is the same p must be assumed also to be the same. Taking $Z = z^{n-1}$ and integrating each term, the equation becomes

$$\frac{p}{q} = C + \frac{x^2}{2f} - \frac{z^n}{n} - \frac{1}{2}\alpha\sin^2\varphi,$$

where C is a constant. At the surface of the ocean, $p = 0$, so that the equation of the surface is

$$0 = C + \frac{x^2}{2f} - \frac{z^n}{n} - \frac{1}{2}\alpha\sin^2\varphi.$$

When Z is at the pole B, $z = 1$, $x = 0$ and $\varphi = 90°$. Hence $C = \frac{1}{n} + \frac{1}{2}\alpha$, so that the equation becomes (§ 10)

$$z^n = 1 + \frac{nx^2}{2f} + \frac{\alpha n x^2}{2z^2}.$$

Now taking Z at A on the equator, $x = z = \mathrm{CA}$, so that

$$z^n = 1 + \frac{nz^2}{2f} + \frac{1}{2}\alpha n.$$

Let $z = 1 + \omega$, so that $z^n = 1 + n\omega$ and $z^2 = 1 + 2\omega$. Then (§11)

$$\omega = \frac{1 + \alpha f}{2(f-1)}.$$

From the ratio given by the geodetic measurements, namely 201:200, it follows that $\omega = 1/200$ approximately, and hence $\alpha = 1/154$, taking $f = 289$, the value first found by HUYGENS (§12). EULER again notes that the value of n does not enter into the result and he also remarks that the maximum value of the lateral force, which is nearly 1/300 and occurs when $\varphi = 45°$, is almost equal to the centrifugal force at the equator (§13).

Since the lateral force

$$S = \frac{\alpha \sin 2\varphi}{2z},$$

it follows that, near the centre of the earth, this force should become infinite. From this EULER draws the conclusion that, if the whole earth were fluid, the surface could never arrive at a state of equilibrium (§15).

LEONHARD EULER'S WORK ON THE TIDES: HISTORICAL COMMENTARY

Geographical exploration on the eve of the seventeenth century scientific Renaissance contributed much new knowledge concerning tidal phenomena in many parts of the world. While this served to establish the tides as a universal phenomenon, systematic observations designed to discover the precise details were first undertaken at the English ports of Bristol, London and Plymouth in the second half of the seventeenth century. These observations led to the publication of accurate descriptions of the principal inequalities in the early volumes of the *Philosophical Transactions of the Royal Society*. In the first two decades of the eighteenth century, several series of systematic observations were instigated at the continental ports of Dunkirk, Le Havre, Brest and Lorient by the Académie Royale des Sciences in Paris. These were reported in the *Histoire* and carefully analysed by JACQUES CASSINI in several memoirs of the Académie, where he developed some empirical rules for the prediction of the tides at these ports. Such rules were of considerable importance in application to practical navigation. Commenting, for example, on the observation that the tide descends more slowly than it rises, CASSINI remarked that a vessel which could enter a port one hour after high water might not always be able to do so one hour before high water[145]). Accurate and reliable prediction, however, could only be reasonably expected on the basis of a knowledge of the causes of the tides. Yet in the early part of the eighteenth century, even such a basic principle as the role of the sun in the production of the tides was not universally recognized, while three types of theory, based on very different principles, each had its supporters. It was against this background that the Académie Royale des Sciences in 1738 proposed the theory of the tides as the subject for a prize essay. Two years later the prize was divided between ANTOINE CAVALLERI, COLIN MACLAURIN, DANIEL BERNOULLI and LEONHARD EULER.

Description of the Phenomena

In his *Almagestum novum* (Bologna, 1651), GIAMBATTISTA RICCIOLI gave a comprehensive account of the history of ideas concerning the tides, together with references to the literature and the observations made possible by expanding geographical exploration[146]). He described the tides as one of three general motions of the sea. The other two were a motion in longitude from east to west and a motion in latitude from north to south. Of these, the first was based on observation of a westerly current in a variety of locations, while the supposed southerly current had been noted by ARISTOTLE[147]) and generally explained

145) *Histoire de l'Académie Royale des Sciences* (Paris), (1712), 1714, pp. 1–3. Cf. *Mémoires*, pp. 86–96.

146) GIAMBATTISTA RICCIOLI, *Almagestum novum* (Bologna, 1651), vol. 1, part 1, pp. 72–74 (Bk. 2, ch. 15) and part 2, pp. 360–380 (Bk. 9, ch. 14).

147) ARISTOTLE, *Meteorologica* 354 a 20–30.

by a hypothetical higher elevation of the earth towards the poles. Dispensing with the southerly current, BERNHARDUS VARENIUS (1622–1650), whose *Geographia generalis* became the standard work on geography for over a century, regarded the general motion of the sea as two-fold; first there was the constant uniform motion from east to west and secondly the tides[148]). FRANCIS BACON (1561–1626) attributed the tides simply to the interruption by the continents of the supposed universal westerly current[149]).

PYTHEAS OF MASSILIA, about 325 B.C., noted the connection between the tides and the moon and in particular the chief monthly inequality in the height, namely the alternation of spring and neap tides with the phases of the moon[150]). The monthly inequality in the time of high water, known as priming and lagging, was described in 1668 by HENRY PHILIPS (?–1677)[151]), who made observations at London Bridge. Between the syzygies and quadratures, the sun crosses the meridian before the moon and the resultant tide is therefore advanced or is said to prime. On the other hand, between the quadratures and syzygies, the sun crosses the meridian after the moon and the resultant tide is delayed or is said to lag. Another monthly inequality, arising from the moon's elliptical orbit, was described in 1670 by JOSHUA CHILDREY (1623–1670)[152]). This inequality is most noticeable when the moon happens to be in perigee at the time of a spring tide. Such a tide was known among seamen as a running-spring, and it was a common observation that two running-spring tides never occurred together. For if the moon was in perigee at the first spring tide, it would be in apogee at the next.

Annual inequalities in the tides arise from the variations in the sun's geocentric distance and declination. The first has the effect of making the spring tides greater and the neap tides smaller in winter than in summer; for in winter the earth is nearer than it is in summer. The effect of the second is that the highest springs and lowest neaps occur respectively in the syzygies and quadratures close to the equinoxes, when the declination of the sun is zero. Both of these inequalities were known to PLINY THE ELDER[153]) and the second was confirmed by the observations of JOSHUA CHILDREY[154]).

There is a difference in the two tides of the same day, depending on the latitude of the place and the declination of the moon. Systematic observations of this diurnal inequality were made at Plymouth by SAMUEL COLEPRESSE[155]) and at Bristol by SAMUEL STURMY (1633–1669)[156]). They noted also that, if in one half of the year the higher tide was in the evening, then in the other half it was in the morning.

148) BERNHARDUS VARENIUS, *Geographia generalis* (Amsterdam, 1650), Sect. 4, ch. 14, prop. 7.

149) FRANCIS BACON, *Works*, ed. T. SPEDDING et al. (London, 1857–1859), vol. 5, p. 449.

150) M. FUHR, *De Pythea Massiliensi* (Darmstadt, 1835), p. 19.

151) *Philosophical Transactions* 3 (1668), 656–659.

152) *Ibid.*, 5 (1670), 2067–2068. In the *Philosophical Transactions*, CHILDREY's name is given, apparently in error, as JOSEPH.

153) PLINY, *Historia naturalis*, Book 2, chapter 99.

154) *Philosophical Transactions* 5 (1670), 2062–2065.

155) *Ibid.*, 3 (1668), 632–634.

156) *Ibid.*, 3 (1668), 813–817.

The observations of the times of high-water at London Bridge made by JOHN FLAMSTEED (1646–1719) and his assistants in 1682 provided the basis of his tide tables for the years 1683 to 1688, published in the *Philosophical Transactions of the Royal Society*[157]). The first series of systematic observations in response to the proposals of the Académie Royale des Sciences were made at Dunkirk and Le Havre for a period exceeding a year during 1701 and 1702[158]). From his analysis of these observations, JACQUES CASSINI found that the highest tides at these ports happened after the syzygies, the mean delay being two days[159]). Observations were carried out at Brest for several years beginning in 1712 and also at Lorient[160]). From his analysis of all these observations, CASSINI deduced the dependence of the tides on the distances of the moon from the equator and the ecliptic (that is, the lunar declination and latitude), noted the reversal of the diurnal inequality previously observed by COLEPRESS and STURMY, and formulated a rule for predicting the true time of the tide for a particular port, given the mean time by a sufficient number of observations[161]).

Theories of the Tides

The theories of GALILEO, DESCARTES and NEWTON may be taken as representatives of the three types of tidal explanation current in the early part of the eighteenth century. GALILEO's first work on the tides was a tract in the form of a letter to Cardinal ORSINI, written in 1616 and entitled, *Discorso sopra il flusso e reflusso del mare*[162]), in which he attributed the tides to the effect of the acceleration and retardation of the parts of the earth arising from the combination of the annual and diurnal motions. The theory was first published in the *Dialogo sopra i due massimi sistemi del mondo* (Florence, 1632)[163]), where GALILEO expanded his earlier account, notably by attempting to explain the monthly and annual inequalities. The spring and neap tides he attributed to a hypothetical monthly inequality in the earth's annual motion, the moon acting like a regulator in a clock[164]). The annual inequality he attributed to the varying angle between the earth's annual and diurnal motions arising from the obliquity of the ecliptic. As RICCIOLI pointed out[165]), the effects

157) *Ibid.*, 13 (1683), 10–15; 14 (1684), 458–462, 821–822; 15 (1685), 1226–1230; 16 (1686–87), 232–235, 428–432.

158) *Histoire de l'Académie Royale des Sciences* (Paris), (1710), 1712, pp. 4–10.

159) *Mémoires de l'Académie Royale des Sciences* (Paris), (1710), 1712, pp. 318–341, especially p. 325, and pp. 366–385.

160) *Histoire de l'Académie Royale des Sciences* (Paris), (1712), 1714, pp. 1–3; (1713), 1716, pp. 1–6; (1714), 1717, pp. 4–7; (1720), 1722, pp. 1–4.

161) *Mémoires de l'Académie Royale des Sciences* (Paris), (1712), 1714, pp. 86–96; (1713), 1716, pp. 14–32 and pp. 267–290; (1714), 1717, pp. 246–258; (1720), 1722, pp. 154–166 and pp. 355–364.

162) GALILEO GALILEI, *Opere*, ed. ANTONIO FAVARO (Florence, 1890–1909), vol. 5, pp. 377–395.

163) GALILEO GALILEI, *Opere*, vol. 7, pp. 21–489, in particular pp. 442–486.

164) See E. J. AITON, "On Galileo and the earth-moon system", *Isis* 54 (1963), 265–266.

165) GIAMBATTISTA RICCIOLI, *Almagestum novum* (Bologna, 1651), vol. 1, part 2, p. 380.

of these causes were not in agreement with the observations. For, according to GALILEO's theory, the tides should be greater at new moon than at full moon and greater in the solstices than in the equinoxes. An even more serious defect of the theory was its failure to show any connection of the two daily tides with the moon.

A modification of GALILEO's theory, designed to explain the dependence of the daily tides on the moon, was introduced in 1650 by GIOVANNI BATTISTA BALIANI (1582–1666), who supposed the earth to revolve about the moon, the moon meanwhile revolving about the sun[166]). This bizarre hypothesis was proposed independently by JACQUES ALEXANDRE (1654–1734) in an essay on the tides that was awarded a prize by the Académie Royale de Bordeaux in 1726[167]). A refutation was published in the following year by JEAN JACQUES D'ORTOUS DE MAIRAN[168]). The theory of BALIANI was also criticized by JOHN WALLIS (1616–1703), who proposed his own modification of GALILEO's theory to explain the dependence of the daily tides on the moon. He held that the earth and the moon were two parts of a single body whose centre of gravity described the orbit around the sun[169]). However, even WALLIS's theory could not explain the inequalities successfully.

GALILEO's theory (and hence also those derived from it) was in effect decisively refuted by DANIEL BERNOULLI in his prize essay. BERNOULLI recognized that, if the earth's diurnal motion is abstracted, both in the case of the motion of the earth in its epicycle about the common centre of gravity of the earth and the moon, and in its motion about the sun, each part of the earth moves in an equal ellipse, so that, at every instant, each point of the earth (and the ocean) experiences the same centrifugal force (equal in magnitude and direction)[170]). Clearly the implication is that the centrifugal force (which is the effect of the earth's orbital motion), acting equally on every part of the earth and the ocean, cannot cause any relative motion of the parts and consequently cannot give rise to the tides.

DESCARTES published his explanation of the tides, based on his theory of celestial vortices, in his *Principia Philosophiae* (Amsterdam, 1644)[171]). The theory was described by VARENIUS in his *Geographia generalis* and also by JACQUES ROHAULT (1620–1675) in his equally influential *Traité de physique* (Paris, 1671)[172]). According to DESCARTES, the tides

166) *Ibid.*, vol. 1, part 1, p. 216 and part 2, p. 381.

167) *Recueil des dissertations qui ont remporté le prix à l'Académie Royale de Bordeaux* (Bordeaux, 1715–1741), vol. 2.

168) *Histoire de l'Académie Royale des Sciences* (Paris), (1727), 1729, pp. 117–131; *Mémoires*, pp. 63–107.

169) *Philosophical Transactions* 1 (1666), 263–289. See also E. J. AITON, "Galileo's theory of the tides", *Annals of Science* 10 (1954), 44–57.

170) DANIEL BERNOULLI, "Traité sur le flux et reflux de la mer", *Die Werke von Daniel Bernoulli*, ed. D. SPEISER et al., vol. 3 (Basel, 1987), pp. 348–349.

171) *Principia Philosophiae*, Part 4, §§48–52. *Oeuvres de Descartes*, ed. C. ADAM and P. TANNERY, vol. 8 (Paris, 1905).

172) In DUGDALE's translation of VARENIUS's work (London, 1734), NEWTON's explanation was also given in the form of footnotes, a device for the dissemination of NEWTON's ideas previously used in the CLARKE editions of ROHAULT's *Traité de physique*.

were an effect of the moon alone, the sun having no role to play. By restricting the circulation of the fluid of the terrestrial vortex, the moon, DESCARTES supposed, caused the fluid to exert extra pressure on the earth directly beneath, displacing the earth slightly from the centre of the vortex until the extra pressure was the same on both sides. Consequently the water was depressed directly under the moon and at the point of the earth diametrically opposite, so that low water should occur at these points. VARENIUS argued that, according to the Cartesian theory, high water should occur directly under the moon and at the point diametrically opposite, though the reasons he gave are very obscure[173]). In fact, neither high water nor low water coincides with the transit of the moon across the meridian.

DESCARTES based his explanation of the spring and neap tides on the lunar inequality known as the variation, which had been discovered observationally by TYCHO BRAHE. This lunar inequality DESCARTES attributed to the elliptical shape of the vortex, the minor axis of which was directed towards the sun. At new moon and full moon, therefore, the moon was in the narrowest parts of the vortex and hence caused greater pressure on the oceans than it did in the quadratures. The annual inequality DESCARTES regarded as a consequence of the inclination of the plane of the equator and the ecliptic. As these planes intersect at the equinoctial points, it followed that, in the syzygies at the equinoxes the moon acted most directly against the earth and thus rendered the tide the greatest of all. DESCARTES also supported the idea of a universal east to west ocean current[174]).

While GALILEO's theory was considerably modified and improved by his disciples, the Cartesians rigidly adhered to the explanations of their master, so that no new ideas were introduced from the time of the first publication of the theory in 1644 until the appearance of CAVALLERI's prize essay in 1740. From his analysis of the observations promoted by the Académie Royale des Sciences, however, JACQUES CASSINI recognized that the Cartesian theory, of which he was himself a supporter, needed some modification. In the course of an historical review, he concluded that the springs and neaps could not be attributed to the different distances of the moon, and it was this phenomenon, he remarked, that led him to believe that the sun contributed to the tides[175]). In his annual summary of the memoirs presented to the Académie, BERNARD LE BOVIER DE FONTENELLE remarked that, if the pressure of the moon was not enough, then some kind of action common to the moon and the sun would have to be found[176]).

NEWTON's theory of the tides, published in his *Philosophiae Naturalis Principia Mathematica* (London, 1687), is based on the principle of universal gravitation. NEWTON in fact proposed two theories of the tides, though he failed to distinguish between them. He began his explanation in Book I by considering a canal encircling the earth along the equator and applying to the particles of water in it the laws he had deduced for the motion of a satellite

173) BERNHARDUS VARENIUS, *A compleat system of general geography*, tr. DUGDALE (London, 1734), p. 248 and p. 255.

174) *Principia Philosophiae*, Part 4, § 53.

175) *Mémoires de l'Académie Royale des Sciences* (Paris), (1713), 1716, pp. 267–290.

176) *Histoire de l'Académie Royale des Sciences* (Paris), (1713), 1716, pp. 5–6.

disturbed by a third body[177]). As the water would be accelerated and retarded in turn, so that its speed was greater than that of the surface of the globe directly under the disturbing body and at the point diametrically opposite but slower at the points a quadrant distant from these, NEWTON concluded that the water would ebb and flow in the canal after the manner of the sea[178]). To satisfy the condition that the total volume of water in the canal is constant, the depth of the water should be greatest where the speed is least, so that low water should occur at the meridian passage of the moon, as in the theory of DESCARTES. By a dynamical investigation, LAPLACE demonstrated that, in a shallow ocean, this inversion of the tides does indeed occur at places near the equator[179]). NEWTON did not state this prediction of his theory but instead located the positions of high and low water by considering the variation of the vertical component of the disturbing forces. Thus he took high water and low water to occur about the octants after the syzygies and quadratures respectively, at which points he estimated the vertical component of the disturbing forces to be zero[180]).

When NEWTON returned to the theory of the tides in Book III, he tacitly made a transition from the kinetic theory described so far to an equilibrium theory, in which he supposed that, at every instant, the ocean assumes the form in which it is in equilibrium under the action of the disturbing forces. In this case high water and low water occur at the points where the vertical component of the disturbing forces is a maximum and a minimum respectively; that is, high water at the points directly under the disturbing body and diametrically opposite, and low water at the points a quadrant distant from these. NEWTON mistakenly attributed the observed delay of the high water after the meridian passage of the moon, and that of the spring tide after the syzygy, to the inertia of the water[181]). The true cause of these delays is to be found in the friction of the water, which of course is not a perfect fluid.

Assuming the figure of the earth to be that which would be taken by a fluid of uniform density in equilibrium under the forces of attraction and rotation, and making use of his calculation of the attraction of a spheroid, NEWTON deduced that the earth was higher at the equator than at the poles by 85472 feet[182]). Then comparing the tidal force due to the sun's attraction with the centrifugal force determining the figure of the earth, he calculated the range of the solar tide[183]). It should be noted, however, that whereas the centrifugal

177) NEWTON, *Principia* (1687), Book I, proposition 66. For the changes in later editions, see *Isaac Newton's Philosophiae Naturalis Principia Mathematica*, ed. by ALEXANDRE KOYRÉ and I. BERNARD COHEN (Cambridge, 1972).

178) *Ibid.*, corollary 19.

179) P. S. LAPLACE (1749–1827), *Oeuvres complètes* (Paris, 1878–1912), vol. 2, p. 212.

180) *Principia*, Book I, proposition 66, cor. 20. In fact the vertical component is zero not in the octants after the syzygies and quadratures but at a distance of $\cos^{-1}(1/\sqrt{3})$ from the syzygies and quadratures.

181) *Principia*, Book III, proposition 24.

182) *Ibid.*, proposition 19.

183) *Ibid.*, proposition 36.

force gives to the earth the form of an oblate spheroid, the tidal force of the sun gives to the ocean the form of a prolate spheroid. NEWTON's comparison has been held to be valid because the effective (that is, the horizontal) components of the forces in the two cases are similarly distributed over the earth's surface[184]). This result is far from obvious, yet NEWTON made no attempt to justify his comparison.

The ratio of the masses of the earth and the moon being unknown, NEWTON could not calculate directly the force of the moon to move the sea. In fact he deduced this force from the ratio of the solar and lunar tides observed at Bristol by STURMY[185]). In arriving at his result, NEWTON made corrections for the declination of the moon in the quadratures and also for the delay of the spring tide after the syzygy. The latter correction was misconceived, as the forces causing the spring tide are those acting at the instant of syzygy and not those at the instant of high water. In fact NEWTON's estimate of the range of the lunar tide is too large by a factor of two.

Finally a remark should be made concerning NEWTON's explanation of the anomalous tides in the Gulf of Tonkin. In effect, NEWTON's explanation is that the semi-diurnal oscillation is destroyed according to the principle of interference of waves arriving by different routes, leaving only the diurnal oscillation, which also vanishes when the moon's declination is zero, so that intermissions in the tides occur at these times[186]).

The Prize Essays

All four prize essays, including the Cartesian essay of ANTOINE CAVALLERI, were influenced by the work of NEWTON. Although CAVALLERI, in his essay entitled, "Dissertation sur la cause du flux et reflux de la mer"[187]), defended the Cartesian theory, he responded to the challenge of CASSINI and FONTENELLE to find some kind of action common to the moon and the sun. Thus accepting that the sun had a role in the production of the tides, he argued that, just as the central force of the terrestrial vortex arising from its rotation acts on the oceans, so does the central force of the solar vortex. Following the ideas of an earlier eighteenth century Cartesian, JOSEPH PRIVAT DE MOLIÈRES (1677–1742), CAVALLERI identified these forces with NEWTON's attraction. For MOLIÈRES[188]), and PHILIPPE VILLEMOT (1651–1713)[189]) before him, deduced from the supposed constitution

184) See J. PROUDMAN, "Newton's work on the theory of the tides", in: W. J. GREENSTREET (ed.), *Isaac Newton* (London, 1927).

185) *Principia*, Book III, proposition 37.

186) *Ibid.*, proposition 24.

187) *Recueil des pièces qui ont remporté les prix de l'Académie Royale des Sciences*, vol. 4 (Paris, 1752), pp. 1–51.

188) *Mémoires de l'Académie Royale des Sciences* (Paris), (1728), 1730, pp. 245–267; (1729), 1731, pp. 235–244; (1733), 1735, pp. 301–312.

189) PHILIPPE VILLEMOT, *Nouveau système ou nouvelle explication du mouvement des planètes* (Lyon, 1707).

of the vortex, that the central force was proportional to the mass of the body in the centre of the vortex and inversely proportional to the square of the distance from the centre. Having similarly demonstrated that the tide-generating forces of the sun and the moon derived from his own theory had those properties that NEWTON had assigned to them, CAVALLERI was able to assimilate the mathematical steps of NEWTON's calculation[190]).

To explain why high water and not low water (as predicted by the Cartesian theory) occurred directly under the moon and at the point diametrically opposite, CAVALLERI argued that the moon deflected and reduced the speed of the fluid of the vortex. Although this fluid then accelerated through the narrower channel between the earth and the moon, this acceleration was not enough to restore the previous loss, so that the surface of the sea experienced less pressure under the moon and at the point diametrically opposite than elsewhere.

COLIN MACLAURIN's prize essay, "De causa physica fluxus et refluxus"[191]), was written in haste but the substance of it was incorporated in a better form in his *Treatise of fluxions* (Edinburgh, 1742, vol. 2). Besides recognizing that NEWTON's kinetic theory was a separate theory, MACLAURIN found that the heights of the tides predicted by this theory were considerably less than NEWTON had predicted on the basis of his equilibrium theory. MACLAURIN's principal contribution to the equilibrium theory was an elegant geometrical demonstration of the theorem that, under the action of the sun or moon, a homogeneous sphere would assume the form of a prolate spheroid having its axis directed towards the disturbing body. NEWTON had assumed this theorem without proof. MACLAURIN was also the first to describe and explain the deflection towards the right in ocean currents along a meridian in the northern hemisphere on a rotating earth[192]).

DANIEL BERNOULLI, in his prize essay entitled, "Traité sur le flux et reflux de la mer"[193]), developed the equilibrium theory to give it a practical value in predicting tides. The tables in the essay show the times and relative heights for the various ages and distances of the moon. Towards the end of the essay he considered the effects of the declinations of the disturbing bodies and the latitude of the port, giving a formula by which the tables could be corrected for these effects[194]). To determine the theoretical range of the solar tide, BERNOULLI considered the equilibrium of a vertical canal under the sun, communicating at the centre of the earth with another canal inclined to it at an angle of 90°, a device NEWTON had employed to determine the figure of the earth. In the case where the earth was supposed

190) In a prize essay of 1730, JOHANN BERNOULLI had also assimilated NEWTON's mathematical results to a Cartesian physical explanation, "Nouvelles pensées sur le système de M. Descartes", *Recueil des pièces qui ont remporté les prix de l'Académie Royale des Sciences*, vol. 2 (Paris, 1752).

191) *Recueil des pièces qui ont remporté les prix de l'Académie Royale des Sciences*, vol. 4 (Paris, 1752), pp. 193–234.

192) *Ibid.*, pp. 227–228 (proposition 7).

193) *Ibid.*, pp. 55–191. Cf. *Die Werke von Daniel Bernoulli*, ed. D. SPEISER et al., vol. 3 (Basel, 1987), pp. 327–438.

194) *Die Werke von Daniel Bernoulli*, pp. 410–425.

to be a homogeneous sphere of the same density as the ocean, he arrived at NEWTON's value, but his conclusions for other laws of density are rendered invalid by the unrecognized differences in the earth's mass implied in his calculations[195]). Observing that the amplitude of a pendulum does not respond immediately to a change in the force causing the motion, whereas the period does so respond, it seemed to BERNOULLI that observations of the duration and intervals of the tides would provide a better basis for the comparison of the lunar and solar tides than the observations of the heights employed by NEWTON. Using this principle he calculated the mean lunar tide to be $2\frac{1}{2}$ times the mean solar tide, a much more accurate value than that of NEWTON[196]).

Near the end of his essay[197]), BERNOULLI recommended the undisturbed level as the fixed point to which observation of the heights of the tides should be referred. According to the equilibrium theory, this should be one-third of the tidal range above low water. LAPLACE later pointed out, however, that when the effect of the earth's rotation is taken into account, the position of the undisturbed level in the tidal range could be considerably changed. At Brest, for example, it was nearly in the middle of the tidal range[198]).

EULER's prize essay, "Inquisitio physica in causam fluxus ac refluxus maris", contains two significant original contributions to the theory of the tides. First, he recognized that the horizontal components of the disturbing forces are the effective tide-generating forces. Secondly, he determined the equilibrium figure of the ocean under the action of the sun by the condition that the resultant force is perpendicular to the surface of the water. EULER's direct analytical approach must be judged superior to NEWTON's argument by analogy with the problem of the figure of the earth. Nevertheless, EULER neglected the attraction of the water in the two high water regions, an approximation that greatly simplifies the problem but the effect of which is not negligible. NEWTON's calculation, it will be recalled, takes account of this attraction, at least in the case of an ocean of the same density as the earth.

Besides the formulation of the equilibrium theory, EULER also attempted a mathematical theory, though in this he was less successful, not having worked out his theory of the motion of fluids at the time of the composition of his essay on the tides. Neglecting the horizontal motion[199]) and considering only the vertical motion, he supposed that each particle of water tends to return to the vertical position of equilibrium with a force proportional to the vertical displacement from this position. Combining this force with the disturbing force of the sun, and thus conceiving the motion as a forced harmonic oscillation, he arrived at a second order differential equation whose solution LAPLACE regarded as the most remarkable thing in the essay. EULER also demonstrated that the solar and lunar tides could be added together (a result NEWTON had assumed without proof) and he

195) *Ibid.*, pp. 357–362.

196) *Ibid.*, pp. 380–382.

197) *Ibid.*, p. 436.

198) P. S. LAPLACE, *Oeuvres complètes* (Paris, 1878–1912), vol. 5, pp. 169–170.

199) The fundamental equations later derived by LAPLACE refer to the horizontal motion. *Oeuvres complètes*, vol. 1, p. 102; vol. 2, p. 184.

emphasized the need for an acceptable theory of the tides to be able to account in detail for the observations that had been collected and systematized, especially those recorded in the first decades of the eighteenth century. The success of his own theory in this respect, he claimed, confirmed its truth.

EULER made a further contribution to the theory of the tides in a memoir presented to the St. Petersburg Academy of Sciences in 1775 [200]). While this added no new results, it greatly simplified the calculation of the equilibrium position of the surface of the ocean under the action of the forces of the sun and the moon. For this purpose EULER used a principle that he had established in a memoir presented to the Berlin Academy of Sciences in 1755 [201]). In effect, he derived the equilibrium state from the condition that the potential is constant over the surface of the water.

Notes on the Edition

In the case of each of LEONHARD EULER's works, the earliest printed text has been edited and reproduced, though in two cases this text is a French translation of a work composed in Latin. However, the version of JOHANN ALBRECHT EULER's essay entered for the prize of the Paris Academy in 1761 and published in 1771 (A 8^2) has been chosen in preference to the first edition printed in Berlin in 1762 (A 8) because it is more carefully written and contains fewer mistakes.

All errors in mathematical formulae have been corrected, though the readings given in the original texts have been indicated in the footnotes, unless they are obvious misprints. Small numerical errors without significant consequences have not always been corrected but in every case the presence of such errors has been noted.

In the cases of E 57, E 850 and E 851, manuscript versions have been used for the correction of the printed texts. Particular attention has been given to the manuscript of the prize essay on the tides (E 57), which was evidently revised by EULER himself after the publication of the printed version. To distinguish different types of emendation, the following conventions have been used. Words interpolated in the manuscript and therefore representing later additions have been included between square brackets [ ] and indicated by the symbol * meaning: In manuscripto interpolatum. Whenever the manuscript has been amended (other than by interpolation), or the manuscript is correct while the printed version is in error, the manuscript has been followed. Emendations of this kind are indicated by the symbol † meaning: Secundum manuscriptum. In all cases of emendation, the reading of the printed version is indicated in footnotes.

200) E 546, pp. 329–347 of this volume.

201) E 225, "Principes généraux de l'état d'équilibre des fluides", *Mémoires de l'Académie Royale des Sciences de Berlin* [11] (1755), 1757, 217–273 = *Opera omnia* II 12, pp. 2–53.

KOMMENTAR ZU EINER SCHRIFT EULERS ÜBER DIE SCHWERKRAFT

ANDREAS KLEINERT

De causa gravitatis ist EULERS erste Veröffentlichung seiner Äthertheorie der Schwerkraft. Sie erschien 1743 im 7. Band der *Miscellanea Berolinensia*, den Abhandlungen der Berliner Akademie, die damals noch «Königliche Sozietät der Wissenschaften» hieß. Weder in der Sekundärliteratur noch im ENESTRÖM-Verzeichnis ist diese anonyme Abhandlung bisher erwähnt worden[1]).

Daß der Artikel von EULER stammt, geht aus seinem Briefwechsel mit dem Genfer Physiker GEORGES-LOUIS LESAGE (1724–1803) hervor[2]). LESAGE war wohl der bekannteste und auch der umtriebigste Gravitationstheoretiker des 18. Jahrhunderts: Er korrespondierte mit Wissenschaftlern in ganz Europa, um sie mit geradezu missionarischem Eifer von der Richtigkeit *seiner* Gravitationstheorie zu überzeugen[3]). Zu den Strategien, die er anwandte, um mit bekannten Gelehrten im Gespräch zu bleiben, gehörte, daß er ihnen Kommentare zu ihren Arbeiten schickte, die irgendwie mit seinem Lieblingsthema zu tun hatten, und so verfuhr er auch mit EULER. In einem Brief von mehr als 13 engzeilig beschriebenen Quartseiten vom 9. August 1765 kritisiert er dessen Gravitationstheorie, wobei er sich unter anderem auf ein «Mémoire anonyme imprimé dans le VIIe Tome des *Mélanges de Berlin*» bezieht. EULERS Reaktion auf diesen Brief ist die Bestätigung, daß er tatsächlich der Verfasser der anonymen Abhandlung ist: In seiner Antwort vom 8. September 1765 nimmt er zu LESAGES Einwänden Stellung und gibt zu, daß er sich in einem wichtigen Punkt geirrt hat[4]).

1) Die einzige größere Studie zu EULERS Gravitationstheorie ist der Aufsatz von CASPAR ISENKRAHE, «Euler's Theorie von der Ursache der Gravitation», in: *Zeitschrift für Mathematik und Physik*, Historisch-literarische Abteilung, 26 (1881), S. 1–19. Die neueste Darstellung der Gravitationstheorien im 18. und 19. Jahrhundert ist FRANS HERBERT VAN LUNTEREN, *Framing hypotheses: Conceptions of gravity in the 18th and 19th centuries*, Dissertation, Utrecht 1991.

2) Der Briefwechsel EULER-LESAGE umfaßt die Briefe R 2065–R 2072 (vgl. *Opera omnia* IV A 1); er wird veröffentlicht in *Opera omnia* IV A 7.

3) Nach LESAGE bewegen sich harte, punktförmige Teilchen, die er als „corpuscules ultramondains" bezeichnet, geradlinig mit großer Geschwindigkeit in alle Richtungen. Einzelne Körper, die sich frei im Raum befinden, werden von diesen Gravitationskorpuskeln nicht beeinflußt, da die Stöße gleichmäßig von allen Seiten erfolgen. Bei zwei benachbarten Massen tritt dagegen ein Abschirmeffekt auf, und sie werden gegeneinander gedrückt, so daß der Eindruck entsteht, als zögen sie sich gegenseitig an. Die älteste Darstellung der Theorie von LESAGE ist PIERRE PREVOST, *Notice de la vie et des écrits de George[s]-Louis Lesage de Genève*, Genf 1805. Vgl. auch VAN LUNTEREN (Fußnote 1), S. 111–121.

4) Es handelt sich um die Briefe R 2069 und R 2070. In den Zitaten aus diesen Briefen wurde die Schreibweise modernisiert.

Es war für Lesage nicht schwer, Euler als Verfasser von *De causa gravitatis* zu identifizieren. Er arbeitete an einer *Histoire critique de la pesanteur*, in der er alle damals bekannten Schwerkraft-Theorien darstellen wollte[5]), und durchsuchte zu diesem Zweck die gesamte Fachliteratur seiner Zeit auf einschlägige Veröffentlichungen. Dabei war er offenbar auch auf die *Dissertatio de magnete* gestoßen, die Euler 1744 der Pariser Akademie als Preisschrift eingereicht hatte und in der er angekündigt hatte, er könne nicht nur den Magnetismus, sondern auch die Schwerkraft mechanisch erklären[6]). Da in *De causa gravitatis* ebenso argumentiert wird wie in der Magnetismus-Schrift und da in beiden Abhandlungen dieselbe Formel für den Ätherdruck auftritt, nahm Lesage zu Recht an, daß beide Texte von Euler stammten.

Eulers Schrift über die Ursache der Schwerkraft hängt mit den Reaktionen der Physiker des 18. Jahrhunderts auf das Werk von Isaak Newton zusammen. Dessen Gravitationstheorie wurde zwar als Hilfsmittel zur exakten Berechnung der Bewegungen der Himmelskörper weitgehend akzeptiert, galt aber dennoch bei vielen als eine Zumutung an den gesunden Menschenverstand.

Unabhängig davon, was Newton selbst geschrieben oder gedacht hatte, unterstellte man ihm und seinen Anhängern einen Rückfall in die finstersten Zeiten des Aristotelismus. So wie Aristoteles gelehrt habe, Ursache der Schwere sei der dem Körper innewohnende Drang, zu seinem natürlichen Ort zu gelangen, schreibe Newton den Körpern eine ähnliche «verborgene Eigenschaft» zu, wenn er von Anziehung spreche. Wie sollte unbelebte Materie durch den leeren Raum hindurch «merken», daß anderswo auch Materie vorhanden ist, und was sollte Anhäufungen von Materie veranlassen, sich gegenseitig anzuziehen? Diese Frage wird in fast jedem Physikbuch des 18. Jahrhunderts erörtert, wie das folgende Zitat zeigen mag:

> Man muß gestehen, daß die Naturlehre unserer Zeiten, welche sich rühmet auf immer gereiniget zu seyn von jenen verborgenen Eigenschaften, welche sie so lächerlich gemacht hatte, nicht ohne Verdruß wahrnehmen muß, daß man wieder in die Materie eine abgezogene Kraft, ein unbekantes, ja gar unbegreiflich Ding hinein setzet, dabey gar nichts auf den Mechanismum ... ankomme[7]).

Dieser «Verdruß» hat zahlreiche Physiker des 18. Jahrhunderts veranlaßt, das nachzuliefern, was der Newtonschen Physik ihrer Meinung nach fehlte, nämlich einen «Mechanismus» als Ursache der Kraft, die die Körper aufeinander zutreibt. Dabei konnten sie an Descartes anknüpfen, der schon 1644 in seinen *Principia philosophiae* eine mechanische Gravitationstheorie entworfen hatte. Nach Descartes sollte eine subtile Materie, die

5) Prevost (Fußnote 3), S. 95–96.

6) E109, *Opera omnia* III 10. Abschnitt 23 dieser Abhandlung trägt die Überschrift «Causa gravitatis indicatur».

7) Jean-Antoine Nollet, *Physicalische Lehrstunden*, Band 6, Erfurt 1766, S. 131. [Deutsche Übersetzung der *Leçons de physique expérimentale*, Paris 1743–1748].

um die Himmelskörper herumwirbelt, bewirken, daß andere Körper zu deren jeweiligem Zentrum gedrückt werden. Die Anhänger dieser Wirbeltheorie mußten sich mit zahlreichen Einwänden auseinandersetzen. So wurde ihnen vorgehalten, daß dadurch allenfalls die Bewegung von Körpern zu einer Achse, aber nicht zu einem Punkt hin erklärt werde; ferner war die Theorie, wenn man sie auf die Planetenbewegung anwandte, nicht mit den KEPLERschen Gesetzen vereinbar – ganz zu schweigen von der Bewegung der Kometen. Viele illustre Physiker und Philosophen – von CHRISTIAAN HUYGENS bis zu CHRISTIAN WOLFF – haben versucht, diese Mängel der Wirbeltheorie DESCARTES' zu beseitigen.

Da keiner dieser Versuche von Erfolg gekrönt war, schlug EULER einen anderen Weg ein. Zwar griff auch er wie DESCARTES auf ein subtiles Fluidum zurück, das er Äther nannte; doch nicht die Zentrifugalkraft im Ätherwirbel sollte die Ursache der Schwerkraft sein, sondern ein im Äther auftretendes Druckgefälle.

Bevor EULER seine Theorie vorstellt, polemisiert er heftig gegen die „attractionis patroni", die behaupten, das Rätsel der Schwerkraft sei gelöst, wenn man einfach allen Körpern eine in die Ferne wirkende anziehende Kraft zuschreibe, ohne weiter zu fragen, woher diese Kraft komme und wie sie übertragen werde. Diese Polemik umfaßt die erste Hälfte der Abhandlung. Viele seiner Argumente gegen die Newtonianer der strikten Observanz stammen aus dem Arsenal des militanten Cartesianismus jener Zeit, doch unter dem Schutz der Anonymität geht EULER noch einen Schritt weiter und erfindet mit viel Phantasie weitere Beispiele für die Schwierigkeiten, die sich ergeben, wenn man die Möglichkeit einer Fernwirkung ohne direkten Kontakt einräumt. Dann, so EULER, müßte es auch möglich sein, den Schall im luftleeren Raum zu übertragen, und man müßte Gegenstände sehen können, die den Sinnesorganen nicht zugänglich sind, weil sie sich in unendlicher Entfernung befinden. Den Gipfel des Spottes bildet die Überlegung, mit der er den polemischen Teil seiner Abhandlung abschließt:

> Die auf dem Planeten Saturn wachsenden Kräuter müßten aufgrund der ihrem Wesen nach unverständlichen Anziehungskraft in meinem Körper einen Brechreiz und andere heftige Bewegungen der Eingeweide hervorrufen, obwohl ich sie mir weder durch die Nase noch durch den Mund jemals zugeführt habe.

Das Druckgefälle im Äther, das nach EULER die Gravitation bewirkt, soll eine Folge der ständigen Bewegung des Äthers um die Erde sein; diese Ätherbewegung zeige sich unter anderem beim Erdmagnetismus. Um das Gravitationsgesetz zu erhalten, muß EULER dann nur noch annehmen, daß die Änderung des Ätherdrucks mit wachsendem Abstand vom Zentrum erfolgt, und zwar nach einer Beziehung, die sich auch dadurch auszeichnet, daß darin das einfachste von allen denkbaren Verhältnissen („ratio ... omnium simplicissima") vorkommt: Der Ätherdruck ist umgekehrt proportional zum Abstand vom Erdmittelpunkt. In EULERS Schreibweise lautet die Formel für den Ätherdruck:

$$C = c - \frac{cg}{x}$$

(C = Druck des Äthers in der Entfernung x vom Erdmittelpunkt; c = Druck des Äthers im Unendlichen; g = konstanter Faktor). Eine einfache Überlegung zeigt, daß daraus tatsächlich das NEWTONsche Gravitationsgesetz folgt.

Etwa zehn Jahre später hat EULER seine Gravitationstheorie weiter ausgebaut; das 19. Kapitel der zwischen 1750 und 1758 entstandenen *Anleitung zur Naturlehre*[8]) lautet „Von der Schwere und den Kräften, so auf die himmlischen Körper wirken". Hier erörtert er einen Einwand, der sich beim Lesen von *De causa gravitatis* zwangsläufig einstellen muß: Nach der oben angegebenen Formel müßte die Schwerkraft nicht nur von der Masse, sondern auch von der Gestalt des Körpers abhängen. Da das nicht der Fall ist, muß EULER zwischen der wahren und der scheinbaren Größe eines Körpers unterscheiden. Die Körper bestehen für ihn aus «grober Materie» und mehr oder weniger vielen Poren, und was wir wahrnehmen, ist immer nur ihre scheinbare Größe, «welche die Poren zugleich mit in sich begreift». Will man die Schwerkraft ermitteln, dann muß man von den Poren des Körpers absehen, «als wenn nur seine grobe Materie allein in einen Klumpen zusammengepresst wäre». Im übrigen gibt EULER offen zu, wo die Grenzen seiner Theorie liegen: Er kann keine Erklärung dafür angeben, wie das Druckgefälle im Äther zustande kommt, und stellt fest, daß wir «kaum hoffen können, jemals die wahre Ursache dieser Verminderung der elastischen Kraft des Aethers zu ergründen». Trotzdem kann er sich mit dieser fragwürdigen Grundlage seiner Theorie «leichter begnügen, als wenn man blosserdings vorgiebt, alle Körper seien von Natur mit einer Kraft begabt, einander anzuziehen».

EULER hat die *Anleitung zur Naturlehre* nie veröffentlicht; sie wurde erst 1862 von seinen Urenkeln PAUL HEINRICH und NIKOLAUS FUSS aus dem Nachlaß herausgegeben. Erst 1768 hat er sich wieder öffentlich zur Gravitationstheorie geäußert, und zwar in den *Briefen an eine deutsche Prinzessin*[9]). Hier wiederholt er seine Ausfälle gegen die «englischen Philosophen», die behaupten, «daß es eine wesentliche Eigenschaft aller Körper sey, sich wechselsweise anzuziehen; und daß alle Körper gleichsam eine gewisse natürliche Neigung gegen einander haben, kraft welcher sie sich bemühen einander näher zu kommen, so als wenn sie eine Empfindung oder Begierde hätten». Er hält es lieber mit denen, «die in der Philosophie helle und begreifliche Grundsätze lieben», und führt die Gravitation auf den Äther zurück, «wenn man auch die Art dieser Wirkung nicht einsieht». In auffälligem Gegensatz zur Gravitationsschrift von 1743 und zum Schwerkraftkapitel aus der *Anleitung zur Naturlehre* verzichtet EULER jetzt darauf, das Gravitationsgesetz aus der Ätherhypothese herzuleiten, und beschränkt sich auf die Feststellung:

> Jetzo kommt es darauf an, die wahre Quelle dieser anziehenden Kraft zu entdecken, welches aber eigentlich mehr für die Metaphysik als Mathematik ge-

8) E 842, *Opera omnia* III 1, S. 16–178; die Zitate sind aus den Abschnitten 141 und 146. Zur Datierung vgl. HELMUT PULTE, *Das Prinzip der kleinsten Wirkung und die Kraftkonzeptionen der rationalen Mechanik: Eine Untersuchung zur Grundlegungsproblematik bei Leonhard Euler, Pierre Louis Moreau de Maupertuis und Joseph Louis Lagrange*, Stuttgart 1989 (= Studia Leibnitiana, Sonderheft 19), S. 164.

9) E 343, *Opera omnia* III 11 und 12; die Zitate sind aus Brief 68. Zitiert wird nach der deutschen Übersetzung von 1769 (E 343 B).

> hört; und ich kann mir nicht schmeicheln, daß mir dieß eben so gut gelingen wird.

Woher diese Bescheidenheit? Der Brief über die Gravitation trägt das Datum «18. Oktober 1760», und wir wissen nicht, ob EULER ihn vor der Veröffentlichung noch einmal überarbeitet hat. Wenn ihm nicht schon Bedenken gekommen waren, als er an die Prinzessin schrieb, so wird ihn spätestens der oben erwähnte Brief von LESAGE davon abgehalten haben, seine Gravitationstheorie von 1743 weiter zu vertreten, und dieser Brief könnte auch dazu beigetragen haben, daß er die *Anleitung zur Naturlehre* nicht veröffentlicht hat. Zumindest das Kapitel über die Schwerkraft hätte er überarbeiten oder weglassen müssen, nachdem LESAGE ihm gezeigt hatte, daß es nicht möglich ist, die Lichtausbreitung und die Gravitation auf denselben Äther zurückzuführen.

Der von LESAGE vorgebrachte Einwand ergab sich aus dem Vergleich der Gravitationsschrift mit zwei Veröffentlichungen EULERS aus dem Jahre 1746, in denen ebenfalls der Äther vorkommt. Die eine ist die *Nova theoria lucis et colorum*[10]), in der der Äther als das Medium der Lichtausbreitung aufgefaßt wird; die andere ist die unmittelbar damit zusammenhängende Abhandlung *De relaxatione motus planetarum*[11]), in der EULER nachweist, daß sein Lichtäther die Planeten, die sich ja darin bewegen, nicht abbremst. Bei der Lektüre dieser Schriften stieß LESAGE auf einen Widerspruch zur in *De causa gravitatis* dargestellten Gravitationstheorie, den er EULER in seinem Brief mit sichtlichem Vergnügen unterbreitete: Vergleicht man die Eigenschaften, die EULER dem Lichtäther zuschreibt, mit denjenigen des Gravitationsäthers, dann ergibt sich, daß die Dichte des Äthers 160 mal größer ist als die Dichte der Luft. Das aber, so LESAGE, ist ein eklatanter Widerspruch zu der unbestreitbaren Tatsache, daß die Planeten bei ihrer Bewegung im Äther nicht abgebremst werden.

EULER sah seinen Irrtum sofort ein. «Vous avez solidement réfuté l'explication de la gravité, que j'avais hazardée autrefois», antwortete er genau einen Monat, nachdem LESAGE ihm geschrieben hatte, und gab zu, daß der Äther als das Medium, in dem sich das Licht ausbreitet, ganz und gar ungeeignet ist, um die Schwerkraft zu verursachen. Freilich hat LESAGE sein Ziel, EULER zu seiner korpuskularen Gravitationstheorie zu bekehren, nicht erreicht. Als Ausweg aus dem Dilemma schlägt EULER vor, man müsse vielleicht noch einen zweiten Äther annehmen, gewissermaßen einen besonderen Gravitationsäther, der sich hinsichtlich seiner Dichte und seiner Elastizität zum Lichtäther ungefähr so verhalte wie dieser zur Luft. Von LESAGES Gravitationsatomen wollte er jedenfalls nichts wissen:

> Je sens encore une très grande répugnance pour vos corpuscules ultramondains et j'aimerais toujours mieux d'avouer mon ignorance sur la cause de la gravité que de recourir à des hypothèses si étranges. Mais j'accorde très volontiers cette liberté à d'autres.

10) E 88, *Opera omnia* III 5, S. 1–45.
11) E 89, S. 195–220 in diesem Band.

Warum hat EULER die Schrift über die Gravitation anonym veröffentlicht, obwohl er, nach dem polemischen und selbstbewußten Ton zu urteilen, seiner Sache sehr sicher war? Ein Brief an CHRISTIAN GOLDBACH vom 14. April 1744 läßt erkennen, daß es wohl nicht in erster Linie die Angst war, seinen guten Ruf mit einer reichlich spekulativen Theorie auf's Spiel zu setzen, sondern daß ein wesentlich handfesterer Grund eine Rolle gespielt hat. In dem Brief heißt es nämlich:

> Ich bin in Verfertigung meiner Piece, welche ich über den Magneten im vorigen Jahr nach Paris geschickt, auf einen Einfall, um die causam gravitatis zu erklären, geraten, welcher mir je länger je gründlicher vorkommt. ... [Ich kann] gewisse Rechnung machen, daß ich diesmal wieder leer ausgegangen. Ich kann auch die Ursach leicht erraten, dann da ich, um meine Erklärung zu bekräftigen, die Meinung der Engelländer von der Attraction als einem attributo essentiali corporum ziemlich stark angegriffen und widerleget, so wird dieses den Herrn Commissariis, welche, wie ich seit der Zeit erfahren, dieser Meinung völlig beistimmen, gar nicht gefallen haben[12]).

Schon 1742 hatte DANIEL BERNOULLI EULER geraten, im eigenen Interesse eine moderate Haltung gegenüber der als «Meinung der Engelländer» bezeichneten Attraktionshypothese einzunehmen, wenn er in Frankreich anerkannt werden wolle, weil «des NEWTONS Reputation in Frankreich nunmehro so groß ist als in England selbst»[13]). Wenn EULER schon den kurzen und zurückhaltend-sachlich formulierten Exkurs in der Magnetismusschrift als «ziemlich starken Angriff» ansah, dann durfte er sicher sein, daß er mit seiner Polemik in der Gravitationsschrift die Pariser Akademiker erst recht massiv gegen sich aufbringen würde. Daran konnte er jedoch aus naheliegenden Gründen kein Interesse haben, denn die Akademiepreise bedeuteten eine erhebliche Aufbesserung seines Gehaltes[14]). Die gegenüber GOLDBACH geäußerte Vermutung war im übrigen falsch: Der Preis der Akademie für die Magnetismus-Frage wurde zu gleichen Teilen an drei Bewerber vergeben, und einer davon war EULER.

---

12) *Leonhard Euler und Christian Goldbach, Briefwechsel 1729–1764*, hrsg. von A. P. JUŠKEVIČ und E. WINTER, Berlin 1965, S. 192–193.

13) PAUL HEINRICH FUSS (Hrsg.), *Correspondance mathématique et physique de quelques célèbres géomètres du XVIIIeme siècle*, Band II, St. Petersburg 1843, S. 480.

14) Für DANIEL BERNOULLI betrug ein Preis der Pariser Akademie etwa zehn Jahresgehälter. Vgl. EDUARD FUETER: *Geschichte der exakten Wissenschaften in der schweizerischen Aufklärung (1680–1780)*, Aarau 1941, S. 146. KNOBLOCH verweist im Zusammenhang mit den Preisaufgaben auf den «materiellen Aspekt, der das kärgliche Einkommen vieler Gelehrter aufbessern konnte», und fügt hinzu: «Für EULER war dieser Aspekt durchaus wichtig: er gewann mit den zwölf Preisen von der Pariser Akademie zugleich 8000 Taler oder 3000 Livres.» (EBERHARD KNOBLOCH, «Die Akademie der Wissenschaften zu Berlin», in: FRIEDRICH RAPP und HANS-WERNER SCHÜTT (Hrsg.), *Philosophie und Wissenschaft in Preußen*, Berlin 1982, S. 115–143, hier S. 127.) Bei der Beurteilung von Preisaufgaben sollten zwar den Gutachtern die Namen der Bewerber nicht bekannt sein, in der Praxis wurde das Anonymitätsprinzip jedoch oft durchbrochen. Bei der Berliner Preisaufgabe von 1746 kannte EULER im voraus die Namen der Verfasser von zwei eingereichten Schriften, darunter denjenigen des

EULERS anonyme Gravitationsschrift hat in der Diskussion um die causa gravitatis keine nachweisbare Rolle gespielt. Obwohl ihr Erscheinen in den *Miscellanea Berolinensia* darauf schließen ließ, daß der Verfasser der Berliner Akademie angehören oder ihr wenigstens nahestehen mußte, wurde sie ebenso ignoriert wie die vielen anderen mechanischen Modellvorstellungen zur Erklärung der NEWTONschen Physik, mit denen sich im 18. Jahrhundert „besonders die zweitklassigen Denker mit leichtfertig vorgetragenen Ideen“ zu profilieren suchten[15]): In der zeitgenössischen Literatur wird sie nirgends erwähnt, und ohne die Aufmerksamkeit von LESAGE, dem emsigen und umtriebigen Gravitationsapostel aus Genf, wüßten wir bis heute nicht, wer sie verfaßt hat.

Preisträgers D'ALEMBERT. Vgl. ANDREAS KLEINERT: «D'Alembert et le prix de l'académie de Berlin en 1746», in: MONIQUE EMERY et PIERRE MONZANI (Ed.), *Jean d'Alembert, savant et philosophe: Portrait à plusieurs voix*, Paris 1989, S. 415–431.

15) RAINER BAASNER, *Das Lob der Sternkunst: Astronomie in der deutschen Aufklärung*, Göttingen 1987, S. 97.

INDEX

Insunt in hoc volumine indicis ENESTROEMIANI commentationes 7, 57, 67, 68, 89, 103, 142, 183, 184, 218, A8, 527, 546, 619, 839, 850, 851, s.n.

INDEX NOMINUM

QUAE AB EULERO CITATA SUNT

INDEX NOMINUM

QUAE AB EDITORE CITATA SUNT

TENTAMEN EXPLICATIONIS PHAENOMENORUM AERIS

Commentatio 7 indicis ENESTROEMIANI
Commentarii academiae scientiarum Petropolitanae 2 (1727), 1729, p. 347–368

1. Quanquam ad intima rerum penetralia et cognitionem ultimae partium structurae aditus non ita patet, ut phaenomenorum, quae inde oriuntur ratio reddi queat: Tamen, ut plerumque a Physicis factum est, a corporum naturalium proprietatibus, quas observavimus, quodammodo ad ipsam eorum structuram concludere licet. Ex qua percepta corporum structura, quo plura phaenomena explicari possunt, eo perfectior ea est; Et, si ex qua Theoria omnes prorsus proprietates, quas quidem cognoscere impossibile est, derivari possent, dubium non est, quin ea vera sit, et re ipsa existat.

2. Aeris quam plurima nota sunt phaenomena, eaque parum a se invicem dependentia; ut is profecto multum praestitisse censendus sit; qui eadem theoria omnibus satisfacere posset. Sed theoriam ita maxime conficere convenit, ut primum excogitetur partium structura, ex qua una tantum proprietas fluat; id quod plerumque pluribus modis fieri potest: et deinceps inquiratur, num ea caeteris quoque proprietatibus explicandis sufficiat? et, si plures primum theoriae conceptae sunt, tum quaeratur, quae maximae parti vel quae omnibus satisfaciat.

3. Inter alias aeris, quas cognoscimus, proprietates ea imprimis idonea visa est, secundum quam structura aeris adornetur, qua aer sese continuo expandere conatur, et reipsa se expandit, si quae impedimento fuerant, removeantur. Haec enim aeris elasticitas prae caeteris proprietatibus maxime explicatu difficilis videtur, ut eius ratione cognita reliqua facile fluere videantur.

4. Nisi velimus hoc aeris phaenomenon occultae cuidam particularum proprietati et vi insitae adscribere, alia via non superest, nisi ut conatus iste a motu materiae cuiusdam subtilis derivetur. Conatus autem seu vis mortua, uti a

LEIBNITIO vocatur, a materia mota ortum trahere potest, si ea in gyrum moveatur; quo fit ut quaevis particula a centro aufugere conetur, atque ita huiusmodi vortex vim sese expandendi acquirat. Hoc usus principio Cel. IOH. BERNOULLI omnem vim elasticam explicare instituit in schediasmate *de communicatione motus*[1]) Lutetiae nuper impresso. Quo vim elasticam a vi centrifuga materiae subtilis oriri asserit.

5. Sequor itaque hac in re, istam, uti mihi videtur, maxime probabilem elasticitatis aeris causam, atque in hac dissertatione examini subiiciam, quantum aeris structura ex hoc fundamento formata reliquis aeris proprietatibus explicandis sufficiat, quantumve minus, ut appareat, utrum aer hanc partium structuram habere possit, an vero non? Quo in casu meliorem oporteret excogitare aeris constitutionem.

6. Suppono igitur aerem constare acervo infinitarum minimarum bullularum, in quibus materia subtilis motu circulari gyratur et vi centrifuga bullulas continuo expandere conatur, easque reipsa semotis obstaculis, expandit. Suppono porro bullulas esse pellicula obductas; quod quidem opus non esset, cum huiusmodi vorticuli sine pelliculis constare possent et tamen mutuo non permiscerentur. Unus enim alterum impedit, quominus extravagetur: Attamen propterea bullulas pelliculis obductas suppono, quod aer nunquam tam purus sit, ut prorsus a vaporibus liber sit. Vapores autem particulas aeris ad instar pellicularum obducere valde probabile est.

7. Constet itaque aer infinito bullularum minimarum numero, quarum crusta exterior sit aquea pro diverso aeris statu maior minorve; intra hanc crustam gyretur materia subtilis certa cum velocitate, quae subinde ab alia subtiliori adhuc materia omnes poros penetrante accelerationes nanciscitur, ne motus tandem consumatur et evanescat. Constat enim aerem calorem semel acceptum sensim amittere, cum autem aer calore rarefiat, sequitur materiam subtilem motu vehementiore agitari; cessante ergo calore, indicio id est, motum materiae esse retardatum.

8. Ex hisce de structura aeris praemissis consequitur, eum in infinitum se expandere debere atque extremum raritatis gradum accipere, quando nihil est quod eius conatum compescat. Sed accedente gravitate, aliter se res habebit, eritque,

1) *Discours sur les loix de la communication du mouvement* (Paris, 1727). EJA.

quod vi aeris elasticae se opponet. Quum enim aer superior inferiorem premat pondere suo, inferior ulterius se expandere nequit, quam quoad eius vis elastica, quae expansione continuo diminuitur, aequalis sit vi incumbentis aeris comprimenti.

9. Patet porro ex concepta aeris constitutione, eum in infinitum comprimi non posse, propter gravitatem specificam, quae in infinitum augeretur. Nam cum in qualibet bullula certa et determinata materiae subtilis quantitas comprehendatur, eaque semper superficiei adhaereat ob vim centrifugam, necesse est ut circa centrum spatium vacuum relinquatur; id quod eo maius esse debet, quo magis aer rarus fuerit: Contra autem continuata aeris compressione, id spatium vacuum continuo diminuetur, donec tandem prorsus evanescat, ultra quem densitatis gradum aerem comprimere impossibile erit.

10. Quod ad velocitatem materiae subtilis attinet, oportet singulis eius particulis eandem attribuere velocitatem, neque quae a centro remotiores sunt, iis maiorem et proprioribus minorem adscribere velocitatem. Praeterea enim, quod hinc theoria nascatur experientiae penitus contraria, ob vim centrifugam in maioribus bullulis maiorem, ex hoc elucere potest, quod bullulam condensando vel expansioni relinquendo velocitas materiae subtilis eadem manere debeat, cum nihil sit, quod eam immutet; Huc enim non pertinet retardatio, de qua § 6 quae non propter immutationem bullulae, sed propter resistentiam quandam contingit. Quare cum velocitas materiae subtilis non a distantia a centro pendere queat necesse est eam ubique constantem statuere.

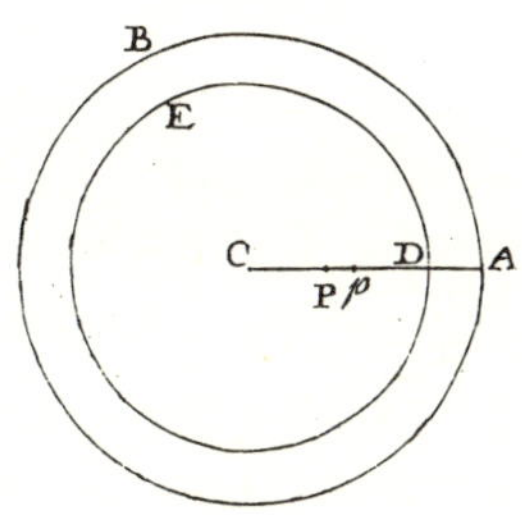

Fig. 1

11. Sit CAB bullula aerea [fig. 1], quoad fieri potest compressa, quae proin est materia subtili vorticosa penitus repleta. Circumdata vero sit crusta aquea $ADEB$, ut ergo reliquum spatium CDE materia subtili impleatur. Sit $AC = g$, $CD = h$. Sumatur pro ratione radii ad peripheriem, $1 : \pi$, pro gravitate specifica materiae subtilis, n et pro gravitate specifica aquae seu crustae m. Erit capacitas

globuli $CAB = \frac{2\pi g^3}{3}$, et capacitas globuli $CDE = \frac{2\pi h^3}{3}$. Ergo soliditas crustae $ADEB = \frac{2\pi}{3}(g^3 - h^3)$. Quamobrem erit massa materiae subtilis spatium CDE implentis $= \frac{2\pi n h^3}{3}$, et massa crustae $= \frac{2\pi m}{3}(g^3 - h^3)$. Et hae massarum quantitates in quantum vis expansis bullulis eaedem manere debent.

12. Exprimat k altitudinem, ex qua grave cadendo velocitatem acquirit, materiae subtilis velocitati aequalem; Unde sequenti modo vis centrifuga, seu vis, qua superficies globuli CDE premitur, invenietur. Sumatur a centro indeterminata $CP = x$ cuius differentiale $Pp = dx$. Erit crusta sphaerica crassitiei Pp et radii $CP = 2\pi xx\,dx$, quae si ducatur in densitatem materiae subtilis, dat massam $2\pi nxx\,dx$, seu pondus. Quum haec materia gyretur velocitate ex altitudine k acquisita, fiat secundum HUGENIUM, ut radius x ad duplam altitudinem, $2k$ ita pondus materiae gyrantis, $2\pi nxx\,dx$ ad pondus vi centrifugae huius crustae aequale, quod ergo erit $= 4\pi nkx\,dx$. Huius ergo integrale $2\pi nkx^2$ exprimit vim centrifugam sphaerae radii CP. Consequenter vis centrifuga bullulae DE erit $= 2\pi nkhh$.

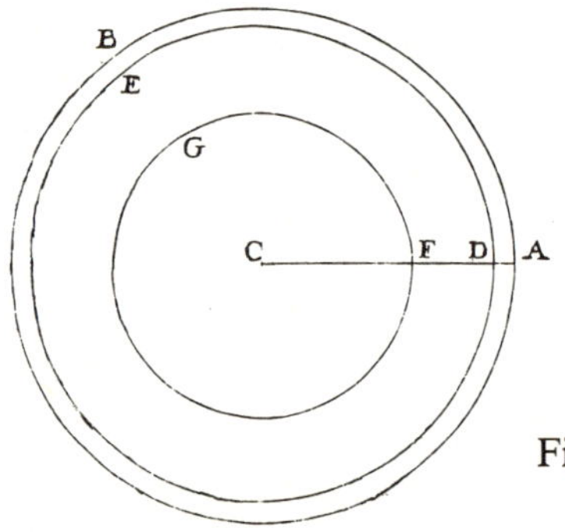

Fig. 2

13. Consideremus nunc [fig. 2] bullulam aeream quomodocunque expansam CAB: Cuius extrema crusta $ADEB$ designet materiam aqueam, media $DFGE$ materiam subtilem circa centrum gyrantem, et tertia seu intima CFG, spatium vacuum, vel quod ad minimum materia gravitatis experti sit repletum. Dicantur $AC = a$, $CD = b$, et $CF = c$. Erit, computo ut supra instituto, soliditas crustae extremae seu aqueae $ADEB = \frac{2\pi}{3}(a^3 - b^3)$. Dein soliditas crustae mediae seu quantitas materiae subtilis $DFGE = \frac{2\pi}{3}(b^3 - c^3)$. Tertio autem capacitas totius

bullulae erit $= \frac{2\pi}{3} a^3$. Sit gravitas specifica aeris seu totius bullulae, i erit pondus eius $\frac{2\pi i}{3} a^3$, id quod aequale est summae ponderum partium, nempe $\frac{2\pi m}{3}(a^3 - b^3) + \frac{2\pi n}{3}(b^3 - c^3)$. Est igitur $ia^3 = ma^3 - mb^3 + nb^3 - nc^3$.

14. Cum et quantitates materiae aquae, et quantitas materiae subtilis aequales esse debeant iis, quae supra erant inventae in casu bullulae maxime compressae, sequentes obtinebuntur aequationes $\frac{2\pi}{3}(g^3 - h^3) = \frac{2\pi}{3}(a^3 - b^3)$ et $\frac{2\pi}{3} h^3 = \frac{2\pi}{3}(b^3 - c^3)$. Quamobrem $g^3 - h^3 = a^3 - b^3$ et $h^3 = b^3 - c^3$. Unde $b = \sqrt[3]{(a^3 - g^3 + h^3)}$ et $c = \sqrt[3]{(b^3 - h^3)} = \sqrt[3]{(a^3 - g^3)}$. Si haec substituantur in superioris § ultima aequatione, reperietur $ia^3 = mg^3 - mh^3 + nh^3$. Unde $h^3 = (ia^3 - mg^3):(n - m)$. Et porro[2]) $b = \sqrt[3]{((i - m + n)a^3 - ng^3):(n - m)}$ ac $c = \sqrt[3]{(a^3 - g^3)}$. Hoc ergo modo ex calculo excluduntur litterae b, c, et h denotantes interiorum bullulae partium a centro distantias.

15. Vis centrifuga materiae subtilis in spatio $DFGE$ gyrantis velocitate ex altitudine k producta ex § 12[3]) inveniri potest hoc modo: Vis centrifuga materiae globum radii x implentis inventa est $= 2\pi nkxx$. Quare se materia subtilis totum spatium CDE impleret, foret eius vis centrifuga $= 2\pi nkbb$, a qua si auferatur, vis centrifuga materiae spatii $CFG = 2\pi nkcc$, restabit vis centrifuga materiae subtilis in spatio $FDEG$ gyrantis, cuius quantitas proin erit $= 2\pi nk(bb - cc)$ et subrogatis loco b et c valoribus § 13 inventis, erit ea

$$= 2\pi nk\left(\left(\frac{(i - m + n)a^3 - ng^3}{n - m}\right)^{2/3} - (a^3 - g^3)^{2/3}\right).$$

Ponatur $h^3 = pg^3$ erit, ob $ia^3 = mg^3 - mh^3 + nh^3$, $ia^3 = (m - mp + np)g^3$ unde $g^3 = ia^3:(m - mp + np)$. Erit ergo pondus vi centrifugae aequivalens

$$= 2\pi nkaa\left(\left(\frac{m - i + pi - mp + pn}{m - pm + pn}\right)^{2/3} - \left(\frac{m - pm + pn - i}{m - pm + pn}\right)^{2/3}\right)$$

2) Intellegimus $\sqrt[3]{((i - m + n)a^3 - ng^3)}:\sqrt[3]{(n - m)}$. EJA.

3) Editio princeps: 11 loco 12. Correxit EJA.

$$= \frac{2\pi nkaa}{\sqrt[3]{(m-pm+pn)^2}}\left(\sqrt[3]{(m-i+pi-pm+pn)^2} - \sqrt[3]{(m-pm+pn-i)^2}\right).$$

16. Cum vi centrifuga efficiatur, ut bullulae aereae sese continuo extendere conentur, erit ea aequalis vi elasticae aeris; ex inventa igitur aequatione, quanta sit aeris elasticitas, inveniri poterit. Verum cum hoc loco primum legem duntaxat, qua aeris vis elastica pro diversis densitatis, humoris et celeritatis gradibus immutetur, persequi conveniat, factor $2\pi n$ utpote constans negligi poterit, eritque vis aeris elastica ut

$$\frac{kaa}{\sqrt[3]{(m-pm+pn)^2}}\left(\sqrt[3]{(m-i+pi-pm+pn)^2} - \sqrt[3]{(m-pm+pn-i)^2}\right).$$

Cum autem sit $ia^3 = (m-pm+pn)g^3$, loco a in computum duco g, ut ea tanquam constans abiici possit; Erit ergo $aa = gg\sqrt[3]{\left(\frac{m-pm+pn}{i}\right)^2}$, quo substituto erit vis elastica aeris ut

$$k\left(\sqrt[3]{\left(\frac{m-i+pi-pm+pn}{i}\right)^2} - \sqrt[3]{\left(\frac{m-i-pm+pn}{i}\right)^2}\right).$$

Caeteris igitur paribus est aeris vis elastica ut altitudo k, velocitatem materiae subtilis in bullulis gyrantis gravi descendenti imprimens.

17. Verum cum vires aeris elasticae inter se comparantur, id fit aeris vim expansivam in eandem basin agentem explorando. Quamobrem, ut mensuram aeris vis elasticae, ut consuetum est, exhibeam, necesse est, ut pressionem aeris in datam basin investigem. Nam, quae hucusque de ista mensura tradidi, huc non quadrant, quia vis tota globuli aeris elastica est supputata, quae propterea in tanto maiorem basin agit, quanto magis bullula est extensa. Sunt autem hae bases ut quadrata radiorum bullularum; Et iis etiam vires elasticae proportionantur. Quocirca assumatur constans quidam sphaerae radius e, fiatque ut a^2 ad e^2 ita vis aeris elastica inventa paragr. praeced. ad vim in datam basin agentem. Multiplicetur ergo oportet formula praecedens per $e^2 : a^2$ at vero est $a^2 = gg\sqrt[3]{\left(\frac{m-pm+pn}{i}\right)^2}$. Quamobrem absoluta divisione, abiectisque e^2 et g^2 tanquam constantibus obtinebitur vis aeris elastica absoluta, quae erit ut

$$k\left(\sqrt[3]{\left(\frac{m-i+pi-pm+pn}{m-pm+pn}\right)^2}-\sqrt[3]{\left(\frac{m-i-pm+pn}{m-pm+pn}\right)^2}\right).$$

18. Evanescat pars bullulae aquea; erit $g = h$ et ideo $p = 1$. Quamobrem vis aeris elastica hoc casu erit

$$k\left(\sqrt[3]{1}-\sqrt[3]{\left(\frac{n-i}{n}\right)^2}\right)$$

seu multiplicato per constantem $\sqrt[3]{n^2}$, erit ea ut $k(\sqrt[3]{n^2}-\sqrt[3]{(n-i)^2})$. Ponatur k seu velocitas constans, ut obtineatur lex elasticitatum pro solis aeris diversis condensationibus, erit tum vis elastica, ut $\sqrt[3]{n^2}-\sqrt[3]{(n-i)^2}$. Et hinc sequentes duco consequentias. Si status aeris quam proxime ad maximam condensationem accedat, erit $n-i$ tantum non $=0$, ergo vis elastica hoc in casu erit ut $\sqrt[3]{n^2}$, i.e. ea erit constans. Aere ergo iam vehementer compresso, vis elastica amplius sensibiliter non immutatur.

19. Deinde si i respectu ipsius n valde parvum sit, seu si densitas aeris ad densitatem materiae subtilis admodum exiguam habuerit rationem erit $(n-i)^{2/3} = n^{2/3} - \frac{2}{3}n^{-1/3}\,i$, consequenter vis aeris elastica erit ut $\frac{2}{3}n^{-1/3}\,i$ sive neglecto $\frac{2}{3}n^{-1/3}$ ut i. Aere ergo valde rarefacto elasticitates erunt ut densitates aeris. Quare cum circa aerem naturalem observemus quantumvis is comprimatur elasticitatem propemodem in eadem ratione crescere, dubium non est, quin aer noster admodum sit dilatatus respectu materiae subtilis, atque rationem specificam aeris ad gravitatem specificam materiae subtilis perquam esse exiguam.

20. Attamen cum ea prorsus negligi nequeat, oportet seriei, inquam $(n-i)^{2/3}$ convertitur, non tantum duos primos, sed tres accipere terminos, qui variationes observatas satis exacte monstrabunt. Hoc ergo pacto erit $(n-i)^{2/3} = n^{2/3} - \frac{2}{3}n^{-1/3}\,i - \frac{1}{9}\,n^{-4/3}\,i^2$. Atque hinc vis elastica erit ut $\frac{2}{3}n^{-1/3}\,i + \frac{1}{9}\,n^{-4/3}\,i^2$ seu multiplicando per $9n^{4/3}$, ut $6ni + i^2$. Dicatur vis elastica v fiatque $fv = 6ni + ii$. Ex hac igitur aequatione ope experimentorum, qua circa aeris incrementum vis elasticae eo continuo magis condensato, instituta sunt a BOYLEO, invenietur ratio $n:i$. Ex quo intelligetur extremus et maximus densitatis gradus, ad quem aerem comprimere possibile est.

21. Consultum ergo esse duxi experimenta BOYLEANA[4]) huc transcribere, ut ex iis de densitate seu gravitate specifica materiae subtilis concludere liceat, et quamnam ad aerem rationem habeat. Aer primo in tubo spatium 12 digit. Angl. replebat postea vero cum columna mercuriali comprimebatur altitudines aeris et mercurii superaffusi in sequenti tabula exhibentur, cuius prior columna A indicat spatium aeris in tubo, et altera B altitudinem mercurii comprimentis aerem: hae vero in digitis Anglic. exprimuntur.

A	B	A	B	A	B
12	0	8	$15\frac{1}{16}$	5	$41\frac{9}{16}$
$11\frac{1}{2}$	$1\frac{7}{16}$	$7\frac{1}{2}$	$17\frac{15}{16}$	$4\frac{3}{4}$	45
11	$2\frac{15}{16}$	7	$21\frac{3}{16}$	$4\frac{1}{2}$	$48\frac{12}{16}$
$10\frac{1}{2}$	$4\frac{6}{16}$	$6\frac{1}{2}$	$25\frac{3}{16}$	$4\frac{1}{4}$	$53\frac{11}{16}$
10	$6\frac{3}{16}$	6	$29\frac{11}{16}$	4	$58\frac{12}{16}$
$9\frac{1}{2}$	$7\frac{14}{16}$	$5\frac{3}{4}$	$32\frac{3}{16}$	$3\frac{3}{4}$	$63\frac{15}{16}$
9	$10\frac{2}{16}$	$5\frac{1}{2}$	$34\frac{15}{16}$	$3\frac{1}{2}$	$71\frac{5}{16}$
$8\frac{1}{2}$	$12\frac{8}{16}$	$5\frac{1}{4}$	$37\frac{15}{16}$	$3\frac{1}{4}$	$78\frac{11}{16}$
				3	$88\frac{7}{16}$

22. Exhibet igitur haec tabula columnam mercurialem, quae pondere suo aerem in datum spatium redigit. Hoc vero pondus non solum aerem comprimit, sed ei insuper adiici debet pondus atmosphaerae, quod simul cum mercurio in aerem agit. Cum ergo summa ponderis mercurii et atmosphaerae ea sit, vis qua aer comprimitur, erit ea aequalis vi aeris elasticae. Unde, si numeris tabulae B addatur altitudo mercurii ponderi atmosphaerae aequalis, quam BOYLEUS se $29\frac{1}{8}$ dig. observasse scribit; habetitur relatio inter densitates aeris et elasticitates. Sed cum ad istud accurate praestandum, exactissime altitudinem mercurii atmosphaeram aequilibrantis observasse necesse sit; idque multis difficultatibus perturbetur: Mallem relicta hac altitudine $29\frac{1}{8}$ dig. ex experimentis ipsis, quum numerus eorum abunde sufficiat, deducere pondus atmosphaerae: Sed quia ad hoc accuratissima requiruntur experimenta, (in quibus praesentia haberi nequeunt) altitudinem $29\frac{1}{8}$ dig. retinere cogor.

4) ROBERT BOYLE (1627–1691), *A defence of the doctrine touching the spring and weight of the air* (London, 1662), pp. 57–68. EJA.

23. Sed densitates aeris sunt reciproce ut volumina eiusdem massae aereae; volumina vero columna A exhibeantur: Ergo densitates erunt reciproce ut numeri columnae A. Si igitur densitas aeris in statu naturali ponatur, 1; reliquae densitates habebuntur si numerus 12 per reliquos respondentes numeros columnae A dividatur. Deinde elasticitates, ut vidimus, sunt ut numeri secundae columnae B aucti numero $29\frac{1}{8}$. Cum vero sit $fv = 6ni + ii$, atque ex observationibus allatis habeantur in quolibet casu et v et i, duae hae literae f et n determinari debent; Id quod duobus quibusvis experimentis praestabitur. Sumatur ad literam f determinandam experimentum primum; Et erit $i = 1$, et $v = 29\frac{1}{8}$, unde $29\frac{1}{8}f = 6n + 1$. Ergo $f = \dfrac{48n + 8}{233}$. Quo valore in aequatione substituto habebitur $48nv + 8v = 1398ni + 233ii$, consequenter $n = \dfrac{8v - 233ii}{1398i - 48v} = \dfrac{233ii - 8v}{48v - 1398i}$.

24. Ut hinc inveniatur n, oportet experimentorum allatorum aliquod adiungere. Sumatur igitur ultimum, erit $i = 12:3 = 4$, et $v = 88\frac{7}{16} + 29\frac{1}{8} = 117\frac{9}{16}$. Unde $n = \dfrac{940\frac{1}{2} - 3728}{5592 - 5643} = \dfrac{2787{,}5}{51}$ hinc erit[5] $n = 54{,}64$. Ut pateat, quantum experimenta inter se conveniant vel disconveniant, accipiatur id quod aer in triplo minus spatium est redactum; Erit ergo $i = 3$, et $v = 58\frac{12}{16} + 29\frac{2}{16} = 87\frac{7}{8}$. Unde habetur $n = \dfrac{2097 - 703}{4218 - 4194} = \dfrac{1394}{24} = 58\frac{1}{12}$. At experimentum quo aer duplo tantum densior exhibetur paulo plus quam 17 pro valore ipsius n exhibet. Ex qua ingenti discrepantia intelligi potest, quam parum accurata haec sint experimenta: Id quod praeterea ex saltibus, qui in iis deprehenduntur, satis colligi potest.

25. Id autem ex reliquis experimentis calculum instituens observavi, inde quo minus aer erat compressus, eo minorem ipsius n valorem inventum. Ex quo intelligi potest, reliquis in numeris saltibus neglectis, vel altitudinem mercurii atmosphaerae aequiponderantis non satis accurate esse assumtam, vel tubum nimis fuisse angustum, ut ne facillime quidem mercurius in eo descendere potuerit. Prius quidem vix credi potest: Sed posterius eo magis verisimile est, quod tanta insit difformitas experimentis: Unde concludi debet, mercurium non successive, sed quasi per saltus descendisse. Eandem difformitatem in BOYLEI experimentis circa rarefactionem aeris advertens, inde quicquam concludere nolui: sed plenius de

5) Valor verus est 54,66. EJA.

densitate materiae subtilis iudicium tamdiu differam, donec vel accuratiora experimenta in manus veniant, vel ipsi instituere vacaverit.

26. Ut autem clarius ob oculos ponatur, qua lege elasticitates aeris pro diversis densitatibus crescant, tota res figura geometrica repraesentari potest. Neglectis pelliculis aqueis inventa est aeris vis elastica proportionalis $\sqrt[3]{n^2} - \sqrt[3]{(n-i)^2}$: Unde patet, id per parabolam cubicalem secundam praestari posse. Sit AMC [fig. 3] parabola cubicalis secunda super axe AB, in qua applica-

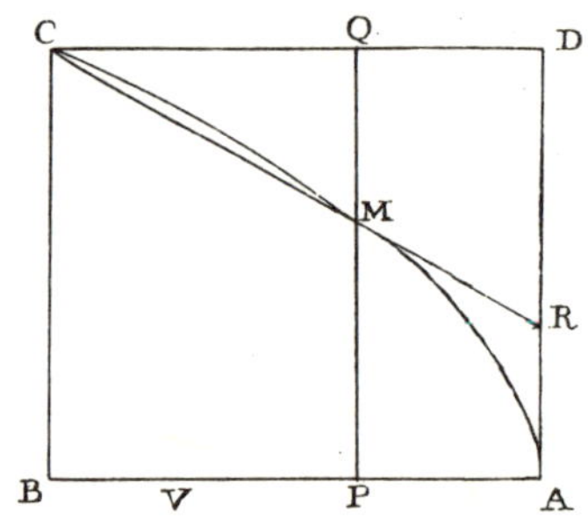

Fig. 3

tae PM serit in ratione subsesquiplicata abscissarum AP. Capiatur $AB = n$ et erecta applicata BC, ducatur axi parallela CD. Dico si in ea capiatur $CQ = i$, applicatam correspondentem QM repraesentare vim aeris elasticam. Nam est $QM = BC - PM$. Sed BC est ut $\sqrt[3]{AB^2}$ seu $\sqrt[3]{n^2}$, et PM ut $\sqrt[3]{AP^2}$, seu, ob $AP = AB - BP[=AB-]\,CQ = n - i$, erit PM ut $\sqrt[3]{(n-i)^2}$. Ut ergo sit QM ut $\sqrt[3]{n^2} - \sqrt[3]{(n-i)^2}$ cui quantitati etiam, ut patet, proportionalis est vis aeris elastica.

27. Si ea accipiatur regula, qua vires aeris elasticae in ratione densitatum ponuntur; Ex hac figura patebit quantum ea a vero, si modo hanc theoriam veram appellare licet, aberret. Ducatur per puncta C et M recta CMR perpendicularem AD ex A in AB ductam secans in R; exprimet haec recta distantiis suis a CD vires elasticas secundum istam regulam aeri iuxta abscissas in linea CD condensato respondentes. Si igitur QM naturalem aeris vim elasticam denotet, regula ista in condensationibus iusto minorem exhibebit vim elasticam, at in rarefactionibus iusto maiorem, donec utraque regula aeri infinite rarefacto elasticitatem nullam attribuat.

28. Si certo constaret ratio quam n ad i habet, quantum haec regula in quovis casu a vero aberret, assignari posset: Nec non aeris vis elastica maxima AD,

seu ratio $AD:QM$. Ob hunc defectum pono saltem $n:i = q:1$ eritque $n = qi$, adeoque vis elastica QM erit ut[6]) $\sqrt[3]{q^2i^2} - \sqrt[3]{(qi-i)^2}$, dividatur per $\sqrt[3]{i^2}$ utpote constantem, erit vis elastica aeris naturalis ut $\sqrt[3]{q^2} - \sqrt[3]{(q-1)^2}$. Assumatur quivis alius condensationis gradus, quo densitas sit ad naturalem ut s ad 1. Erit ea densitas si, adeoque vis elastica respondens erit $\sqrt[3]{q^2i^2} - \sqrt[3]{(qi-si)^2}$, erit ea igitur ut $\sqrt[3]{q^2} - \sqrt[3]{(q-s)^2}$. Unde sequitur, elasticitatem aeris naturalis esse ad elasticitatem aeris s vicibus densioris ut 1 ad $\dfrac{\sqrt[3]{q^2} - \sqrt[3]{(q-s)^2}}{\sqrt[3]{q^2} - \sqrt[3]{(q-1)^2}}$ sed secundum regulam vulgarem oporteret esse ut 1 ad s, si $s = q$ tum erit $DR = q \cdot QM$, et $AD = \dfrac{\sqrt[3]{q^2}}{\sqrt[3]{q^2} - \sqrt[3]{(q-1)^2}} \cdot QM$. Quia vero q valde est magnum respectu 1, erit $\sqrt[3]{q^2} - \sqrt[3]{(q-1)^2} = \dfrac{2}{3\sqrt[3]{q}}$; erit itaque $AD = \frac{3}{2}q \cdot QM$. Regula ergo ea plus dimidio nunquam a vero aberrare potest.

29. Cognita pro quovis condensationis gradu aeris elasticitate, poterit inde inveniri quanta esse debeat aeris densitas in data quacunque altitudine. Cum enim aer naturalis comprimatur a pondere aeris superincumbentis, necesse est, ut, quo altius ascendatur, aer ob imminutum ibi atmosphaerae pondus rarior fiat. Nam ubique eo usque aer dilatatur, quoad pressio aequalis fit eius elasticitati. Sit igitur curva BMV [fig. 4] scala densitatum aeris, cuius nimirum applicatae PM

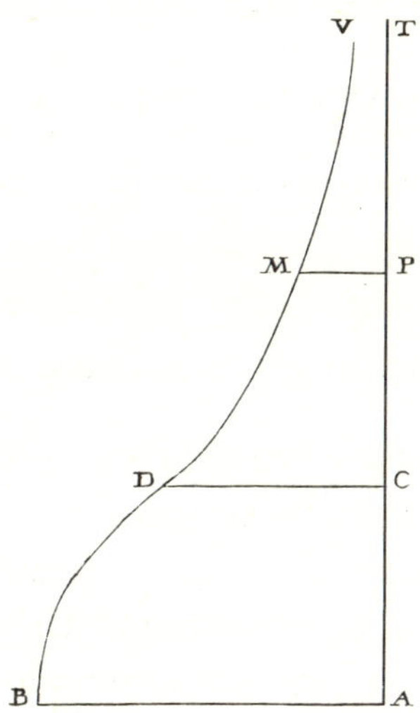

Fig. 4

6) Intellegimus $\sqrt{(q^2i^2)}$. EJA.

exprimant aeris densitates in altitudinibus P. Sit A is locus, quo densitas aeris est maxima, adeoque ubi $AB = n$. Accipiatur locus quicunque p, cuius altitudo AP super A dicatur x; densitas vero ibi seu $PM = y$, erit ibi aeris vis elastica ut $\sqrt[3]{n^2} - \sqrt[3]{(n-y)^2}$, cui proportionalis esse debet pressio ab aere superiore PT orta. Pressiones autem sunt ut densitates et altitudines coniunctim: Quamobrem erit pressio aeris superioris ut area $MPTV$, i.e. ut $-\int y\,dx$. Est itaque $a\int y\,dx = \sqrt[3]{n^2} - \sqrt[3]{(n-y)^2}$, adeoque $ay\,dx = \dfrac{2\,dy}{3\sqrt[3]{(n-y)}}$; unde $a\,dx = \dfrac{2\,dy}{3y\sqrt[3]{(n-y)}}$ seu posito $a = \frac{2}{3}$, erit $dx = \dfrac{dy}{y\sqrt[3]{(n-y)}}$ quae hoc modo integrari debet, ut posito $x = 0$, y fiat $= n$.

30. Si fiat $n = y$ erit tum dx infinities maius quam dy, ergo tangens in B parallela erit axi verticali AT. Propterea haec curva alicubi punctum flexus contrarii habere videtur; id quod hoc modo invenietur. Quia est $dx = \dfrac{dy}{y\sqrt[3]{(n-y)}}$; erit $dy = y\,dx\sqrt[3]{(n-y)}$. Assumto dx pro constante, erit

$$ddy = dy\,dx\sqrt[3]{(n-y)} - \tfrac{1}{3}y\,dx\,dy\,(n-y)^{-2/3} = 0\,.$$

Unde $3n - 3y = y$. Consequenter $y = \frac{3}{4}n$. Quam ob rem punctum flexus contrarii eo erit loco, quo densitas aeris est ad maximam ut 3 ad 4. Applicetur igitur $CD = \frac{3}{4}AB$, erit in puncto D punctum flexus contrarii. Est deinde subtangens huius curvae $\dfrac{y\,dx}{dy} = \dfrac{1}{\sqrt[3]{(n-y)}}$. Unde colligitur si y fuerit respectu ipsius n valde parvum, tum esse subtangentem constantem; Ut adeo hoc in casu haec curva cum logarithmica confundatur.

31. Potest quidem aequatio pro ista curva $dx = \dfrac{dy}{y\sqrt[3]{(n-y)}}$ ad quadraturam circuli et logarithmos reduci: sed inde multo difficilior enascitur eius curvae constructio, quam si per quadraturas construatur. Designet igitur AMC [fig. 5] parabolam cubicalem secundam, ut in Fig. 3 sitque $CD = n$. Assumatur aeris densitas quaevis in CD, puta CQ, ponaturque $CQ = y$. Cuius applicata respondens QM erit $\sqrt[3]{n^2} - \sqrt[3]{(n-y)^2}$, cui proportionalis accipi $-\int y\,dx$ debet. Dicatur $QM = z$, brevitatis ergo: eritque $-y\,dx = dz$ et $dx = \dfrac{-dz}{y}$, atque $x = \int\dfrac{-dz}{y}$.

Ducatur PM, quae erit y, et in ea producta, si opus est, capiatur $PM = \frac{1}{y}$, erit area $PBEN = \int \frac{-dz}{y}$. Quapropter in MQ prolongata accipiatur QL, quae sit ut area $PBEN$. Erit punctum L in curva quaesita. Est enim in ea, ducta LH, $CH = LQ = \int \frac{-dz}{y} = x$, et $HL = CQ = y$. Hoc igitur modo curva DLV determinabitur.

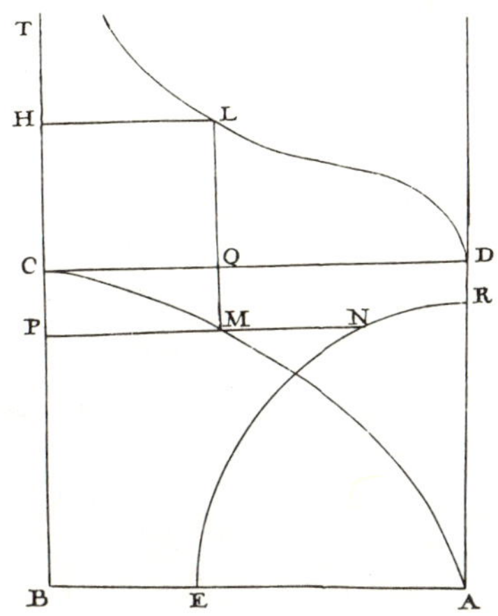

Fig. 5

32. Quae hucusque aeris proprietates ex theoria exposita derivatae sunt, eae nihil absoluti in se continent, sed tantum rationem dant, secundum quam elasticitas aeris pro diversis densitatibus, humiditatibus et materiae subtilis celeritatibus existimari debeat. Verum nunc absoluti quid tradam altitudinem columnae mercurialis determinaturus, quam datus aereus globulus sustinere valet. Sit itaque AB [fig. 6] diameter horizontalis bullulae aereae, de qua intelligi debent, quae § 14

Fig. 6

inventa sunt. Incumbat ei columna mercurialis $ABED$ altitudinis $AD = f$, quae tanta sit, ut in aequilibrio consistat cum vi, quam bullula habet, sese expandendi. Haec autem columna in singulis bullulae punctis perpendiculariter agit in eius superficiem, idque vi, quae est ut altitudo columnae f, et basis seu superficies bullulae, quam premit, atque gravitas specifica coniunctim. Cum autem semidiameter AC sit $= a$; erit circulus maximus bullulae $\frac{\pi aa}{2}$, adeoque semisuperficies eius $= \pi aa$, quae est basis, quae a columna mercuriali premitur. Exprimatur porro gravitas specifica mercurii, respectu habito ad reliquas gravitates specificas, litera r, erit pressio, quam columna mercurialis in bullulam exercet $= \pi aarf$.

33. Haec autem pressio destrui debet pressione a vi centrifuga materiae subtilis orta, quae etiam in singula superficiei puncta aequaliter agit. Quamobrem vis, qua vis centrifuga in haemisphaerium agit, idque extendere annititur, aequalis esse debet vi comprimenti columnae mercurialis. Vis autem ea est dimidium vis elasticae totius bullulae, cuius aequale pondus § 15 [7]) inventum est,

$$\frac{2\pi nkaa}{\sqrt[3]{(m-pm+pn)^2}}\left(\sqrt[3]{(m-i+pi-pm+pn)^2}-\sqrt[3]{(m-i-pm+pn)^2}\right);$$

huius ergo dimidio aequari debet pondus columnae mercurialis $\pi aarf$. Unde sequens enascitur aequatio:

$$rf\sqrt[3]{(m-pm+pn)^2} = nk\left(\sqrt[3]{(m-i+pi-pm+pn)^2}-\sqrt[3]{(m-i-pm+pn)^2}\right)$$

seu

$$f = \frac{nk}{r}\left(\sqrt[3]{\left(\frac{m-i+pi-pm+pn}{m-pm+pn}\right)^2}-\sqrt[3]{\left(\frac{m-i-pm+pn}{m-pm+pn}\right)^2}\right).$$

34. Ut haec aequatio tractatu facilior evadat saltem pro naturali aeris statu, pono i admodum parvum respectu n; et propterea erit

$$\sqrt[3]{(m-i+pi+pm+pn)^2} = \sqrt[3]{(m-pm+pn)^2} + \frac{2(pi-i)}{3\sqrt[3]{(m-pm+pn)}}.$$

Atque eodem modo

$$\sqrt[3]{(m-i-pm+pn)^2} = \sqrt[3]{(m-pm+pn)^2} - \frac{2i}{3\sqrt[3]{(m-pm+pn)}}.$$

Quibus valoribus substitutis, orietur haec aequatio

$$rf(m - pm + pn) = \frac{2pnik}{3}, \quad \text{unde} \quad f = \frac{2pnik}{3r(m - pm + pn)}.$$

Sed si ponatur humiditas in aere evanescens, erit $p = 1$. Tum igitur erit $f = \frac{2ik}{3r}$. Si autem aer vaporibus fuerit infectus p eo minor erit unitate, quo plus vaporum in aere hospitatur; ponatur itaque hoc in casu $p = 1 - q$, erit

$$f = \frac{2nik(1-q)}{3r(qm + n(1-q))} = \frac{2ik}{3r} - \frac{2qmik}{3r(qm + n(1-q))} = \frac{2ik}{3r}\left(1 - \frac{qm}{qm + n(1+q)}\right).$$

35. Cum autem f indicet altitudinem columnae mercurialis in aequilibrio consistens, exprimet eadem litera f altitudinem mercurii in barometro. Ex inventa igitur aequatione, datis velocitate materiae subtilis in bullulis gyrantis, aeris et materiae subtilis gravitatibus specificis, atque quantitate aquae in aere versantis, inveniri poterit altitudo mercurii in barometro. Nam r gravitas specifica mercurii, ut et m gravitas specifica aquae aliunde iam constant. Percurram itaque casus, quibus mercurius ascendere, et quibus descendere debet; ut inde pateat, quid in aere acciderit et ascendente et descendente mercurio in barometro. Ad hoc cum tantum ratione opus sit, negligo factorem $\frac{2}{3r}$ tanquam constantem, eritque

$$f \quad \text{ut} \quad ik\left(1 - \frac{qm}{qm + n(1-q)}\right).$$

36. Hinc igitur consequitur manente facto, ik mercurium in barometro ascendere decrescente fractione

$$\frac{qm}{qm + n(1-q)}.$$

Haec vero fractio crescente q etiam crescit, decrescente vero q decrescit: Nam crescente q elemento dq, fractio crescet elemento

$$\frac{mn\,dq}{(qm + n(1-q))^2}.$$

7) Editio princeps: 14 loco 15. Correxit EJA.

Quamobrem manente facto ik mercurius in barometro ascendere decrescente aeris humiditate; ea vero aucta mercurius descendere debebit. Atque hanc puto esse rationem, cur ascensus mercurii in barometro plerumque coelum serenum, descensus vero pluviam adversamque tempestatem praenunciet; Illo enim casu aer maximam partem a vaporibus vacuus est, hoc vero iis magis infectus.

37. Possunt quidem aliae concurrere rationes, ob quas mercurius ascendere vel descendere queat immutata vaporum quantitate: Quando scilicet factum ik crescit vel decrescit. Sed fortasse hoc factum sensibiliter neque crescere neque decrescere potest, propter alterutram literam eadem fere ratione auctam, qua altera diminuitur. Nam velocitas materiae subtilis, cuius quadratum est ut k, augmenta accipit aucto calore, sed idem calor aerem rarefacit, et quantitatem i minorem efficit, ut ergo factum ik quasi semper constans permaneat. Ex quo intelligitur tute semper ascensum vel descensum mercurii diminutae vel auctae vaporum in aere versantium quantitati attribui posse, quanquam negari non possit et factum ik quodammodo effectum humiditatis et augere et diminuere posse.

38. Neque vero hinc inferre licet, barometrum idem ac hygrometrum praestare oportere; cum et hoc humiditatem aeris monstret. Sed id considerandum est, barometri effectus a tota aeris massa seu totius atmosphaerae statu pendere; hygrometri autem a solo aere id ambiente. Quamobrem altitudo mercurii in barometro incrementa accipit, si universus aer a vaporibus liberatur, decrementa vero si is vaporibus impraegnatur. Unde colligitur hygrometrum fere summam siccitatem ostendere posse, cum altitudo mercurii minima sit; et similiter hygrometrum humiditatem indicare posse, cum mercuriis summam altitudinem attigerit. Plus enim immensa aeris altitudo in barometrum valet, quam infima haec regio, quae sola in hygrometrum agit.

39. Si humiditas aeris evanescat, habetur iuxta § 33 haec aequatio $f = \frac{2ik}{3r}$, quam n non ingreditur. Ex ea igitur cum experimentis ratio $r:i$, ut et altitudo f constet, inveni[ri] potest altitudo k, ex qua grave cadendo velocitatem acquirit ei, qua materia subtilis in bullulis aeris gyratur, aequalem; Est enim $k = \frac{3rf}{2i}$. Circa quam expressionem observo eam excepto coefficiente numero $\frac{3}{2}$ eandem esse cum altitudine generante velocitatem, qua sonus per aerem promovetur, ut ergo velocitas materiae subtilis constantem habeat rationem ad velocitatem soni. Hic

autem animum abducere oportet humiditate aeris, qua accedente k alio modo exprimetur.

40. Observatur altitudo mercurii in barometro a 22 usque ad 24 et ultra dig. Pedis Rhenani [8]). Cum autem aerem a vaporibus vacuum supponam, attribuo literae f maximam, quam habere potest altitudinem, nempe 2460 scrup. Ped. Rhenani. Dein rationem r ad i pono ut 10000 ad 1. Quemadmodum ex experimentis de gravitate aeris concluditur. Quibus positis erit $k = \frac{30000 \cdot 2460}{2} = 36900$ pedibus. Adeoque materia subtilis velocitate movetur tanta, quanta a gravi ex altitudine 36900 ped. in vacuo descendente acquiritur. Si ergo haec materia sua velocitate in directum pergeret, conficeret uno minuto secundo $1518\frac{1}{2}$ ped. Rhenanos.

41. Isto hanc dissertationem finio, cum desint accurata experimenta, ex quibus reliqua adhuc desiderata determinentur, et quibus Theoria haec plenius confirmetur. Incerta est adhuc ratio n ad i, seu quam habet gravitas specifica materiae subtilis ad gravitatem specificam aeris. Ad hanc vero investigandam accuratis experimentis ad id facientibus instituendis operam studiumque adhibebo. Quantitas autem n si haberetur, facile formulae inventae ad praxin applicarentur; atque aliis idoneis instrumentis adhibendis quovis tempore, quantum aquae in aere contineatur, assignari posset. Et forsitan multa insuper alia, ad quae, iis quae sufficiunt cognitis, quasi manu duceremur.

8) Pes aequatur 10 digitis vel 1000 scrupulis. HCI.

PIECES
QUI ONT REMPORTÉ
LE PRIX
DE L'ACADÉMIE ROYALE
DES SCIENCES,
EN M. DCC. XL.

Sur le Flux & Reflux de la Mer.

Selon la fondation faite par feu M. ROUILLÉ DE MESLAY, ancien Conseiller au Parlement.

A PARIS, rue Saint Jacques.

Chez G. MARTIN, J. B. COIGNARD, & les Freres GUERIN, Libraires.

M. DCC. XLI.

INQUISITIO PHYSICA
IN CAUSAM
FLUXUS AC REFLUXUS
·MARIS·

A. D. D. EULER, *Mathematicarum Profeſſore, è Societate Academiæ Imperialis Sancti-Petersburgenſis.*

Cur nunc declivi nudentur littora Ponto,
Adverſis tumeat nunc Maris unda fretis;
Dum veſtro monitu naturam conſulo rerum:
Quàm procul à Terris abdita cauſa latet!
In Solem Lunamque feror. Si plauditis auſo;
Sidera ſublimi vertice ſumma petam.

Commentatio 57 indicis ENESTROEMIANI
Pièces qui ont remporté le prix de l'académie royale des sciences de Paris en 1740 (Paris, 1741), p. 235–350

INQUISITIO PHYSICA IN CAUSAM FLUXUS AC REFLUXUS MARIS

CAPUT PRIMUM

De Causa Fluxus ac Refluxus Maris in genere

1. Omnem mutationem, quae in corporibus evenit, vel ab ipsa motus conservatione proficisci, vel a viribus motum generantibus, hoc quidem tempore, quo qualitates occultae causaeque imaginariae penitus sunt explosae, nulla indiget probatione. Hoc autem discrimen quovis oblato Phaenomeno diligentissime considerari oportet, ne tam motus conservationi eiusmodi effectus tribuatur, qui sine viribus oriri nequit, quam vires investigentur, quae motum sua natura conservandum producant. Quo quidem in negotio, si debita attentio adhibeatur, errori vix ullus relinquitur locus: cum ex legibus naturae satis superque constet, cuiusmodi motus vel per se conserventur, vel viribus externis debeantur. Corpus scilicet in motu positum propria vi hunc motum uniformiter in directum retinet: atque corpus, quod circa axem convenientem per centrum gravitatis transeuntem motum rotatorium semel est consecutum, eodem motu rotari perpetuo sua sponte perget: neque huiusmodi motuum causam in ulla re alia, nisi in ipsa corporum natura, quaeri oportet. Quocirca si huius generis Phaenomenon fuerit propositum, alia causa investigari non potest, nisi quae a principio tales motus procreaverit.

2. Huius generis foret quaestio, si quaereretur causa motus vertiginis Planetarum ac Solis; hic enim sufficeret eam causam assignasse, quae initio hos motus produxisset, cum Sol aeque ac Planetae talem motum semel consecuti eundem propria vi perpetuo conservare debeant, neque ad hoc Phaenomenon explicandum vis ulla externa etiam nunc durans requiratur. Longe aliter se res habet, si motus proponatur neque uniformis, neque in directum procedens, cuiusmodi est motus Planetarum periodicus circa Solem: hoc enim casu minime sufficit ea vis, quae initio Planetas ad istiusmodi motus impulerit, sed perpetuo novae virium actiones requiruntur, a quibus tam celeritas quam directio continuo immutetur: quae vires,

quam primum cessarent, subito Planetae orbitas suas desererent, atque in directum motu aequabili avolarent. Quod si igitur Phaenomenon quodcunque naturae proponatur, ante omnia sollicite est inquirendum, ad quodnam genus id pertineat, atque utrum causa in viribus externis sit quaerenda, an in ipso subiecto corpore? Quinetiam saepenumero usu venire potest, ut effectus utriusque generis in eodem Phaenomeno multum sint inter se permixti; quo casu summo studio ii a se invicem discerni ante debebunt, quam causarum investigatio suscipiatur.

3. His rite perpensis explicatio Galilei, quam in suis Dialogis de aestu Maris[1]) assignare est conatus, mox concidit; putavit enim Fluxum ac Refluxum Maris tantum a motibus Terrae rotatorio circa axem et periodico circa Solem oriri, neque aliis viribus tribui oportere, nisi quae hos motus cum producant, tum conservent. Namque si ponamus Terram solo motu diurno esse praeditam, iste motus Mare aliter non afficiet, nisi id sub Aequatore attollendo, ex quo figura Terrae sphaeroidica compressa nascitur, motus vero reciprocus in Mari omnino nullus hinc generari poterit. Quod si autem Terrae insuper motum aequabilem in directum tribuamus, priora Phaenomena nullo modo afficientur, sed prorsus eadem manebunt, quemadmodum ex principiis mechanicis clarissime perspici licet, quibus constat motum uniformem in directum omnibus partibus Systematis cuiuscunque corporum aequaliter impressum nullam omnino mutationem in motu et situ partium relativo inferre. Abeat nunc motus iste aequabilis Terrae in directum impressus in circularem vel ellipticum per vires quibus Terra perpetuo ad Solem urgeatur; ac ne hoc quidem casu ullus motus reciprocus in Mari produci poterit; quod cum per se est perspicuum, tum etiam ab ipso Galileo non statuitur: ipse enim non tam ex mixtione motus vertiginis et periodici aestum Maris proficisci est arbitratus, quam ex motu quocunque progressivo sive rectilineo sive curvilineo, si is cum motu rotatorio combinetur.

4. Quanquam autem motus Terrae periodicus circa Solem cum motu rotatorio circa axem coniunctus nullum in Mari motum reciprocum generare valet, tamen Mare, quod si motus esset aequabilis in directum in quiete persisteret, aliquantum turbari debebit. Quod si autem ad vim qua Terra in orbita sua continetur attendamus, non difficulter mutationem, quam Mare ab ea patietur, colligere poterimus. Nam cum partes Terrae a Sole remotiores minori vi, propiores vero maiori sollici-

1) Galileo Galilei, *Dialogo sopra i due massimi sistemi del mondo, Tolemaico e Copernicano* (Firenze, 1632), pars 4. EJA.

tentur, illae ad maius tempus periodicum, hae vero ad minus absolvendum cogentur, ex quo partibus Terrae fluidis, ut pote mobilibus, motus ab Oriente versus Occidentem secundum ecclipticam inducetur, hancque veram esse causam existimo ac praecipuam cur tam Oceanus quam aer sub Aequatore perpetuo habeat Fluxum ab ortu versus occasum. Possem etiam ex eodem principio clare ostendere tam Maris, si omnino liberum esset, quam aeris celeritatem tantam fore, qua tempore viginti-quatuor horarum spatium circiter viginti graduum absolvatur; sed cum haec inquisitio ad praesentem quaestionem proprie non pertineat, atque inclyta Academia fortasse alia occasione quaestiones huc spectantes sit propositura, uberiorem explicationem huius insignis Phaenomeni eo usque differendam esse censemus; hoc quidem tempore tantum indicasse contenti, motum Terrae periodicum coniunctim cum motu diurno Mari [quidem]* motum aliquem imprimere posse, sed neutiquam motum reciprocum, uti GALILEUS est arbitratus.

5. Uti in omnibus omnino quaestionibus physicis multo facilius est, quae non sit causa Phaenomeni cuiuspiam oblati, quam quae sit, ostendere; ita etiam praesens quaestio de Fluxu ac Refluxu Maris est comparata, ut non difficulter causas falso assignatas possimus refellere. Ac primo quidem post eversam GALILEI sententiam, explicatio aestus Maris CARTESIANA[2]) pressioni Lunae innixa tot tantisque laborat difficultatibus, ut omnino subsistere nequeat. Praeterquam enim quod istiusmodi pressio aliunde probari nequeat, atque ad hoc solum Phaenomenon explicandum gratuito assumatur, observationibus etiam minime satisfacit. In aperto enim ac libero Oceano [idque sub ipso aequatore]* aquam mox post transitum Lunae per Meridianum elevari observamus, cum secundum CARTESII sententiam eodem tempore deprimi deberet; neque praeterea hoc modo satis distincte explicatur, cur Luna sub Terra latens eundem fere effectum exerat, ac si super Horizonte versatur [et quae sunt alia huius generis phaenomena]*. Deinde hoc idem negotium non feliciori successu aggressus est WALLISIUS[3]) causam in communi centro gravitatis Terrae et Lunae quaerens, cuius explicatio mox satis dilucide est subversa. Superest denique NEWTONI[4]) theoria, quae nemine contradicente

---

* In Manuscripto interpolatum. Vide p. LXXXVI. EJA.

2) RENÉ DESCARTES, *Principia Philosophiae* (Amsterdam, 1644), pars 4, §§49–56. EJA.

3) "An essay of Dr. JOHN WALLIS, exhibiting his hypothesis about the Flux and Reflux of the Sea", *Philosophical Transactions* 1 (1666), 263–289. EJA.

4) ISAAC NEWTON, *Philosophiae Naturalis Principia Mathematica* (Londini, 1687), Liber 1, propositio 66; Liber 3, propositio 24, 36 et 37. EJA.

Phaenomenis multo magis est consentanea: at in ea id ipsum quod hoc loco quaeritur, causa scilicet physica, non assignatur, sed potius ad qualitates occultas referri videtur; interim tamen ne haec quidem theoria satis est evoluta, ut de eius sive consensu sive dissensu cum observationibus iudicium satis tutum ferri queat.

6. Cum igitur dubium sit nullum, quin Fluxus ac Refluxus Maris causa in viribus externis et realibus sit posita, quae si cessarent, simul aestus Maris mox evanesceret, ubi lateant hae vires et quomodo sint comparatae potissimum nobis erit explicandum, hoc enim est id ipsum, quod celeberrima Academia Scientiarum Regia in quaestione proposita requirit. Neque vero vires tantummodo indicasse sufficiet, verum praeterea id maxime erit monstrandum, quomodo istae vires agant, atque hos ipsos effectus, quos observamus, non vero alios producant; in hoc enim totius quaestionis cardo, explicationis scilicet confirmatio, vertitur. Quoniam autem plerumque pluribus viribus excogitandis idem Phaenomenon explicari potest, studium adhibendum est summum in hac indagatione, ne ad vires inanes atque imaginarias delabamur, quae in mundo neque sunt neque locum habere possunt. Parum enim scientiae naturali consulunt, qui quovis Phaenomeno oblato sibi pro arbitrio mundi structuram peculiarem effingunt, neque sunt solliciti, utrum ea compages cum aliis Phaenomenis consistere queat, an vero secus. Quod si enim iam aliunde constet existere in mundo eiusmodi vires, quae oblato effectui producendo sint pares, frustra omne studium in conquisitione virium novarum collocabitur.

7. Quoniam autem ad causam cuiusque Phaenomeni detegendam, ad singulas circumstantias sedulo attendere necesse est, ante omnia mirificum consensum aestus Maris cum motu Lunae contemplari conveniet. Non solum enim insignis harmonia inter aestum Maris ac Lunae motum diurnum deprehenditur, sed etiam revolutio synodica respectu Solis ingentem affert varietatem. Omnes denique observationes abunde declarant rationem Fluxus et Refluxus Maris a situ cum Lunae tum etiam Solis coniunctim pendere: ex quo statim prono ratiocinio consequitur, vires illas aestum Maris producentes, quaecunque etiam sint, cum Lunam potissimum, tum vero etiam Solem respicere debere. Quamobrem imprimis nobis erit inquirendum, utrum eiusmodi vires Solem et Lunam respicientes, quae in aquis talem effectum, qualis est aestus Maris, producere queant, iure ac ratione statui possint, an secus. Ac si pluribus modis istiusmodi vires animo concipere liceat, diligenter erit dispiciendum, quaenam cum aliis Phaenomenis consistere possint nec ne. Quantumvis enim explicatio quaepiam cum Phaenomenis conspiret, nisi virium, quae assumuntur, existentia aliunde comprobetur, labili ea omnino innititur funda-

mento. Quod si autem contra effectus eiusmodi viribus tribuatur, quas in mundo revera existere alia Phaenomena clare docuerunt, atque summus explicationis cum experientia consensus deprehendatur, dubium erit nullum, quin ista explicatio sit genuina et sola vera.

8. Quamvis autem certis viribus Lunae ac Soli tribuendis Phaenomenon aestus Maris commode explicari posset, tamen ob hanc solam causam istiusmodi vires statuere nimis audax videtur: quamobrem imprimis erit dispiciendum, num aliae rationes eiusmodi vires non solum admittant, sed etiam actu existere manifesto indicent. Perlustremus igitur vires, quas iam aliunde in mundo vigere novimus, sciscitemurque paucis an ad motum reciprocum Oceano inducendum sint idoneae: tales enim vires si in mundo iam extent, omnis labor in aliis inquirendis impensus irritus foret ac ridiculus. Ac primo quidem si Solem spectamus, motus Terrae annuus omnino declarat Terram perpetuo versus Solem urgeri et quasi attrahi, idque fortius in minori distantia, debilius vero in maiori; atque adeo hanc Solis vim in Terram rationem tenere reciprocam duplicatam distantiarum: ex quo sponte sequitur non solum universam Terram, sed etiam singulas eius partes perpetuo versus Solem urgeri. Tota quidem Terra aeque fortiter ad Solem sollicitatur, ac si omnis materia in eius centro esset congesta; interim tamen partes circa superficiem sitae vel magis vel minus ad Solem allicientur, quam totum Terrae corpus, prouti vel minus vel magis sint remotae a Sole, quam centrum Terrae. Hinc igitur fit, ut haec eadem vis ad Solem tendens aquam modo magis modo minus trahat, ex qua alterna actione motus reciprocus in Fluidis necessario oriri debet. Quocirca ista Solis vis in praesenti negotio neutiquam negligi poterit, cum ea, si forte sola causam aestus Maris non constituit, certe effectum aliarum virium necessario afficere ac turbare debeat.

9. Quemadmodum autem Terra cum omnibus suis partibus versus Solem sollicitatur, ita eorum sententia non multum a veritate abhorrere videtur, qui in Luna similem vim collocant. Observationes quidem huiusmodi vim in Luna non demonstrant sicuti in Sole; cum motus Terrae in orbita sua a Luna omnino non affici deprehendatur: sed si docuerimus eandem vim ad Lunam respicientem, quae aestui Maris producendo sit par, in motu Terrae nullam sensibilem anomaliam producere valere, audacia, quae forte in talis vis admissione consistere videatur, multum mitigabitur. Huiusmodi autem vis existentia aliis rationibus, nullo ad aestum Maris habito respectu, satis clare evinci potest; quia enim nullum est dubium, quin Luna ad Terram constanter feratur, ob aequalitatem actionis et

reactionis, Terram quoque versus Lunam pelli necesse est. Namque si ponamus Sole penitus sublato, Terrae ac Lunae omnem motum subito adimi, Luna utique ad Terram accedet; nemo autem non concedet, probe perpensis principiis mechanicis, Terram interea non prorsus esse quieturam, sed Lunae obviam ituram, concursumque in communi gravitatis centro contingere: hoc autem evenire non poterit, nisi Terra actu ad Lunam sollicitetur. Deinde in ipsa Luna gravitatem dari similem huic, quam in Terra sentimus, negari non potest; nisi enim talis vis in Luna vigeret, partes Lunae fluidae, cum ob gravitatem in Terram, tum ob motum Lunae circa proprium axem, etsi sit admodum lentus, et tempori periodico aequalis, iam dudum avolassent, partesque solidae consistentiam suam amisissent. Pluribus denique aliis rationibus ex natura vorticum petitis, magis confirmari posset tale corpus mundanum, cuiusmodi est Luna, subsistere non posse, nisi vortice sit cinctum, quo gravitas in id generetur. Quod si autem gravitationem versus Lunam concedamus, cur eius actionem non ad nos usque [porrigi]* admittamus, nulla omnino ratio suadet: quin potius eiusmodi vim similem statui conveniet, reliquis in mundo deprehensis [viribus]*, quae quasi in infinitum porriguntur, atque inversam duplicatam tenent distantiarum rationem.

10. His expositis manifestum est, et quasi experientia convictum, Terram cum singulis suis partibus tam versus Lunam quam versus Solem perpetuo sollicitari, atque utramque vim proportionalem esse reciproce quadratis distantiarum. Hae igitur vires, cum actu existant, constanterque effectum suum exerant, in praesenti negotio, quo in causam aestus Maris inquirimus, praeteriri omnino nequeunt; nisi dilucide ante sit probatum, eas non solum Fluxum ac Refluxum non generare, se ne quidem quicquam afficere. Si enim istae vires ullum duntaxat motum reciprocum Mari inducere valeant, quantumvis is etiam sit exiguus, atque adeo aestui Maris fortasse contrarius, earum tamen ratio necessario erit habenda, cum sine illis vera causa, quaecumque sit, neque investigari neque cognosci possit. Neque praeterea sanae rationis praecepta permittunt alias vires excogitare, in iisque causam aestus Maris collocare, antequam evidenter sit demonstratum, binas istas vires Solem Lunamque spectantes, quas non gratuito assumsimus, sed ex certissimis Phaenomenis in mundo existere novimus, ad Fluxum ac Refluxum Maris producendum non esse sufficientes. In sequentibus autem capitibus clarissime sumus ostensuri, ab his duabus viribus non solum in Oceano motum reciprocum generari debere, sed etiam eum ipsum, qui aestus marini nomine insigniri solet: atque hanc ob rem firmiter iam affirmamus veram Fluxus ac Refluxus causam in solis illis duabus viribus, quarum altera ad Solem est directa, altera ad Lunam, esse

positam; hocque simul omnium eorum sententias funditus evertimus, qui vel aliis omnino viribus idem Phaenomenon adscribere, vel cum his ipsis alias vires coniungere conantur.

11. Quaestio igitur de causa Fluxus ac Refluxus Maris, prouti ea ab Illustrissima Academia Regia est proposita, ad hanc deducitur quaestionem, ut binarum illarum virium, quibus singulae Terrae partes cum ad Solem tum ad Lunam perpetuo urgentur, idque in distantiarum ratione reciproca duplicata, causa assignetur Physica. Ex quo tractationem nostram bipartitam esse oportebit. Primo scilicet ex principiis Mechanicis dilucide erit ostendendum, a binis illis viribus Solem Lunamque respicientibus cum Fluxum ac Refluxum Maris generatim oriri debere, tum etiam hoc modo singula Phaenomena distincte explicari posse: hac enim parte absoluta nullum superest dubium, quin origo aestus Maris his ipsis viribus, quas actu iam in mundo existere docuimus, debeatur. Deinde vero harum virium causa Physica indicari debet, cum id sit praecipuum, quod Inclyta Academia requirit. Quod quidem ad illam partem attinet, in eius explicatione minime haesitamus; et clarissimis certissimisque demonstrationibus evincere pollicemur, per istas vires omnia omnino aestus Maris Phaenomena absolutissime explicari posse; qua in re nulli dubitationi ullus relinquetur locus, cum tota ad Geometriam et Mechanicam sublimiorem pertineat, calculoque analytico sit subiecta. Altera vero pars, in scientiam naturalem imprimis incurrens, maiori difficultati videtur obnoxia, nec tantae evidentiae capax; verum cum ista res occasione plurium quaestionum ab Academia Celeberrima antehac propositarum iam tanto studio sit investigata atque absoluta, eam non minori certitudine expedire confidimus.

12. Explosis hoc saltem tempore qualitatibus occultis, missaque Anglorum quorumdam renovata attractione, quae cum saniori philosophandi modo nullatenus consistere potest, omnium virium quae quidem in mundo observantur, duplex statuendus est fons atque origo. Nempe cum viribus tribuatur vel motus generatio vel immutatio, iste effectus semper vel ab allisione corporum, vel a vi centrifuga proficiscitur, quarum actionum utraque facultati, qua omnia corpora sunt praedita in statu suo sive quietis sive motus aequabilis in directum perseverandi, debetur. Ob hanc enim ipsam facultatem corpus in motu positum alia corpora, quae vel ipsius motui directe sunt opposita, vel eius directionem mutare cogunt, ad motum sollicitat; atque priori casu regulae collisionis corporum, posteriori vero vis centrifugae indoles et proprietates oriuntur ac demonstrantur. Cum igitur omnia corpora terrestria tam versus Solem, quam versus Lunam perpetuo sollicitentur, causa

huius sollicitationis vel continuo appulsui materiae cuiusdam subtilis, vel vi centrifugae similis materiae tribui debebit. Priori igitur casu materiam subtilem statui oporteret, quae constanter summa rapiditate cum ad Solem tum ad Lunam ferretur: huiusmodi vero hypothesis ob maximas difficultates, quibus est involuta, admitti minime potest. Primo enim perpetuo novis viribus esset opus, quae materiam subtilem indesinenter versus Solem Lunamque pellerent, qua quidem re quaestio non maiorem lucem assequeretur. Deinde talis motus per se diu consistere non posset, propter perpetuum materiae subtilis ad eadem loca affluxum nullumque refluxum, ut taceamus alia maxima incommoda cum istiusmodi positione permixta.

13. Exclusa igitur materiae subtilis continua allisione, tanquam ad vires cum ad Solem tum Lunam tendentes producendas minime idonea, alia harum virium causa non relinquitur, nisi quae in vi centrifuga consistat. Quemadmodum autem materia subtilis in gyrum acta ac vorticem formans non solum animo concipi, sed etiam in mundo persistere queat, iam satis superque est expositum, cum in dissertationibus, quae cum quaestio de causa gravitationis agitaretur, laudes Illustrissimae Academiae merebantur, tum etiam in aliis operibus; quibus in locis simul dilucide est ostensum, quomodo eiusmodi vortices comparatos esse oporteat, ut vires centrifugae fiant quadratis distantiarum a centro vorticis reciproce proportionales. Quae res cum meo quidem iudicio iam tam plana sit facta, ut vix quicquam ad praesens institutum attinens adiici queat, vorticum ulteriori examini sine ulla haesitatione supersedemus; idque eo magis, quod Celeberrima Academia eiusmodi amplam atque adeo iam confectam digressionem postulare haud videatur. Quoniam enim quaestio de causa gravitatis cum versus Terram tum etiam versus Solem et Planetas iam satis est investigata ac diremta; nunc quidem, si cuiuscunque Phaenomeni causa eo fuerit perducta, ibidem acquiescendum videtur, neque actum agendo denuo in causa gravitatis investiganda nimium immorari conveniret. Denique in praesenti negotio sufficere posset, si aestus Maris causa adhuc tantis tenebris obvoluta ad alia maxime aperta Phaenomena reducatur, quorum causa non solum habetur probabilis, sed etiam quae sola sit veritati consentanea, cuiusmodi est gravitatio tam versus Solem quam Lunam.

14. Causam igitur Fluxus ac Refluxus Maris proximam in binis vorticibus materiae cuiusdam subtilis collocamus, quorum alter circa Solem alter vero circa Lunam ita circumagatur, ut in utroque vires centrifugae decrescant in duplicata ratione distantiarum a centro vorticis; quae lex vis centrifugae obtinebitur, si

materiae subtilis vorticem constituentis celeritas statuatur tenere rationem reciprocam sub duplicatam distantiarum a centro vorticis. Quaecunque igitur corpora in istiusmodi vortice posita ad eius centrum pellentur vi acceleratrice, quae pariter ac vis centrifuga quadratis distantiarum reciproce est proportionalis. Vis absoluta autem qua corpus quodpiam in data distantia a centro vorticis collocatum eo urgetur, pendet a celeritate materiae subtilis absoluta. Ac primo quidem quod ad vorticem circa Solem rotatum attinet, eius vis absoluta ex tempore Terrae periodico cum distantia eiusdem a Sole comparato tanta colligitur, ut corpus, cuius distantia a centro Solis aequalis est semidiametro Terrae, eo sollicitetur vi, quae sit 227 512 vicibus maior, quam est gravitas naturalis in superficie Terrae. Metiemur autem hanc ipsam vim absolutam cuiusque vorticis, per vim, quam idem vortex exerit in distantia a suo centro semidiametro Terrae aequali: ex quo si vis gravitatis terrestris designetur per 1, erit vis absoluta Solis $= 227\,512$, cuius numeri loco brevitatis gratia utemur littera S. Simili modo vim vorticis Lunam cingentis absolutam indicabimus littera L, cuius valorem NEWTONUS recte cum ex ipso Fluxu ac Refluxu Maris, tum etiam ex praecessione Aequinoctiorum constituisse videtur circiter $\frac{1}{40}$. Quare si, posita Terrae semidiametro $= 1$, corporis cuiusdam a centro Solis vel Lunae distantia fuerit x, erit vis, qua id corpus vel ad Solem sollicitatur vel ad Lunam, vel $= \frac{S}{xx}$ vel $= \frac{L}{xx}$, uti ex indole horum vorticum prona consequentia fluit. In his quidem litterarum S et L determinationibus assumsimus mediam Solis a Terra distantiam 20 620 semidiametrorum Terrae, quae ex parallaxi horizontali 10″ sequitur, Lunae vero a Terra distantiam mediam 60 semid. Terrae; interim tamen vires ad Mare movendum hinc ortae ab his hypothesibus non pendent, uti ex sequentibus patebit.

15. Quoniam igitur aestum Maris per binas vires, quarum altera Solem respicit, altera Lunam, sumus expósituri, facile videri possemus eandem omnino explicationem suscipere, quam NEWTONUS dedit in suis Principiis Mathematicis Philosophiae Naturalis. Primum autem notandum est, quod si NEWTONUS veram causam huius Phaenomeni assignasset, summopere absurdum atque absonum foret, novitatis studio aliam causam, quae certo falsa futura esset, excogitare. Deinde vero NEWTONUS ne vestigium quidem reliquit, ex quo causa harum virium attractivarum, quas Soli Lunaeque tribuit, colligi posset, sed potius de causae Physicae inventione, qualem Academia Regia potissimum requirit, desperasse videtur; id quod eius asseculae aperte testantur, qui attractionem omnibus corporibus propriam esse, neque ulli causae externae deberi firmiter asserunt, atque adeo ad

qualitates occultas confugiunt. Denique NEWTONUS deductionem et expositionem omnium Phaenomenorum ad aestum Maris pertinentium minime perfecit, sed quasi tantum adumbravit; plena enim explicatio tot tamque difficilium Problematum solutionem postulat, quae NEWTONUS non est aggressus: cum enim huius quaestionis enodatio amplissimos calculos requirat, ipse analysin vitans pleraque tantum obiter indicasse contentus fuit; ob quem defectum plurimis adhuc dubiis circa ipsius explicationem locus est relictus. Neque enim in his viribus veram aestus Maris causam contineri ante certum esse potest, quam absoluto calculo perfectus consensus Phaenomenorum cum Theoria fuerit declaratus.

CAPUT SECUNDUM

De viribus Solis et Lunae ad Mare movendum

16. Effectus, quos vires cum Solis tum Lunae ante stabilitae in Terram exerunt, ad duo genera sunt referendi: quorum alterum eos complectitur effectus quos Sol ac Luna in universam Terram tanquam unum corpus consideratam exercet; alterum vero eos, quos singulae Terrae partes a viribus Solis ac Lunae patiuntur. Ad effectus prioris generis investigandos, omnis Terrae materia tanquam in unico puncto, centro scilicet gravitatis, collecta consideratur, ac tam ex motu insitu quam viribus sollicitantibus motus Terrae progressivus in sua orbita determinari solet. Ex hocque principio innotuit vim hanc Solis efficere, ut Terra circa Solem in orbita elliptica circumferatur, vim Lunae autem tam esse debilem, ut vix ac ne vix quidem ullam sensibilem perturbationem in motu Terrae annuo producere valeat. Contra autem docebitur, vim Lunae ad partes Terrae inter se commovendas ac Mare agitandum multo esse fortiorem vi Solis; ex quo plerisque primo intuitu summe paradoxon videatur, quod vis Lunae in priori casu respectu vis Solis evanescat, cum tamen eadem casu posteriori multum excedat vim Solis. Sed mox, cum effectus utriusque generis diligentius evolvemus et perpendemus, satis dilucide patebit, eos inter se maxime discrepare, atque a vi, quae in universam Terram minimum exerat effectum, maximam tamen agitationem partium Terrae inter se oriri posse et vicissim.

17. Ad illum autem harum virium effectum, qui in commotione partium Terrae inter se consistit, diiudicandum, ante omnia probe notari oportet, si singulae Terrae partes viribus aequalibus et in directionibus inter se parallelis sollicitentur,

eo casu nullam omnino commotionem partium oriri, etiamsi sint maxime fluidae nulloque vinculo invicem connexae, sed totum virium effectum in integro tantum corpore movendo consumtum iri; perinde ac si totum Terrae corpus vel in unico puncto esset conflatum, vel ex materia firmissime inter se connexa constaret. Ex quo manifestum est partes Terrae saltem fluidas, quae viribus cedere queant, inter se commoveri non posse, nisi a viribus dissimilibus urgeantur: atque hanc ob rem non magnitudo virium partes Terrae sollicitantium, sed potius dissimilitudo, qua cum quantitatis tum directionis ratione inter se discrepant, eum effectum, quo situs partium mutuus perturbetur, producit. Ita vis Solis, etsi est maxima, tamen ob insignem distantiam partes Terrae fere aequaliter afficit, contra vero vis Lunae ob propinquitatem admodum inaequaliter: unde a Luna multo maior agitatio Oceani resultat, quam a Sole, quamvis ea vis, quae ad Solem tendit, insigniter maior sit altera Lunam respiciente. Atque hoc pacto dubium ante allatum funditus tollitur, hocque adhuc planius fiet, si utriusque vis effectus ad calculum revocabimus.

18. Ad inaequalitatem igitur virium quibus singulae Terrae partes vel a Sole vel a Luna sollicitantur, definiendam, ante omnia vim, qua universa Terra, si in suo centro gravitatis esset concentrata, afficeretur, determinari oportet, haecque est ea ipsa vis, quae Terrae motum progressivum in sua orbita respicit et turbat; deinde dispiciendum est, quantum vires, quibus singulae Terrae partes urgentur, tam ratione quantitatis quam directionis ab illa vi totali discrepent. Quod si enim nulla deprehendatur differentia, partes quoque singulae situm suum relativum inter se retinebunt; at quo maior erit differentia inter vires illas singulas partes sollicitantes, eo magis eae inter se commovebuntur, situmque relativum permutabunt. In hac autem investigatione, simul gravitatis naturalis, qua omnia corpora versus centrum Terrae tendunt, ratio est habenda; haec enim vis in causa est, quod quantumvis vires Solis et Lunae in diversis Terrae regionibus sint inaequales, aequilibrii tamen status detur, in quo partes tandem singulae conquiescant, neque perpetuo inter se agitari pergant. Atque hanc ob rem singulae Terrae partes a tribus viribus sollicitatae considerari debebunt, primo scilicet a propria gravitate, qua directe deorsum nituntur; tum vero a vi, qua ad Solem urgentur, ac tertio a vi versus Lunam directa; haeque tres vires, cuiusmodi Phaenomena quovis tempore in partibus Terrae fluidis gignant, erit investigandum.

19. Quo igitur vim totalem, qua Terra vel a Sole vel a Luna urgetur, definiamus, consideremus primum peripheriam circuli MN [fig. 1] tanquam ex materia homogenea conflatam, cuius centro P verticaliter immineat Sol vel Luna in S, ita

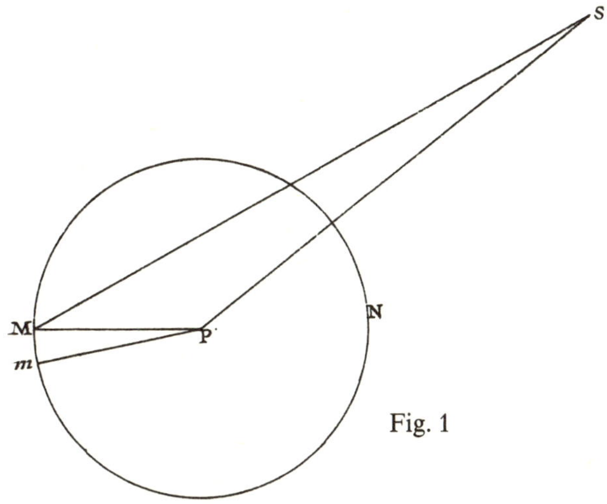

Fig. 1

ut recta PS ad planum circuli MN sit perpendicularis. Sit circuli huius radius $PM = y$, et distantia $SP = x$, ac vis sive Solis sive Lunae absoluta $= S$. His positis elementum peripheriae Mn pelletur ad S in directione MS vi acceleratrice $= \frac{S}{MS^2} = \frac{S}{xx+yy}$, posita cum vi gravitatis naturalis in superficie Terrae $=1$, tum etiam semidiametro Terrae $=1$: atque hanc ob rem elementum Mm versus S nitetur vi $= \frac{S \cdot Mm}{xx+yy}$. Resolvatur haec vis in binas laterales, quarum alterius directio cadat in MP, alterius vero sit parallela directioni PS; atque evidens erit vires omnes MP per totam peripheriam se mutuo destruere, alterarum vero mediam directionem cadere in PS, ac vim his omnibus aequivalentem iisdem coniunctim sumtis fore aequalem. Trahetur autem elementum Mm in directione ipsi PS parallela vi $= \frac{Sx \cdot Mm}{(xx+yy)^{3/2}}$, unde posita ratione radii ad peripheriam $= 1:\pi$ tota circuli MN peripheria, quae erit $= \pi y$, urgebitur seu quasi gravitabit versus S in ipsa directione PS vi $= \frac{\pi Sxy}{(xx+yy)^{3/2}}$. Vis autem acceleratrix qua haec peripheria circuli versus S sollicitabitur, prodibit, si vis motrix inventa dividatur per massam movendam, quae est $= \pi y$, eritque $= \frac{Sx}{(xx+yy)^{3/2}}$.

20. Hoc praemisso, contemplemur superficiem sphaericam genitam conversione circuli AMB [fig. 2] circa diametrum AB; sitque semidiameter $AC = BC = r$; erit ipsa superficies $= 2\pi rr$. Iam attrahatur haec superficies ad

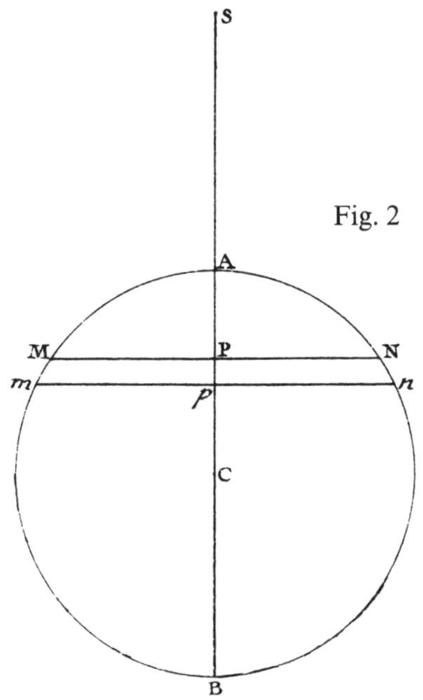

Fig. 2

Solem Lunamve in S, existente distantia $SC = a$; atque ad vim totalem seu conatum quo integra superficies ad S tendet, inveniendum, concipiatur annulus genitus conversione elementi Mm circa diametrum AB, quae protensa per S transeat. Positis igitur $Sp = x$, $PM = y$, erit per § praec. conatus huius annuli in directione $PS = \frac{\pi Sxy \cdot Mm}{(xx+yy)^{3/2}}$. At posito $Pp = dx$, erit $Mm = \frac{r\,dx}{y}$, et $xx + yy = 2ax - aa + rr$, unde annuli conatus versus S erit $= \frac{\pi Srx\,dx}{(2ax - aa + rr)^{3/2}}$, cuius integrale est $= \mathrm{C} + \frac{\pi Sr(ax - aa + rr)}{\sqrt{(2ax - aa + rr)}}$, ex quo conatus portionis superficiei sphaericae conversione arcus AM ortae prodibit $= \frac{\pi Srr}{aa} + \frac{\pi Sr(ax - aa + rr)}{\sqrt{(2ax - aa + rr)}}$. Quare si ponatur $Sp = SB$ seu $x = a + r$, emerget conatus totius superficiei sphaericae $= \frac{2\pi Srr}{aa}$: hincque cum ipsa superficies sit $= 2\pi rr$, erit vis acceleratrix qua superficies sphaerica actu versus S tendet $= \frac{S}{aa}$, ideoque tanta, quanta foret, si tota superficies in centro C esset collecta.

21. Cum igitur superficies sphaerica perinde ad Solem sive Lunam in S sollicitetur, ac si tota in ipso centro esset conflata, haec proprietas ad omnes superficies sphaericas, ex quibus integra Sphaera composita concipi potest, patebit,

dummodo singulae hae superficies ex materia homogenea constent, sive quod eodem redit, ipsa Sphaera in iisdem a centro distantiis sit aeque densa. Hanc ob rem eiusmodi Sphaera quoque perinde ad S in directione PS urgebitur, ac si tota ipsius materia in centro C esset concentrata; haecque proprietas non solum in eiusmodi Sphaeras competit, quae totae ex materia uniformi sunt confectae, sed etiam ut iam indicavimus, in tales, quae ex materia constant difformi, dummodo in aequalibus a centro distantiis, materia circumquaque sit homogenea seu saltem eiusdem densitatis. Cum igitur Terram sibi repraesentare liceat tanquam Sphaeram, si non ex uniformi materia conflatam, tamen sine ullo errore ita comparatam, ut in aequalibus circa centrum intervallis materiam aeque densam includat, Terra quoque universa tam a Sole quam a Luna aeque sollicitabitur, ac si omnis eius materia in centro esset collecta. Quanquam enim nunc quidem accuratissimis ab Illustrissima Academia Regia institutis passim mensuris satis est demonstratum, Terrae figuram ad polos esse compressam, tamen tantilla a perfecta Sphaera aberratio, in aliis quidem negotiis maximi momenti, in hoc instituto tuto negligi potest, [ut infra usu fieri ostendemus]*. Parique ratione, etiamsi Terra in aequalibus a centro distantiis non sit aeque densa, tamen differentia certe non est tanta, ut error sensibilis inde sit metuendus.

22. Ut igitur vires inveniantur, quae tendant ad situm partium Terrae relativum immutandum, definienda est vis acceleratrix, qua centrum Terrae sive ad Solem sive ad Lunam urgeatur: qua cognita, si comperiantur omnes Terrae partes aequalibus viribus acceleratricibus et in directionibus parallelis urgeri, nulla omnino situs mutatio, nullaque proinde Maris agitatio orietur. Sed Terra in se spectata omnium partium situm mutuum invariatum conservabit. At si vires, quibus singulae partes a Sole aut Luna urgentur, discrepent a vi centrum Terrae afficiente, tam ratione quantitatis quam directionis, tum nisi firmissime inter se sint connexae in situ suo mutuo perturbari debebunt. Hocque casu aquae, quae ob fluiditatem vi etiam minimae cedunt, sensibiliter agitabuntur, atque affluendo defluendoque aliis locis elevabuntur, aliis deprimentur. Cum autem iste motus, qui in singulis Terrae partibus generatur, a differentia inter vires centrum Terrae et ipsas partes sollicitantes proficiscatur, propria vis, qua quaeque particula agitabitur, innotescet, si a vi acceleratrice illam particulam sollicitante auferatur vis acceleratrix, quam centrum Terrae patitur: haecque subtractio ita instituitur, ut cuique particulae praeter vim actu eam sollicitantem alia vis aequalis illi, quam centrum perpetitur, in directione contraria applicata concipiatur: tum enim vis quae ex compositione harum duarum oritur, erit vera vis particulam illam de loco suo deflectens.

23. Consentanea est haec reductio principiis Mechanicis, quibus statuitur motum relativum in systemate quotcunque corporum et a quibuscunque viribus sollicitatorum manere invariatum, si non solum toti systemati motus aequabilis in directum simul imprimatur, sed etiam singulis partibus vires aequales, quarum directiones sint inter se parallelae, applicentur. Nostro igitur casu motus intestinus partium Terrae non turbabitur, si singulis particulis vires aequales in directionibus parallelis applicemus ut fecimus: quod si autem istae vires aequales sint illi, qua tota Terra seu centrum sollicitatur, et contrariae, hoc ipso Terrae motum curvilineum et inaequabilem, quippe qui ab iisdem viribus oritur, adimemus. Quare si insuper toti Terrae motum aequalem et contrarium illi, quo actu fertur, impressum concipiamus, obtinebimus totam Terram quiescentem, atque etiamnunc partes perinde agitabuntur et inter se commovebuntur, ac si nullas istiusmodi mutationes intulissemus. Quilibet autem facile percipiet, quantum ex hac reductione subsidium assequamur; multo enim facilius erit mutationes, quae in ipsa Terra accidunt, percipere atque explicare, si centrum Terrae constituatur immotum, quam si totalis motus singularum partium motibus esset permixtus. Hanc ob rem ista reductione qua centrum Terrae in quietem redigitur, perpetuo utemur, quo Phaenomena aestus Maris, prouti in Terra immota sentiri debent, eliciamus; quippe qui est casus naturalis, ad quem omnes observationes sunt accomodatae, omnes vero theoriae accomodari debent.

24. Concipiatur nunc Terra tota tanquam globus $ADBE$ [fig. 3] urgeri ad Solem Lunamve in S existentem cuius vis absoluta seu ea, quam in distantia a

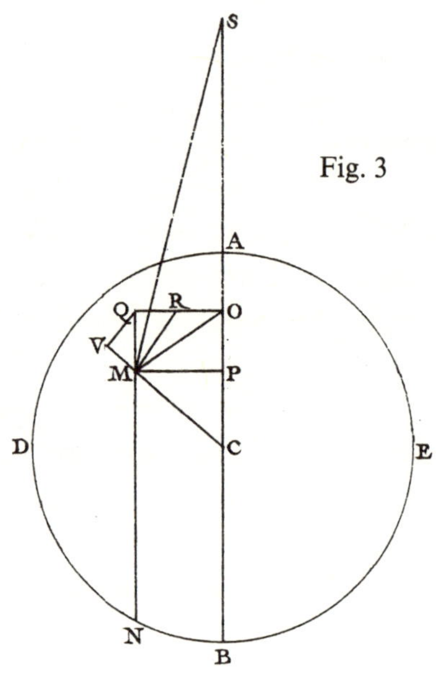

Fig. 3

centro suo S semidiametro Terrae aequali exerit, sit $=S$, distantia vero centri Terrae C ab S seu CS ponatur $=a$; eritque vis acceleratrix, qua tota Terra tanquam in C collecta sollicitabitur in directione CS, $=\frac{S}{aa}$. Contemplemur iam particulam Terrae quamcunque M cuius situs ita sit definitus, ut sit $CP=x$ et $PM=y$, existente MP normali ad CS; hinc igitur habebitur $SP=a-x$ et $SM=\sqrt{((a-x)^2+y^2)}$. Vis igitur acceleratrix, qua particula M versus S pelletur, erit $=\frac{S}{(a-x)^2+y^2}$; a qua cum auferri debeat vis, qua tota Terra versus S nititur, concipienda est particulae M applicata vis $=\frac{S}{aa}$ in directione MN ipsi CS parallela et opposita; quae duae vires particulam M aeque afficient ac si universa Terra quiesceret vel uniformiter in directum moveretur, qui casus ab illo non differt. Ex his igitur ambabus viribus conatus innotescet, quo particula M a vi ad S directa de loco suo recedere annitetur: ad ipsum autem motum definiendum insuper vis gravitatis erit respicienda: et quia haec particula non est libera, sed quaquaversus materia terrestri circumdata, investigari oportet, quantum ista materia effectum viribus sollicitantibus concedat.

25. Quoniam autem in hoc capite nobis nondum est propositum in ipsum effectum ab his viribus oriundum inquirere, sed tantum conatum evolvere atque explorare; diligentius perpendemus, cuiusmodi vires ex combinatione harum potentiarum particulam M sollicitantium resultent. Hunc in finem resolvatur vis MS in duas laterales, quarum alterius directio parallela sit ipsi CS, altera vero in MP cadat: ex quo reperietur vis illa particulam M in directione MQ urgens $=\frac{S(a-x)}{((a-x)^2+y^2)^{3/2}}$; altera vero vis in directione MP trahens $=\frac{Sy}{((a-x)^2+y^2)^{3/2}}$. Cum autem particula M insuper trahatur in directione MN vi $=\frac{S}{aa}$, tres istae vires a Sole Lunave in S existente reducentur ad duas, quarum altera in directione MQ urgens erit $=\frac{S(a-x)}{((a-x)^2+y^2)^{3/2}}-\frac{S}{a^2}$, altera vero directionem habens MP $=\frac{Sy}{((a-x)^2+y^2)^{3/2}}$. Quare si rectae MQ et MP his viribus proportionales capiantur, et rectangulum $MQOP$ compleatur exprimet diagonalis MO tam directionem quam quantitatem vis ex tribus praecedentibus ortae: erit

autem anguli OMP tangens $=\frac{a-x}{y}-\frac{((a-x)^2+y^2)^{3/2}}{a^2y}$; quo cognito, si fiat ut MP ad MO ita $\frac{Sy}{((a-x)^2+y^2)^{3/2}}$ ad quartam, haec ipsa quarta proportionalis erit vis particulam M in directione MO sollicitans, quae oritur a vi ad S tendente.

26. Ut autem istae vires facilius cum gravitate naturali, cuius directio est MC, coniungi queant, resolvantur eae in binas, quarum altera in ipsam directionem MC cadat, alterius vero directio sit MR normalis ad MC. Ad hoc commodissime praestandum, resolvatur vis MS primum in duas, quarum altera ut ante directionem habeat ipsi CS parallelam, alterius vero directio in ipsam MC incidat. Cum igitur sit $MC=\sqrt{(x^2+y^2)}$, erit prior vis $=\frac{Sa}{((a-x)^2+y^2)^{3/2}}$, posterior vero $=\frac{S\sqrt{(x^2+y^2)}}{((a-x)^2+y^2)^{3/2}}$, qua vis gravitatis augebitur. At si a priori auferatur vis $=\frac{S}{aa}$, remanebit vis particulam M in directione MQ sollicitans

$$=\frac{Sa}{((a-x)^2+y^2)^{3/2}}-\frac{S}{a^2}.$$

Iam ex Q in CM productam demittatur perpendiculum QV, eritque ob similitudinem triangulorum QVM et MPC vis gravitati contraria secundum directionem MV agens ex vi MQ orta

$$=\frac{Sax}{((a-x)^2+y^2)^{3/2}\sqrt{(x^2+y^2)}}-\frac{Sx}{a^2\sqrt{(x^2+y^2)}};$$

unde omnino particula M a vi ad S tendente versus C urgebitur vi

$$=\frac{Sx}{a^2\sqrt{(x^2+y^2)}}-\frac{S(ax-xx-yy)}{((a-x)^2+y^2)^{3/2}\sqrt{(x^2+y^2)}}.$$

Praetera vero eadem particula M in directione MR ad MC normali sollicitabitur vi

$$=\frac{Say}{((a-x)^2+y^2)^{3/2}\sqrt{(x^2+y^2)}}-\frac{Sy}{a^2\sqrt{(x^2+y^2)}}.$$

27. Tametsi istae expressiones tantopere sint compositae, ut parum ex iis ad usum deduci posse videatur, tamen si consideremus distantiam Lunae a Terra, multo magis autem distantiam Solis, vehementer excedere quantitatem Terrae, ac propterea quantitates x et y; per approximationem satis commodas formulas ex iis derivare licebit. Cum enim sit proxime

$$\frac{1}{((a-x)^2+y^2)^{3/2}} = (a^2-2ax+x^2+y^2)^{-3/2}$$
$$= \frac{1}{a^3} + \frac{3(2ax-xx-yy)}{2a^5} + \frac{15(2ax-xx-yy)^2}{8a^7},$$

loco $\frac{1}{((a-x)^2+y^2)^{3/2}}$ satis tuto substitui poterit $\frac{1}{a^3}+\frac{3x}{a^4}+\frac{3(4xx-yy)}{2a^5}$. Ex his autem obtinebitur vis, qua particula M praeter gravitatem a vi Solis sive Lunae in S existentis ad centrum Terrae C in directione MC urgetur,

$$= \frac{S(yy-2xx)}{a^3\sqrt{(x^2+y^2)}} + \frac{3Sx(3yy-2xx)}{2a^4\sqrt{(x^2+y^2)}}.$$

Praeterea autem eadem particula M sollicitabitur in directione MR ad MC normali, vi[5])

$$= \frac{3Sxy}{a^3\sqrt{(x^2+y^2)}} + \frac{3Sy(4xx+yy)}{2a^4\sqrt{(x^2+y^2)}} = \frac{3Sy}{a^3\sqrt{(x^2+y^2)}}\left(x+\frac{4xx+yy}{2a}\right).$$

Atque cum in his formulis termini primi posteriores multis vicibus excedant, rem crassius inspiciendo, particula M a vi Solis Lunaeve secundum MC urgebitur vi $=\frac{S(yy-2xx)}{a^3\sqrt{(x^2+y^2)}}$, in directione vero MR vi $=\frac{3Sxy}{a^3\sqrt{(x^2+y^2)}}$.

28. Ex his igitur postremis formulis intelligitur ab actione Solis sive Lunae in S existentis gravitatem particulae M augeri, si eius situs respectu rectae SC ita fuerit comparatus, ut sit $yy > 2xx$ hoc est tangens anguli $MCP > \sqrt{2}$ posito sinu toto $=1$, contra vero gravitatem diminui, si fuerit $yy < 2xx$. Quare cum

5) Editio princeps et Manuscriptus: $\frac{4xx-yy}{2a}$ loco $\frac{4xx+yy}{2a}$. Correxit EJA.

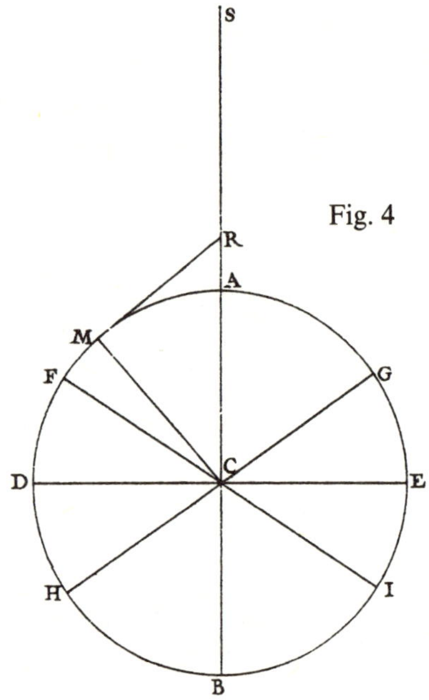

Fig. 4

angulus cuius tangens est $=\sqrt{2}$ contineat $54^\circ\,45'$ circiter, si concipiatur circulus Terrae maximus quicunque $ADBE$ [fig. 4], cuius planum per punctum S transeat, in eoque ducantur rectae FCI et GCH, quae cum recta SAB angulos constituant $54^\circ\,45'$; tum omnes Terrae particulae in spatiis FCH et GCI sitae gravitatis naturalis augmentum accipient, reliquae vero particulae in spatiis FCG et HCI positae decrementum gravitatis patientur. Atque hinc, quacumque Terrae particula proposita, definiri poterit, quantum eius gravitas a Sole Lunave in S existente vel augeatur vel diminuatur. Altera vero vis, qua particula M [fig. 3] in directione horizontali MR urgetur, affirmativa erit, in eamque plagam, quae in figura repraesentatur, verget, si quantitates x et y ambae fuerint vel affirmativae vel negativae: contrariumque eveniet, si earum altera sit affirmativa, altera negativa. Quare si particula M [fig. 4] sita fuerit vel in quadrante ACD vel ACE, tum vis horizontalis ad rectam CA tendet; contra vero haec vis ad radium CB dirigetur, si particula M sit vel in quadrante BCD vel BCE constituta. Ex quibus perspicitur effectus vel Solis vel Lunae in ambo hemisphaeria, superius scilicet DAE et inferius DBE, inter se esse fere similes; quae similitudo quoque in ipso aestu Maris observatur.

29. Ponamus nunc particulam M in ipsa Terrae superficie esse constitutam, eritque $\sqrt{(x^2+y^2)}=1$ ob Terrae semidiametrum $=1$. Quare si particula M

fuerit posita in M, existente anguli ACM sinu $= y$ et cosinu $= x$, eius gravitas naturalis acceleratrix a Sole Lunave in S augebitur vi $= \frac{S(yy - 2xx)}{a^3}$, secundum horizontem autem in directione MR urgebitur vi $= \frac{3Sxy}{a^3}$. Gravitas igitur maxime augebitur, si particula M posita fuerit in D vel E, quibus in locis punctum S in horizonte apparet; ibi vero gravitatis augmentum erit $= \frac{S}{a^3}$. In punctis autem A et B, quae punctum S vel in suo zenith vel nadir positum habent, maximum deprehendetur gravitatis decrementum, quod scilicet erit $= \frac{2S}{a^3}$: ita ut maximum gravitatis decrementum duplo maius sit quam maximum incrementum. Vis autem horizontalis $\frac{3Sxy}{a^3}$ maxima evadet, si angulus ACM fuerit semirectus, id quod accidit in iis Terrae regionibus, in quibus punctum S conspicitur vel 45° gradibus supra horizontem elevatum, vel tantundem sub horizonte depressum latet: his igitur casibus ob $xy = \frac{1}{2}$ fiet vis horizontalis $= \frac{3S}{2a^3}$. Huius ergo vis effectus in hoc consistet, ut directio gravitatis mutetur, atque versus rectam SC inclinetur angulo cuius tangens est $= \frac{3S}{2a^3}$, existente sinu toto $= 1$, quia gravitatem unitate designamus.

30. Hae itaque vires si satis essent magnae, in ponderibus utique sentiri deberent, ac prior quidem gravitatem naturalem vel augens vel diminuens in oscillationibus pendulorum animadverti deberet, eorum motum vel accelerando vel retardando; posterior vero vis situm pendulorum quiescentium verticalem de hoc situ deflecteret, atque ad horizontem inclinatum efficeret. Quoniam autem huiusmodi perturbationes non observamus, operae pretium erit dilucide monstrare vires illas tam esse exiguas, ut hi effectus sensus nostros omnino effugiant. Primum igitur cum pro Sole sit $S = 227\,512$ atque $a = 20\,620$, erit $\frac{S}{a^3} = \frac{1}{38\,535\,570}$; pro Luna autem quia est $S = \frac{1}{40}$ et $a = 60$, erit $\frac{S}{a^3} = \frac{1}{8\,640\,000}$; ex quo vis Lunae plus quam quater maior est vi Solis, ceteris paribus; atque si Solis et Lunae vires prorsus conspirent, erit ex iis coniunctim $\frac{S}{a^3} = \frac{1}{7\,057\,700}$ seu proxime $= \frac{1}{7\,000\,000}$. Hinc

maxima gravitatis diminutio, quae quidem oriri poterit, erit $= \frac{1}{3\,500\,000}$, maximum vero incrementum $= \frac{1}{7\,000\,000}$: unde numerus oscillationum eiusdem penduli eodem tempore editarum, illo casu erit ut $\sqrt{\left(1 - \frac{1}{3\,500\,000}\right)}$ seu $1 - \frac{1}{7\,000\,000}$, hoc vero casu ut $\sqrt{\left(1 + \frac{1}{7\,000\,000}\right)}$ seu $1 + \frac{1}{14\,000\,000}$. Numeri ergo oscillationum ab eodem pendulo eodem tempore absolutarum, cum gravitas maxime est diminuta, et cum maxime est aucta, tenebunt rationem ut 13 999 998 ad 14 000 001, hoc est ut 4 666 666 ad 4 666 667; ex quo satis perspicitur differentiam hanc minime percipi posse. Similis autem omnino est ratio alterius Phaenomeni declinationis scilicet a situ verticali comparata, quae numquam ad 5‴ exsurgere potest.

CAPUT TERTIUM

De Figura, quam vires cum Solis tum Lunae Terrae inducere conantur

31. Cum igitur in capite praecedente vires tam a Sole quam a Luna oriundas determinaverimus, quibus singulae Terrae particulae ad situm relativum cum inter se tum respectu centri, quod in hoc negotio tanquam quiescens consideratur, immutandum sollicitantur; ordo requireret, ut iam in ipsum motum, quo singulae particulae inter se commoveri debeant, inquireremus. Verum cum haec investigatio sit altioris indaginis, atque opus habeat principiis mechanicis ad motum partium inter se respicientibus, qualia vix usquam adhuc reperiuntur; in hoc capite rem secundum principia statica ulterius persequi pergamus, ac figuram determinemus, quam vires Solis et Lunae cum seorsim tum etiam coniunctim inducere conantur. Hunc in finem Terram undequaque materia fluida seu aqua cinctam contemplabimur, quo sollicitationibus obedire ac figuram iis convenientem actu induere queat. In hoc scilicet negotio Solem et Lunam pariter ac ipsam Terram quiescentes concipimus, ita ut inter se perpetuo eundem situm relativum conservent, quo pacto Terrae ab actionibus Solis ac Lunae figura permanens mox induetur, quam tandiu retinebit, quoad idem situs relativus duret. Perspicuum autem est cognitionem huius figurae magno futuram esse adiumento ad eiusdem figurae transmutationem definiendam, si tam Soli quam Lunae motus tribuatur.

32. Consideremus igitur primum Terram in statu suo naturali, in quem se sola vi gravitatis composuit; in quo, cum habitura sit figuram sphaericam, reprae-

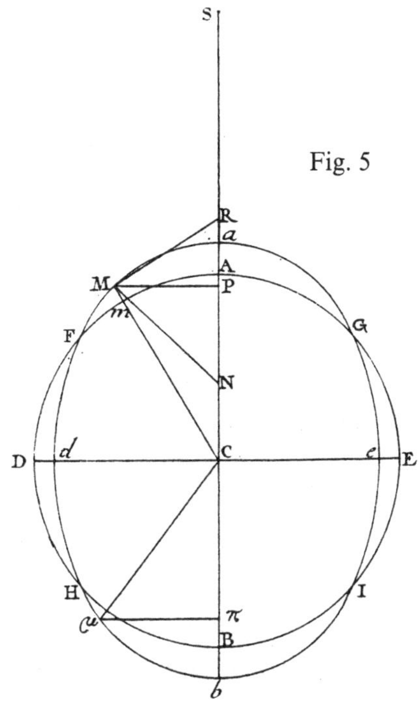

Fig. 5

sentet circulus $ADBE$ [fig. 5] seu potius globus eius rotatione ortus Terram, quam praeterea undique aqua circumfusam ponimus. Versetur iam Sol vel Luna in S, a cuius vi cum gravitas naturalis tam in A quam in B diminuatur, in D vero et E augeatur, manifestum est Terram seu potius aquam illi circumfusam elevatum iri in A et B, contra vero in D et E deprimi, idque eousque, quoad sollicitationes a Sole Lunave in S oriundae cum vi gravitatis ad aequilibrium fuerint redactae. Sit itaque curva *adbe* ea figura, quae circa axem *ab* rotata generet Terrae formam, quam a vi ad S directa tandem recipiet, atque cum aquae nunc ponantur in aequilibrio constitutae, necesse est ut directio media omnium sollicitationum, quibus singulae Terrae particulae in suprema superficie sitae urgentur, ad ipsam superficiem sit normalis. Quare si particulam quamcunque M spectemus; ea primum a gravitate naturali in directione MC urgetur deorsum, idque vi, quam constanter ponimus $=1$; quippe quae est ipsa gravitas in superficie Terrae, eo quod elevatio vel depressio huius particulae distantiam eius a centro Terrae, a qua variatio gravitatis pendet, sensibiliter non immutet. Deinde vero eadem particula M a vi in S existente sollicitatur duplici vi, quarum alterius directio in ipsam MC incidit, alterius vero in MR normalem ad MC. Quocirca trium harum virium mediam directionem incidere oportet in rectam MN normalem ad curvam aMd, quo ipso natura huius curvae determinabitur.

33. Dubium hic subnasci posset, quod cum ad praesens institutum omnium virium, quibus singulae particulae sollicitantur, ratio haberi debeat, eam hic negligamus, quae a vi centrifuga motus Terrae diurni oritur, quippe quae non solum non est infinite parva, sed multis vicibus maior, quam vires quae vel a Sole vel Luna resultant: sed quia haec vis constantem producit effectum, Terrae scilicet figuram sphaeroidicam ad polos compressam, mutationem, quae in Fluxu ac Refluxu Maris observatur, sensibiliter afficere nequit. Deinde quamvis hic figuram Terrae sphaericam ponamus, tamen in aberrationem praecipue ab hac figura tam a Sole quam Luna oriundam inquirimus: manifestum autem est, quantum figura aquae ob vires Solis Lunaeve a sphaerica recedat, tantundem aquae figuram admisso motu diurno Terrae a figura sphaeroidica esse discrepaturam. Quapropter in hoc negotio sufficere potest, si, Terra instar sphaerae perfectae considerata [sublato motu vertiginis]*, definiamus quantam differentiam in aquae figura vires cum Solis tum Lunae producant: hac enim determinata, si Terrae motus vertiginis restituatur, perspicuum erit totam figuram sub aequatore intumescere, sub polis autem subsidere; ita tamen ut ubique eadem vel elevatio vel depressio aquae a viribus Solis Lunaeve maneat. Namque si ulla etiam varietas in aestu Maris a motu vertiginis Terrae proficiscatur, ea calculo monstrante nusquam maior esse potest parte $\frac{1}{289}$ aestus totalis; tantilla autem differentia notari non meretur, neque ob eam causam operae praetium est tam complicatos et abstrusos calculos inire, ad quos perveniretur, si Terrae figura naturalis a sphaerica diversa poneretur, atque insuper vis centrifuga a motu vertiginis Terrae in computum duceretur.

34. Ad curvam igitur $aMdb$, cui ea quae ex altera parte axis ab similis est et aequalis, determinandam, ponatur vis absoluta sive Solis sive Lunae in S existentis $= S$, distantia $CS = a$, ac ducta semiordinata MP vocetur $CP = x$, et $PM = y$. Ex praecedenti igitur capite habebitur vis, qua punctum M vel a Sole vel Luna versus C urgebitur $= \frac{S(yy-2xx)}{a^3\sqrt{(xx+yy)}}$, insuper autem idem punctum M sollicitabitur in directione MR normali ad MC vi[6])

$$= \frac{3Syx}{a^3\sqrt{(xx+yy)}} + \frac{3Sy(4xx-yy)}{2a^4\sqrt{(xx+yy)}}.$$

6) Secundum Manuscriptum. Editio princeps: $a^3\sqrt{(xx-yy)}$ loco $a^3\sqrt{(xx+yy)}$. EJA.

Praeter has vero vires punctum M gravitate naturali deorsum pellitur vi $=1$ secundum directionem MC, ita ut punctum M ab omnibus his viribus coniunctim in directione MC deorsum urgeatur vi $=1+\dfrac{S(yy-2xx)}{a^3\sqrt{(xx+yy)}}$ ubi ob 1 sequens terminus tuto negligi potest, et in directione MR vi

$$=\frac{3Syx}{a^3\sqrt{(xx+yy)}}+\frac{3Sy(4xx-yy)}{2a^4\sqrt{(xx+yy)}};$$

quarum duarum virium si MN ponatur media directio, prodibit per regulas compositionis motus anguli CMN tangens[7])

$$=\frac{3Sy(2ax+4xx-yy)}{2a^4\sqrt{(xx+yy)}+2Sa(yy-2xx)},$$

quae divisione actu instituta, iisque terminis neglectis in quorum denominatoribus a plures quam quatuor obtinet dimensiones, abit in hanc expressionem

$$\frac{3Sxy}{a^3\sqrt{(xx+yy)}}+\frac{3Sy(4xx-yy)}{2a^4\sqrt{(xx+yy)}},$$

quae est ea ipsa formula, qua vis MR exprimebatur. Quocirca angulus CMN prorsus non pendet ab aucta minutave gravitate, sed tantum a vi horizontali singulis particulis in Terrae superficie sitis impressa.

35. Quoniam vero haec ipsa media directio MN debet esse ad curvam aMd in puncto M normalis, erit subnormalis $PN=\dfrac{-y\,dy}{dx}$ et $CN=\dfrac{x\,dx+y\,dy}{dx}$. Cum igitur sit anguli MNP tangens $=\dfrac{-dx}{dy}$ et anguli MCP tangens $=\dfrac{y}{x}$, erit horum angulorum differentiae, hoc est anguli CMN tangens $=\dfrac{y\,dy+x\,dx}{y\,dx-x\,dy}$, quae superiori expressioni, qua haec eadem tangens designabatur, aequalis posita pro curva quaesita $aMdb$ sequentem praebebit aequationem

7) Editio princeps: $2S(yy-2xx)$ loco $2Sa(yy-2xx)$. Correxit EJA.

$$\frac{y\,dy + x\,dx}{y\,dx - x\,dy} = \frac{3Sxy}{a^3\sqrt{(xx+yy)}} + \frac{3Sy(4xx-yy)}{2a^4\sqrt{(xx+yy)}},$$

ad quam integrandam ponimus $\sqrt{(xx+yy)} = z = MC$, et anguli MCA cosinum $\frac{x}{\sqrt{(xx+yy)}} = u$, unde fiet $x = uz$ et $y = z\sqrt{(1-uu)}$, atque

$$y\,dx - x\,dy = \frac{zz\,du}{\sqrt{(1-uu)}}, \quad \text{itemque} \quad x\,dx + y\,dy = z\,dz.$$

Hac autem facta substitutione, aequatio inventa abit in hanc

$$\frac{dz}{zz} = \frac{3Su\,du}{a^3} + \frac{3Sz\,du(5uu-1)}{2a^4},$$

cuius postremus terminus, qui ob parvitatem prae reliquis fere evanescit, si abesset, foret integrale $\frac{1}{c} - \frac{1}{z} = \frac{3Suu}{2a^3}$ seu $z = c + \frac{3Sccuu}{2a^3}$ proxime. Ponamus itaque completum integrale esse

$$z = c + \frac{3Sc^2u^2}{2a^3} + \frac{3Sc^3V}{2a^4},$$

ac facta applicatione reperietur $V = \frac{5u^3 - 3u}{3}$, ita ut habeatur

$$z = c + \frac{3Sccuu}{2a^3} + \frac{Sc^3u(5uu-3)}{2a^4},$$

quod autem integrale proxime tantum satisfacit; at mox alia via aperietur verum ipsius z valorem per u commodius et propius definiendi.

36. Cum autem soliditas sphaeroidis, quod generatur ex conversione curvae *adb* circa axem *ab*, aequalis esse debeat soliditati Sphaerae radio $CA = 1$ descriptae, hinc constans quantitas c quae per integrationem est ingressa, definietur: id quod commodissime praestabitur, si utraque sphaeroidis semissis, superior scilicet versus S directa, atque inferior seorsim investigetur. Quoniam igitur pro semissi superiori est

$$CP = x = zu = cu + \frac{3Sccu^3}{2a^3} + \frac{Sc^3u^2(5uu-3)}{2a^4},$$

et

$$MP^2 = y^2 = z^2(1-uu) = (1-uu)\left(cc + \frac{3Sc^3u^2}{a^3} + \frac{Sc^4u(5uu-3)}{a^4}\right),$$

erit $\int yy\,dx$, cui soliditas genita conversione spatii $dCPM$ est proportionalis,

$$= c^3u - \frac{c^3u^3}{3} + \frac{5Sc^4u^3}{2a^3} - \frac{3Sc^4u^5}{2a^3} - \frac{3Sc^5u^2}{a^4} + \frac{21Sc^5u^4}{4a^4} - \frac{5Sc^5u^6}{2a^4}.$$

Posito igitur $u=1$, prodibit superioris semissis ut $\frac{2}{3}c^3 + \frac{Sc^4}{a^3} - \frac{Sc^5}{4a^4}$. Simili modo cum pro inferiori semissi sit[8])

$$C\mu = z = c + \frac{3Sc^2u^2}{2a^3} - \frac{Sc^3u(5u^2-3)}{2a^4},$$

erit eius soliditas ut $\frac{2}{3}c^3 + \frac{Sc^4}{a^3} + \frac{Sc^5}{4a^4}$; ex quibus totius sphaeroidis soliditas erit ut $\frac{4}{3}c^3 + \frac{2Sc^4}{a^3}$. Quare cum Sphaerae radio $=1$ descriptae soliditas pari modo definita, sit ut $\frac{4}{3}$, fiet $1 = c^3 + \frac{3Sc^4}{2a^3}$; hincque $c = 1 - \frac{S}{2a^3}$. Quamobrem pro curva quaesita habebitur, hoc valore loco c substituto, ista aequatio

$$z = 1 + \frac{S(3u^2-1)}{2a^3} + \frac{Su(5uu-3)}{2a^4};$$

ex qua natura istius curvae luculenter cognoscitur.

37. Hinc igitur perspicitur a Sole vel Luna in S existente aquam, cuius superficies ante erat in A, attolli in a, ita ut sit elevatio $Aa = \frac{S}{a^3} + \frac{S}{a^4}$; atque in regione opposita B, aquam pariter elevari per spatium $Bb = \frac{S}{a^3} - \frac{S}{a^4}$: unde patet

8) Editio princeps: Cac loco $C\mu$, vide fig. 5. EJA.

aquas in A et B, ad eandem fere altitudinem elevari, cum excessus superioris elevationis super inferiorem sit tantum $\frac{2S}{a^4}$, quod discrimen respectu totius elevationis vix est sensibile. Contra vero in regionibus lateralibus D et E, aqua circumquaque aequaliter deprimetur, et quidem per intervallum $Dd = Ee = \frac{S}{2a^3}$; ex quo ista depressio duplo minor est, quam elevatio quae in A et B accidit. In punctis praeterea F, G, H et I, quae a cardinalibus A et B distant angulo 54° 45′, quippe pro quo est $3uu - 1 = 0$, neque elevabitur aqua neque deprimetur, sed naturalem tenebit altitudinem. In loco autem Terrae quocumque M cognoscetur aquae vel elevatio vel depressio ex angulo ACM, cuius cosinus u est sinus altitudinis sub qua Sol vel Luna in S existens super horizonte conspicitur ab observatore in M constituto; hoc enim in loco aqua elevata erit supra naturalem altitudinem intervallo $= \frac{S(3uu-1)}{2a^3} + \frac{Su(5uu-3)}{2a^4}$: quae expressio, si fit negativa, maris depressionem indicat. Hic autem annotare non est opus, quod si punctum S sub horizonte lateat, tum sinus depressionis maneat quidem u, sed negative accipi debeat.

38. Definiamus igitur primum cum elevationem tum depressionem, quae a sola vi Solis ubique terrarum produci deberet, si, uti ponimus, omnia in statu aequilibrii essent constituta. Quoniam itaque est $S = 227\,512$ atque $a = 20\,620$ semid. Terrae, si una Terrae semidiameter assumatur 19 695 539 pedum Paris. erit $\frac{S}{a^3} = 0{,}5072$ ped.[9]) seu pauxillum excedet semipedem: valor autem $\frac{S}{a^4}$ omnino erit quantitas evanescens et imperceptibilis. Hanc ob rem in regionibus sub Sole verticaliter sitis, quae habeant Solem vel in Zenith vel Nadir, aqua ultra altitudinem naturalem attolletur ad semipedem cum pollicis parte decima circiter; depressio autem maxima cadet in loca, quae Solem in horizonte conspicient, ubi aqua ad quadrantem pedis tantum deprimetur; ex quo totum discrimen, quod a Sole in altitudine aquae naturali oritur, ad tres quartas pedis partes circiter assurget. Iste Solis effectus autem distantiae tantum mediocri Solis a Terra est tribuendus: quod si enim Sol versetur vel in apogaeo vel perigaeo, eius effectus vel diminui vel augeri debebit in ratione reciproca triplicata distantiarum Solis a Terra, quia pendet a

9) Valor verus est 0,5111. Talia errata parva in sequentibus non correcta sunt. EJA.

valore $\frac{S}{a^3}$. Cum igitur orbitae Terrae excentricitas sit $= \frac{163}{10000}$, erit intervallum Aa vel Bb, dum Sol in perigaeo versatur, $= 0{,}5332$ ped. sin autem Sol in apogaeo sit constitutus, $= 0{,}4825$ pedum; quorum differentia ad vicesimam pedis partem ascendit: valor autem medius est $= 0{,}5072$, quem pro mediocri distantia Solis a Terra invenimus.

39. Problema hoc, quod hucusque dedimus solutum, quodque maximi est momenti ad effectus cum Solis tum Lunae in mari elevando et deprimendo definiendos, NEWTONUS ne attigit quidem, sed aliam viam secutus, non solum indirectam sed etiam erroneam, invenit Mare a sola vi Solis ad altitudinem duorum fere pedum elevari debere; cum tamen tam eandem vim Soli absolutam quam eandem distantiam a Terra assumsisset, quibus nos sumus usi. Conclusit autem hunc enormem effectum ex comparatione vis Solis seu valoris $\frac{S}{a^3}$ cum vi Terrae centrifuga a motu diurno orta, qua Terra sub aequatore extenditur ac crassior redditur quam sub polis; atque assumit elevationem aquae a vi Solis ortam eandem tenere debere rationem ad incrementum Terrae sub aequatore a vi centrifuga factum, quam teneat vis Solis ad vim centrifugam. Sed praeterquam quod hoc ratiocinium nimis infirmo superstructum sit fundamento, nostra via directa, qua sumus usi, statim evertitur: ex ipsa enim rei natura, nullis precariis assumtis principiis, elevationem aquarum a vi Solis oriundam directe et luculenter determinavimus; ac si ullum etiam dubium ob integrationem per approximationes tantum institutum restaret, id mox tolletur, cum infra idem problema alia methodo prorsus diversa sumus resoluturi, congruentemque solutionem exhibituri.

40. Quamvis autem iste Solis effectus in Mari tam elevando quam deprimendo non adeo certus et planus esse videatur ob parallaxin Solis, quam $10''$ assumsimus, nondum accuratissime definitam; a qua tam distantia Solis a Terra a, quam aestimatio vis absolutae S, pendet; tamen si rem attentius perpendamus, comperiemus expressionem $\frac{S}{a^3}$ perpetuo eundem retinere valorem, quaecumque Soli parallaxis tribuatur: mutata enim parallaxi, valor litterae S praecise in eadem ratione, in qua cubus distantiae a^3, mutabitur. Per leges enim motus firmissime stabilitas patebit quantitatem $\frac{S}{a^3}$ a solo tempore periodico Terrae circa Solem determinari, cuius quantitas accuratissime est definita. Quod ut clarius appareat,

consideremus planetam quemcunque circa Solem in orbita elliptica revolventem, cuius semiaxis transversus seu distantia a Sole media sit $=a$, vis autem Solis absoluta $=S$, erit tempus periodicum semper ut $\frac{a\sqrt{a}}{\sqrt{S}}$; quod si igitur tempus periodicum sit $=t$, erit t ut $\frac{a\sqrt{a}}{\sqrt{S}}$ et $\frac{S}{a^3}$ uti $\frac{1}{tt}$. Ad valorem autem fractionis $\frac{S}{a^3}$ absolute inveniendum, exprimatur a in semidiametris Terrae, atque in minutis secundis dato tempore periodico t, erit semper $t=\frac{5064\frac{1}{2}a\sqrt{a}}{\sqrt{S}}$; ex quo prodit $\frac{S}{a^3}=\frac{5064\frac{1}{2}\cdot 5064\frac{1}{2}}{tt}$, posita unitate cum pro gravitate naturali, tum pro una Terrae semidiametro. At si tempus Terrae periodicum seu annus sidereus in minutis secundis exponatur, fiet $t=31\,558\,164$, atque $\frac{S}{a^3}=0{,}50723$ ped. posita semidiametro Terrae per observationes exactissimas $19\,695\,539$ ped. Paris. Reg. omnino uti ante invenimus.

41. Simili modo ex superiori aequatione elevatio aquae a vi Lunae oriunda determinabitur; posita enim vi Lunae absoluta $=L$, poni oportet $S=L$, eiusque valor proxime erit $=\frac{1}{40}$, quem a NEWTONO repertum tantisper retinebimus, quod verus valor per alia Phaenomena accuratius definiatur. Quoniam itaque Lunae a Terra mediocris distantia est $60\frac{1}{2}$ semid. Terrae, erit $\frac{S}{a^3}=L\cdot 88{,}94$ ped. $=2{,}223$ ped. et $\frac{S}{a^4}=L\cdot 1{,}47=0{,}037$ ped. Cum autem Lunae excentricitas sit quasi $\frac{550}{10\,000}$, erit dum Luna in perigaeo versatur $\frac{S}{a^3}=L\cdot 104{,}44$ ped. $=2{,}611$ ped. et $\frac{S}{a^4}=L\cdot 1{,}82=0{,}045$ pedum. At si Luna fuerit in apogaeo, prodibit $\frac{S}{a^3}=L\cdot 75{,}74$ ped. $=1{,}893$ ped. et $\frac{S}{a^4}=L\cdot 1{,}19=0{,}030$ pedum. Ex his igitur si Luna a Terra mediocriter distet, erit aquae elevatio $Aa=L\cdot 90{,}41$ ped. $=2{,}260$ ped. elevatio autem $Bb=L\cdot 87{,}47$ ped. $=2{,}187$ pedum: ac depressio ad latera $Dd=Ee=L\cdot 44{,}47$ pedum $=1{,}112$ ped. Pro perigaeo vero Lunae fiet $Aa=L\cdot 106{,}26$ ped. $=2{,}656$ pedum; $Bb=L\cdot 102{,}62$ ped. $=2{,}565$ pedum; atque $Dd=Ee=L\cdot 52{,}22=1{,}305$ pedum. Pro apogaeo denique Lunae habebitur

$Aa = L \cdot 76{,}93$ ped. $= 1{,}923$ pedum, et $Bb = L \cdot 74{,}55$ ped. $= 1{,}864$ pedum, atque $Dd = Ee = L \cdot 37{,}87$ ped. $= 0{,}947$ pedum.

42. Tametsi autem hac methodo non difficulter tam elevatio Maris quam depressio quae vel a Sole vel Luna seorsum gignitur, sit determinata, si quidem omnia ad statum quietis redacta concipiantur; tamen nimium foret difficile eiusdem methodi ope easdem res definire, si Sol et Luna coniunctim agant. Quamobrem aliam methodum exponamus, cuius usus pro utroque casu aeque pateat; quae cum a priori penitus sit diversa, simul ea, quae iam sunt eruta atque a NEWTONIANIS diversa deprehensa, maxime confirmabit. Petita vero est haec altera methodus ex ea aequilibrii proprietate, qua requiritur, ut omnes columnae aqueae a superficie Terrae ad centrum pertingentes sint inter se aequiponderantes. Existente igitur vel

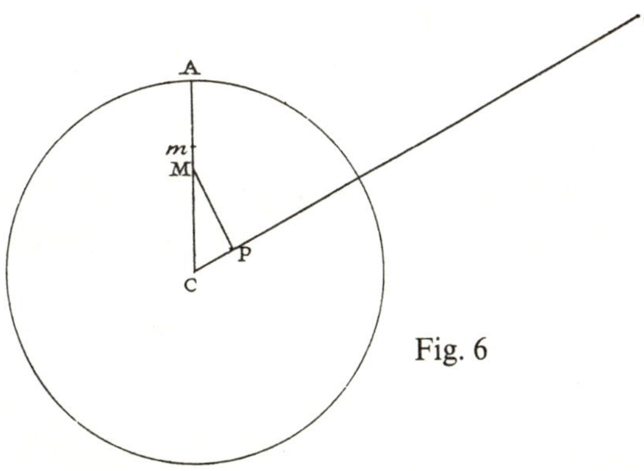

Fig. 6

Sole vel Luna in S [fig. 6], cuius vis absoluta ponatur $= S$, et distantia $SC = a$, sit AC columna aquea a superficie Terrae A ad centrum C usque pertingens, quae altitudo AC sit $= h$. Ponatur anguli ACS cosinus $= u$, qui simul erit sinus altitudinis sub qua punctum S a spectatore in A constituto super horizonte elevatum conspicitur; sumaturque intervallum quodcunque $CM = z$, et consideretur totius columnae elementum $Mm = dz$. Hoc igitur elementum primo a gravitate deorsum versus C urgebitur, cuius effectus, cum intra Terram pro variis distantiis non satis constet, ponatur dignitati cuicunque distantiarum a centro puta ipsi z^n proportionalis: mox enim planum fiet exponentem n nil omnino determinationes esse turbaturum. Urgebitur ergo elementum Mm versus centrum C vi $= z^n dz$; ex quo totius columnae AC nisus deorsum a gravitate oriundus, erit $= \dfrac{h^{n+1}}{n+1}$.

43. Praeterea autem elementum $Mm = dz$ a vi S sollicitabitur duplici modo, altero deorsum in directione MC, altero in directione ad illam MC

normali, quae posterior vis, cum pondus columnae nequaquam afficiat, tuto negligetur, solaque prior considerabitur. Demisso autem ex M in CS perpendiculo MP, positisque $CP = x$ et $PM = y$, erit $\sqrt{(x^2+y^2)} = z$, et $x = uz$ atque $y = z\sqrt{(1-uu)}$. At ex §27 vis, qua particula Mm deorsum sollicitatur, est

$$= \frac{S(yy-2xx)}{a^3\sqrt{(xx+yy)}} + \frac{3Sx(3yy-2xx)}{2a^4\sqrt{(xx+yy)}} = \frac{Sz(1-3uu)}{a^3} + \frac{3Suz^2(3-5uu)}{2a^4}.$$

Quae expressio per dz multiplicata, tumque integrata facto $z = h$, praebebit totius columnae AC nisum a vi S oriundum $= \dfrac{Sh^2(1-3uu)}{2a^3} + \dfrac{Sh^3u(3-5uu)}{2a^4}$. Quocirca totus columnae AC nisus deorsum tendens erit

$$= \frac{h^{n+1}}{n+1} + \frac{Sh^2(1-3uu)}{2a^3} + \frac{Sh^3u(3-5uu)}{2a^4};$$

qui cum in omnibus columnis debeat esse idem, aequabitur conatui, quo columna aequalis semidiametro Terrae 1 in statu naturali a sola gravitate deorsum nititur, quae vis est $= \dfrac{1}{n+1}$. Hinc igitur sequens emergit aequatio

$$1 = h^{n+1} + \frac{(n+1)Sh^2(1-3uu)}{2a^3} + \frac{(n+1)Sh^3u(3-5uu)}{2a^4};$$

ex qua elicitur

$$h = 1 + \frac{S(3uu-1)}{2a^3} + \frac{Su(5uu-3)}{2a^4},$$

quae est ea ipsa expressio, quam supra §36 altera methodo invenimus.

44. Agant nunc vires ambae ad Solem Lunamque directae coniunctim; ac primo quidem designet S Solis vim absolutam, a eius distantiam a Terra, et u sinum anguli, quo Sol supra horizontem est elevatus. Deinde sit simili modo pro Luna L eius vis absoluta, b eius distantia a Terra, atque v sinus altitudinis Lunae super horizonte. Ex his igitur columna aquea $AC = h$ tam vi propriae gravitatis quam a viribus Solis ac Lunae coniunctim in centrum C urgebitur vi

$$= \frac{h^{n+1}}{n+1} + \frac{Sh^2(1-3uu)}{2a^3} + \frac{Lh^2(1-3vv)}{2b^3} + \frac{Sh^3u(3-5uu)}{2a^4} + \frac{Lh^3v(3-5vv)}{2b^4},$$

quae aequalis esse debebit vi $\frac{1}{n+1}$. Ex hac autem aequatione resultat

$$h = 1 + \frac{S(3uu-1)}{2a^3} + \frac{L(3vv-1)}{2b^3} + \frac{Su(5uu-3)}{2a^4} + \frac{Lv(5vv-3)}{2b^4}.$$

Quocirca aqua in A supra situm naturalem, quem a sola gravitate sollicitata obtineret, a viribus Solis ac Lunae coniunctim sollicitantibus, elevabitur per intervallum

$$= \frac{S(3uu-1)}{2a^3} + \frac{L(3vv-1)}{2b^3} + \frac{Su(5uu-3)}{2a^4} + \frac{Lv(5vv-3)}{2b^4},$$

ex qua expressione status aquae vel elevationis vel depressionis ubique terrarum cognoscetur.

45. Hanc posteriorem viam secuti, non solum actiones Solis ac Lunae commode coniungere potuimus, sed etiam nunc nobis licebit motus vertiginis Terrae, et vis centrifugae inde ortae, rationem habere; id quod methodo priore opus fuisset insuperabile. Ponamus enim altitudinem columnae naturalem AC, quam habitura esset a vi gravitatis et vi centrifuga simul, seu quod eodem redit, in figura Terrae sphaeroidica compressa, esse $=f$, altitudinem autem quam habebit accedentibus viribus Solis ac Lunae esse $=h$; atque manifestum est quantitates f et h quam minime ab 1 discrepare. Cum igitur utriusque columnae f et h idem debeat esse nisus deorsum, columnae autem f in quam sola gravitas et vis centrifuga agunt nisus sit $=\frac{f^{n+1}}{n+1} - \alpha ff$, denotante α quantitatem a vi centrifuga in A pendentem, columnae vero h nisus sit

$$= \frac{h^{n+1}}{n+1} - \alpha h^2 + \frac{Sh^2(1-3uu)}{2a^3} + \frac{Lh^2(1-3vv)}{2b^3} + \frac{Sh^3u(3-5uu)}{2a^4} + \frac{Lh^3v(3-5v^2)}{2b^4};$$

erit aequalitate facta[10])

$$f^{n+1} - (n+1)\alpha ff = h^{n+1} - (n+1)\alpha h^2 + \frac{(n+1)Sh^2(1-3uu)}{2a^3}$$
$$+ \frac{(n+1)Lh^2(1-3vv)}{2b^3} + \frac{(n+1)Sh^3u(3-5uu)}{2a^4} + \frac{(n+1)Lh^3v(3-5vv)}{2b^4}.$$

10) Editio princeps et Manuscriptus: Lh^3 loco $(n+1)Lh^3$. Correxit EJA.

Ponatur $h = f + s$, erit ob α quantitatem vehementer parvam, a vero et b maximas,

$$0 = f^n s + \frac{Sf^2(1-3uu)}{2a^3} + \frac{Lf^2(1-3vv)}{2b^3} - 2\alpha fs + \frac{Sfs(1-3uu)}{a^3} + \frac{Lfs(1-3vv)}{b^3} + \frac{Sf^3u(3-5uu)}{2a^4} + \frac{Lf^3v(3-5vv)}{2b^4},$$

neglectis terminis in quibus s plures obtinet dimensiones, ob summam ipsius s parvitatem respectu ipsius f. Hinc itaque fiet

$$s = \frac{\dfrac{S(3uu-1)}{2a^3} + \dfrac{L(3vv-1)}{2b^3} + \dfrac{Sfu(5uu-3)}{2a^4} + \dfrac{Lfv(5vv-3)}{2b^4}}{f^{n-2} - \dfrac{2\alpha}{f} + \dfrac{S(1-3uu)}{a^3 f} + \dfrac{L(1-3vv)}{b^3 f}}.$$

Quod si porro ponatur semiaxis Terrae per polos transiens $= 1$, erit ob aequilibrium $\frac{f^{n+1}}{n+1} - \alpha ff = \frac{1}{n+1}$ et $f = 1 + \alpha$, ex quo denominator praecedentis fractionis ab unitate quam minime discrepabit; sub ipso enim aequatore est $\alpha = \frac{1}{578}$, ubi quidem est maximum: unde erit omnino ut ante elevatio aquae a viribus Solis ac Lunae orta supra altitudinem naturalem

$$s = \frac{S(3uu-1)}{2a^3} + \frac{L(3vv-1)}{2b^3} + \frac{Su(5uu-3)}{2a^4} + \frac{Lv(5vv-3)}{2b^4};$$

discrimen enim quod revera aderit, sensus omnino effugiet, pendebitque simul a valore exponentis n.

CAPUT QUARTUM

De Fluxu ac Refluxu Maris si aqua omni inertia careret

46. Quae in capite praecedente sunt tradita respiciunt hypothesin assumtam, qua Solem ac Lunam respectu Terrae perpetuo eundem situm tenere posuimus; ibique praecipue statum aequilibrii, ad quem oceanus a viribus Solis et Lunae perducatur, determinavimus. Longe aliter autem se res habet, si tam Luna et Sol

quam Terra in motum collocentur, quo casu ob perpetuam situs relativi mutationem nunquam aequilibrium adesse poterit; cum enim tempore opus sit, quo data vis datum corpus ad motum perducat, duplici modo status oceani assignatus a vero discrepabit. Namque primo aqua quovis momento in eum aequilibrii situm, quem vires sollicitantes intendunt, pervenire non poterit, sed tantum ad eum appropinquabit continuo; deinde etiamsi in ipsum aequilibrii situm perveniat, in eo tamen non acquiescet, sed motu iam concepto ulterius feretur, uti ex natura motus abunde constat. Huius autem utriusque aberrationis ratio in inertia aquae est posita, qua fit ut aqua nec subito in eum situm se conferat, in quo cum viribus datur aequilibrium, nec cum hunc aequilibrii situm attigerit, ibi quiescat. Quocirca ne difficultatum multitudine obruamur, aquam omni inertia carentem assumamus, hoc est istius indolis, ut non solum quovis momento se in statum aequilibrii subito recipiat, sed ibi etiam omnem motum insitum deponendo permaneat, quamdiu iste situs viribus sollicitantibus conveniat. Hac itaque facta hypothesi, perspicuum est aquam quovis temporis momento in eo ipso statu fore constitutam, qui secundum praecepta capitis praecedentis positioni cum Solis tum Lunae respondeat.

47. Ut igitur in hac hypothesi, qua Mare vis inertiae expers ponimus, pro quovis loco ad quodvis tempus statum Maris quam commodissime definiamus, primum solam Lunam considerabimus, cum in ea praecipua aestus Maris causa contineatur, atque tam Fluxus quam Refluxus Maris a transitu Lunae per meridianum computari soleat: quod si enim Lunae effectus innotuerit, non solum Solis effectus quoque mutatis mutandis colligetur, sed etiam effectus, qui ab ambobus luminaribus simul agentibus proficiscitur. Propositus igitur sit Terrae locus quicunque, cuius in coelo Zenith sit Z [fig. 7], horizon HQO et P polus borealis, ita ut

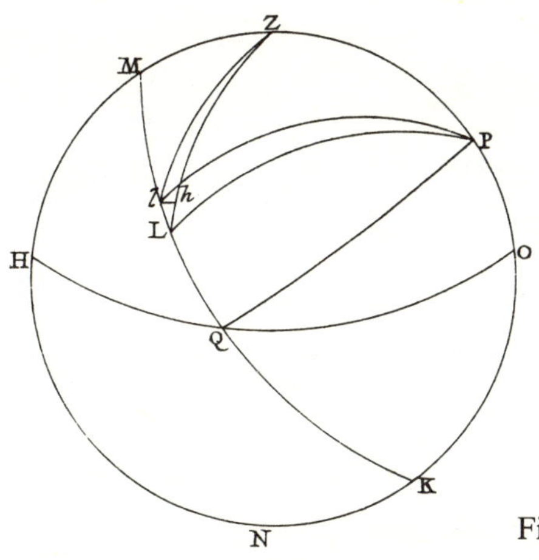

Fig. 7

arcus PO sit huius loci elevatio poli, et circulus $PZHNO$ meridianus. Sit porro MLK parallelus aequatori, in quo Luna iam motu diurno circumferatur, atque hoc momento reperiatur Luna in L; eritque tempus, quo Luna vel ex L ad meridianum M appellet, vel vicissim a meridiano ad L pertigit, ut angulus MPL, sive hoc tempus se habebit ad tempus unius revolutionis Lunae, quod est 24 horarum 48′, uti se habet angulus MPL ad quatuor rectos. Sit igitur anguli MPL cosinus $= t$, cosinus[11]) elevationis poli PO seu sinus arcus $PZ = p$, cosinus [arcus PZ] $= P$, ac sinus declinationis Lunae borealis $= Q$, qui idem est cosinus[12]) distantiae Lunae a polo PL, huius vero ipsius arcus sinus sit $= q$, cui simul cosinus declinationis Lunae aequatur, atque ob sinum totum constanter positum $= 1$, erit $Q^2 + q^2 = 1$. Cum iam in triangulo sphaerico ZPL dentur arcus PZ et PL cum angulo ZPL, reperietur per Trigonometriam sphaericam arcus ZL cosinus $= tpq + PQ$, qui simul est sinus altitudinis Lunae supra horizontem, quem ante posuimus $= v$. Ex quibus erit $v = tpq + PQ$, et $3vv - 1 = 3(tpq + PQ)^2 - 1$, atque $5vv - 3 = 5(tpq + PQ)^2 - 3$; qui valores in formulis praecedentis capitis substituti praebebunt statum Maris, hoc est vel elevationem vel depressionem, pro loco proposito at tempus assignatum.

48. Quod si ergo Lunae vis absoluta ponatur $= L$, eiusque a Terra distantia $= b$, erit intervallum, quo aqua supra statum naturalem elevabitur,

$$= \frac{L(3(tpq + PQ)^2 - 1)}{2b^3} + \frac{L(tpq + PQ)(5(tpq + PQ)^2 - 3)}{2b^4},$$

quae expressio si fit negativa, indicat aquam infra statum naturalem esse depressam. Ponamus Lunam super horizonte seu versus austrum per meridianum transire, quo casu erit $t = 1$; hoc igitur tempore aqua supra statum naturalem erit elevata intervallo

$$= \frac{L(3(pq + PQ)^2 - 1)}{2b^3} + \frac{L(pq + PQ)(5(pq + PQ)^2 - 3)}{2b^4}.$$

Contra vero dum Luna sub horizonte vel versus boream ad meridianum appellit, fiet elevatio aquae supra statum naturalem per intervallum

11) Editio princeps et Manuscriptus: sinus loco cosinus. Correxit EJA.
12) Editio princeps et Manuscriptus: sinus loco cosinus. Correxit EJA.

$$= \frac{L(3(PQ-pq)^2-1)}{2b^3} + \frac{L(PQ-pq)(5(PQ-pq)^2-3)}{2b^4},$$

quoniam hoc casu sit $t = -1$. Tempore autem intermedio inter binos hos appulsus ad meridianum loci propositi, seu eo tempore quo angulus ZPL fit rectus, hoc est 6 horis 12′ vel ante vel post transitum per meridianum ob $t = 0$, erit intervallum, quo aqua elevabitur,

$$= \frac{L(3P^2Q^2-1)}{2b^3} + \frac{LPQ(5P^2Q^2-3)}{2b^4};$$

quae expressio semper est negativa, ideoque indicat aquam infra statum naturalem consistere. Namque cum P ubique sit minor unitate nisi sub ipsis polis, ac declinatio Lunae nunquam ad 30° assurgere possit, ex quo $Q < \frac{1}{2}$ et $QQ < \frac{1}{4}$, erit $3P^2Q^2$ perpetuo unitate minor; ideoque illa expressio negativa.

49. De ratione autem elevationis aquae in genere iudicari licebit ex formula

$$\frac{L(3vv-1)}{2b^3} + \frac{Lv(5vv-3)}{2b^4},$$

seu cum posterior terminus vix sit sensibilis, ex solo priore

$$\frac{L(3vv-1)}{2b^3}.$$

Ex hac autem expressione intelligitur aquae elevationem a sola elongatione Lunae ab horizonte pendere, sive Luna sit super sive sub horizonte, retinet enim $3vv-1$ eundem valorem sive v sit affirmativum sive negativum. Deinde quia fit $3vv-1=0$ si Luna ab horizonte distet arcu 35° 16′, tum aqua in ipso statu naturali erit constituta, neque elevata neque depressa. Elevabitur ergo aqua, cum Luna ultra 35° 16′ vel supra vel infra horizontem versetur, e contrario autem deprimetur quando Lunae ab horizonte distantia minor est quam 35° 16′. Omnino autem aqua maxime erit depressa dum Luna ipsum horizontem occupat, hocque tempore infra situm naturalem subsidet intervallo $\frac{L}{2b^3} = 1{,}111$ pedum (§41); atque de hoc situ elevabitur recedente Luna ab horizonte sive super sive sub Terra. Hinc iis in regionibus, in quibus Luna oritur et occidit, tempore 24 hor. 48′ Mare bis maxime erit depressa, bisque elevata; status scilicet depressionis incidet in appulsus Lunae ad horizontem, status autem elevationis in appulsus Lunae ad meridianum. At quibus in regionibus Luna nec oritur nec occidit, quoniam ibi Luna altero appulsu ad meridianum maxime, altero minime ab horizonte distat, spatio 24 h. 48′ aqua semel tantum elevabitur, semelque deprimetur: sub ipsis autem polis aestus Maris omnino erit nullus, diurnus scilicet; nam variatio declinationis sola statum Maris turbabit.

50. Cum igitur sub polis Terrae nullus sit Fluxus ac Refluxus Maris, sed aqua tantum aliquantulum ascendat descendatque, prout Luna vel magis ab aequatore recedit vel ad eum accedit; videamus etiam quomodo aestus Maris in aliis Terrae regionibus secundum nostram hypothesin debeat esse comparatus. Considerabimus autem praecipue tres regiones, quarum prima posita sit sub ipso aequatore, secunda habeat elevationem poli 30 graduum, tertia vero 60 graduum. Quia igitur in his omnibus regionibus Luna oritur atque occidit, maxima depressio aquae ubique erit eadem, scilicet per intervallum $\frac{L}{2b^3}$ infra situm naturalem, eaque continget bis, quando nimirum Luna in ipso horizonte versatur. Ab hoc itaque statu maximae depressionis elevationes Maris indicabimus et computabimus, spatiis assignandis, per quae aqua attolletur dum Luna vel supra horizontem in M vel infra in K ad meridianum appellit, itemque dum ab utroque meridiano aequaliter distat, qui locus sit L existente angulo MPL recto. Praeterea tres quoque Lunae situs in sua orbita contemplabimur, quorum primus sit, cum Luna in ipso aequatore versatur, secundus cum Luna habet declinationem borealem 20 graduum, tertius vero cum Luna declinationem habet australem pariter 20 graduum. Denique in tabella sequente adscripsimus quantitatem anguli MPQ, ex quo tempus tam ortus quam occasus Lunae, quo aqua maxime est depressa, atque elevatio existit nulla, innotescit.

In locis sub Aequatore sitis, est elevatio Maris[13]), *dum Luna versatur in*

		M	*L*	*K*	*MPQ*
☾ Declinatio	0°	$\frac{3L}{2b^3}+\frac{2L}{2b^4}$	0	$\frac{3L}{2b^3}-\frac{2L}{2b^4}$	90° 0′
☾ Decl. boreal.	20°	$\frac{2{,}649L}{2b^3}+\frac{1{,}549L}{2b^4}$	0	$\frac{2{,}649L}{2b^3}-\frac{1{,}549L}{2b^4}$	90° 0′
☾ Decl. austr.	20°	$\frac{2{,}649L}{2b^3}+\frac{1{,}549L}{2b^4}$	0	$\frac{2{,}649L}{2b^3}-\frac{1{,}549L}{2b^4}$	90° 0′

Sub elevatione Poli 30°, erit Maris elevatio

☾ Declinatio 0°	$\frac{2{,}250L}{2b^3}+\frac{1{,}082L}{2b^4}$	0	$\frac{2{,}250L}{2b^3}-\frac{1{,}082L}{2b^4}$	90° 0′
☾ Decl. boreal. 20°	$\frac{2{,}909L}{2b^3}+\frac{1{,}880L}{2b^4}$	$\frac{0{,}087L}{2b^3}-\frac{0{,}156L}{2b^4}$	$\frac{1{,}239L}{2b^3}-\frac{0{,}154L}{2b^4}$	102° 8′
☾ Decl. austr. 20°	$\frac{1{,}239L}{2b^3}+\frac{0{,}154L}{2b^4}$	$\frac{0{,}087L}{2b^3}+\frac{0{,}156L}{2b^4}$	$\frac{2{,}909L}{2b^3}-\frac{1{,}880L}{2b^4}$	77° 52′

Sub elevatione Poli 60°, erit Maris elevatio

☾ Declinatio 0°	$\frac{0{,}740L}{2b^3}-\frac{0{,}125L}{2b^4}$	0	$\frac{0{,}740L}{2b^3}+\frac{0{,}125L}{2b^4}$	90° 0′
☾ Decl. boreal. 20°	$\frac{1{,}760L}{2b^3}+\frac{0{,}582L}{2b^4}$	$\frac{0{,}263L}{2b^3}-\frac{0{,}514L}{2b^4}$	$\frac{0{,}092L}{2b^3}+\frac{0{,}158L}{2b^4}$	129° 5′
☾ Decl. austr. 20°	$\frac{0{,}092L}{2b^3}-\frac{0{,}158L}{2b^4}$	$\frac{0{,}263L}{2b^3}+\frac{0{,}514L}{2b^4}$	$\frac{1{,}760L}{2b^3}-\frac{0{,}582L}{2b^4}$	50° 55′

51. Si quis iam ex hac tabula elevationem Maris supra statum maximae depressionis in mensuris cognitis definire voluerit, is loco fractionum $\frac{L}{b^3}$ et $\frac{L}{b^4}$ earum valores in pedibus parisinis ex § 41 substituat, habita ratione distantiae Lunae a Terra, prout ibidem est expositum. Consequuntur autem ex hac tabula multa egregia consectaria, quae vero nondum summo cum rigore ad experientiam examinari possunt, etiamsi iam insignis convenientia deprehendatur. Aquam enim adhuc omnis inertiae expertem ponimus, perspicuum autem est, si aquae inertia tribuatur, tum diversa omnino Phaenomena oriri oportere. Quod si igitur hi assignati effectus iam cum observationibus plane consentirent, id potius theoriam everteret quam confirmaret, cum aquam extra statum suum naturalem simus contemplati. Interim tamen satis tuto iam status Maris sub ipsis polis poterit definiri, qui etsi ad experientiam examinari non potest, tamen ipsa ratione confirmabitur. Ac primo quidem sub polis nulla erit Maris mutatio diurna, cum Luna per totum diem eandem teneat ab horizonte distantiam, id quod ipsa quoque ratio dictat, quia ibi

13) Valores terminorum posteriorum minime sunt accurati. EJA.

non datur meridianus, a cuius appulsu aestus Maris alibi aestimari solet. Dabitur tamen his locis mutatio menstrua, atque aqua maxime erit humilis cum Luna in ipso aequatore versatur, quo quippe tempore perpetuo horizontem occupabit. Hinc porro aqua sensim elevabitur prout Lunae declinatio sive versus boream sive versus austrum augetur, donec tandem si declinatio fit maxima, per spatium 10 pollicum tantum elevetur; quae mutatio cum sit perquam lenta, ab inertia aquae vix turbabitur.

52. Ex his vero iisdem formulis effectus a Sole oriundus non difficulter colligetur; tantum enim quantitates S et a, loco L et b substitui oportet, quo facto effectus Solis circiter quater minor reperietur quam is qui a Luna oritur. Seorsim autem cum Solis tum Lunae effectibus definitis, per coniunctionem simplicem effectus, quem ambo luminaria coniunctim producunt, determinabitur. Ponamus itaque primum Solem Lunamque in coniunctione versari, id quod fit tempore novilunii; tum igitur neglecta Lunae latitudine, Sol et Luna in eodem eclipticae loco versabuntur, atque simul ad meridianum aeque ac ad horizontem appellent. Quocirca manentibus superioribus denominationibus, erit quoque Solis declinationis sinus $= Q$, cosinus $= q$, ac pro angulo MPL cuius cosinus est $= t$, erit sinus altitudinis Solis pariter uti Lunae $= tpq + PQ$. Ex quo dum ambo luminaria per meridianum versus austrum transeunt, aquae elevatio, quae tum erit maxima, altitudinem naturalem superabit intervallo

$$= \left(\frac{S}{2a^3} + \frac{L}{2b^3}\right)(3(pq + PQ)^2 - 1) + \frac{L(pq + PQ)}{2b^4}(5(pq + PQ)^2 - 3),$$

neglecto altero termino a vi Solis oriundo, cum sensus omnino effugiat. At dum ambo luminaria infra horizontem ad meridianum pertingunt, erit elevatio aquae

$$= \left(\frac{S}{2a^3} + \frac{L}{2b^3}\right)(3(PQ - pq)^2 - 1) + \frac{L(PQ - pq)}{2b^4}(5(PQ - pq)^2 - 3).$$

Maxima denique aquae depressio incidet, quando luminaria vel oriuntur vel occidunt, eaque minor erit quam altitudo aquae naturalis intervallo $= \dfrac{S}{2a^3} + \dfrac{L}{2b^3}$. Cum igitur $\dfrac{S}{2a^3}$ sit circiter subquadruplum ipsius $\dfrac{L}{2b^3}$, in novilunio omnes effectus Lunae supra recensiti quarta sui parte augebuntur.

53. In plenilunio omnia eodem se habere modo deprehenduntur, quo in novilunio, quia enim tum Sol et Luna in oppositione versantur, erit declinatio Solis aequalis et contraria declinationi Lunae, unde quidem pro Sole fit $-Q$, quod in novilunio erat $+Q$; at cum Sol secundum ascensionem rectam a Luna distet 180°, erit hoc casu $-t$, quod ante erat $+t$, ex quo pro plenilunio habetur sinus altidudinis Solis $= -tpq - PQ$, qui pro novilunio erat $= tpq + PQ$, ex quo quadratum huius sinus utroque casu est idem, ideoque etiam eadem Phaenomena in novilunio atque plenilunio. Deinde etiam hoc tempore aqua maxime deprimetur, cum luminaria ambo in horizonte versantur, tumque aqua humilior erit quam in statu naturali intervallo $= \frac{S}{2a^3} + \frac{L}{2b^3}$. Ex hoc itaque situ donec Luna ad meridianum supra Terram appellit, aqua elevabitur per intervallum $= 3(PQ + pq)^2\left(\frac{S}{2a^3} + \frac{L}{2b^3}\right)$, tantoque iterum subsidet usque ad Lunae obitum; tum vero rursus elevabitur usque ad appulsum Lunae ad meridianum infra horizontem, idque per spatium $3(PQ - pq)^2\left(\frac{S}{2a^3} + \frac{L}{2b^3}\right)$, neglecto termino sequente quippe fere insensibili. Cum igitur sint $PQ + pq$ et $PQ - pq$ sinus distantiae Lunae ab horizonte dum in meridiano versatur, erunt spatia per quae aqua tempore pleniluniorum ac noviluniorum supra statum maxime depressum elevatur, in ratione duplicata sinuum distantiarum Lunae ab horizonte, dum per meridianum transit. Nisi ergo vel Luna in ipso aequatore existat, vel Terrae locus sub aequatore sit situs, Fluxus Maris diurni ac nocturni erunt inaequales; luminaribus autem in aequatore extantibus, utraque aquae elevatio fiet per spatium $= 3pp\left(\frac{S}{2a^3} + \frac{L}{2b^3}\right)$.

54. Ut nunc in effectus, quos Sol et Luna in quadraturis siti coniunctim producunt, inquiramus; ponamus, ne calculus nimium fiat prolixus, Solem in ipso aequatore versari, quoniam tum plerumque minimus aestus observatur. Hoc itaque casu Solis declinatio erit nulla, Lunae vero maxima, quam neglecta latitudine assumamus 23° 29′, cuius sinus sit $= Q$, cosinus $= q$, posita hac declinatione boreali. Iam ponamus Lunam in meridiano in M versari, quo tempore Sol erit in horizonte; unde cum aqua supra statum naturalem elevetur a Luna intervallo $\frac{L(3(pq + PQ)^2 - 1)}{2b^3}$, a Sole vero deprimatur intervallo $\frac{S}{2a^3}$, ab utraque vi coniunctim elevabitur per spatium $\frac{L(3(pq + PQ)^2 - 1)}{2b^3} - \frac{S}{2a^3}$: at dum

Luna sub horizonte ad meridianum appellit, aqua elevabitur per spatium $\frac{L(3(PQ-pq)^2-1)}{2b^3}-\frac{S}{2a^3}$. Sumatur inter has ambas elevationes inaequales more solito medium, eritque elevatio aquae media hac quadratura eveniens

$$=\frac{L(3p^2q^2+3P^2Q^2-1)}{2b^3}-\frac{S}{2a^3}.$$

Refluxus vero continget, cum Luna horizontem attinget, quo tempore Sol in meridiano proxime versabitur, ex quo depressio totalis aquae in Refluxu infra statum naturalem proxime erit $=\frac{L}{2b^3}-\frac{S(3pp-1)}{2a^3}$: quare à Fluxu usque ad subsequentem Refluxum aqua subsidet per intervallum

$$=\frac{3L(p^2q^2+P^2Q^2)}{2b^3}-\frac{3Spp}{2a^3}.$$

55. Quamvis motus Maris hoc modo assignatus ab inertia aquae multum immutetur, tamen quia eandem fere mutationem tam maioribus aestibus quam minoribus infert, satis tuto assumere posse videmur spatia, per quae aqua circa aequinoctia cum tempore plenilunii sive novilunii, tum etiam tempore quadraturarum actu ascendit, expressionibus inventis esse proportionalia. Quamobrem si in dato Terrae loco ex pluribus observationibus determinetur spatium medium, per quod Mare a Refluxu ad Fluxum ascendit, tempore aequinoctiorum, tam in pleniluniis noviluniisve quam in quadraturis, eorum ratio ad eam quae ex formulis consequitur, proxime accedere debebit. Atque hinc ex definita hac ratione per observationes ratio poterit inveniri inter vires Solis et Lunae absolutas S et L, quae est ipsa via qua NEWTONUS est usus ad vim Lunae absolutam definiendam, cum vis Solis sit cognita: quod negotium, cum a NEWTONO non satis accurate sit pertractatum, nos id ex istis principiis expediemus. Exprimat igitur $m:n$ rationem intervallorum eorum, per quae Oceanus in dato Terrae loco, cum in syzygiis luminarium quam quadraturis tempore aequinoctiorum, ascendendo descendendoque oscillatur; eritque

$$m:n=3pp\left(\frac{S}{2a^3}+\frac{L}{2b^3}\right):\frac{3L(p^2q^2+P^2Q^2)}{2b^3}-\frac{3Spp}{2a^3};$$

ex qua elicitur ista proportio

$$m\left(q^2 + \frac{P^2Q^2}{p^2}\right) - n : m + n = \frac{S}{a^3} : \frac{L}{b^3};$$

ex qua cum data sit vis a Sole orta $\frac{S}{a^3}$, deducetur vis a Luna oriunda $\frac{L}{b^3}$ saltem proxime. Instituamus calculum pro observationibus in Portu Gratiae (*Havre de Graces*) factis, ex quibus diligenter inter se collatis pro ratione $m:n$ prodit ista $17:11$. Cum igitur huius loci elevatio poli sit circiter 50°, erit $P = \sin 50°$, et $Q = \sin 23° \, 29'$; hincque $qq + \frac{P^2Q^2}{pp} = 1{,}0668$: ex quo prodibit

$$\frac{S}{a^3} : \frac{L}{b^3} = 7{,}1356 : 28;$$

ita ut vis Lunae $\frac{L}{b^3}$ sit fere quadrupla vis Solis $\frac{S}{a^3}$, uti iam NEWTONUS ex aliis observationibus conclusit: atque hanc ob rem ipsius determinationem vis Lunae absolutae L retinuimus.

56. Si haec, quae de combinatione virium Lunam Solemque respicientibus sunt allata, attentius considerentur, mox patebit maximos aestus menstruos in novilunia ac plenilunia incidere debere; his enim temporibus tam elevatio aquae quam depressio a Luna oriunda a vi Solis maxime adiuvatur, cum eodem tempore, quo Luna aquam maxime vel elevat vel deprimit, simul quoque Solis vis aquam maxime vel elevet vel deprimat. In quadraturis autem hae duae vires fere perpetuo dissentiunt, ac dum Luna aquam maxime vel elevat vel deprimit, eodem tempore Sol contrarium exerit effectum, aquamque maxime vel deprimit vel elevat, ex quo minimum discrimen inter quemque Fluxum ac subsequentem Refluxum observabitur, aestusque erunt minimi. Quamobrem circa alias Lunae phases aestus Maris medium teneat inter maximum minimumque necesse est, quia tum vires Solis ac Lunae nec omnino conspirant, nec sibi invicem adversantur. Per totum autem annum quibus noviluniis pleniluniisque maximus eveniat aestus, quibusque quadraturis minimus aestus respondeat, absolute sine respectu ad situm loci habito definiri nequit. Sub aequatore quidem ubi Luna, cum est in aequatore, maxima vi gaudet, dubium est nullum, quin aestus maximi in aequinoctia incidat, quando ambo luminaria in aequatore sunt posita, quae eadem proprietas etiam in loca ab aequatore non multum dissita competit: at in locis ab aequatore magis remotis

aestus Maris, cum Luna maximam habet declinationem, dantur quidem maiores ex Tabula § 50, verum aestus mox subsequentes multo sunt minores. Quod si autem inter binos aestus a Luna oriundos consequentes medium capiatur, patebit in regionibus 30° ab aequatore remotis, quibus aestus est $\frac{2{,}250}{2b^3}L$ si Lunae declinatio sit nulla, aestum Maris medium, cum Luna habet declinationem 20 graduum, fore $= \frac{2{,}074}{2b^3}L$, ideoque adhuc minorem quam cum Luna aequatorem tenet. Contra vero sub elevatione poli 60 graduum, est aestus Maris, Luna versante in aequatore, $= \frac{0{,}740}{2b^3}L$, aestus autem medius, cum Lunae declinatio est 20°, est $= \frac{0{,}926}{2b^3}L$, ideoque maior. Ex quo consequitur in regionibus polis vicinioribus aestus maximos, non in aequinoctia, sed potius circa solstitia, incidere debere, qua quidem in re theoria nostra per experientiam mirifice confirmatur.

CAPUT QUINTUM

De tempore Fluxus ac Refluxus Maris in eadem hypothesi

57. Quanquam in praecedente capite, quo in quantitatem aestus Maris praecipue inquisivimus, etiam tempora, quibus tam Fluxus quam Refluxus eveniat, iam indicavimus; tamen hoc capite istud argumentum fusius atque ad observationes accommodate persequemur. Observationes enim, quae circa aestum Maris institui solent, ad tria genera commodissime referuntur; ad quorum primum pertinet Maris cum elevatio maxima tum maxima depressio, atque indicatur quantum quovis aestu aqua cum ascendat tum descendat. Ad secundum observationum genus numerari convenit eas, quae ad tempus respiciunt, quibusque definitur, quonam temporis momento ubivis terrarum aqua cum summam teneat altitudinem tum minimam. Tertium denique genus observationum ad ipsum motum Maris reciprocum spectat, iisque determinatur quanta celeritate quovis temporis momento alterna Maris elevatio ac depressio absolvatur, sive momentanea mutatio, dum Mare a Fluxu ad Refluxum transit et vicissim, investigatur. Quibus tribus rebus cum observationes convenientissime instituantur, iisdem theoria atque explicatio phaenomenorum commodissime tractabitur. Ac primae quidem et tertiae parti pro nostra hypothesi in precedentibus capitibus abunde satisfactum videtur.

58. Quoniam autem a Maris inertia aliisque circumstantiis Maris motum turbantibus omnes cogitationes adhuc abstrahimus, manifestum est ubique terrarum, si sola Lunae vis Mare agitaret, aquam maxime elevari debere cum Luna ab horizonte longissime fuerit remota, hoc est iis ipsis momentis quibus Luna per meridianum dati loci tam supra quam infra Terram transit: sunt enim elevationes aquae in duplicata ratione sinuum distantiarum Lunae ab horizonte, ex quo simul successiva Maris commotio cognoscitur. Excipiuntur autem hinc, ut iam notavimus, loca polis Terrae proxima, quibus Luna vel non oritur vel non occidit; ibi enim altero Lunae ad meridianum appulsu aqua debet esse summa, altero ima. Verum de his locis non admodum erimus solliciti; cum tam observationes sufficientes, quibus theoria probetur, deficiant, quam ipse Maris motus indicatus rationi sit consentaneus, neque confirmatione indigeat. In Terrae locis ergo a polis satis remotis seu extra circulos polares sitis, quibus Luna intervallo 24 h. 48′ tam oritur quam obit, elevabitur Mare eodem temporis intervallo bis, totiesque deprimetur; atque utraque maxima Maris altitudo continget, cum Luna ad meridianum illius loci pervenit, minima vero cum Luna horizontem attingit. Hinc igitur temporis intervallum inter binos aquae Fluxus seu summas elevationes interiectum constanter erit 12 h. 24′, ab anomaliis Lunae mentem abstrahendo; at tempus summae depressionis, cum respondeat appulsui Lunae ad horizontem, inter binas elevationes aequaliter non interiacebit, sed alteri elevationi eo erit propius, quo maior fuerit cum loci propositi elevatio poli tum Lunae declinatio, hoc est quo maius fuerit discrimen inter ortum obitumve Lunae et circulum horarium sextum.

59. Sed coniungamus cum Luna vim Solis, ut nostrae conclusiones magis ad observationes perducantur. Ac primo quidem manifestum est tempore tam novilunii quam plenilunii aquam maxime fore elevatam, quando Luna per meridianum loci transit, quippe quo momento etiam Sol ad eundem meridianum appellit, si quidem syzygia ipso meridie vel media nocte celebratur. Quamobrem si novilunium pleniluniumve in ipsum meridiem incidat, ipso quoque meridiei momento maxima habebitur aquae elevatio; pariterque si id eveniat media nocte, eodem ipso momento aqua maximam obtinebit elevationem. Verum si coniunctio vel oppositio luminarium meridiem vel praecedat vel sequatur, tum Fluxus non in ipsum meridiem incidet, sed vel tardius vel citius veniet, quia Luna his casibus tanquam primaria aestus causa vel post vel ante meridiem ad meridianum pertingit. Atque hinc eo die, in quem sive plenilunium sive novilunium incidit, facile poterit definiri acceleratio vel retardatio Fluxus respectu meridiei. Ponamus enim novilunium seu plenilunium celebrari n horis ante meridiem, unde cum motus Lunae medius a

Sole diurnus sit 12° circiter, ipso meridie Luna a meridiano iam distabit angulo horario $\frac{n}{2}$ grad. versus ortum, ex quo Luna post meridiem demum per meridianum transibit, elapsis $\frac{n}{30}$ horis seu $2n$ minutis primis. Sin autem novilunium pleniluniumve accidat n horis post meridiem, tum Maris maxima elevatio $2n$ minutis ante meridiem eveniet. Haec autem momenta accuratissime cognoscentur, si ad singulos dies transitus Lunae per meridianum computentur; ac praeterea tam ortus quam occasus notetur, quippe quibus momentis maxima aquae depressio respondet; maiorem autem huiusmodi tabula afferet utilitatem, si insuper quovis die distantia Lunae a Terra indicetur, quippe a qua Lunae effectus praecipue pendet.

60. Congruunt haec iam apprime cum observationibus, quibus constat, diebus novilunii vel plenilunii aestum Maris accelerari si novilunium pleniluniumve post meridiem accidat, contra vero retardari. Quamvis enim ob aquae inertiam maxima Maris elevatio non respondeat appulsui Lunae ad meridianum, sed tardius eveniat, uti post docebitur, tamen similibus casibus aequaliter retardabitur; pro termino igitur fixo, si ad observationes respiciatur, non sumi debet momentum meridiei, sed id momentum, quo si Lunae cum Sole coniunctio vel oppositio in ipsum meridiem incidit, summa aquae elevatio observatur. Hoc igitur momento notato, uti ab iis qui huiusmodi observationes instituunt fieri solet, si plenilunium noviluniumve vel ante vel post meridiem incidat, summa Maris elevatio vel tardius vel citius continget: et quidem si syzygia vera n horis vel ante meridiem eveniat vel post, tum Fluxus $2n$ minutis vel tardius vel citius observari debebit [quam si syzygia in ipsum meridiem incidisset]*. Atque haec est ea ipsa regula quam Celeb. CASSINI in Mem. Academiae Regiae[14]) pro Anno 1710, ex quamplurimis observationibus inter se comparatis derivavit; iubet scilicet numerum horarum, quibus coniunctio sive opposito luminarium verum meridiem vel praecedit vel sequitur, duplicari, totidemque minuta prima ad tempus medium notatum, quo fluxus evenire solet, vel addi vel ab eo subtrahi, quo verum Fluxus momentum obtineatur. Quoniam autem haec correctio nititur motu Lunae medio, perspicuum est eam correctione ulteriori opus habere a vero Lunae motu petita, quae vero plerumque erit insensibilis, cum summa aquae elevatio non subito adsit, sed per tempus satis notabile duret.

14) *Mémoires de l'Académie Royale des Sciences* (Paris), (1710), 1712, pp. 318–341. EJA.

61. Nisi autem luminaria proxima sint vel coniunctioni vel oppositioni, maxima Maris elevatio non in ipsum Lunae transitum per meridianum incidet. Quoniam enim Luna dum prope meridianum versatur, per aliquod tempus eandem altitudinem conservat, tantisper etiam Mare eandem elevationem retinebit; et hanc ob rem si Sol interea sensibiliter vel ab horizonte recedat, vel ad eundem accedat, vis Solis ad Mare elevandum vel crescet sensibiliter, vel decrescet; ex quo dum Luna prope meridianum existit, fieri potest, ut tamen mare etiamnum elevetur, vel adeo iam deprimatur a Sole. Ex his igitur perspicuum est summam Maris altitudinem tardius seu post transitum Lunae per meridianum accidere debere, si eo tempore Sol ab horizonte recedat, id quod evenit diebus novilunium et plenilunium praecedentibus. Contra autem si Luna post Solem per meridianum transeat, idque vel ante Solis ortum vel ante occasum; tum, quia Mare in transitu Lunae per meridianum a vi Solis iam deprimitur, maximam habuit altitudinem ante appulsum Lunae ad meridianum, id quod contingit diebus novilunium pleniluniumve sequentibus. Quando autem Sol ipsum horizontem occupat, dum Luna in meridiano versatur, tum etiamsi distantia Solis horizonte perquam sit mutabilis, tamen cum elevationis vis quadrato sinus altitudinis Solis sit proportionalis, quod [quadratum]* omnino evanescit, etiam hoc casu maxima aquae elevatio in ipsum Lunae per meridianum transitum incidet, hicque casus circa quadraturas luminarium locum habet.

62. Ut igitur innotescat, quantum vires cum Solis tum Lunae ad Mare elevandum dato tempore vel crescant vel decrescant, dum ab horizonte aliquantillum vel recedunt, vel ad eundem accedunt, ponamus Solem Lunamve in L [fig. 7] versari, atque inde ad punctum meridiani M progredi. Tempusculo ergo per angulum $LPl = d\theta$ repraesentato progredietur Luna vel Sol ex L in l atque ab horizonte removebitur intervallo Lh: ad quod inveniendum sit ut ante anguli MPL cosinus $= t$, et sinus $= T$, eritque ipse angulus $LPl = d\theta$ $= \dfrac{dt}{\sqrt{(1-tt)}} = \dfrac{dt}{T}$, ex quo orietur anguli MPl cosinus $= t + dt = t + T\,d\theta$. Si iam ponatur sinus elevationis poli $= P$, sinus declinationis borealis puncti $L = Q$, nam si declinatio sit australis, sinus Q sumi debet negative, cosinus vero respondentes sint p et q, reperietur sinus altitudinis L supra horizontem $= v = tpq + PQ$: punctique l sinus altitudinis $v + dv = tpq + PQ + Tpq\,d\theta$. Quocirca si Luna ponatur in L, cum eius vis ad Mare attollendum sit $= \dfrac{L(3vv-1)}{2b^3}$, erit huius vis incrementum tempusculo $d\theta$ ortum $= \dfrac{3Lv\,dv}{b^3}$

$=\frac{3L(tpq+PQ)\,Tpq\,d\theta}{b^3}$. At si Sol ponatur in L, eius vis ad Mare elevandum tempusculo $d\theta$ capiet incrementum $=\frac{3S(tpq+PQ)\,Tpq\,d\theta}{a^3}$. Quamvis autem pro Sole et Luna eidem angulo $d\theta$ non aequalia tempora respondeant, tamen quia ea proxime ad rationem aequilitatis accedunt, sunt enim ut 24 ad $24\frac{3}{4}$ seu ut 32 ad 33, sine sensibili errore pro aequalibus haberi poterunt. Interim tamen si res accurate definiri debeat, et vis Solis incrementum angulo $d\theta$ acquisitum sit $=\frac{3S(tpq+PQ)\,Tpq\,d\theta}{a^3}$, erit vis Lunae incrementum eodem tempusculo acceptum $=\frac{32L(tpq+PQ)\,Tpq\,d\theta}{11b^3}$. Ex his intelligitur haec incrementa tribus casibus evanescere, quorum primus evenit sub polis, quia ibi est $p=0$; secundus, si punctum L in meridiano sit situm, tum enim fit $T=0$; tertius denique locum habet, si punctum L in horizonte existat, ubi est $tpq+PQ=0$.

63. Ponamus nunc Solem in L versari ac Lunam per meridianum iam transiisse, hocque momento maxime aquam esse elevatam; iam enim ostendimus dum Sol ab horizonte recedit, aquam summam incidere post transitum Lunae per meridianum. Hoc ergo momento necesse est, ut decrementum vis Lunae, quod tempusculo $d\theta$ patitur, aequale sit incremento vis Solis eodem tempore accepto. Sit igitur anguli horarii ad polum sumti quo Luna iam a meridiano recessit, cosinus $=n$, sinus $=N$, atque sit Lunae declinationis borealis sinus $=R$, cosinus $=r$, ex quibus orietur decrementum vis Lunae tempusculo $d\theta$ ortum $=\frac{3L(npr+PR)\,Npr\,d\theta}{b^3}$, quod cum aequale esse debeat incremento vis Solis eodem tempusculo nato $=\frac{3S(tpq+PQ)\,Tpq\,d\theta}{a^3}$, denotante Q sinum declinationis borealis Solis, et q eius cosinum, habebitur haec aequatio

$$\frac{L(npr+PR)\,Nr}{b^3}=\frac{S(tpq+PQ)\,Tq}{a^3},$$

neglecta fractione $\frac{32}{33}$, per quam incrementum vis Lunae multiplicari deberet. Quoniam autem Luna a meridiano non procul distabit, poni poterit $n=1$, atque cum sit proxime $\frac{L}{b^3}=\frac{4S}{a^3}$, obtinebitur iste valor $N=\frac{Tq(tpq+PQ)}{4r(pr+PR)}$; qui in tempus conversus dabit temporis spatium, quo aqua post transitum Lunae per meridianum maximam altitudinem attingit. Sub aequatore ergo erit $N = \frac{Ttqq}{4rr}$, ob $P = 0$ et $p = 1$; quare si declinationes Luminarium vel negligantur vel aequales assumantur, ita ut sit $qq = rr$, fiet $N = \frac{Tt}{4}$, cuius expressionis valor extat maximus si angulus MPL sit 45°, quo casu erit $N = \frac{1}{8}$, et angulus respondens $= 7°\,11'$, qui indicat aquam summam 30 minutis post transitum Lunae per meridianum contingere debere: totidemque minutis aqua ante transitum Lunae per meridianum maxime erit elevata, si Sol tum versus occasum versetur angulo MPL = semirecto. Quamobrem si Luna ad meridianum appellat hora nona sive matutina sive pomeridiana, Fluxus demum post semihoram eveniet; at si hora tertia appellat Luna ad meridianum, aqua summa 30′ ante observabitur: in aliis vero Terrae regionibus ista aberratio magis est irregularis; interim tamen satis prope ex formula data per solam aestimationem potest definiri.

64. Quod si autem hanc rem curatius investigare velimus, amborum Luminarium declinationes non pro arbitrio fingere licet, pendent enim a se mutuo maxime ob angulum horarium MPL inter ea interiectum datum: ut igitur pro data Lunae phasi aberrationem maximae aquae elevationis a transitu Lunae per meridianum determinemus, repraesentet nobis circulus $ZBNC$ [fig. 8] verticalem primarium, BC horizontem, ZN meridianum per dati loci Zenith Z et Nadir N ductum, atque aequator sit BAC, polus australis p, et ecliptica ♊♎♐. Constitutus nunc sit Sol in S et Luna in L, quae modo per meridianum transierit, quo tempore ponimus aquam maxime esse elevatam. Ponamus porro longitudinis Solis

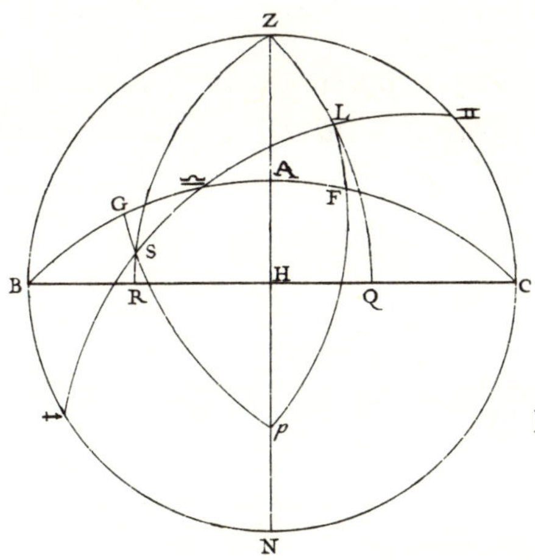

Fig. 8

ab aequinoctio verno computatae sinum esse $=F$, cosinum $=f$; Lunae vero longitudinis sinum esse $=G$, cosinum $=g$; sitque inclinationis eclipticae B♎♐ sinus $=M$, cosinus $=m$. Ex his definientur declinationes cum Solis tum Lunae, quarum sinus ante erant positi Q et R; erit scilicet $Q=FM$, $R=GM$; hincque $q=\surd(1-F^2M^2)$ et $r=\surd(1-G^2M^2)$. Deinde angulus SpL aequalis est angulo cuius tangens est $\frac{mF}{f}$ demto angulo cuius tangens est $\frac{mG}{g}$; huius vero eiusdem anguli ob angulos SpZ et LpZ datos, quorum sinus sunt positi T et N, tangens quoque est $\frac{nT+Nt}{nt-NT}$, quae tangens propter sinum N valde parvum proxime est $=\frac{T}{t}+\frac{N}{tt}$. Ponatur autem K pro sinu anguli qui excessus est anguli habentis tangentem $=\frac{mF}{f}$ super angulum cuius tangens est $\frac{mG}{g}$, et k pro cosinu [eiusdem]*, reperietur $T=K-Nk$ et $t=k+NK$ scripto 1 pro n: quibus valoribus substitutis prodibit

$$N=\frac{Kq(kpq+PQ)}{4r(pr+PR)+(2k^2-1)pq^2+kPQq},$$

ex aequatione $N=\frac{Tq(tpq+PQ)}{4r(pr+PR)}$, paragr. praeced.

65. Ponamus nunc Lunam in quadraturis versari ac primo quidem in primo post novilunium quadrante, ita ut arcus LS futurus sit 90°, erit $G=f$ et $g=-F$, unde $Q=MF$ et $R=Mf$, ex quibus prodibit[15]) $K=\sin\left(\mathrm{A\,tang}\frac{mF}{f}-\mathrm{A\,tang}\frac{-mf}{F}\right)$ atque k eiusdem anguli cosinui aequabitur. Quare his tempestatibus aqua maxime erit elevata post transitum Lunae per Meridianum, intervallo temporis quod in arcum aequatoris conversum dabit angulum cuius sinus erit

$$N=\frac{Kq(kpq+PQ)}{4r(pr+PR)+(2k^2-1)pq^2+kPQq}.$$

Pro posteriore vero quadratura post novilunium, erit $G=-f$ et $g=F$, unde erit

15) A tang denotat arcum tangentis. EJA.

$Q = MF$ et $R = -Mf$, ex quibus fit ut ante $K = \sin\left(\mathrm{A\,tang}\frac{mF}{f} - \mathrm{A\,tang}\frac{-mf}{F}\right)$ et $k =$ cosinui respondenti. Ne autem hic signa $+$ et $-$ calculum confundant, notari convenit K esse sinum arcus, qui restat, si ascensio recta Lunae subtrahatur ab ascensione recta Solis; atque k esse eiusdem arcus cosinum. Ponamus exempli causa Solem in initio Arietis versari, erit longitudo Solis $= 0°$, seu $360°$, et longitudo Lunae $=$ vel $90°$ vel $270°$, unde fiet $F = 0$, $f = 1$, $G = \mp 1$, et $g = 0$, atque $Q = 0$. Praeterea ascensio recta Solis est $360°$, et ascensio recta Lunae vel $90°$ vel $270°$; utroque casu ergo fit $k = 0$, unde etiam prodit $N = 0$; quod idem evenit, si Sol versetur in initio Librae. In utroque igitur aequinoctio, dum Luna in quadraturis versatur, aqua maxime erit elevata eo ipso momento, quo Luna ad meridianum appellit.

66. Sit porro Sol in solstitio aestivo, Luna vero in ultimo quadrante, erit longitudo Solis $90°$, Lunae vero $= 0°$, unde fit $F = 1$, $f = 0$; $G = 0$, $g = 1$, indeque $Q = M$ et $R = 0$; itemque $q = m$ et $r = 1$. Solis vero ascensio recta habebitur $90°$, Lunae vero $= 0°$, ex quo $K = 1$ et $k = 0$. Hinc ergo fit $N = \frac{mMP}{(4 - m^2)p}$. Pro prima autem quadratura est longitudo Lunae $180°$, unde $G = 0$, $g = -1$, at ut ante $F = 1$, $f = 0$; ergo $Q = M$, $R = 0$, itemque $q = m$ et $r = 1$. Cum igitur Lunae ascensio recta sit $180°$, erit $K = \sin -90° = -1$, et $k = 0$, ex quibus fit $N = \frac{-mMP}{(4 - m^2)p}$. Quoniam autem est $4 > m^2$, dum Sol in solstitio aestivo versatur maxima aquae elevatio in ultima quadratura continget post Lunae transitum per meridianum supra Terram, priore vero quadratura ante hunc transitum, haecque aequatio eo erit maior, quo maior fuerit elevatio poli; sub aequatore enim omnino evanescit. Sit poli elevatio $45°$, fietque his regionibus $N = \pm\frac{Mm}{4 - m^2}$; quare cum sit M sinus $23°\,29'$, prodibit $N =$ sinui anguli $6°\,33'$, qui in tempus conversus dat $26'$. In prima igitur quadratura totidem minutis ante transitum Lunae per meridianum aqua maxime erit elevata, in ultima vero quadratura tot minutis post transitum. Contrarium evenit si vel Luna sub Terra ad meridianum appellat, vel Sol in solstitio hyemali versetur. Ex his igitur formulis, si tabulae adhibeantur, non erit difficile pro quovis loco Terrae ad quodvis tempus definire, quantum maxima aquae elevatio transitum Lunae per meridianum vel praecedere vel sequi debeat; cuiusmodi supputationes maximam etiam afferent utilitatem, quando etiam inertiae aquae ratio habebitur.

67. Quoniam igitur satis est expositum, quo momento Mare maxime sit elevatum, maximam quoque Maris depressionem definire aggrediamur. Ac primo quidem manifestum est, si sola Luna Mare agitaret, tum minimam aquae altitudinem observatum iri, eo ipso momento, quo Luna in horizonte versetur: atque hinc perspicuum est, idem usu venire debere, si Sol eodem momento quoque in horizonte existat, id quod accidit cum noviluniis tum pleniluniis. Praeterea vero etiam ima aqua respondebit situi Lunae in horizonte, si eo tempore Sol meridianum occupet, quia tum vis Solis per notabile temporis intervallum neque augetur nec diminuitur, etiamsi tum aqua non tantum deprimatur, quam circa novilunia ac plenilunia. Ponamus igitur, quo reliquos casus evolvamus, dum Luna horizontem occupat, Solem ab horizonte removeri; hoc ergo casu aqua iam elevabitur, ex quo necesse est imam aquam ante adventum Lunae ad horizontem extitisse, contra vero si dum Luna in horizonte versatur, Sol ad horizontem appropinquet, aqua tardius scilicet post appulsum Lunae ad horizontem continget. Ponamus itaque Lunam ante ortum sub horizonte *Hh* [fig. 9] in ☾ adhuc versari, solemque in ⊙ esse

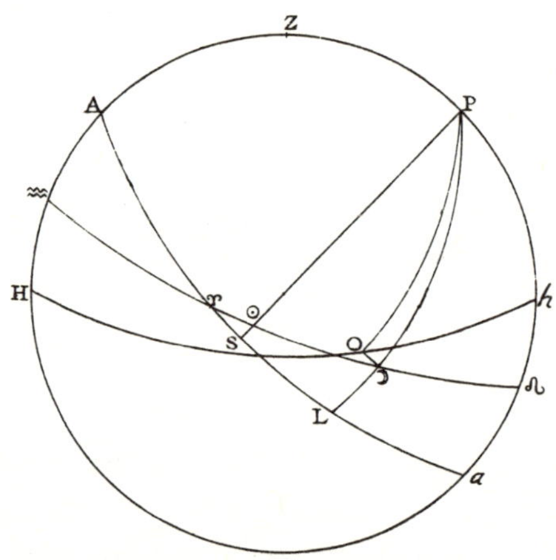

Fig. 9

positum, unde ad meridianum *PZH* progrediatur, hocque ipso momento aquam maxime esse depressam. Necesse igitur est, ut decrementum momentaneum vis Lunae ad Mare movendum aequale sit incremento momentaneo vis Solis. Ad hanc aequalitatem declarandam sit anguli ☾*PO* ad polum sumti, distantiam Lunae a suo ortu *O* indicantis, sinus $= V$ et cosinus $= v$, qui ob angulum ☾*PO* valde parvum tuto sinui toti 1 aequalis concipi potest. Invento ergo angulo hoc ☾*PO* seu arcu aequatoris illi respondente, eoque in tempus converso, constabit quanto temporis intervallo ima aqua appulsum Lunae ad horizontem praecedat: idem vero calculus tam ad Lunae occasum quam ad accessionem Solis ad horizontem facile accommodabitur.

68. Positis nunc $A\Upsilon a$ aequatore ac ♒♈♌ ecliptica, sit elevationis poli Ph sinus $= P$, cosinus $= p$; sinus declinationis Lunae borealis ☾$L = R$, cosinus $= r$; ex quibus fiet anguli APO cosinus $= \frac{-PR}{pr}$, quia Lunae, cum in horizontem O pervenit, altitudo evanescit. Cum igitur anguli APO sinus sit $= \frac{\sqrt{(p^2r^2 - P^2R^2)}}{pr}$ $= \frac{\sqrt{(1-P^2-R^2)}}{pr} = \frac{\sqrt{(pp-RR)}}{pr}$, erit anguli AP☾ sinus $= \frac{v\sqrt{(pp-RR)} - VPR}{pr}$, et cosinus $= \frac{-vPR - V\sqrt{(pp-RR)}}{pr}$, unde emergit decrementum momentaneum vis Lunae

$$= \frac{3LV\sqrt{(pp-RR)}\,(\sqrt{(pp-RR)} - VPR)\,d\theta}{b^3} = \frac{3LV(pp-RR)\,d\theta}{b^3},$$

ob $v = 1$ et V valde exiguum. Sit porro Solis declinationis borealis ☉S sinus $= Q$ et cosinus $= q$, atque anguli AP☉ sinus $= T$, cosinus $= t$, erit vis Solis incrementum momentaneum

$$= \frac{3S(tpq + PQ)\,Tpq\,d\theta}{a^3},$$

quod illi vis Lunae decremento aequale est ponendum, siquidem Maris altitudo hoc tempore est mimima. Quare cum sit fere $\frac{L}{b^3} = \frac{4S}{a^3}$, ista habebitur aequatio

$$4V(pp - RR) = Tpq\,(tpq + PQ),$$

quae praebet $V = \frac{Tpq\,(tpq + PQ)}{4(pp - RR)}$: cum igitur hoc pacto innotescat angulus OP☾, is in tempus conversus dabit temporis spatium, quo summa Maris depressio ante ortum Lunae contingit. At si punctum O designet Lunae occasum, idem angulus praebebit tempus post Lunae occasum, quo Mare maxime deprimetur. Intelligitur ex formula inventa quibus casibus ima aqua in ipsum appulsum Lunae ad horizontem incidat; hoc scilicet primo evenit, si $T = 0$, hoc est si Sol in meridiano versetur, deinde si $tpq + PQ = 0$, id est si Sol quoque horizontem occupet; quos binos casus iam notavimus.

69. Sit locus noster Terrae sub aequatore situs, seu elevatio poli nulla, erit $P = 0$, et $p = 1$, unde efficitur $V = \frac{Ttqq}{4(1 - RR)} = \frac{Ttqq}{4rr}$; in qua formula cum q et r denotent cosinus declinationum Solis ac Lunae, non multum inter se discrepabunt; ponamus enim alteram declinationem esse maximam, alteram vero minimam seu $= 0$, erit tamen cosinuum ratio minor quam $1 : \sqrt{\frac{3}{4}}$, ex quo fractio $\frac{qq}{rr}$ semper intra hos limites $\frac{4}{3}$ et $\frac{3}{4}$ continebitur. Quod si ergo hanc ab aequalitate aberrationem negligamus, id quod tuto facere possumus, quia rem tantum prope definire conamur, habebitur $V = \frac{Tt}{4} = \frac{2Tt}{8}$. Denotat autem $2Tt$ sinum dupli anguli horarii quo Sol a meridiano distat, et hanc ob rem ad momentum maximae depressionis aquae assignandum, videndum est qua diei hora Luna ad horizontem appellat, huiusque temporis vel a meridie vel media nocte intervallum capiatur, atque in arcum aequatoris convertatur. Huius deinde arcus vel anguli sumatur duplum, huiusque dupli sinus, cuius pars octava praebebit sinum anguli, qui in tempus conversus dabit temporis intervallum, quo ima aqua Lunae appulsum ad horizontem vel praecedit vel sequitur; id quod ex notatis circumstantiis discernere licet. Sic si Luna hora 9 matutina oriatur[16]), erit tempus usque ad meridiem 3 horarum, angulusque respondens 45°, cuius dupli sinus est ipse sinus totus, cuius pars octava fit sinus anguli 7° 11′, cui tempus respondet fere 30 minutorum, tantum itaque ima aqua ortum Lunae praecedet.

70. Ut haec ad datum Lunae cum Sole aspectum accommodari queant, ponamus longitudinis Solis ♈☉ sinum esse $= F$, cosinum $= f$; longitudinis vero Lunae ♈☾ sinum esse $= G$, cosinum $= g$; atque inclinationis eclipticae ☊♈a sinum $= M$, cosinum $= m$. His positis erit $Q = MF$, et $R = MG$; atque ascensionis rectae Solis ♈S tangens reperietur $= \frac{mF}{f}$, Lunae vero ascensionis rectae ♈L tangens $= \frac{mG}{g}$. Subtrahatur ascensio recta Solis ab ascensione recta Lunae, et differentiae sinus sit $= K$, cosinus $= k$. Cum igitur anguli ☉P☾ sit sinus $= K$ et cosinus $= k$, anguli vero AP☾ sinus $= \frac{\sqrt{(pp - RR)} - VPR}{pr}$

16) Editio princeps: adoriatur loco oriatur. EJA.

ob $v = 1$, et cosinus $= \dfrac{-PR - V\sqrt{(pp-RR)}}{pr}$, erit anguli $AP\odot$ sinus

$$= T = \frac{(k+KV)\sqrt{(pp-RR)} - kPRV + KPR}{pr}$$

et cosinus

$$= t = \frac{(K-kV)\sqrt{(pp-RR)} - KPRV - kPR}{pr};$$

quibus valoribus substitutis, simulque sinu V tanquam valde parvo considerato, reperietur sinus V

$$= \frac{(KPR + k\sqrt{(pp-RR)})\,q\,(Kq\sqrt{(pp-RR)} - kPRq + PQr)}{4rr(pp-RR)}.$$

Sub aequatore autem, quo fit $P = 0$, $V = \dfrac{Kkqq}{4rr}$: ex quo pro aequatore regula superior a distantia Solis a meridiano petita simul ad differentiam ascensionalem Solis et Lunae potest accommodari, ita ut maneat invariata. Sed ad praesens institutum, quo tantum veritatem causae Fluxus ac Refluxus Maris exhibitae declarare annitimur, non opus est haec pluribus persequi, quippe quae potissimum ad accuratissimas aestus marini Tabulas supputandas pertinent, quae res in proposita quaestione illustrissimae Academiae non contineri videtur.

CAPUT SEXTUM

De vero aestu Maris, quatenus a Terris non turbatur

71. Quae hactenus ex viribus Solis ac Lunae circa aestum Maris fusius deduximus, ea hypothesi nituntur assumta, qua aquam inertiae expertem posuimus: quamobrem non est mirandum si plerique effectus assignati cum Phaenomenis minus congruant, atque adeo pugnare videantur; quod si enim inter se prorsus convenirent, theoria non solum non eo consensu confirmaretur, sed potius omnino subverteretur, cum quilibet facile agnoscat ob aquae inertiam determinationibus exhibitis ingentem mutationem inferri debere. Quae autem ex deductis conclusionibus maxime ab experientia dissentiunt, potissimum quantitatem elevationis aquae ac temporis momentum, quo tam summa Maris elevatio quam ima depressio

contingere solet, respiciunt. Nusquam enim ubi quidem Mare est liberum atque apertum, tam exiguum discrimen inter Fluxum ac Refluxum in aquae altitudine observatur, quale in praecedentibus definivimus, quatuor scilicet pedum tantum; quae elevatio insuper tamen maxima est deprehensa, ac tum solum oriunda, quando tam regio prope aequatorem est sita, quam vires luminarium inter se maxime conspirant. Experientia namque constat, plerisque in locis, si aestus contingat maximus, aquam non solum ad altitudinem duplo maiorem, sed etiam quadruplam, imo nonnullis in locis adeo decuplam attolli; quanquam haec enormis elevatio non soli inertiae aquae, sed maximam partem vicino continenti ac littorum situi est tribuenda, uti in sequenti capite clarissime monstrabitur. Deinde etiam quod ad tempus attinet, nusquam illis ipsis momentis, quae assignavimus, Fluxus ac Refluxus unquam contingunt, nec etiam tempestatibus hic definitis Fluxus maximi vel minimi, sed ubique tardius evenire constanter observantur; cuius quidem retardationis causa in ipsa aquae inertia posita esse prima etiam fronte perspicitur.

72. Quantumvis autem agitatio Maris in praecedentibus capitibus determinata ab observationibus dissentiat, tamen complures circumstantiae sese iam praebuerunt experientiae tantopere consentaneae, ut amplius dubitare omnino nequeamus, quin in viribus Solem Lunamque respicientibus, quas non temere assumsimus, sed aliunde existere demonstravimus, vera et genuina aestus Maris causa contineatur. Hanc ob rem iam merito suspicari licet, dissensiones quae inter theoriam nostram, quatenus eam assumtae hypothesi superstruximus, et experientiam intercedunt, ab aquae inertia aliisque circumstantiis, quarum nullam adhuc rationem habuimus, proficisci. Quocirca si omnia inertiae ratione habita ad observationes propius accedant, id quidem nostrae theoriae maximum afferet firmamentum, atque simul omnes alias causas, quae praeter has vel sunt prolatae vel proferri possunt, excludet, irritasque reddet. Cum igitur consensum huius theoriae cum Phaenomenis, mox simus evidentissime ostensuri, quaestioni ab Inclyta Academia propositae ex asse satisfecisse iure nobis videbimur: cum non solum nullas vires imaginarias effinxerimus, sed etiam virium Lunam Solemque respicientium existentiam aliunde dilucide evicerimus. Neque vero in hoc negotio cum plerisque Anglorum ad qualitates occultas sumus delapsi, verum potius causam istarum virium modo rationali et legibus motus consentaneo in vorticibus constituimus, quorum formam atque indolem luculenter explicare possemus; idque fecissemus, nisi ab aliis cum iam satis esset expositum, tum etiam ab Illustrissima Academia in praesente quaestione non requiri videatur.

73. Dum igitur hactenus aquae omnem inertiam cogitatione ademimus, ipsi eiusmodi qualitatem affinximus, qua viribus sollicitantibus subito obsequeretur, seque in instanti in eum statum reciperet, in quo cum viribus in aequilibrio consisteret; hocque pacto aquam non solum subito omnis motus capacem posuimus, sed etiam ita comparatam, ut quovis momento omnem pristinum motum amittat. Longe aliter autem res se habet, si inertiae ratio in computum ducatur; haec enim efficit ut primo aqua non subito se ad eum situm componat, quem vires intendunt, sed pedetentim per omnes gradus medios ad eum accedat; deinde vero eadem inertia in causa est, quod aqua, cum in statum aequilibrii pervenerit, ibi non acquiescat, sed ob motum insitum ultra progrediatur, quoad omnem motum a potentiis renitentibus amittat. Ex quo perspicuum est, admissa inertia aquae, a potentiis sollicitantibus motum omnino diversum actu imprimi debere ab eo, quem reciperet, si inertia privata esset; cuius discriminis ratio exemplo corporis penduli commode ob oculos poni potest. Ponamus enim corpus pendulum OC [fig. 10] ob

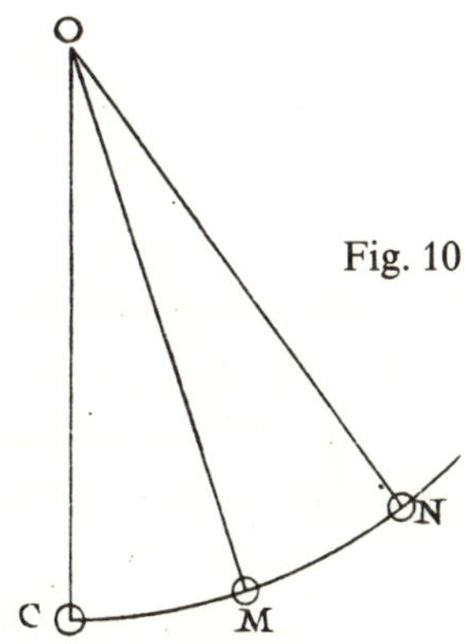

Fig. 10

gravitatem situm tenens verticalem, a vi quapiam in latus secundum directionem CM sollicitari. Si nunc hoc pendulum inertia careret, seu eiusmodi esset indolis, cuius aquam hactenus sumus contemplati, tum subito situm OM acciperet, in quo haec vis cum gravitate aequilibrium teneret. At cum pendulum inertia praeditum consideratur, post aliquod demum tempus elapsum ad situm OM perveniet: ac deinde quia motu accelerato eo pertingit, ibi non quiescet, sed ultra excurret, puta in N usque, ita ut spatium CN fere sit duplo maius spatio CM, prouti calculus clare indicat. Propter inertiam igitur pendulum primum tardius vi sollicitanti obtemperat, atque a situ aequilibrii recedit; deinde vero etiam magis recedit, maioremque excursionem conficit, quam si inertia careret; quae sunt eae ipsae duae res, in quibus theoria ante exposita ab experientia maxime dissentire deprehensa est.

74. Si nunc istud penduli exemplum ad nostrum casum aestus Maris transferamus, primo ingens similitudo in situ penduli verticali ac statu Maris naturali, quem obtinet remotis potentiis externis, observatur. Nam quemadmodum pendulum, si in quamcunque plagam de situ verticali declinetur, propria vi gravitatis se in eundem recipit, ita etiam aqua, si ex situ suo aequilibrii depellatur, vi gravitatis se ad eundem componit, ac praeterea pariter ac pendulum oscillationes peragit, cuiusmodi oscillationum casus in aqua observati passim inveniuntur expositi. Deinde etiam simili modo, quo pendulum, Mare quo magis ex situ suo naturali fuerit deturbatum, eo maiorem habebit vim sese in situm aequilibrii restituendi. Quod si igitur Mare a viribus externis, Solis scilicet ac Lunae, mox elevetur mox deprimatur, necesse est ut inde motus oscillatorius seu reciprocus oriatur aestui Maris omnino similis, qui autem per leges motus difficulter definiri queat acurate quidem; nam vero proxime, hoc non adeo erit difficile. Duae autem sunt res, quae absolutam ac perfectam totius motus determinationem summopere reddunt difficilem, quarum altera physicam spectat, atque in ipsa fluidorum natura consistit, quorum motus difficulter ad calculum revocatur, praecipue si quaestio sit de amplissimo Oceano, qui aliis in locis elevetur, aliis vero deprimatur. Altera autem difficultas in ipsa analysi est posita, eo quod iste motus Maris reciprocus prorsus sit diversus ab omnibus oscillationibus a Mathematicis adhuc consideratis: vires enim Lunae ac Solis Mare sollicitantes neque a situ corporis oscillantis, neque ab eius celeritate pendent, uti id usu venit in omnibus oscillationum casibus etiam nunc expositis, sed eae vires a situ luminarium respectu Terrae, ideoque a tempore determinantur, cuiusmodi oscillationes nemo adhuc, quantum quidem constat, calculo subiecit.

75. Quod quidem ad priorem difficultatem physicam attinet, res hoc quidem tempore fere desperata videtur; quamquam enim ab alique tempore theoria motus aquarum ingentia sit assecuta incrementa, tamen ea potissimum motum aquarum in vasis et tubis fluentium respiciunt, neque vix ullum commodum inde ad motum Oceani definiendum derivari potest. Quamobrem in hoc negotio aliud quicquam praestare non licet, nisi ut hypothesibus effingendis, quae a veritate quam minime abludant, tota quaestio ad considerationes pure geometricas et analyticas revocetur: alteram autem difficultatem mathematicam, etiamsi difficillimis integrationibus sit involuta, tamen feliciter superare confidimus. Considero scilicet superficiem aquae RS [fig. 11], quae hoc in situ aequilibrium teneat cum reliqua aqua, remotis viribus externis; his vero accedentibus alternis vicibus attollatur in A, deprimaturque in B. Quod si igitur aqua in M usque sit depressa, atque externae vires

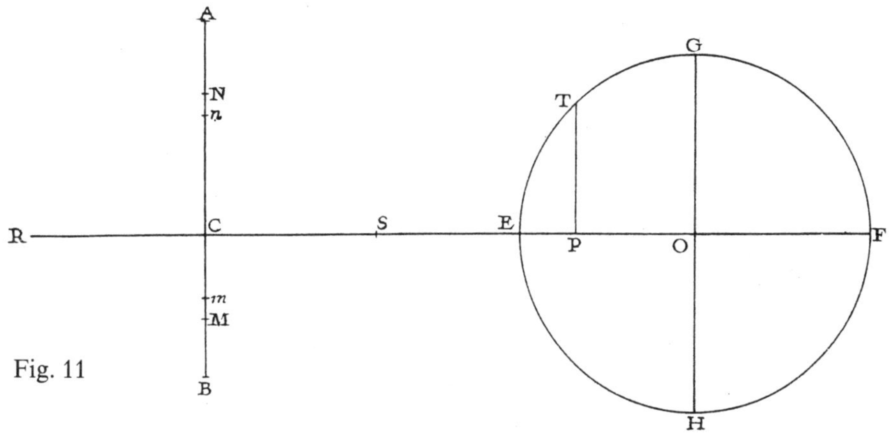

Fig. 11

Solis ac Lunae subito cessarent, tum vi gravitatis propriae conaretur sese elevare usque in situm RS naturalem, isteque conatus eo erit maior, quo maius fuerit spatium CM quo a situ naturali distat. A veritate itaque non multum recedemus, si hanc vim ipsi spatio MC ponamus proportionalem: quamobrem posito spatio $MC = s$, erit vis, quae aquae superficiem in M usque depressam attollet $= \frac{s}{g}$, quae hypothesis ad veritatem eo propius accedit, quod sponte indicat, si aquae superficies supra C iam sit elevata, tum vim fieri negativam, adeoque aquam deprimere. Praeterea vero eadem hypothesis confirmatur pluribus phaenomenis aquae nisum respicientibus, ita ut de eius veritate amplius nullum dubium supersit.

76. Ponamus iam aquam in M constitutam urgeri a sola Luna, atque ut calculus per se molestus minus habeat difficultatis, sit locus C sub ipso aequatore situs, Lunaeque declinatio nulla, ex quo Luna in circulo maximo per loci zenith transeunte aequatore scilicet circumferetur: sit $EGFH$ iste circulus, cuius radius ponatur $= 1$, atque EF repraesentet horizontem, et G zenith. Positis his, sit Luna in T dum Maris superficies versatur in M, ita ut $PT = y$ exprimat sinum altitudinis Lunae super horizonte; unde vis Lunae Mare attollens erit $= \frac{L(3yy-1)}{2b^3} = \frac{3yy-1}{h}$, posito brevitatis gratia h pro $\frac{2b^3}{L}$. Hanc ob rem ergo superficies Maris in M duplici vi attolletur scilicet vi $= \frac{s}{g} + \frac{3yy-1}{h}$. Quod si ergo ponamus aquam in M iam habere motum sursum directum, cuius celeritas tanta sit quanta acquiritur lapsu gravis ex altitudine v, atque spatium $Mm = -ds$ tempusculo infinite parvo absolvatur, habebitur per principia motus

$$dv = -ds\left(\frac{s}{g} + \frac{3yy-1}{h}\right).$$

Ponamus porro tempus ab ortu lunae in E iam elapsum, quod arcui ET est proportionale, esse $=z$, quae littera ipsum arcum ET simul denotet, erit $y=\sin z$ scilicet sinui arcus z, hoc enim modo sinus ac cosinus arcuum sumus indicaturi: unde orietur $1-2yy=\cos 2z$, atque $3yy-1=\frac{1}{2}-\frac{3}{2}\cos 2z$, hincque

$$dv = -ds\left(\frac{s}{g} + \frac{1}{2h} - \frac{3}{2h}\cos 2z\right).$$

77. Cum igitur elementum temporis sit $=dz$, erit ex natura motus $dz = -\frac{ds}{\sqrt{v}}$, atque $v=\frac{ds^2}{dz^2}$; unde sumto elemento dz pro constante, fiet

$$dv = \frac{2ds\,dds}{dz^2} = -ds\left(\frac{s}{g} + \frac{1}{2h} - \frac{3}{2h}\cos 2z\right),$$

atque

$$2dds + \frac{s\,dz^2}{g} + \frac{dz^2(1-3\cos 2z)}{2h} = 0,$$

quae aequatio duas tantum continet variabiles s et z, et propterea si debito modo integretur, indicabit situm seu statum aquae ad quodvis tempus. Quoniam autem haec aequatio est differentialis secundi gradus, atque insuper arcus et sinus arcuum continet, facile intelligitur eius integrationem minus esse obviam; interim tamen cum alterius variabilis s plus una dimensione nusquam adsit, ea per methodos mihi familiares tractari poterit. Soleo autem, quoties eiusmodi occurrunt, [aequationes]* initio eos terminos in quibus altera variabilis s omnino non inest, reiicere; unde haec consideranda venit aequatio

$$2dds + \frac{s\,dz^2}{g} = 0,$$

quae per ds multiplicata fit integrabilis, existente integrali

$$ds^2 + \frac{ss\,dz^2}{2g} = c\,dz^2$$

ob dz constans. Hinc porro elicitur[17]) $dz = \frac{ds\sqrt{2g}}{\sqrt{(2cg - ss)}}$, atque $\frac{z}{\sqrt{2g}}$ = arcui cuius sinus est $\frac{s}{\sqrt{2cg}}$, ex quo obtinetur $s = \sqrt{2cg} \cdot \sin \frac{z}{\sqrt{2g}}$. Cognito autem hoc valore, idonea nascitur substitutio facienda pro aequatione proposita

$$2dds + \frac{s\,dz^2}{g} + \frac{dz^2(1 - 3\cos 2z)}{2h} = 0;$$

fiat enim $s = u \sin \frac{z}{\sqrt{2g}}$, erit $ds = du \sin \frac{z}{\sqrt{2g}} + \frac{u\,dz}{\sqrt{2g}} \cos \frac{z}{\sqrt{2g}}$, atque

$$dds = ddu \sin \frac{z}{\sqrt{2g}} + \frac{2du\,dz}{\sqrt{2g}} \cos \frac{z}{\sqrt{2g}} - \frac{u\,dz^2}{2g} \sin \frac{z}{\sqrt{2g}}.$$

Quibus valoribus substitutis emerget ista aequatio

$$2ddu \sin \frac{z}{\sqrt{2g}} + \frac{4du\,dz}{\sqrt{2g}} \cos \frac{z}{\sqrt{2g}} + \frac{dz^2(1 - 3\cos 2z)}{2h} = 0,$$

in qua hoc commode accidit, ut ipsa variabilis u non insit, sed tantum eius differentialia.

78. Quod si ergo ponatur $du = p\,dz$, erit $ddu = dp\,dz$, et aequatio nostra transibit in sequentem differentialem primi gradus tantum

$$2dp \sin \frac{z}{\sqrt{2g}} + \frac{4p\,dz}{\sqrt{2g}} \cos \frac{z}{\sqrt{2g}} + \frac{dz(1 - 3\cos 2z)}{2h} = 0$$

quae integrabilis reddi invenitur, si multiplicetur per quantitatem quampiam ex z et constantibus compositam, eo quod p plures una dimensiones habet nusquam. Ad integrationem autem absolvendam notandum est huius aequationis $dp + pZ\,dz = \Sigma\,dz$, in qua Z et Σ functiones quascunque ipsius z denotent, integrale esse $e^{\int Z dz} p = \int e^{\int Z dz} \Sigma\,dz$. Reducta autem nostra aequatione ad hanc formam, habetur

17) $\sqrt{2g}$ denotat $\sqrt{(2g)}$ passim. EJA.

$$dp + \frac{2p\,dz\cos\dfrac{z}{\sqrt{2g}}}{\sqrt{2g}\sin\dfrac{z}{\sqrt{2g}}} = \frac{dz(3\cos 2z - 1)}{4h\sin\dfrac{z}{\sqrt{2g}}},$$

ideoque

$$Z\,dz = \frac{2dz\cos\dfrac{z}{\sqrt{2g}}}{\sqrt{2g}\sin\dfrac{z}{\sqrt{2g}}} = \frac{2\,\text{diff.}\sin\dfrac{z}{\sqrt{2g}}}{\sin\dfrac{z}{\sqrt{2g}}};$$

atque hinc $\int Z\,dz = 2\log\sin\dfrac{z}{\sqrt{2g}}$; et $e^{\int Z\,dz} = \left(\sin\dfrac{z}{\sqrt{2g}}\right)^2$. Ex his sequitur integrale nostrae aequationis

$$p\left(\sin\frac{z}{\sqrt{2g}}\right)^2 = \frac{1}{4h}\int dz\sin\frac{z}{\sqrt{2g}}(3\cos 2z - 1)$$
$$= \frac{3}{4h}\int dz\sin\frac{z}{\sqrt{2g}}\cos 2z - \frac{1}{4h}\int dz\sin\frac{z}{\sqrt{2g}},$$

ad quas integrationes perficiendas notetur esse $\int dz\sin\alpha z = C - \dfrac{1}{\alpha}\cos\alpha z$, atque[18])

$$\int dz\sin\alpha z\cos\beta z = C - \frac{\beta\sin\alpha z\sin\beta z + \alpha\cos\alpha z\cos\beta z}{\alpha^2 - \beta^2}:$$

ex his itaque conficietur

$$p\left(\sin\frac{z}{\sqrt{2g}}\right)^2 = C + \frac{\sqrt{2g}}{4h}\cos\frac{z}{\sqrt{2g}} - \frac{3\left(2\sin\dfrac{z}{\sqrt{2g}}\sin 2z + \dfrac{1}{\sqrt{2g}}\cos\dfrac{z}{\sqrt{2g}}\cos 2z\right)}{\left(\dfrac{1}{2g} - 4\right)4h}$$

atque

18) Editio princeps et Manuscriptus: Expressiones mathematicae usque ad finem paragraphes errata parva continent. Correxit EJA.

$$p = \frac{C}{\left(\sin\frac{z}{\sqrt{2g}}\right)^2} + \frac{\sqrt{2g}\cos\frac{z}{\sqrt{2g}}}{4h\left(\sin\frac{z}{\sqrt{2g}}\right)^2} - \frac{3\left(4g\sin\frac{z}{\sqrt{2g}}\sin 2z + \sqrt{2g}\cos\frac{z}{\sqrt{2g}}\cos 2z\right)}{4h(1-8g)\left(\sin\frac{z}{\sqrt{2g}}\right)^2}.$$

79. Cum autem posuissemus $du = p\,dz$, erit

$$u = \int p\,dz = \int\frac{C\,dz}{\left(\sin\frac{z}{\sqrt{2g}}\right)^2} + \int\frac{dz\sqrt{2g}\cos\frac{z}{\sqrt{2g}}}{4h\left(\sin\frac{z}{\sqrt{2g}}\right)^2}$$
$$-\frac{3}{4h}\int\frac{dz\left(4g\sin\frac{z}{\sqrt{2g}}\sin 2z + \sqrt{2g}\cos\frac{z}{\sqrt{2g}}\cos 2z\right)}{(1-8g)\left(\sin\frac{z}{\sqrt{2g}}\right)^2}.$$

Hae autem formulae omnes sunt absolute integrabiles, prodibitque

$$u = D - \frac{C\sqrt{2g}\cos\frac{z}{\sqrt{2g}}}{\sin\frac{z}{\sqrt{2g}}} - \frac{g}{2h\sin\frac{z}{\sqrt{2g}}} + \frac{3g\cos 2z}{2h(1-8g)\sin\frac{z}{\sqrt{2g}}};$$

ex quo tandem resultat

$$s = u\sin\frac{z}{\sqrt{2g}} = D\sin\frac{z}{\sqrt{2g}} + C\cos\frac{z}{\sqrt{2g}} - \frac{g}{2h} + \frac{3g\cos 2z}{2h(1-8g)},$$

quae est aequatio generalis ad quodvis tempus z statum aquae, seu distantiam eius supremae superficiei a C indicans, ubi constantes C et D ex dato Maris statu ad datum tempus definiri oportet. Quod si igitur ponamus motum aquae iam ad uniformitatem esse deductum, ita ut aqua omnibus diebus, quando Luna in T versatur, in eodem loco M versetur, necesse erit ut valor ipsius s maneat idem, etsi arcus z integra peripheria 2π vel eius multiplo augeatur. At posito $z + 2\pi$ loco z, terminus $\cos 2z$ manet quidem invariatus, at $D\sin\frac{z}{\sqrt{2g}} + C\cos\frac{z}{\sqrt{2g}}$ fit

$= D \sin \frac{z+2\pi}{\sqrt{2g}} + C \cos \frac{z+2\pi}{\sqrt{2g}}$, quae aequalitas adesse non potest nisi vel $\frac{1}{\sqrt{2g}}$ sit numerus integer, vel C et $D = 0$. Cum itaque g determinari non liceat, quia iam est datum, ponendum erit $C = 0$ et $D = 0$, ita ut ista habeatur aequatio $s = \frac{-g}{2h} + \frac{3g \cos 2z}{2h(1-8g)}$, ex qua facillime ad quodvis tempus status Maris cognoscetur: valores scilicet affirmativi ipsius s dabunt situm aquae infra situm naturalem C, negativi vero supra C.

80. Cognito autem spatio s per tempus z, celeritas quoque Maris qua in M ascendit reperietur ex aequatione $dz = \frac{-ds}{\sqrt{v}}$, erit enim $\sqrt{v} = \frac{-ds}{dz} = \frac{3g \sin 2z}{h(1-8g)}$, quae expressio ipsi celeritati, qua aquae superficies, dum in M versatur, elevatur, est proportionalis: haec ergo celeritas aquae semper est ut sinus dupli arcus ET, vel etiam ut sinus dupli temporis, quo Luna a transitu per meridianum abest, tempore scilicet in arcum aequatoris converso. Hinc igitur celeritas aquae erit nulla si Luna fuerit vel in E vel in G vel in F vel in H, hoc est, vel in horizonte vel in meridiano: quare cum his temporibus aqua vel maxime sit elevata vel maxime depressa, una Lunae revolutione aqua bis elevabitur, bisque deprimetur, ideoque bini Fluxus binique Refluxus contingent. Aqua quidem maxime erit depressa iis ipsis momentis, quibus Luna ad horizontem appellit, tum enim fit $\cos 2z = 1$; atque spatium CB erit[19]) $= s = \frac{g(1+4g)}{h(1-8g)}$; at maxima elevatio incidet in ipsos Lunae transitus per meridianum, quibus est $\cos 2z = -1$: ac tum altitudo CA erit $= -s = \frac{g(2-4g)}{h(1-8g)}$. Quanquam autem haec momenta cum experientia non satis conveniunt, tamen ea hypothesi assumtae plane congruunt, qua posuimus Lunam solam agere, ac perpetuo in ipso aequatore versari, ex quo aestus se tandem ad summam regularitatem componat necesse est. Quod si enim Lunae declinatio ponatur variabilis, atque Sol insuper agat, aestus iam formati perpetuo turbabuntur, ex quo ob aequabilitatem continuo sublatam effectus tardiores necessario consequi debebunt. Praeterea quoque nullam adhuc motus Maris horizontalis habuimus rationem, cum enim aqua ad aestum formandum motu horizontali progredi debeat, persipicuum est [et]* hinc retardationem in aestu oriri oportere.

19) Editio princeps: 2 loco h. EJA.

81. Si aqua, uti in praecedentibus capitibus posuimus, inertia careret, tum foret ex aequatione prima $dv = -ds\left(\frac{s}{g} + \frac{3yy-1}{h}\right)$ perpetuo $s = \frac{g(1-3yy)}{h}$, quia aqua tum quovis momento cum viribus sollicitantibus in aequilibrio consisteret. Maxima igitur depressio etiam tum Lunae horizontali responderet, cum est $y = 0$, foretque spatium depressionis $CM = \frac{g}{h}$; maxima vero elevatio, quae circa Lunae appulsum ad meridianum continget, fiet per spatium $CN = \frac{2g}{h}$ ob $y = 1$. Quare si aqua inertia careret, foret spatium MN, per quod aqua motu reciproco agitaretur, $= \frac{3g}{h}$; inertia autem admissa agitationes perficientur in spatio maiore $AB = \frac{3g}{h(1-8g)}$, cuius excessus super spatium MN erit $= \frac{24gg}{h(1-8g)}$. Quantitas itaque aestus pendet a valore litterae g, qui quidem semper est affirmativus; nam si foret $g = 0$, quod evenit si gravitatis vis esset infinite magna respectu virium Lunae et Solis, tum etiam nullus aestus oriretur; deinde quo magis $8g$ ad 1 accedit, eo maior prodibit aestus, qui adeo in infinitum excrescere posset si foret $8g = 1$, hoc quippe casu vis Lunae gravitatem superaret, omnesque aquas ad Lunam attraheret; quod autem fieri non potest, multo minus vero[20]) esse potest $8g > 1$, quod tamen si [quidem]* eveniret, maxima elevatio appulsui Lunae ad horizontem, maximaque depressio Lunae meridianum occupanti responderet.

82. Cum igitur aqua, si inertia careret, agitetur per spatium $MN = \frac{3g}{h}$, supra autem §41 eadem hac hypothesi, qua tam locus quam Luna in aequatore ponitur, [reverta sit]* aqua[21]) elevari supra libellam per spatium 2,260 pedum, infra eam vero deprimi spatio 1,112 pedum, erit $\frac{3g}{h} = 3{,}372$ pedum, ideoque $\frac{g}{h} = 1{,}124$ pedum $= 1\frac{1}{8}$ pedum. Quod[22]) vero valor ipsius g cum unitate comparatur, id inde[23]) venit, quod tempus per ipsum arcum circuli cuius radius est $= 1$ expressimus: hinc itaque valor ipsius g respectu unitatis definietur tempore eodem

20) Editio princeps: autem loco vero. EJA.
21) Editio princeps: aquam loco aqua. EJA.
22) Editio princeps: Quoniam loco Quod. EJA.
23) Editio princeps: ideo loco id inde. EJA.

modo expresso, quo aqua in M usque depressa sola vi gravitatis se in C restitueret, quod tempus ex circumstantiis facile poterit aestimari: prodibit autem per calculum tempus huius restitutionis $=\frac{\pi}{2}\sqrt{2g}$, denotante π semiperipheriam circuli radium $=1$ habentis, seu tempus duodecim horarum Lunarium. Quod si igitur restitutio ponatur actu fieri tempore $\frac{12}{n}$ horarum, erit $\frac{\pi}{n}=\frac{\pi\sqrt{2g}}{2}$, et $g=\frac{2}{nn}$, ex quo perspicuum est, quo citius aqua se propria sua vi restituere valeat, eo minus excessurum esse spatium AB spatium MN. Cum autem de hac restitutione non satis tuto iudicare queamus, praestabit ex observationibus rationem spatii AB ad MN proxime assumere. Si enim ponamus esse $AB=2MN$, erit $\frac{3}{1-8g}=6$, erit[24]) $g=\frac{1}{16}$; sin autem sit $AB=3MN$, fiet $\frac{3}{1-8g}=9$ et $g=\frac{1}{12}$: at posito $AB=4MN$, erit $g=\frac{3}{32}$. Quoniam igitur aqua ob inertiam fere duplo maius spatium absolvere poni potest, assumamus $g=\frac{2}{36}$ seu $n=6$, ita ut aqua propria vi gravitatis tempore circiter 2 horarum in statum naturalem se restituere valeat. Posito autem $g=\frac{1}{18}$, fiet $\frac{3}{1-8g}=5{,}4$; spatiumque $AB=6$ ped. proxime. Ne autem tractatio nimis fiat specialis, retineamus litteram n, cuius valorem esse circiter 6 vel 5 notasse sufficiet, qui valor satis prope ad aestimationem accedit: ita ut sit $g=\frac{2}{nn}$ et $AB=\frac{3nn}{nn-16}\cdot\frac{9}{8}$ pedum: unde satis patet n necessario esse debere >4, eritque adeo vel 5 vel 6.

83. Tentemus nunc idem hoc problema in sensu latiori, ac ponamus regionis C elevationis poli sinum esse $=P$, cosinum $=p$; Lunae vero declinationis borealis sinum esse $=Q$, cosinum $=q$; Lunamque super Terra iam per meridianum transiisse, ab eoque distare angulo horario $=z$, ita ut z ut ante tam tempus quam arcum circuli radii $=1$ designet; quod si nunc arcus z cosinus ponatur $=t$, erit sinus altitudinis Lunae super horizonte $=tpq+PQ$; ideoque vis Lunae Mare elevans

$$=\frac{L}{2b^3}\left(3(tpq+PQ)^2-1\right)=\frac{3p^2q^2tt+6pqPQt+3P^2Q^2-1}{h},$$

24) Editio princeps usque ad finem paragraphes errata continet. EJA.

posito ut ante $\frac{L}{2b^3} = \frac{1}{h}$. Quoniam vero est $t = \cos z$ erit $2tt - 1 = \cos 2z$ et $tt = \frac{1 + \cos 2z}{2}$, ex quo vis Lunae ad Mare elevandum habebitur

$$= \frac{3p^2q^2 \cos 2z}{2h} + \frac{6pqPQ \cos z}{h} + \frac{3p^2q^2 + 6P^2Q^2 - 2}{2h}.$$

Ponamus nunc superficiem aquae in M versari, existente $CM = s$, et celeritatem eius qua actu ascendit debitam esse altitudini v, erit $dv = -ds\left(\frac{s}{g} + \text{vi Lunae}\right)$, cum vero sit $dz = \frac{-ds}{\sqrt{v}}$ seu $\sqrt{v} = \frac{-ds}{dz}$ = ipsi celeritati ascensus; erit $v = \frac{ds^2}{dz^2}$ atque[25]) $dv = \frac{2ds\,dds}{dz^2}$, posito dz constante: hinc igitur emerget ista aequatio[26])

$$2dds + dz^2\left(\frac{s}{g} + \frac{3p^2q^2 + 6P^2Q^2 - 2}{2h} + \frac{6pqPQ \cos z}{h} + \frac{3p^2q^2 \cos 2z}{2h}\right) = 0,$$

relationem inter tempus z et statum Maris s continens.

84. Quod si nunc haec aequatio eodem modo tractetur, quo superior, ea pariter bis integrari posse deprehendetur, integrationibus autem singulis debito modo absolutis, et constantibus ita determinatis ut motus aquae fiat uniformis, reperietur

$$s = \frac{-g(3p^2q^2 + 6P^2Q^2 - 2)}{2h} - \frac{6gpqPQ \cos z}{h(1-2g)} - \frac{3gp^2q^2 \cos 2z}{2h(1-8g)}$$

ac celeritas ascensus

$$\sqrt{v} = \frac{-ds}{dz} = \frac{-6gpqPQ \sin z}{h(1-2g)} - \frac{3gp^2q^2 \sin 2z}{h(1-8g)}.$$

Cum autem sit $\sin 2z = 2 \sin z \cos z$, celeritas duobus casibus evanescit, quorum primus est si $\sin z = 0$, alter si $\cos z = \frac{-PQ(1-8g)}{pq(1-2g)}$; illi casus dabunt aquam

25) Editio princeps: dz loco dz^2. EJA.
26) Editio princeps: $PQ \cos 2z$ loco $PQ \cos z$. EJA.

summam, hi vero imam. Hinc igitur patet aquam summam contingere debere iis ipsis momentis, quibus Luna per meridianum transit, imam vero non tum, cum Luna horizontem attingit; namque Luna horizontem attingit, si est $\cos z = \frac{-PQ}{pq}$, aqua vero est ima si est $\cos z = \frac{-PQ\,(1-8g)}{pq(1-2g)} = \frac{-5PQ}{8pq}$ posito $g = \frac{1}{18}$. Hic autem idem est notandum quod supra, scilicet nos posuisse motum aquae esse uniformem seu quotidie sui similem, Lunamque in ecliptica locum tenere fixum, seu saltem suam declinationem non variare. Quoniam vero ob variabilitatem declinationis Lunae, itemque ob actionem Solis, iste motus perpetuo turbatur, atque insuper motus Maris horizontalis nulla adhuc habita est ratio, facile intelligitur, tam Fluxus quam Refluxus tardius venire debere, quam quidem ex his formulis sequitur.

85. Bini ergo una Lunae revolutione contingent Fluxus, alter si Luna super horizonte ad meridianum appellit, alter si sub Terra; priori casu est $\cos z = 1$, et $\cos 2z = 1$, hoc itaque tempore Mare supra libellam C elevabitur per spatium

$$\frac{g(3p^2q^2 + 6P^2Q^2 - 2)}{2h} + \frac{3gp^2q^2}{2h(1-8g)} + \frac{6gpqPQ}{h(1-2g)}.$$

Dum autem Luna sub horizonte meridianum attingit, tum aqua elevabitur per spatium[27])

$$= \frac{g(3p^2q^2 + 6P^2Q^2 - 2)}{2h} + \frac{3gp^2q^2}{2h(1-8g)} + \frac{6gpqPQ}{h(1-2g)},$$

propter $\cos z = -1$ at $\cos 2z = 1$ hoc casu: harum igitur altitudinum differentia est $= \frac{12gpqPQ}{h(1-2g)}$: atque Mare in transitu Lunae per meridianum supra horizontem altius elevatur, si declinatio Lunae sit borealis; contra vero si declinatio fuerit australis, maior Maris elevatio respondebit appulsui Lunae ad meridianum infra horizontem. Luna vero in ipso aequatore versante, ambo Fluxus inter se erunt aequales. Ratione autem elevationis poli, horum binorum Fluxuum successivorum inaequalitas erit maxima sub elevatione poli 45°, pro his enim regionibus fit

27) Editio princeps et Manuscriptus: $3p^2q^2$ loco $3gp^2q^2$. Correxit EJA.

pP maximum; atque in aliis regionibus eo minor erit inaequalitas, quo magis fuerint a latitudine 45° remotae. Mare autem maxime deprimetur, si fuerit $\cos z = \frac{-PQ\,(1-8g)}{pq(1-2g)}$; quo valore substituto, reperietur aqua infra libellam C subsidere per spatium

$$= \frac{3gp^2q^2}{2h(1-8g)} - \frac{g(3p^2q^2+6P^2Q^2-2)}{2h} + \frac{3gP^2Q^2\,(1-8g)}{h(1-2g)^2};$$

omnino igitur aqua in aestu movebitur per spatium

$$= \frac{3gp^2q^2}{h(1-8g)} \pm \frac{6gpqPQ}{h(1-2g)} + \frac{3gP^2Q^2(1-8g)}{h(1-2g)^2},$$

quorum signorum ambiguorum superius $+$ valet si Luna super horizonte, altero vero $-$ si Luna sub horizonte in Fluxu meridianum attingit.

86. Si aqua inertia careret, tum superiore Lunae transitu per meridianum elevaretur supra libellam C per spatium $= \frac{3(pq+PQ)^2-1}{h}g$, inferiori vero transitu per meridianum elevaretur ad altitudinem $\frac{3(pq-PQ)^2-1}{h}g$, quarum altitudinum discrimen est $= \frac{12gpqPQ}{h}$; ita ut discrimen admissa inertia maius sit parte circiter octava, quam idem discrimen si inertia tollatur. Maxime autem deprimetur aqua sublata inertia, si fuerit $\cos z = \frac{-PQ}{pq}$, tumque infra libellam erit constituta intervallo $= \frac{g}{h}$; ex quo spatium, per quod aestus Maris fit sublata inertia, prodit $= \frac{3p^2q^2+3P^2Q^2 \pm 6pqPQ}{h}g$; cum igitur idem spatium concessa inertia, sit

$$= \frac{3gp^2q^2}{h(1-8g)} \pm \frac{6gpqPQ}{h(1-2g)} + \frac{3gP^2Q^2(1-8g)}{h(1-2g)^2},$$

erit excessus huius spatii super illud

$$= \frac{24g^2p^2q^2}{h(1-8g)} - \frac{12g^2P^2Q^2(1+g)}{h(1-2g)^2} \pm \frac{12g^2pqPQ}{h(1-2g)}.$$

Fieri ergo potest ut spatium, in quo aestus Maris continetur, maius sit sublata inertia, quam si ea aquae tribuatur, id quod eveniet si[28]) $\frac{P^2Q^2(1+g)}{(1-2g)} > \frac{2p^2q^2}{1-8g}$ vel $\frac{PQ}{pq} > \frac{(1-2g)\sqrt{2}}{\sqrt{(1+g)(1-8g)}}$, hoc est, $\frac{PQ}{pq} > \sqrt{\frac{256}{95}}$, posito $g = \frac{1}{18}$; quod vero si evenit, Luna ne quidem ad horizontem in cursu diurno attingit, ac propterea aquam non deprimit. Ex quo sequitur aestum ubique ab inertia aquae augeri: erit autem ad usum magis accommodate spatium AB, per quod Mare agitatur, ita expressum ut sit

$$AB = \frac{3g}{h(1-8g)}\left(pq \pm \frac{PQ(1-8g)}{1-2g}\right)^2,$$

ubi signorum ambiguorum superius transitum Lunae per meridianum super horizonte, inferius vero sub horizonte respicit.

87. Cum sit $\frac{3g}{h} = 3{,}372$ pedum, Luna mediocrem a Terra distantiam tenente, atque g sit circiter $\frac{2}{25}$ vel $\frac{1}{18}$; erit posito $g = \frac{2}{25}$ spatium $AB = \frac{25}{9}(pq \pm \frac{3}{7}PQ)^2 \cdot 3{,}372$ ped. at facto $g = \frac{1}{18}$ erit spatium $AB = \frac{9}{5}(pq \pm \frac{5}{8}PQ)^2 \cdot 3{,}372$ ped. Ex his colligitur aestum fore maximum pro eadem elevatione poli, si fuerit tangens declinationis Lunae $= \frac{3}{7}\frac{P}{p}$ casu $g = \frac{2}{25}$, vel $= \frac{5}{8}\frac{P}{p}$ casu $g = \frac{1}{18}$: horum autem casuum prior veritati magis videtur consentaneus, atque hanc ob rem valorem $g = \frac{2}{25}$ retineamus: hinc igitur sequitur sub aequatore aestum fore maximum si Luna nullam habeat declinationem, atque simul pro quaque regione declinatio Lunae poterit assignari, cui maximus aestus respondeat: uti ex adiecto laterculo apparet:

Elevatio Poli	Declinatio ☾	Elevatio Poli	Declinatio ☾	Elevatio Poli	Declinatio ☾
0°	0° 0′	30°	13° 54′	60°	———
5°	2° 8′	35°	16° 42′	65°	———
10°	4° 19′	40°	19° 46′	70°	———
15°	6° 33′	45°	23° 11′	75°	———
20°	8° 52′	50°	27° 3′	80°	———
25°	11° 18′	55°	maxima	85°	———

In locis ergo ultra 45° ab aequatore remotis aestus erit maximus, si Luna maximam obtineat declinationem, si quidem fuerit $g = \frac{2}{25}$: ac si per observationes constet cuinam Lunae declinationi maximus aestus respondeat, tum inde valor litterae g innotescet: quoniam autem sub elevatione poli 50° aestus maximi nondum maximae declinationi respondere observantur, ponamus id evenire sub elevatione poli 60°, reperietur $\frac{1-8g}{1-2g} = \frac{1}{4}$ atque $g = \frac{1}{10}$, unde ipsius g tuto hi limites constitui posse videntur $\frac{1}{10}$ et $\frac{1}{18}$; ex hac vero hypothesi valor $\frac{1}{10}$ multo propius ad veritatem accedit; interim tamen etiamnum nil definimus, sed observationes hunc in finem sollicite institutas expectamus.

88. Quod si autem ponamus $g = \frac{1}{10}$, tum bini aestus successivi, dum Luna in maxima declinatione versatur, eo magis ad aequalitatem perducentur, quo ipso theoria ad experientiam propius accedit; cum enim sit horum binorum aestuum maior ad minorem uti $\left(pq + \frac{PQ(1-8g)}{1-2g}\right)^2$ ad $\left(pq - \frac{PQ(1-8g)}{1-2g}\right)^2$, haec ratio eo propius ad aequalitatem accedet, quo minor fuerit fractio $\frac{1-8g}{1-g}$, fit autem haec fractio $= \frac{1}{4}$ si ponatur $g = \frac{1}{10}$. Hac itaque hypothesi erit quantitas aestus maioris $= (pq + \frac{1}{4}PQ)^2 \cdot 16{,}86$ ped. minoris vero $= (pq - \frac{1}{4}PQ)^2 \cdot 16{,}86$ ped. At inter hos binos aestus aqua humillima non medium interiacet, sed minori est vicinior, neque tamen tanta inaequalitate binos Fluxus dirimit, quam fieret, si ima aqua Lunae horizontali responderet. Si enim tempus medium inter binos Fluxus ponatur z, erit $\cos z = 0$, at temporis, quo Refluxus Fluxum maiorem insequitur, cosinus est $= \frac{-PQ}{4pq}$, eiusque ergo intervalli a tempore medio sinus est $= \frac{PQ}{4pq}$, quae expressio adeo sub elevatione poli 60°, pro maxima Lunae declinatione 28°, tantum fit $= \sin 13°$, unde Refluxus a tempore inter Fluxus medio circiter 54′ aberrabit: minor vero erit aberratio, quo propius cum regio Terrae tum Luna ad aequatorem versentur, id quod cum experientia mirifice convenit. Quoniam autem haec ex valore

28) $\surd(1+g)(1-8g)$ denotat $\surd((1+g)(1-8g))$. EJA.

ipsius g assumto consequuntur, imprimis notari oportet, litteram g non posse absolute determinari, sed eius quantitatem, quippe quae mobilitatem totius oceani spectat, cum ab extensione tum etiam profunditate Maris pendere; ex quo variis in locis haec eadem littera g, varias significationes sortietur.

89. Ex solutione horum duorum problematum, quae quidem in se spectata non solum sunt attentione digna, sed etiam cum analysin tum etiam motus scientiam amplificant, quamvis ea casum propositum non penitus exhauriant, tamen motus in praecedentibus capitibus definitus multo magis cum experientia conciliatur, id quod theoriae nostrae iam insigne addit firmamentum. Simili autem modo vis a Sole profecta cum inertia aquae potest coniungi, atque aestus Maris definiri, quatenus a sola vi Solis oritur, quibus duobus effectibus coniungendis iudicare licebit, quantus aestus quovis tempore et quovis loco debeat evenire. In hoc quidem capite cogitationes adhuc ab omnibus obstaculis a Terra et littoribus oriundis prorsus abstrahimus, atque universam Terram undiquaque aqua circumfusam ponimus; ex quo regulas hinc natas praecipue cum eiusmodi observationibus, quae in amplissimo oceano apud exiguas insulas sunt institutae, conferri conveniet. Quoniam autem nondum motus aquae progressivi, quo alternative ad loca, in quibus Fluxus et Refluxus accidit, progreditur et recedit, rationem habuimus, necesse est ut etiam hunc motum et Phaenomena inde orta contemplemur. Ac primo quidem facile intelligitur, cum ob inertiam aquae tum etiam alia impedimenta motui opposita, aquam tam tardius elevari quam deprimi oportere, quam ex allatis hactenus consequitur: unde Fluxus non ad transitus Lunae per meridianum contingent, sed aliquanto serius evenient, omnino uti experientia testatur.

90. Haec autem retardatio praecise ad calculum revocari non potest, quia a motu aquae eiusque profunditate plurimum pendeat, prouti etiam videmus in diversis locis eam vehementer esse diversam, atque aliis locis Fluxum contingere post Lunae culminationem tribus horis nondum elapsis, aliis vero locis plus quam duodecim horis tardius venire, quae quidem insignis retardatio terrarum positioni est adscribenda; interim tamen hinc sufficienter constat motum Maris admodum posse impediri. Pro eodem vero loco satis luculenter perspicitur, quo maior atque altior Fluxus evenire debeat, eo tardius eundem accidere oportere. Quod si enim aestus contingat infinite parvus, dubium est nullum, quin is stato tempore adveniat, cum impedimentis hoc casu ne locus quidem concedatur agendi: unde dilucide sequitur aestus eo tardius advenire debere, quo sint maiores. Atque hoc ipsum experientia confirmat, qua constat aestus maiores, qui circa novilunia ac plenilunia

contingunt, tardius insequi transitum Lunae per meridianum, quam aestus minores, qui circa quadraturas contingunt. Cum enim Luna in quadraturis circiter 6 horis tardius respectu Solis per meridianum transeat, quam in syzygiis, aestus tamen non 6 horis tardius, sed tantum circiter $5\frac{1}{4}$ horis tardius accidit. Videtur vero etiam calculus, qui pro utraque vi Solis ac Lunae coniunctim institui potest simili modo, quo pro sola vi Lunae fecimus, eiusmodi retardationem maiorem in syzygiis quam in quadraturis indicare, etiamsi eum ob summas difficultates ad finem perducere non valuerimus; interim tamen satis planum est praecipuam eius causam in ipsa natura aquae esse quaerendam. Haec autem allata ratio retardationis a FLAMSTEDIO[29]) maxime probatur, quippe qui observavit maximam retardationem non tam syzygiis luminarium, neque minimam quadraturis respondere, sed iis tempestatibus, quibus aestus soleant esse maximi et minimi, id quod demum post syzygias et quadraturas contingit.

91. Ad hanc autem Fluxuum a syzygiis ad quadraturas accelerationem, respectu transitus Lunae per meridianum, ac retardationem a quadraturis ad syzygias, plurimum quoque vis Solis conferre videtur. Supra enim iam indicavimus post syzygias Fluxum transitum Lunae per meridianum antecedere debere, ob Solem tum iam versus horizontem declinantem; unde etiam, stabilita inertia, diebus novilunia ac plenilunia sequentibus aestus Maris citius insequi debet transitum Lunae per meridianum, quam in ipsis syzygiis, id quod etiam observationes mirifice confirmant; inter Fluxum enim quintum et sextum post syzygias retardatio respectu Solis tantum 17 minut. deprehenditur, cum tamen Luna 24′ retardetur. Hanc ob rem a Sole determinatur aestus ad actionem virium magis exacte sequendam, quae determinatio cum duret usque ad quadraturas, mirum non est, quod aestus tum respectu Lunae citius contingant, magisque ad calculum accedant. Contrarium evenit in progressu a quadraturis ad syzygias, quo tempore aestus a Sole continuo retardantur; hocque necessario efficitur, ut tandem in ipsis syzygiis Fluxus tardius insequatur Lunae culminationem quam in quadraturis. Hanc autem rationem cum magnitudine aestus coniungendam esse putamus ad haec phaenomena perfecte explicanda, saepissime enim in hac quaestione plures causae ad eundem effectum producendum concurrunt; hoc autem est id ipsum quod calculus ille summopere implicatus et molestus quasi per transennam ostendere visus est.

29) JOHN FLAMSTEED, «Table of times of high water at London Bridge», *Philosophical Transactions* 13 (1683), 10–15, p. 13. EJA.

92. Quo autem tam de his Phaenomenis quam reliquis certius et solidius iudicare queamus, ipsum motum progressivum, quem aqua ab aestu recipit, investigabimus. Cum enim aqua eodem loco nunc elevetur nunc subsidat, necesse est ut priori casu aqua aliunda affluat, posteriori vero ab eodem loco defluat, unde nomina Fluxus ac Refluxus originem traxerunt. Repraesentet igitur tempore quocunque figura *ADBE* [fig. 12] statum aquae totam Terram ambientis, ita ut in locis

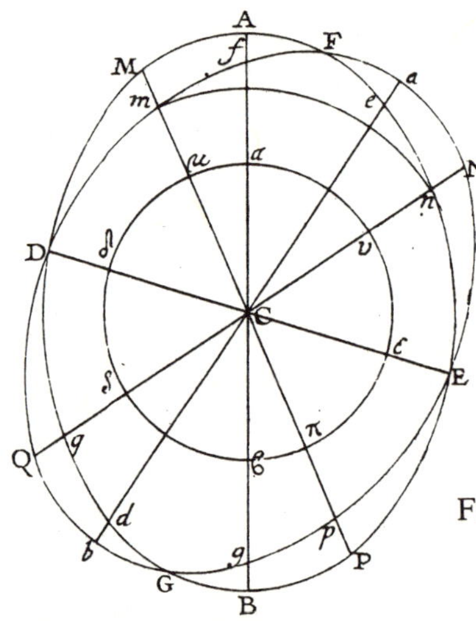

Fig. 12

A et *B* aqua maxime sit elevata, in locis vero mediis ab *A* et *B* aequidistantibus, maxime depressa. Post aliquod tempus transferatur aestus summus ex *A* et *B* in *a* et *b*, sitque *aDbE* figura aquae Terram circumdantis: hoc igitur tempore necesse est, ut a parte oceani *DF* defluxerit aquae copia *FAMDmf*, in partem vero *FE* tantundem aquae affluxerit, portio scilicet *FaNEne*: simili modo portio *EG* decrevit copia aquae *EPBGgp*, portioque *GD* augmentum accepit *GbQDqd*. Si nunc ponamus portionem *FMm* transire in locum *FNn*, ac portionem *EPp* in *ENn* deferri, satis clare motum aquae progressivum intelligere licebit. Cum enim motus aquae summae *A* fiat ab ortu in occasum, aqua quae circa *A* versus orientem scilicet ab *M* ad *N* usque est sita, in occasum movebitur; similiterque ea quae huic e diametro est opposita et spatium *PQ* occupat. Contra vero reliqua aqua in *MQ* et *NP* contenta in ortum promovebitur. Verum celeritas ubique non erit eadem; in punctis enim *M*, *N*, *P* et *Q* quippe limitibus inter motus versus ortum et obitum, celeritas erit nulla, deinde ab *M* usque ad *F* crescet ubique ita ut incrementa celeritatis in punctis mediis ut *A* sint differentiis *Af* proportionalia: ab *F* vero usque ad *N* celeritas decrescere debet, et decrementum celeritatis in *e* erit ut ae; similique modo comparatus erit motus in reliquis portionibus figurae propositae.

93. Si haec diligentius prosequamur ac punctum a ipsi A proximum ponamus, reperiemus in loco quocunque M fore intervallum Mm sinui dupli anguli MCA proportionale. Quare si anguli ACM sinus ponatur $= x$, cosinus $= y$, ac celeritas quam aqua in M habet, versus occasum $= u$, erit du ut $2xy$. Cum autem elementum arcus AM sit ut $\frac{dx}{y}$; nam figuram instar circuli considerari licet: erit du ut $2x\,dx$, atque u proportionale erit ipsi $2xx - 1$ eiusmodi adiecta constante, ut ubi Mm est maximum, ibi celeritas evanescat. Hanc ob rem erit celeritas in loco quocunque M, quam aqua versus occidentem habebit, uti cosinus dupli anguli MCA. Maxima igitur aquae celeritas versus occidentem erit in iis locis, in quibus aqua maxime est elevata; huicque celeritati aequalis est ea, qua aqua in locis ubi maxime est depressa, versus orientem promovetur; si quidem haec in circulo fieri concipiamus, nam in sphaera motus aliquantum diversus erit, sed tamen hinc intelligi poterit. At in locis quae ab A et B 45 grad. distant, ob cosinum dupli anguli $= 0$, aqua omnino nullum habebit motum horizontalem. Ex his igitur non solum motus aquae progressivus cognoscitur, quo alterna elevatio ac depressio producitur, sed etiam luculenter perturbationes, quae a Terris, littoribus atque etiam a fundo Maris proficisci possunt, perspiciuntur. Ceterum quanquam sectio nostra plana $ADBE$ aequatorem solum denotare videtur, tamen eadem ad parallelum quemvis significandum satis commode adhiberi potest: quin etiam motus pro sphaera hinc satis distincte colligi poterit, operae enim pretium non iudicamus, per solidorum introductionem hanc rem cognitu tanto difficiliorem reddere.

94. Eo minus autem huius [negotii]* accuratae inquisitioni insistemus, quod celeritas progressiva insuper a profunditate maris pendeat. Quod si enim ponamus mn iam esse Maris fundum, ita ut profunditas Maris in M maior non esset quam Mm, tum isti aquae tantus motus inesse deberet, quo ea, dum fluxus ex A in a transit, ex situ $nFMm$ in situm $mFNn$ transferri posset. Hic autem motus quamvis sit difformis et per totam massam inaequabilis, tamen si tota translatio spectetur, totus motus ex spatio a centro gravitatis interea percurso est aestimandus. Hoc igitur casu, quo Terrae superficiem solidam ad mn usque pertingere ponimus, reperietur centrum gravitatis massae $nFMm$ fere aeque celeriter promoveri debere

ac punctum A, ex quo eius celeritas tanta esse deberet, qua tempore unius horae spatium fere 15 graduum percurrere posset, quae celeritas utique foret enormis ac stupenda. At si Mari profunditatem maiorem tribuamus, scilicet ad $\mu\nu$ usque, tum illa celeritas multo fiet minor, decrescet namque in eadem ratione in qua profunditas crescit. Cum igitur celeritas Maris, quae ante in se spectata inventa est cosinui dupli anguli MCA proportionalis, eo fiat minor, quo maiorem Mare habeat profunditatem, tenebit ea in quoque loco rationem compositam ex ratione directa cosinus dupli anguli MCA atque ex inversa profunditatis.

95. Datur autem alius modus celeritatem Maris horizontalem, posita scilicet ubique profunditate eadem, determinandi, qui tamen etiam ad diversas profunditates patet, si cum ratione invenienda coniungamus reciprocum [rationem]* profunditatum uti fecimus; deduciturque hic modus ex motu Maris verticali, quo modo ascendit modo descendit, qui iam supra est definitus. Primo enim manifestum est, si Mare ubique eadem celeritate, (posita profunditate ubique aequali) in eandem plagam promoveretur, tum etiam altitudinem mansuram esse eandem ubique, neque ullam mutationem in elevatione aquae orturam esse. At si aqua motu inaequabili progrediatur, manifestum est iis in locis, ubi celeritas diminuitur, aquam turgescere atque adeo elevari debere, quoniam plus aquae affluit quam defluit; contra vero ubi celeritas aquae crescat, ibi aquam subsidere oportere. Quare cum elevatio et depressio Maris a motus progressivi horizontalis inaequalitate pendeat, licebit pro quovis loco hanc inaequalitatem definire, ex motu ascensus et descensus cognito. Cum enim celeritas ascensus sit decremento celeritatis progressivae proportionalis[30]), celeritas descensus vero incremento celeritatis progressivae, ex dato motu verticali ratio motus horizontalis definiri poterit. Invenimus autem supra § 84, si Luna a meridiano versus occasum iam recessit angulo z, hoc est cum regio proposita ab ea, in qua aqua est summa, versus orientem secundum longitudinem distet angulo z, fore celeritatem qua aqua ascendit

$$= \frac{-6gpqPQ\sin z}{h(1-2g)} - \frac{3gp^2q^2\sin 2z}{h(1-8g)}.$$

Quare cum huic celeritati ascensus proportionale sit decrementum motus horizontalis, erit ipsa celeritas horizontalis versus occasum ut

30) Editio princeps: aequalis loco proportionalis. EJA.

$$\frac{g(3p^2q^2+6P^2Q^2-2)}{2h}+\frac{6gpqPQ\cos z}{h(1-2g)}+\frac{3gp^2q^2\cos 2z}{2h(1-8g)};$$

huius enim differentiale negative sumtum et per dz divisum dat ipsam celeritatem ascensus. Quoniam autem haec expressio simul exhibet spatium, quo Mare supra libellam elevatur, erit celeritas Maris in quovis loco versus occidentem proportionalis elevationi supra libellam, et inverse profunditati Maris, quae est vera regula pro motu Maris, tam verticali quam horizontali, definiendo; atque ita priori modo insufficienti supersedere potuissemus.

96. Consideremus ergo motum, quo aqua tam verticaliter quam horizontaliter promovetur a Fluxu usque ad Refluxum, indeque ad sequentem Fluxum, idque sub aequatore, dum Luna pariter in aequatore versatur: erit itaque celeritas ascensus ut $-\sin 2z$, celeritas autem horizontalis versus occasum ut $15\cos 2z+1$ posito $g=\frac{1}{10}$, cui expressioni simul altitudo aquae supra libellam est proportionalis. Quod si ergo superficies Terrae seu perimeter aequatoris in 24 partes aequales dividatur, atque in locis A et B [fig. 13] aqua sit maxime elevata, in C et D vero minime, numeri 1, 2, 3, etc. designabunt ea Terrae loca in quibus ante unam vel duas vel tres vel etc. horas lunares aqua maxime fuit elevata, tribuendo uni horae Lunari 62 minuta. In Tabula ergo annexa exhibetur motus tam verticalis, quam horizontalis, ad singulas horas post Fluxum elapsas.

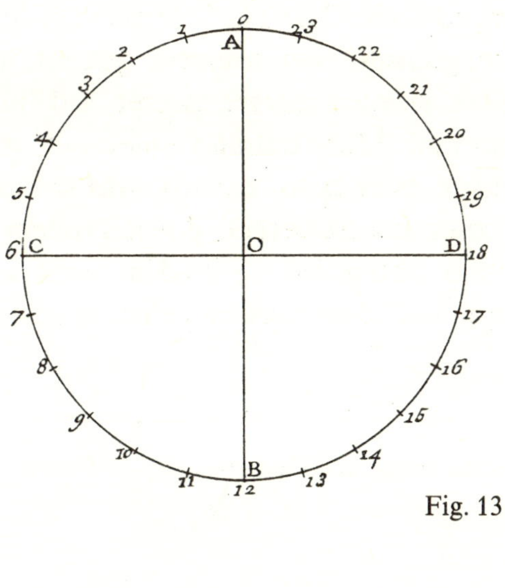

Fig. 13

Horae post Fluxum	Celeritas Maris verticalis	Celeritas Maris horizontalis[31])
0	0,000 descendit.	1,067 in occasum.
1	0,500 descendit.	0,927 in occasum.
2	0,860 descendit.	0,567 in occasum.
3	1,000 descendit.	0,067 in occasum.
4	0,860 descendit.	0,432 in ortum.
5	0,500 descendit.	0,792 in ortum.
6	0,000 ascendit.	0,932 in ortum.
7	0,500 ascendit.	0,792 in ortum.
8	0,860 ascendit.	0,432 in ortum.
9	1,000 ascendit.	0,067 in occasum.
10	0,860 ascendit.	0,567 in occasum.
11	0,500 ascendit.	0,927 in occasum.
12	0,000 descendit.	1,067 in occasum.

Facile autem intelligitur pro regionibus ab aequatore remotis, praecipue si Luna habeat declinationem, tum utrumque motum magis fore irregularem, atque mox ascensum citius absolvi mox vero descensum; totus autem motus facilius ex ipsis formulis datis cognoscetur. Hic denique profunditatem ubique eandem posuimus; quod si enim esset diversa, motus horizontalis simul rationem inversam profunditatis tenebit.

97. Denique antequam hoc caput finiamus, notari oportet, neque maximos aestus iis ipsis temporibus evenire posse, quibus vires Solis et Lunae maxime vigent, nec minimos aestus tum, cum vis a Luna et Sole nata est debilissima, sed aliquanto tardius. Aestus enim magnitudo non solum a quantitate virium sollicitantium pendet, uti id usu veniret, si aqua inertia careret, sed insuper a motu iam ante concepto. Quod si enim ante Mare omnino quievisset, tum primus certe aestus oriundus admodum futurus esset exilis, etiamsi vires sollicitantes essent maximae; sequentes vero aestus continuo crescerent, donec tandem post tempus infinitum magnitudinem assignatam obtinerent, si quidem vires sollicitantes idem robur perpetuo servarent: atque hoc idem evenire debet, si aestus praecedentes tantum

31) Pro calculo, celeritas horizontalis ponitur $= \frac{1}{15}(15\cos 2z + 1)$. Ponitur autem $\sin 60° = 0{,}860$. EJA.

fuerit minores, quam is qui viribus sollicitantibus convenit. Quare cum aestus novilunia ac plenilunia praecedentes sint minores, ii quidem his temporibus ab auctis viribus augebuntur, non vero subito totam suam quantitatem consequentur, atque hanc ob rem aestus etiamnum post syzygias augmenta accipient, donec ob tum secutura virium decrementa, aestus iterum decrescere incipiant. Ita tempore noviluniorum et pleniluniorum non tam ipsi aestus quam incrementa eorum censenda sunt maxima, quatenus scilicet aestus praecedentes maxime deficiunt, ab iis qui sequi deberent; ex quo manifestum est non illos aestus, qui in ipsis syzygiis luminarium contingunt, esse maximos, sed sequentes esse maiores. Hocque idem intelligendum est de aestibus minimis, qui non in ipsas quadraturas incidunt, sed tardius sequuntur: unde ratio luculenter perspicitur, cur aestus tam maximi quam minimi non ipsis syzygiarum et quadraturarum tempestatibus respondeant, sed serius observentur, tertii scilicet demum vel quarti post haec tempora.

CAPUT SEPTIMUM

Explicatio praecipuorum Phaenomenorum circa Aestum Maris observatorum

98. In praecedentibus capitibus fusius exposuimus effectus, qui in Mari a viribus illis duabus, quarum altera versus Lunam est directa, altera versus Solem, produci debent; eosque cum per calculum analyticum, tum per solida ratiocinia ita determinavimus, ut de eorum existentia dubitari omnino non liceat, si quidem illae vires admittantur. At vero istas vires in mundo existere non solum per alia phaenomena evidentissime probavimus, sed etiam earum causam physicam assignavimus, quam in binis vorticibus, quorum alter circa Solem, alter circa Lunam sit constitutus, posuimus, quippe quae est unica ratio cum gravitatem tum etiam vires, quibus planetae in suis orbitis circa Solem continentur, explicandi. Quin etiam haec ipsa phaenomena internam vorticum structuram et indolem commonstrarunt; ob eaque vortices ita comparatos esse statuimus, ut vires centrifugae decrescant in duplicata ratione distantiarum a centris eorundem. Quare cum in his viribus nihil gratuito assumserimus, si effectus ex iis oriundi cum phaenomenis aestus Maris conveniant, certissime nobis persuadere poterimus, in assignatis viribus veram aestus Maris causam contineri; absonumque omnino fore, si causam aestus Maris in aliis viribus imaginariis anquirere vellemus. Quamobrem in hoc capite constituimus omnes effectus, qui in superioribus capitibus sparsim sunt eruti, coniunctim et ordine

proponere, summumque eorum consensum cum experientia declarare. Quoniam autem nondum impedimentorum a littoribus terrisque oriundorum rationem habuimus, facile intelligitur, hinc excludi adhuc debere eiusmodi anomalias aestus Maris, quae evidentissime a Terris contingentibus ortum habeant, cuiusmodi sunt aestus vel vehementer enormes vel vix sensibiles, uti in Mari Mediterraneo, vel insignes retardationes eorum, quibus rebus explicandis sequens caput ultimum destinavimus: ita in hoc capite tantum ea aestus Maris phaenomena explicanda suscipimus, quae in portubus amplissimum oceanum respicientibus vel insulis observari solent in oceano sitis.

99. Si omnes proprietates, quibus Fluxus ac Refluxus Maris praeditus esse observatur, distincte enumerare atque exponere velimus, deprehendemus eas ad tres classes revocari debere. Ad primam scilicet classem referenda sunt phaenomena, quae in uno aestu in se spectato conspiciuntur, cum ratione temporis tum etiam ratione quantitatis; haecque phaenomena commodissime sub varietatibus diurnis comprehendi possunt, quatenus ea se offerunt observatori, qui per integrum tantum diem observationes instituit, neque ea cum aliis phaenomenis aliis temporibus occurrentibus comparat. Secunda classis complectitur varietates menstruas, quae sese observatori per integrum mensem aestum Maris contemplanti offerunt, quorsum pertinent aestus maximi minimique, item retardationes modo maiores modo minores. Tertia denique classis comprehendit varietates annuas ac plusquam annuas, quae sequuntur vel varias Lunae a Terra distantias, vel Solis; vel etiam luminarium declinationem. Hanc ob rem phaenomena uniuscuiusque classis recensebimus, atque quomodo singula cum theoria tradita congruant, ostendemus. Hic vero, ut iam est monitum, a perturbationibus quae a Terris ac littoribus provenire possunt, animum prorsus abstinemus, eas sequenti capiti reservantes. Multo minus vero ad ventum hic respicimus, quo aestus Maris cum ratione magnitudinis tum temporis plurimum affici observatur; sed tantum eiusmodi phaenomena explicare hic conabimur, quae memoratis perturbationibus minime sint obnoxia.

100. Quod igitur ad primam classem attinet, praecipuum Phaenomenum in hoc consistit, quod ubique in amplissimo oceano quotidie bini Maris Fluxus seu elevationes, binique Refluxus seu depressiones observentur, atque tempus inter binos Fluxus successivos circiter 12 h. 24′ deprehendatur. Huic vero Phaenomeno, si ulli alii, per theoriam nostram plenissime est satisfactum, ubi ostendimus maximam aquae elevationem deberi transitui Lunae per meridianum tam supra quam infra

Terram: ex quo cum Luna una revolutione diurna bis ad eiusdem loci meridianum appellat intervallo temporis circiter 12 hor. 24′, necessario sequitur una revolutione Lunae circa Terram binos Fluxus tanto tempore a se invicem dissitos oriri debere, quemadmodum hoc ipsum calculus tam pro hypothesi aquae inertia carentis, quam admissa inertia, clarissime indicavit. Simul autem ex iisdem determinationibus intelligitur sub ipsis polis nullum omnino aestum dari diurnum, in regionibus vero a polis non procul remotis, ubi luminaria vel non oriuntur vel non occidunt, quotidie unum tantum Fluxum unicumque Refluxum contingere debere; quae consequentia theoriae, etsi observationibus nondum satis est comprobata, tamen quia ex iisdem principiis sequitur quae institutis observationibus satisfaciant, nulli amplius dubio subiecta videtur. In locis autem aequatori propioribus, quibus quotidie bini Fluxus totidemque Refluxus eveniunt, momentum, quo aqua maxime deprimitur non satis exacte medium interiacere observatur inter Fluxuum momenta, sed mox priori mox posteriori est propius, quod Phaenomenum cum nostra theoria apprime congruit; ostendimus enim momentum Refluxus non exacte tempori medio inter Fluxus respondere, nisi vel locus situs sit sub aequatore, vel Lunae declinatio fuerit nulla, sed modo priori modo posteriori Fluxui esse propius.

101. Secundum Phaenomenum huc redit, ut ubique locorum Fluxus post transitum Lunae per meridianum venire observetur, idque aliquot horarum spatio, in portubus versus apertum oceanum patentibus. Nam in regionibus quae cum oceano non liberrime communicantur, sed ad quas aqua iuxta littora deferri debet, multo tardius aestus advenit, quae retardatio si fere ad 12 horas ascendit, in causa esse solet, ut huiusmodi in locis Fluxus ante transitum Lunae per meridianum venire videatur. Ita ad Portum Gratiae videri posset Fluxus 3 horis Lunae culminationem antecedere, cum tamen, re bene considerata, a praecedente culminatione oriatur, atque adeo eam 9 fere horis demum sequatur, uti apparebit si aestuum momenta, quae successive ad littora Britanniae minoris et Normanniae observantur continuoque magis retardantur, attentius inspiciantur. Deberet quidem ubique Fluxus in ipsos Lunae transitus per meridianum incidere, imo quandoque ob Solem praecedere, non solum demta inertia, sed etiam ea posita, si tantum aquae motus verticalis spectetur; at si etiam motus horizontalis ratio habeatur, tum dilucide ostendimus Fluxum perpetuo retardari, ac demum post Lunae transitum per meridianum evenire debere. Tempus quidem huius retardationis, cum sit admodum variabile pluribusque circumstantiis subiectum, non definivimus, interim tamen id ex § 82 colligi poterit, remotis externis impedimentis: cum enim invenerimus aquam

propria vi gravitatis sese in situm aequilibrii recipere tempore $\frac{12}{n}$ horarum, ac numerum n esse circiter 5 vel 6, manifestum est tanto etiam tempore opus esse, quo aqua eum situm quem vires intendunt, induat, ex quo Fluxus circiter 2 horas vel $2\frac{1}{2}$ hor. post transitum Lunae per meridianum contingere debebit, id quod cum observationibus in oceano libero institutis egregie convenit; hancque idcirco praecipuam huius retardationis causam merito assignamus.

102. Tertium Phaenomenon suppeditat aestus magnitudo, quae autem tam diversis locis quam diversis tempestatibus maxime est mutabilis. Interim tamen exceptis enormibus illis aestubus, qui nonnullis in portubus observari solent, reliqui cum nostra Theoria egregie consentiunt; inertia enim sublata, invenimus sub aequatore maximum aestum fore per spatium circiter 4 pedum, ab inertia autem hoc intervallum augeri ita ut duplo, vel triplo, vel etiam quadruplo et plus fiat maius, prout valor ipsius g (vid. §82) minor fuerit vel maior, quippe qui a facultate oceani sese propria sua vi in statum aequilibrii restituendi pendet; ex quo sub aequatore spatium per quod maximus aestus agitatur ad 8, 12, 16 et plures pedes exsurgere potest. In regionibus autem ab aequatore remotis invenimus magnitudinem aestus tenere rationem duplicatam cosinuum elevationis poli, unde sub elevatione poli 45°, magnitudo aestus circiter duplo erit minor quam sub ipso aequatore; cuius veritas in locis a littoribus aliquot milliaria remotis per experientiam eximie comprobatur. Deprehenditur enim ubique in locis a littoribus remotis aestus multo minor quam ad littora; cuius discriminis causa in sequenti capite dilucide indicabitur. Quin etiam in medio Mari plerumque aestus adhuc minor observatur, quam haec regula requirit; id autem ostendetur a non satis ampla oceani extensione secundum longitudinem proficisci, quemadmodum in oceano Atlantico qui versus occidentem littoribus Americae, versus orientem vero littoribus Africae et Europae terminatur, quae amplitudo non est satis magna, ut integram aestus quantitatem suscipere queat.

103. Quartum Phaenomenon varietates menstruas respicit, atque ostendit aestus, qui circa plenilunia et novilunia contingunt, inter reliquos eiusdem mensis esse maximos, aestus vero circa quadraturas luminarium minimos; quae inaequalitas cum theoria nostra ad amussim quadrat. Cum enim aestus Maris non solum ab ea vi, quae vortici Lunam ambienti competit, oriatur, sed etiam a vi Solem spectante pendeat, quae ceteris paribus circiter quadruplo minor est vi Lunae, manifestum est aestum Maris maximum esse debere, si ambae vires inter se conspirent,

atque aquam simul vel elevent vel deprimant, id quod accidere ostendimus tam pleniluniis quam noviluniis. Deinde simili modo, quoniam istae vires inter se maxime discrepant in quadraturis, quibus temporibus dum aqua a Luna maxime elevatur, simul a Sole maxime deprimitur ac vicissim, perspicuum est iisdem temporibus aestum minimum esse debere. Praeterea vero ipsum discrimen cum theoria exacte convenit; in pluribus enim portubus aestus maximos et minimos ad calculum revocavimus, atque ex relatione eorum relationem inter vires Lunae ac Solis investigavimus; hincque perpetuo eandem fere rationem inter vires Solis ac Lunae absolutas elicuimus, quemadmodum id fecit NEWTONUS[32]) ex observationibus Bristolii et Plymouthi, nos vero in Portu Gratiae institutis[33]), conclusionibus mirifice inter se congruentibus: qualis consensus profecto expectari non posset, si theoria veritati non esset consentanea. Neque etiam aliae theoriae adhuc productae, cuiusmodi sunt GALILAEI, WALLISII atque CARTESII, qui causam in pressione Lunae collocavit, huic phaenomeno perfecte satisfaciunt, sed potius prorsus evertuntur.

104. Quintum Phaenomenon in hoc consistat, quod unius mensis intervallo maximi aestus non sint ii, qui novilunia ac plenilunia proxime insequuntur, sed sequentes tertii scilicet circiter vel quarti, similique intervallo aestus minimi demum post quadraturas contingunt[34]). Huius autem Phaenomeni ratio in §97 fusius est exposita, ubi ostendimus, cum aestus ante syzygias incidentes essent minores, maximam vim a Sole et Luna ortam non subito aestum maximum producere valere, sed tantum Mare ad eum statum sollicitare. Cum igitur post syzygias vis aestum efficiens sensibiliter non decrescat, aestus etiamnum post hoc tempus incrementa capiet, atque ideo demum post syzygias fiet maximus; similisque est ratio diminutionis aestuum, quae etiamnum post quadraturas contingere debet, ita ut aestus minimi demum post quadraturas eveniant. Huiusmodi autem retardationes effectuum a viribus in mundo existentibus provenientium quotidie abunde experimur: ob similem enim rationem singulis diebus maximum calorem non in ipso meridie sentimus, etiamsi hoc tempore vis Solis calefaciens sine dubio sit maxima, sed demum aliquot horis post meridiem, atque propter eandem causam neque solstitii aestivi momento maximus calor annuus sentitur, neque tempore solstitii hyberni frigus summum, sed utrumque notabiliter tardius.

32) NEWTON, *Principia,* Liber III, propositio 37. EJA.

33) Vide *Histoire de l'Académie Royale des Sciences* (Paris), (1710), 1712, p. 4. EJA.

34) Vide *Histoire de l'Académie Royale des Sciences* (Paris), (1710), 1712, p. 8 et (1712), 1714, p. 2. EJA.

105. Sextum Phaenomenon in hoc ponimus, quod momenta Fluxuum tempore syzygiarum multo strictius ordinem tenere observantur, quam circa quadraturas. Hic vero ante omnia animadvertendum est praecipuam sensibilem anomaliam in momentis aestuum inde originem trahere, quod haec momenta ex tempore solari atque a vero meridie seu transitu Solis per meridianum soleant computari, cum ea potius a transitu Lunae per meridianum pendeant. Quod si autem ad has observationes tempus lunare a transitu Lunae per meridianum computandum adhibeatur, irregularitates apparentes maximam partem evanescent, hoc vero multo magis in fluxubus circa syzygias quam quadraturas: in quadraturis enim quoniam, dum Luna per meridianum transit, Sol non semper in horizonte versatur, sed vel ad horizontem demum accedit vel iam ab eo recedit, necesse est ut illo casu Fluxus citius, hoc vero tardius contingat: quod discrimen cum partim ab elevatione poli partim a declinatione luminarium pendeat, momenta Fluxuum in quadraturis magis irregularia reddit: interim tamen habita harum circumstantiarum ratione satis prope definire potest. Circa tempora Fluxuum autem, qui in noviluniis ac pleniluniis incidunt, haec sola correctio seu reductio ad transitum Lunae per meridianum omnem fere anomaliam tollit, quorsum spectat regula a celeb. CASSINO[35]) in Mem. 1710 tradita, qua pro totidem horis, quibus plenilunium seu novilunium vel ante meridiem vel post incidit, totidem bina minuta ad tempus Fluxus medium vel addere vel ab eo subtrahere iubet, quippe quae ex motu Lunae est petita. Interim tamen hac correctione adhibita aliqua anomalia superesse deprehenditur, cuius autem ratio ex nostra theoria sponte sequitur. Quando enim syzygia ante meridiem celebratur, tum dum Luna per meridianum transit, Sol iam ante eum est transgressus, atque ideo iam horizonti appropinquant, ex quo necesse est ut fluxus citius eveniat, quam prima regula sola adhibita indicat. Atque etiam idem in tabulis Fluxuum Dunkerquae[36]) et in Portu Gratiae[37]) observatorum, Mem. 1710 insertis, manifesto conspicitur: quando enim novilunium pleniluniumve pluribus horis ante meridiem accidit, tum Fluxus citius advenisse observatur, quam calculus CASSINIANUS indicabat; contra vero tardius si syzygiae demum pluribus horis post meridiem inciderint, cuius maioris retardationis causa in Sole tum adhuc ab horizonte recedente est quaerenda.

35) JACQUES CASSINI, «Reflexions sur les observations du flux et du reflux de la mer, faites à Dunquerque», *Mémoires de l'Académie Royale des Sciences* (Paris), (1710), 1712, pp. 318–341, vide p. 322. EJA.

36) *Mémoires de l'Académie Royale des Sciences* (Paris), (1710), 1712, p. 330 et p. 332. EJA.

37) J. CASSINI, «Reflexions sur les observations du flux et du reflux de la mer, faites au Havre de Grace», *Mémoires de l'Académie Royale des Sciences* (Paris), (1710), 1712, pp. 366–379. EJA.

106. Septimum Phaenomenon suppeditat diversa retardatio Fluxuum in syzygiis luminarium et quadraturis respectu appulsus Lunae ad meridianum; tardius scilicet ubique locorum Fluxus, qui in syzygiis contingunt, insequuntur culminationem Lunae, quam ii, qui circa quadraturas veniunt. Huius autem Phaenomeni duplex causa potest assignari, quarum prima a sola quantitate aestuum petitur, quia enim aestus syzygiarum multo sunt maiores quam aestus quadraturarum, consentaneum videtur illos tardius venire quam hos. Altera vero causa quae hoc Phaenomenon multo distinctius explicat, nullique dubio locum relinquit, nostrae theoriae omnino est propria, priorique longe est praeferenda. Ponamus enim t esse tempus, quo in noviluniis ac pleniluniis Fluxus post appulsum Lunae ad meridianum venire solet; sequentibus igitur diebus hoc tempus t continuo diminuetur, quia tum Sol, dum Luna in meridiano versatur, Mare iam deprimit; quae diminutio cum duret fere usque ad quadraturas, necesse est ut his temporibus Fluxus multo citius post culminationem Lunae sequantur, viribusque sollicitantibus magis obtemperent, uti hoc fusius §91 explicavimus, unde tempus retardationis in quadraturis tantum erit $t-\theta$. Post quadraturas autem Sol exerit contrarium effectum, atque adventum Fluxus continuo magis retardat, idque aequali modo, quo ante acceleraverat, ex quo usque ad sequentem syzygiam intervallum $t-\theta$ iterum at t usque augebitur. Huiusque Phaenomeni solius explicatio sufficere posset ad veritatem theoriae nostrae evincendam, cum id omnibus aliis theoriis explicatu sit insuperabile; neque a nemine adhuc saltem probabilis eius causa sit assignata.

107. Octavum Phaenomenon petamus ex inaequalitate duorum Fluxuum sese immediate insequentium, quorum alter transitui Lunae superiori per meridianum respondet, alter inferiori, quae inaequalitas maxime observatur in regionibus ab aequatore multum remotis, ac tum, cum Lunae declinatio est maxima. Theoria quidem declarat Lunam, etiamsi in ipso aequatore versetur, tamen maiori vi gaudere ad Mare movendum, quando super horizonte meridianum attingit, quam infra horizontem; at discrimen adeo sub aequatore tam est exiguum, ut vix in sensus occurrere queat, integrum enim digitum non attingit (§41); atque in regionibus ab aequatore remotis fit multo minus. Vera igitur huius Phaenomeni ratio in altitudine Lunae meridiana seu distantia ab horizonte continetur; hinc enim sequitur quo maior fuerit differentia inter distantias Lunae ab horizonte, dum per meridianum transit tum super horizonte tum sub horizonte, eo maiorem esse debere differentiam inter binos Fluxus successivos, ex quo perspicuum est istam differentiam versus polos continuo crescere debere, si quidem Luna habeat declinationem. Quod

si ergo Luna habuerit declinationem borealem, tum in regionibus septentrionalibus Fluxus erit maior qui transitum Lunae per meridianum superiorem sequitur, alter vero sequens, qui transitui inferiori respondet, minor. Contra autem si Lunae declinatio fuerit australis, appulsui Lunae ad meridianum superiori Fluxus succedet minor, inferiori vero maior; hancque differentiam FLAMSTEDIUS observavit diligenter, nullumque est dubium, quin ea per copiosissimas observationes, quas Academia Celeberrima Regia Parisina collegit, omnino confirmetur. In hoc autem negotio indoles Fluxuum probe est inspicienda, quoniam aliquibus in portibus tantopere retardantur, ut sequentibus Lunae transitibus per meridianum sint propiores, quam illi, cui suam originem debent; ita Dunkerquae circa syzygias Fluxus circiter meridie observari solet, neque vero illi ipsi transitui Lunae per meridianum est tribuendus qui eodem tempore fit, sed praecedenti, prouti successiva retardationis incrementa ad littora Galliae et Belgii borealia evidentissime testantur. Quare si verbi gratia Dunkerquae quis huiusmodi observationes perlustrare voluerit, is quemque Fluxum non cum transitu Lunae per meridianum proximo comparet, sed cum eo qui propemodum 12 horis ante contigit; alioquin enim contraria Phaenomena esset deprehensurus.

108. Commodus hic nobis praebetur locus explicandi transitum a binis aestubus, qui quotidie in regionibus extra circulos polares sitis eveniunt, ad singulos aestus, qui secundum theoriam nostram in regionibus polaribus contingere debent. Quoniam enim theoria nostra monstrat, in zonis temperatis et torrida quotidie duos Fluxus observari debere, in zonis frigidis autem unum tantum, transitio subitanea a binario ad unitatem maxime mirabilis ac paradoxa videri posset. Sed quia si Fluxus bini successivi inter se sunt inaequales, Refluxus aquae seu maxima depressio Fluxui minori est vicinior, bini aestus quoque successivi ratione temporis inter se erunt inaequales, si quidem voce aestus intelligamus motum aquae a summa elevatione ad imam depressionem usque, ac vicissim. Quo magis itaque ab aequatore versus polos recedatur, eo maior deprehendetur inter binos aestus successivos inaequalitas, cum ratione magnitudinis tum temporis, maior enim diutius durabit quam minor, ambo vero simul ubique absolventur tempore 12 horarum, cum 24′ circiter: quod si itaque in eas regiones usque perveniatur, in quibus Luna utraque vice vel super horizonte vel sub horizonte meridianum attingit, aestus minor omnino evanescet, solusque maior supererit, qui tempus 12 h. 24′ adimplebit. Ex quibus perspicuum est, si Luna habeat declinationem, inaequalitatem binorum aestuum successivorum ad polos accedendo continuo fieri

maiorem, atque tandem minorem omnino evanescere debere, quod cum evenit, bini aestus in unum coalescunt.

109. Explicatis anomaliis aestus Maris menstruis, pervenimus ad anomalias annuas vel plusquam annuas, ac nonum quidem Phaenomenon desumimus ex variatione aestus, quae a diversis Lunae a Terra distantiis proficiscitur. Observantur enim aestus ubique maiores ceteris paribus, in iisdem scilicet luminarium aspectibus iisdemque declinationibus, si Luna in suo perigaeo versetur, minores vero, Luna in apogaeo existente. Egregie autem haec conveniunt cum nostra theoria, qua demonstravimus Lunae vires ad Mare movendum decrescere in triplicata ratione distantiarum Lunae a Terra: quod si igitur Luna versetur in perigaeo Fluxus debebunt esse maiores, quam si Luna apogaeum occupat. Praeterea etiam tabula quam Celeb. CASSINI[38]) in Mem. 1713 pro diversis Lunae a Terra distantiis ex plurimis observationibus Brestiae institutis collegit, satis accurate cum theoria nostra conspirat, etiamsi enim pro Luna perigaea minorem elevationem aquae tribuat, quam ista ratio requireret, tamen discrimen valde est exiguum: quin etiam facile concedetur Lunam perigaeam totum suum effectum non tam cito consequi posse, quem tandem consequeretur, si Luna perpetuo in perigaeo versaretur. Aliter autem Luna apogaea est comparata, quae ad diminuendum aestum Maris tendit, cum enim Mare ob inertiam et impedimenta ipsum ad diminutionem aestus sit proclive, sine ulla resistentia Luna in apogaeo constituta effectum suum exeret. Huc etiam pertinet, quod pariter Celeb. CASSINI se observasse testatur, similem differentiam etsi multo minorem a variis Solis a Terra distantiis produci, id quod nostrae theoriae non solum est consentaneum, sed inde etiam ipsa quantitas huius differentiae potest definiri.

110. Denique decimum Phaenomenon sese nobis contemplandum offert, quo vulgo statui solet aestus tam noviluniorum quam pleniluniorum, qui contingant circa aequinoctia, ceteris esse maiores, etiamsi observationes hanc regulam non penitus confirment; quamobrem videamus quomodo aestus ceteris paribus comparatus esse debeat pro diversis Lunae declinationibus. Ac primo quidem ex nostra theoria constat (§ 87) aestus dum Luna in aequatore versatur, maximos esse non posse, nisi in locis sub ipso aequatore sitis; atque eodem loco tabellam adiecimus,

---

38) J. CASSINI, «Reflexions sur des nouvelles observations du flux et reflux de la mer, faites au Port de Brest», *Mémoires de l'Académie Royale des Sciences* (Paris), (1713), 1716, pp. 14–32. EJA.

ex qua patet, cuinam Lunae declinationi maximi aestus respondeant. Ita pro elevatione poli 50°, aestus maximi incidunt Lunae declinationi 27°, si quidem g ponatur $= \frac{2}{25}$; at posito $g = \frac{1}{10}$, quod probabilius videtur, prodit Lunae declinatio maximum aestum producens circiter 16°, id quod mirifice convenit cum observationibus ad Littora Galliae Septentrionalia institutis, quibus constat maximos syzygiarum aestus mensibus Novembri et Februario accidere solere, quibus temporibus Luna fere assignatam obtinet declinationem. At quod forte illi regulae, qua Lunae in aequatore versanti maximi aestus adscribi solet, ansam praebuisse videtur, est modus aestuum quantitates definiendi peculiaris ac satis perversus; cum enim crederent plerique observatores causis alienis tribuendam esse inaequalitatem, quae inter binos aestus successivos intercedat, veram aquae elevationem accuratius definire sunt arbitrati, si sumerent medium inter binos Fluxus successivos. Quod si autem hoc modo quique aestus aestimentur, tum utique maximi aestus in aequinoctia incidere observabuntur, id quod etiam nostrae theoriae maxime est conforme, exceptis tantum regionibus polis vicinioribus. Cum enim positis sinu elevationis poli $= P$, cosinu $= p$, sinu declinationis Lunae $= Q$, cosinu $= q$, maior aestus fiat per spatium

$$= \frac{3g}{h(1-8g)}\left(pq + \frac{PQ(1-8g)}{1-2g}\right)^2,$$

minor vero per spatium

$$= \frac{3g}{h(1-8g)}\left(pq - \frac{PQ(1-8g)}{1-2g}\right)^2,$$

(§86) erit per hunc aestum Maris mensurandi modum quantitas aestus

$$= \frac{3g}{h(1-8g)}\left(p^2q^2 + \frac{(1-8g)^2P^2Q^2}{(1-2g)^2}\right)$$

$$= \frac{3g}{h(1-8g)}\left(p^2 - p^2Q^2 + \frac{(1-8g)^2P^2Q^2}{(1-2g)^2}\right);$$

ex qua expressione perspicitur maximos aestus ubique, si quidem modo recensito mensurentur, Lunae in ipso aequatore degenti respondere, nisi sit $\frac{(1-8g)^2P^2}{(1-2g)^2} > p^2$, hoc est nisi tangens elevationis poli maior sit quam $\frac{1-2g}{1-8g}$: his scilicet regionibus etiam Luna declinans ab aequatore maiores aestus producet. At si ponatur $g = \frac{2}{25}$, prodit elevatio poli, ubi regula prolata fallere incipit, 66°; sin autem ponatur $g = \frac{1}{18}$, fit elevatio poli maior quam 58°; at posito $g = \frac{1}{10}$, provenit poli elevatio 76°. Cum igitur in locis polis tam vicinis observationes institui non soleant, satis tuto affirmare licet, maximos aestus menstruos accidere circa aequinoctia, si quidem quantitas aestus quotidie mensuretur per medium arithmeticum inter spatia, quae duo aestus successivi conficiunt.

111. Quid nunc aliud de theoria nostra sit sentiendum, nisi eam veram et genuinam aestus Maris causam, qualis ab Illustrissima Academia Regia in proposita quaestione desideratur, in se complecti, non videmus? Non solum enim omnia Phaenomena, quae in aestu Maris observantur, clare et distincte explicavimus, sed etiam existentiam actualem earum virium, quibus hos effectus adscribimus evidentissime demonstravimus; ex quo efficitur causam a nobis assignatam, non tantum omnibus Phaenomenis satisfacere, sed etiam esse unicam quae cum vera consistere queat. Quod si enim quispiam alias vires excogitet, quibus aeque omnia Phaenomena explicare posset, etiamsi hoc fieri posse minime concedamus, eius certe explicatio subito concideret et everteretur a viribus nostrae theoriae, quas aliunde in mundo existere abunde constat; quoniam ab illis viribus imaginariis hisque realibus coniunctim effectus duplicatus consequi deberet, quem experientia aversatur. Nunc igitur nobis summo iure asserere posse videmur, veram aestus Maris causam in duobus vorticibus esse positam, quorum alter circa Solem, alter circa Lunam agitetur, atque uterque eius sit indolis, ut vires centrifugae decrescant in duplicata ratione distantiarum a centris utriusque vorticis: quae proprietas obtinetur, si celeritas materiae subtilis gyrantis in quoque vortice teneat rationem reciprocam subduplicatam distantiarum. Neque vero hi duo vortices ad libitum sunt excogitati, sed ille qui Solem circumdat est is ipse, qui omnes planetas in suis orbitis continet; alter vero Lunam circumdans, etsi eius vis nisi in aestu Maris non sentitur, tamen sine ulla haesitatione admitti potest, cum certo constet Terram, Iovem ac Saturnum similibus vorticibus esse cinctas, unde eiusmodi vortices nulli omnino corpori mundano denegari posse videntur. Parcius quidem hic materiam de vorticibus tractavimus, etiamsi in illis veram aestus Maris causam ponamus; hoc autem de industria fecimus, cum hoc argumentum iam toties sit tractatum ac fere exhaustum; neque nobis persuadere possumus, si hac occasione doctrinam de vorticibus etiam melius, quam etiamnum a quoquam est factum, expediremus, ob eam rem praemium nobis tributum iri.

CAPUT OCTAVUM

De Aestus Maris perturbatione a Terris ac littoribus oriunda

112. Pervenimus tandem ad ultimam nostrae disquisitionis partem, quae praecipua est, in qua Theoriam expositam ad statum telluris, in quo revera reperitur, debito modo accommodabimus. Hactenus enim, quo ardua ista disquisitio facilior redderetur, ab omnibus circumstantiis externis quibus effectus a viribus Solis ac Lunae oriundis vel turbari vel determinatu difficiliores reddi possent, cogitationem abstraximus. Primo scilicet non solum totam Terram ex aqua conflatam posuimus, sed etiam inertiam aquae mente sustulimus, ut eo pauciores res in computum ducendae superessent. Deinde inertiae quidem habuimus rationem, ac praecedentes determinationes debito modo correximus; verum totam Terram aqua undiquaque circumfusam assumsimus, seu etiamnum anomalias a Terris oriundas negleximus. Nunc itaque nostra theoria eo est perducta, ut nihil amplius adiicere necesse foret, si quidem aestus Maris a Terris littoribusque sensibiliter non afficeretur; nisi forte anomaliae quaedam a ventis oriundae commemorari deberent, quae autem motu aquae perspecto facile diiudicantur, atque ad omnes theorias aeque pertinent. Quamobrem ultimum hoc caput destinavimus explicationi Phaenomenorum quorumdam singularium, quorum causa non tam in ipsa aqua viribusque eam sollicitantibus, quam in Terra continenti littoribusque est quaerenda: hac enim parte absoluta nihil amplius restare videtur, quod vel ad Theoriae nostrae confirmationem, vel ad omnium Phaenomenorum adaequatam explicationem desiderari queat. Quamvis enim Illustrissima Academia totum hoc argumentum non penitus exhauriri iubeat, cum adhuc nonnullas quaestiones de eodem in posterum proponere constituisset, tamen quia hoc tempore vera causa physica desideratur, veritatem nostrae theoriae non satis confirmari arbitramur, nisi eius convenientiam cum omnibus Phaenomenis dilucide ostenderemus, cum si vel unicum Phaenomenon refragaretur, eo ipso tota theoria subverteretur; quam ob causam prolixitatem nostrae tractationis, atque transgressionem limitum praescriptorum nobis sine difficultate condonatum iri confidimus.

113. Primum autem perspicuum est motum Maris horizontalem quo vel versus orientem vel occidentem progreditur, ob Terram interpositam non solum perturbari, verum etiam quandoque prorsus impediri debere. Supra enim ostendimus, si tota Terra aqua esset circumfusa, tum ubique ad Fluxum formandum aquam ab oriente advehi debere, ante refluxum autem versus ortum defluere. Quod

si ergo oceanus versus orientem Terris terminetur, fieri omnino nequit tempore Fluxus ad haec littora aqua ab oriente affluat, quo ipso cursus aquae naturalis penitus impedietur. Quoniam autem vires Solis ac Lunae nihilominus his in regionibus Mare attollere conantur, effectum consequi non poterunt, nisi aqua ab occidente afferatur: sic quando ad littora Europae aqua a viribus Solis ac Lunae elevatur, aqua ab occidente eo deferatur necesse est, ab iis scilicet regionibus, ubi aqua eodem tempore deprimitur; quod idem fieri debet ad littora Africae et Americae occidentalia. Contra vero ad littora Asiae et Americae orientalia aqua naturali motu feretur, atque in Fluxu ab oriente adveniet, in Refluxu vero versus orientem recedet. Vires namque Solis ac Lunae motum aquae horizontalem non plane[39]) determinant, sed eatenus tantum, quatenus aliis in locis aquam attollunt, aliis vero eodem tempore deprimunt; atque aqua ob propriam gravitatem eum seligit motum, quo facillime a locis quibus deprimitur, ad loca quibus attollitur promoveatur: quamobrem iste motus maxime a Terris oceanum includentibus determinetur necesse est. Hinc igitur perspecta positione littorum cuiusvis Maris facile definiri poterit, a quanam plaga aqua in Fluxu venire, quorsumque in Refluxu decedere debeat, si modo elevationes et depressiones aquae per totum Mare attente considerentur: tota enim haec quaestio pertinebit ad hydrostaticam.

114. Cum igitur ad littora Europae aqua elevari nequeat, nisi affluxus ab occidente fiat copiosus, ad littora quae versus occidentem respiciunt aqua directe ab occidente adveniet, quae autem littora ad aliam plagam sunt disposita, aquae cursus versus orientem directus inflectetur iuxta littora, priusquam eo pertingat, omnino uti inspectio mapparum docebit. Quoniam vero iste aquae iuxta littora Fluxus tantam celeritatem, quantam habet Luna, recipere nequit, necesse est, ut Fluxus ad littora magis ad orientem sita tardius advehatur. Haec autem versus littora orientaliora retardatio maxime perspicua est in portubus Galliae, Belgii, Angliae et Hiberniae; cum enim ad ostia fluviorum Garumnae et Ligeris, quae versus oceanum amplissimum patent, tempore pleniluniorum ac noviluniorum Fluxus adveniat hora tertia pomeridiana, quae retardatio naturalis censeri potest, neque littoribus adhuc turbata; hinc aqua demum ad littora Britanniae minoris ac Normanniae progreditur; atque idcirco his in regionibus Fluxus tardius evenire observantur. Sic ad Portum S. Malo tempore syzygiarum Fluxus demum hora sexta sequitur, ad ostia vero Sequanae usque ad horam nonam retardatur: atque ita

39) Secundum Manuscriptum. Editio princeps: per se *loco* plane. EJA.

porro retardatio augetur, donec tandem in freto Gallico Dunkerquae et Ostendae media nocte incidat. Ex hac vero retardatione innotescit celeritas aquae, qua iuxta littora progreditur, eaque tanta deprehenditur qua una hora spatium circiter 8 milliarium conficiat. Denique aqua tantam fere viam absolvere debet usque ad Dublinum, quantam ad fretum Gallicum, ex quo Fluxus etiam Dublini hora circiter decima pomeridiana observari solet. Atque simili modo retardatio Fluxuum ad littora aliarum regionum sine ulla difficultate explicari poterit.

115. Quod autem ad quantitatem aestus Maris ad littora attinet, facile intelligitur aestum Maris ad littora maiorem esse debere, quam in medio mari. Primo enim aqua cum impetu ad littora allidit, ex quo allapsu solo iam intumescentia oriri debet. Deinde quoniam aqua eadem celeritate, quam habebat in oceano, ubi maxima est profunditas, progredi conatur, ad littora locaque vadosa vehementer inturgescet, tantum enim fere aquae ad littora affertur, quantum sufficeret ad spatium, quod Terra occupat, inundandum. Tertio iste aquae affluxus in sinibus vadosis multo adhuc magis increscere debet, eo quod aqua his in locis iam multum appulsa ad latera diffluere nequit, si quidem sinus directe versus eam plagam pateat, unde aqua advehitur. Ex his igitur non solum ratio patet, cur aqua fere ubique ad littora ad multo maiorem altitudinem elevetur, quam in medio Mari, sed etiam cur Bristolii tam enormis Fluxus circa syzygias luminarium observetur; cum enim in hac regione littus sit valde sinuosum ac vadosum, aqua maxima vi appellitur, neque ob sinuositatem tam cito diffluere potest. Atque ex his principiis non erit difficile rationem inconsuetorum aestuum, qui passim in variis portubus animadvertuntur, indicare atque explicare; quamobrem huius generis Phaenomenis explicandis diutius non immoramur, cum consideratio littorum et Fluxus aquae eo sponte quasi manuducat.

116. Quamvis autem tam Affluxus aquae ex oceano Atlantico, quam Refluxus per fretum Galliam ab Anglia dirimens, ingenti fiat celeritate, tamen cum versus Belgium foederatum Mare mox vehementer dilatetur, ab isto alterno Fluxu ac Refluxu altitudo Maris in oceano Germanico sensibiliter mutari nequit. Atque hanc ob causam statui oportet, in hoc Mari aestum proficisci maximam partem ab affluxu et refluxu aquae circa Scotiam, ubi communicatio huius Maris cum oceano Atlantico multo maior patet; quam sententiam magnopere confirmat ingens aestuum retardatio ad littora Belgii et Angliae orientalia observata, ad Ostia scilicet Thamisis pertingit Fluxus elapsis iam duodecim horis post transitum Lunae per meridianum, atque ad Londinum usque tribus fere horis tardius defertur; quod

Phaenomenon consistere non posset si aqua per fretum Gallicum solum moveretur, cum iam in ipso freto duodecim horis retardetur Fluxus. Interim tamen negari non potest quin communicatio Maris Germanici cum oceano Atlantico per fretum Gallicum aestum quodammodo afficiat, atque Fluxum qui circa Scotiam advehitur vel adiuvet vel turbet, prout hi ambo motus ad Mare elevandum ac deprimendum vel magis inter se conspirent vel minus. Simul autem hinc intelligitur aestum Maris ex oceano Atlantico neque cum Mari Mediterraneo neque cum Mari Baltico communicari posse, cum intervallo sex horarum per freta Herculea et Oresundica tantum aquae in haec maria neque affluere queat neque inde refluere, ut sensibilis mutatio in altitudine aquae oriri queat. Quamobrem in istiusmodi maribus quae a vasto oceano tantum angustis fretis separantur, aestus omnino nullus contingere potest, nisi forte talia maria Terris inclusa ipsa tam sint ampla, ut vires Solis ac Lunae aestum peculiarem in iis producere queant; qua de re mox videbimus.

117. Quemadmodum autem vidimus in Mari Germanico duplicem extare aestum, quorum alter, qui quidem longe est minor, per fretum Gallicum, alter circa Scotiam advehitur ex eodem oceano Atlantico: ita propter singularem littorum quorumdam situm mirabilia Phaenomena in aestu Maris evenire possunt. Quod si enim littus quodpiam ita fuerit comparatum, ut aestus in id duplici via vel ex eodem oceano, vel ex diversis communicetur, ratione temporis, quo bini isti aestus adveniunt insignes discrepantiae oriri poterunt. Nam si per utramque viam Fluxus eodem tempore advehatur, atque adeo simul Refluxus congruant, aestus multo maiores existere debebunt. Sin autem eo tempore, quo per alteram viam Fluxus advenit, ex altera via Refluxus incidat, tum aestus omnino destruetur si quidem per utramque viam aqua aequali [copiose]* vel affluat vel defluat. Ad hoc vero non sufficit ut ambae viae sint aequales, sed etiam requiritur ut bini aestus successivi sint aequales, id quod evenit si Luna vel non habeat declinationem, vel littus in aequatore fuerit positum. Quod si autem eadem duplici communicatione posita, tam Luna habeat declinationem, quam littus notabiliter ab aequatore sit remotum, tum ob inaequalitatem binorum aestuum sese insequentium, Fluxus maiores ex altera via advenientes, superabunt Refluxus minores eodem tempore per alteram viam factos, atque hoc modo in tali littore singulis diebus non bini Fluxus, sed unus tantum accidet; hancque rationem allegat NEWTONUS aestus illius singularis Tunquini observati, ubi si Luna in aequatore versatur nullus aestus deprehenditur, sin autem Luna habeat declinationem unicus tantum una Lunae revolutione circa Terram [observatur]*. Nos autem mox huius mirabilis Phaenomeni aliam magis naturalem nostraeque theoriae conformem indicabimus causam.

118. Hactenus aestum Maris, quemadmodum in amplissimo oceano a viribus ad Lunam ac Solem tendentibus producatur, atque vario littorum situ cum ratione quantitatis tum retardationis diversimode turbetur, sumus contemplati, neque necesse esse duximus ventorum Marisque cursuum propriorum rationem habere, cum satis pronum sit perspicere, quomodo his rebus aestus Maris tam augeri vel diminui, quam accelerari vel retardari debeat. Superest igitur ut exponamus, quomodo in satis amplo tractu Maris, qui ab oceano vel omnino est seiunctus, vel per angustum tantum canalem coniunctus, peculiaris aestus a viribus Lunae ac Solis produci queat. Perspicuum enim est si talis tractus secundum longitudinem ultra 90 gradus pateat, aestum pari modo generari debere, ac in amplissimo oceano, qui totam tellurem ambire ponitur. Nam quoniam extensio tanta est, ut vires Lunae et Solis in eo tractu simul maximam ac minimam aquae altitudinem inducere queant, necesse est etiam, ut aqua alio in loco tantum elevetur, inque alio tantum deprimatur, quantum fieret, si iste tractus omnino non esset terminatus. At si iste tractus tam fuerit parvus ut singulae partes aequalibus fere viribus simul vel attollantur vel deprimantur, nulla sensibilis mutatio oriri poterit. Aqua enim uno in loco attolli nequit nisi in alio subsidat et contra, si quidem eadem aquae copia in eo tractu perpetuo conservetur. Atque haec est ratio ut in Mari Baltico, Caspio, Nigro, aliisque minoribus lacubus nullus omnino aestus deprehendatur.

119. Quod si autem istiusmodi Maris tractus tantum spatium occupet, ut vires attollentes et deprimentes in extremitatibus sensibiliter differant, tum necesse est ut non solum aqua in altero extremo elevetur in alteroque deprimatur, sed etiam ut differentia inter aquae altitudines tanta sit, quanta in aperto oceano eidem virium differentiae respondet. Quamobrem definiri conveniet, quanta differentia in diversis Terrae locis eodem tempore in altitudinibus aquae a viribus Lunae ac Solis produci queat. Ne autem calculus nimium fiat prolixus, solam Lunae vim in computum ducemus, quippe quae vim Solis multum excedit; et quoniam effectu Lunae cognito facile est Solis effectum aestimando vel adiicere vel auferre. Repraesentet ergo [fig. 14] $PLpl$ superficiem Terrae cuius poli sint P et p, atque M et N sint duo termini in eodem Maris tractu assumti, in quibus quantum Maris altitudo quovis tempore differat, sit investigandum. Repraesentet porro Ll parallelum, in quo Luna moveatur hoc tempore, sitque Luna in L; atque exprimet angulus LPM tempus, quod post Lunae transitum per meridianum termini M est praeterlapsum, angulus vero LPN tempus post transitum Lunae per meridianum alterius termini N. Ductis autem circulis maximis PM et PN, erit arcus PM complementum

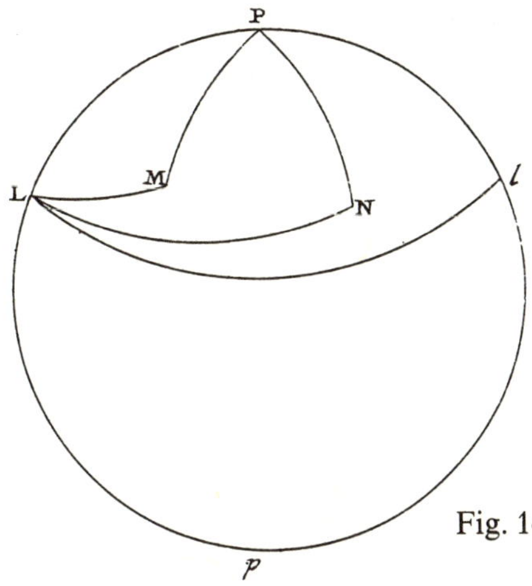

Fig. 14

latitudinis loci M, arcus PN vero loci N; angulus vero MPN dabit differentiam longitudinis locorum M et N; quae proinde omnia ponuntur cognita.

120. Ducantur iam ex loco Lunae L ad terminos M et N circuli maximi LM et LN, exhibebuntque isti arcus complementa altitudinum, quibus hoc tempore Luna in locis M et N supra horizontem elevata conspicitur. Ponatur arcus PL sinus $= q$, cosinus $= Q$, erit Q sinus declinationis borealis Lunae, si quidem Q habeat valorem affirmativum, ac P polum borealem denotet. Deinde ponatur arcus PM sinus $= p$, cosinus $= P$, erit P sinus elevationis poli pro loco M; similique modo sit arcus PN sinus $= r$ et cosinus $= R$, ita ut R sit sinus elevationis poli loci N: denique sit anguli MPN sinus $= M$ et cosinus $= m$, anguli vero LPM sinus $= T$, cosinus $= t$; unde erit anguli LPN cosinus $= mt - MT$. Ex his per trigonometriam sphaericam reperietur sinus altitudinis Lunae supra horizontem loci M seu cosinus arcus $LM = tqp + QP$: pro loco N vero erit altitudinis Lunae sinus $= (mt - MT)qr + QR$. Quare si ut supra vis absoluta ad Lunam urgens ponatur $= L$ et distantia Lunae a Terra $= b$, erit altitudo ad quam aqua in M elevari deberet $= \dfrac{L(3(tpq + PQ)^2 - 1)}{2b^3}$, et altitudo ad quam aqua in N elevari deberet $= \dfrac{L(3((mt - MT)qr + QR)^2 - 1)}{2b^3}$, utroque casu supra libellam naturalem. Si ergo illa expressio hanc excedat, aqua in M altius erit elevata quam in N intervallo

$$\frac{3L}{2b^3}\left((tpq + PQ)^2 - ((mt - MT)qr + QR)^2\right),$$

haecque expressio, quando fit negativa, indicabit, quanto aqua in N altius consistat quam in M. In hoc vero negotio inertiam aquae negligimus, quoniam tantum proxime Phaenomena huiusmodi casibus oriunda indicare annitimur; si enim hanc materiam perfecte evolvere vellemus, integro tractatu foret opus.

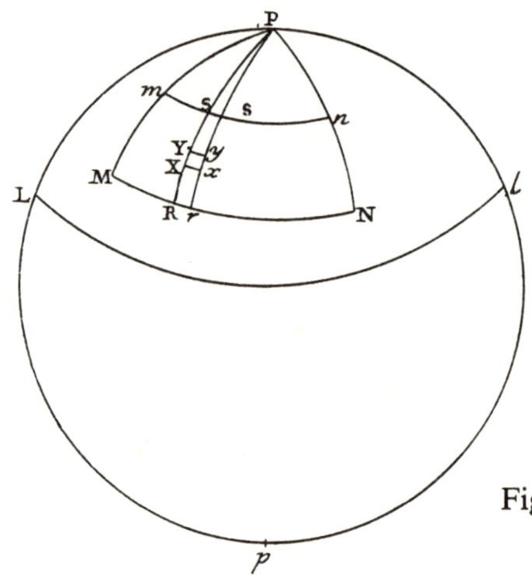

Fig. 15

121. Ponamus tractum nostrum Maris ab oriente N [fig. 15] versus occidentem M sub eodem parallelo extendi, ita ut elevatio poli in locis M et N sit eadem; erit adeo $R = P$, et $r = p$. Transeat nunc Luna per meridianum loci M supra Terram ita ut sit $T = 0$, $t = 1$; hoc ergo tempore [aqua]* magis erit elevata in M quam in N intervallo

$$\frac{3L}{2b^3}((pq + PQ)^2 - (mpq + PQ)^2) = \frac{3L}{2b^3}(M^2p^2q^2 + 2(1 - m)pqPQ).$$

At quando Luna per meridianum loci N supra Terram transit, aqua tantundem magis erit elevata in N quam in M. Ex quo sequitur, dum Luna a meridiano loci N ad meridianum loci M progreditur, aquam in M sensim elevari per spatium

$$\frac{3Lpq}{2b^3}(M^2pq + 2(1 - m)PQ),$$

interea vero in N tantundem subsidere. Sin autem Luna infra Terram a meridiano loci N ad meridianum loci M progrediatur, aqua in M elevabitur interea per spatium

$$= \frac{3Lpq}{2b^3}\left(M^2pq - 2(1-m)PQ\right),$$

per tantumque spatium aqua in N subsidet. Ponamus nunc angulum LPM esse 90 graduum, seu quaestionem institui, cum Luna iam ante sex horas meridianum loci M sit transgressa, atque obtinebitur differentia inter aquae altitudines in locis[40]) M et N

$$= \frac{3L}{2b^3}\left(P^2Q^2 - (PQ - Mpq)^2\right) = \frac{3Lpq}{2b^3}(2MPQ - M^2pq).$$

Sex autem horis, antequam Luna ad meridianum loci M appellit, aqua in N magis erit elevata quam in M intervallo

$$= \frac{3Lpq}{2b^3}(2MPQ + M^2pq).$$

Sequuntur haec si inertia aquae negligatur; at inertia admissa ex praecedentibus satis clarum est, cum has differentias maiores esse debere, tum tempora mutationum tardius sequi debere.

122. Quoniam vero in hoc Maris tractu perpetuo eadem aquae quantitas contineri debet, necesse [est] ut quantum aquae una parte supra libellam attollatur, tantundem ea in reliqua parte infra libellam deprimatur. Quo igitur hinc altitudinem Maris quovis loco exacte determinemus, ponamus tractum nostrum secundum longitudinem terminari binis meridianis PM et PN, secundum latitudinem vero binis parallelis MN et mn, positaque Luna in L sit sinus $PL = q$, cosinus $= Q$; sinus $LPM = T$, cosinus $= t$. Porro sit sinus arcus $PM = p$, cosinus $= P$, sinus $Pm = r$, cosinus $= R$, atque anguli MPN sinus $= M$ et cosinus $= m$. Praeterea sit elevatio in M dum Luna in L versatur, supra libellam $= \alpha$, ita ut hoc loco suprema aquae superficies a centro Terrae distet intervallo $= 1 + \alpha$, unde cum sinus altitudinis Lunae in M sit $= tpq + PQ$, erit gravitatio totius columnae aquae ab M ad centrum Terrae

$$= \frac{(1+\alpha)^{n+1}}{n+1} + \frac{L(1 - 3(tpq + PQ)^2)}{2b^3} = \frac{1}{1+n} + \alpha + \frac{L(1 - 3(tpq + PQ)^2)}{2b^3},$$

40) Editio princeps: $(PQ - Mpq)$ loco $(PQ - Mpq)^2$. EJA.

prouti supra §§43 et 44 demonstravimus. Consideretur iam locus quicunque X in nostro tractu, in quo aqua supra libellam sit elevata spatio $=\varphi$; ac ducto per hunc locum meridiano PR, sit anguli LPR sinus $=X$, cosinus $=x$; arcus PX sinus $=z$ et cosinus $=Z$, unde gravitatio columnae aqueae ex X ad centrum Terrae pertingentis erit

$$=\frac{1}{1+n}+\varphi+\frac{L(1-3(xqz+QZ)^2)}{2b^3}.$$

Cum igitur haec gravitatio aequalis esse debeat illi, orietur

$$\varphi=\alpha+\frac{3L}{2b^3}\left((xqz+QZ)^2-(tpq+PQ)^2\right),$$

ex qua formula si modo constaret elevatio aquae in M, simul innotesceret elevatio vel depressio in quovis loco X.

123. Cum ergo in X aqua supra libellam elevetur spatio φ, in elemento tractus infinite parvo $XYyx$, plus inerit aquae, quam in statu naturali, et quidem quantitas $XY\cdot Xx\cdot\varphi$, cuius elementi integrale per totum tractum sumtum debet esse $=0$, ex quo valor ipsius α innotescet. Erit autem angulus $RPr=\frac{dX}{x}$, hincque arculus $Xx=\frac{z\,dX}{x}$, at elementum $XY=\frac{dZ}{z}$, ex quo infinite parvum rectangulum $XYyx=\frac{dX\,dZ}{x}$, in quo ergo excessus aquae supra statum naturalem est

$$=\frac{\varphi\,dX\,dZ}{x}=\frac{dX}{x}\left(\alpha\,dZ+\frac{3L\,dZ}{2b^3}\left((xqz+QZ)^2-(tpq+PQ)^2\right)\right),$$

quae formula bis debet integrari. Ponatur primo X constans, et integratione absoluta reperietur in elemento $RSsr$ excessus aquae supra statum naturalem

$$=\frac{dX}{x}\Bigg(\alpha(R-P)+\frac{3L}{2b^3}\Bigg(q^2x^2(R-P)-\frac{x^2q^2}{3}(R^3-P^3)-\frac{2xQq}{3}(r^3-p^3)$$
$$+\frac{Q^2}{3}(R^3-P^3)-(tpq+PQ)^2(R-P)\Bigg)\Bigg).$$

Integretur haec formula denuo ut integrale ad totum tractum $MNnm$ extendatur, prodibitque incrementum aquae, quod toti tractui accessisse oporteret[41]),

$$\begin{aligned}
&= \alpha(R-P)\,\mathrm{A}\sin M \\
&+ \frac{3L}{2b^3}\left(\frac{q^2(3(R-P)-(R^3-P^3))}{6}(Mm(1-2TT)-2M^2Tt)\right. \\
&\quad + \frac{2Qq(r^3-p^3)}{3}(T-Mt-mT)+\frac{q^2(R-P)}{2}\mathrm{A}\sin M \\
&\quad \left. + \frac{(3Q^2-1)(R^3-P^3)}{6}\mathrm{A}\sin M-(tpq+PQ)^2(R-P)\,\mathrm{A}\sin M\right),
\end{aligned}$$

quae adeo quantitas debet esse $=0$: unde oritur

$$\begin{aligned}
\alpha &= \frac{3L(tpq+PQ)^2}{2b^3}+\frac{L(1-3Q^2)(R^2+PR+P^2)}{4b^3}-\frac{3Lq^2}{4b^3} \\
&+ \frac{3L}{2b^3(R-P)\,\mathrm{A}\sin M}\left(\frac{q^2(3(R-P)-(R^3-P^3))}{6}(2M^2Tt-Mm(1-2TT))\right. \\
&\quad \left. + \frac{2Qq(p^3-r^3)}{3}(T-Mt-mT)\right).
\end{aligned}$$

124. Cognita igitur vera elevatione aquae in M supra libellam, quam ante posuimus $=\alpha$, hinc intelligetur vera aquae elevatio supra libellam in loco quocunque X. Ponatur enim sinus anguli $MPX = S$ et cosinus $=s$, erit $\sin LPR = X = Ts + tS$ et $x = ts - TS$, manentibusque arcus PX sinu $=z$ et cosinu $=Z$, erit elevatio aquae in X

$$= \varphi = \alpha + \frac{3L}{2b^3}((ts-TS)qz+QZ)^2 - \frac{3L}{2b^3}(tpq+PQ)^2;$$

quare loco α valore invento substituto, reperietur aqua in X supra libellam attoli actu per spatium

41) Asin denotat arcum sinus. EJA.

$$= \frac{3L}{2b^3}((ts-TS)qz+QZ)^2 + \frac{L(1-3Q^2)(R^2+PR+P^2)}{4b^3} - \frac{3Lq^2}{4b^3}$$

$$+ \frac{3L}{2b^3(R-P)\mathrm{A}\sin M}\left(\frac{q^2(3(R-P)-(R^3-P^3))}{6}(2M^2Tt-Mm(1-2TT)) + \frac{2Qq(p^3-r^3)}{3}(T-Mt-mT)\right).$$

Quod si ergo ponatur tractus noster ita augeri ut totam tellurem ambiat, orietur casus iam supra tractatus; quoniam enim fit $MN = 360°$, seu $\mathrm{A}\sin M = 2\pi$ denotante $1:\pi$ rationem diametri ad peripheriam, erit $M = 0$ et $m = 1$: praeterea vero quia M in polum australem p, m vero in borealem P incidit, erit $p = 0$, $P = -1$, $r = 0$ et $R = +1$: si hi valores substituantur, prodibit elevatio aquae in X

$$= \frac{L}{2b^3}(3((ts-TS)qz+QZ)^2-1),$$

quae expressio, quia $ts - TS$ denotat cosinum anguli LPX atque $(ts-TS)qz+QZ$ sinum altitudinis Lunae supra horizontem in X, cum superioribus formulis exactissime convenit: si quidem terminus $\frac{L}{b^4}$ negligatur. Haec vero eadem ipsa expressio quoque emergit, si tantum alterum hemisphaerium vel boreale vel australe ponatur aqua totum circumfusum, manent enim omnia ut ante, nisi quod fiat $p = 1$ et $P = 0$: utroque enim casu fit $R^2 + PR + P^2 = 1$; ultimusque terminus ob $M = 0$ utroque casu evanescit.

125. Ponamus nunc tractum Maris secundum longitudinem MN usque ad 180 gradus extendi, erit $M = 0$ et $m = -1$ et $\mathrm{A}\sin M = \pi$, denotat enim $\mathrm{A}\sin M$ semper arcum circuli, qui mensura est anguli MPN: hinc si brevitatis gratia ponatur sinus anguli, quo Luna in X supra horizontem elevata apparet, $= v$, erit aquae elevatio in X supra libellam

$$= \frac{3Lv^2}{2b^3} + \frac{L(1-3QQ)(R^2+PR+P^2)}{4b^3} - \frac{3Lqq}{4b^3} + \frac{2LTQq(p^3-r^3)}{(R-P)b^3\pi}.$$

Ponamus porro integrum hemisphaerium $LPlp$ aqua esse circumfusum, fiet $p = 0$, $P = -1$, $r = 0$ et $R = 1$; unde elevatio aquae in X erit $= \frac{L(3v^2-1)}{2b^3}$, omnino ac si tota Terra aqua cincta esset, uti in praecedentibus capitibus posuimus, vel quod eodem redit, dummodo omnis aqua super Terra mutuam habeat communicationem satis amplam. Quod si autem tractus noster Maris tantum ad aequatorem usque porrigatur a polo P, ita ut quartam superficiei terrestris partem solum obtegat, tum erit $p=1$, $P=0$, $r=0$ et $R=1$, hoc itaque casu aqua in X elevabitur ad altitudinem $=\frac{L(3v^2-1)}{2b^3}+\frac{2LTQq}{\pi b^3}$: ex quo perspicitur hoc casu elevationem in X maiorem fore, quam si tota Terra aqua esset circumdata, si expressio TQq habeat valorem affirmativum, minorem vero si TQq habeat valorem negativum. Sed limites huic quaestioni praescripti non permittunt hinc plura consectaria deducere, cum debita evolutio satis amplum tractatum requirat, neque theoria ulteriori confirmatione indigeat. Quocirca coronidis loco duos tantum casus evolvemus, quorum altero latitudo tractus ponetur infinite parva, altero vero longitudo: quippe qui ad phaenomena quaedam singularia explicanda inservire poterunt.

126. Ponamus igitur latitudinem Mm infinite esse parvam, seu $R=P$ et $r=p$, reperietur aquae in X elevatio supra libellam

$$=\frac{3Lv^2}{2b^3}+\frac{3L(P^2-q^2-3P^2Q^2)}{4b^3}$$
$$+\frac{3Lpq}{2b^3\operatorname{A}\sin M}\left(\frac{pq}{2}(2M^2Tt-Mm(1-2TT))+2PQ(T-Mt-mT)\right).$$

Consideremus autem elevationem in M, ubi cum sit $v=tpq+PQ$, erit ea

$$=\frac{3Lpq(2ttpq+4tPQ-pq)}{4b^3}$$
$$+\frac{3Lpq}{4b^3\operatorname{A}\sin M}\left(pq\,(2M^2Tt-Mm(1-2TT))+4PQ(T-Mt-mT)\right).$$

Transeat nunc Luna per meridianum loci M supra Terram, erit $T=0$, et $t=1$, atque elevatio in M prodibit[42])

42) Editio princeps et Manuscriptus: Expressiones mathematicae usque ad finem paragraphes errata continent. Correxit EJA.

$$= \frac{3Lpq(pq+4PQ)}{4b^3} - \frac{3Lpq}{4b^3\,\mathrm{A}\sin M}(Mmpq+4MPQ);$$

at si per eundem meridianum infra Terram transeat, erit aquae elevatio

$$= \frac{3Lpq(pq-4PQ)}{4b^3} - \frac{3Lpq}{4b^3\,\mathrm{A}\sin M}(Mmpq-4MPQ).$$

Quod si autem Luna versus ortum a meridiano distet angulo horario 90 graduum, seu circiter 6 horis ante appulsum Lunae ad meridianum in M superiorem, erit $T=-1$ et $t=0$, unde elevatio erit

$$= -\frac{3Lp^2q^2}{4b^3} + \frac{3Lpq}{2b^3\,\mathrm{A}\sin M}\big(pqMm - 4PQ(1-m)\big);$$

sex vero horis post transitum Lunae per meridianum loci M versus occasum, erit altitudo aquae in M supra libellam

$$= -\frac{3Lp^2q^2}{4b^3} + \frac{3Lpq}{2b^3\,\mathrm{A}\sin M}\big(pqMm + 4PQ(1-m)\big).$$

127. Tribuamus huic tractui longitudinem 90 graduum, ut sit $M=1$, $m=0$, et $\mathrm{A}\sin M = \frac{\pi}{2}$, unde oritur elevatio aquae in M

$$= \frac{3Lpq(2ttpq+4tPQ-pq)}{4b^3} + \frac{3Lpq}{2\pi b^3}\big(2pqTt+4PQ(T-t)\big).$$

Quae si etiam declinatio Lunae ponatur $=0$, fiet

$$= \frac{3Lp^2q^2(2tt-1)}{4b^3} + \frac{3Lp^2q^2Tt}{\pi b^3}$$

existente $q=1$, unde apparet maximam elevationem non accidere cum Luna per meridianum loci M transit, sed tardius, et quidem si dupli anguli LPM tangens[43])

43) Editio princeps et Manuscriptus: sinus loco tangens. Correxit EJA.

fuerit $= \frac{2}{\pi}$, hoc est fere una hora post transitum Lunae per meridianum, hoc igitur casu Fluxus in M una fere hora tardius observetur, quam si tota Terra aqua esset circumfusa. Dum autem Luna per meridianum superius transit, erit elevatio $= \frac{3Lpp}{4b^3}$, quae etiam valet si Luna infra Terram meridianum attingat; at sex horis vel ante vel post, quando Luna in horizonte versatur, erit aquae depressio $= -\frac{3Lpp}{4b^3}$. Unde intelligitur in tali Maris tractu pariter quotidie binos Fluxus totidemque Refluxus accidere debere, atque aestum propemodum fore similem aestui generali, nisi quod maioribus anomaliis sit obnoxius, praecipue si Luna habeat declinationem.

128. Hinc explicari potest ratio aestus, qui in Mari Mediterraneo observatur, et qui in ipso hoc Mari generatur. Cum enim longitudo huius Maris ne 60 quidem gradus attingat, aestus erunt multo minores; decrescunt enim si cum longitudo diminuatur, tum elevatio poli augeatur. Quod si ergo in his formulis angulus MPN ponatur fere 60 graduum, atque elevatio poli debita introducatur, reperientur quidem aestus bini quotidie evenire debere, qui autem futuri sint multo minores, quam in medio Mari, et pluribus anomaliis subiecti, quas quidem omnes ex formulis traditis definire licebit. Quoniam ergo tam exigui aestus a ventis et cursu aquae, qui in hoc Mari notabilis deprehenditur, vehementer turbantur, ad pleraque littora huius Maris vix usquam aestus regularis observabitur. Excipi autem debet Mare Adriaticum, quod cum sinum formet amplum, advenientem aquam melius colliget, atque elevationem multo sensibiliorem patietur, a quo aestus Maris Venetiis observatus originem habet. Tametsi enim Mare Mediterraneum non solum satis amplam habeat latitudinem, sed etiam vehementer inaequalem, tamen eiusmodi marium aestus admodum exquisite ex praesenti casu, quo latitudinem omnino negligimus, colligi potest, quia extensio Maris in longitudinem praecipuam causam aestuum binorum singulis diebus evenientium continet, neque extensio latitudinis multum conferat.

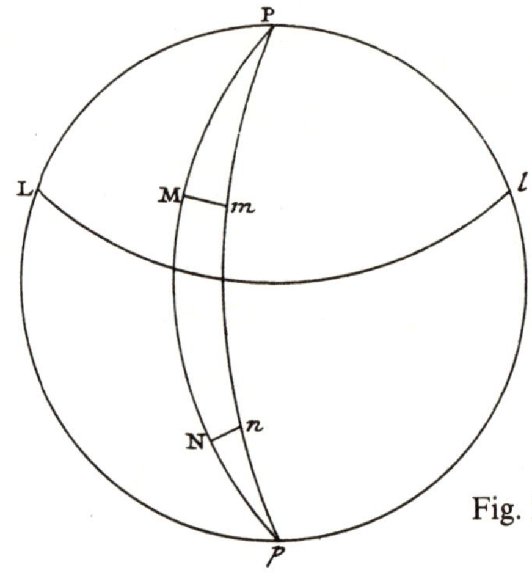

Fig. 16

129. Ponamus nunc tractus nostri Maris longitudinem evanescere, totumque tractum in eodem meridiano Pp [fig. 16] ab M usque ad N extendi, ita ut sit $M = 0$, $m = 1$; sinus autem elevationis poli in M sit $= P$, cosinus $= p$, in N vero sit sinus elevationis poli $= R$, cosinus $= r$. Ex his si Luna in L versetur, ob $\mathrm{A}\sin M = M$, erit in M elevatio aquae supra libellam

$$= \frac{3L(tpq + PQ)^2}{2b^3} + \frac{L(1 - 3Q^2)(P^2 + PR + R^2)}{4b^3} - \frac{3Lq^2}{4b^3}$$

$$+ \frac{L}{4b^3}\left(q^2(3 - P^2 - PR - RR)(2TT - 1) - \frac{4Qqt(p^3 - r^3)}{R - P}\right)$$

$$= \frac{L}{2b^3}\left((ttqq - QQ)(R^2 + PR - 2P^2) + \frac{2Qqt(3PpR + r^3 - 3P^2p - p^3)}{R - P}\right).$$

Quod si nunc ponatur alter terminus N ultra aequatorem versus austrum situs, ita ut sinus elevationis poli australis in N duplo maior sit quam sinus elevationis borealis in M, seu $R = -2P$ et $r = \surd(1 - 4P^2)$, erit $R^2 + PR - 2P^2 = 0$, atque elevatio aquae in M supra libellam erit $= \dfrac{LQqt}{3b^3P}(9P^2p + p^3 - r^3)$. Ex hac igitur formula sequitur, si Lunae declinatio sit nulla seu $Q = 0$, tum nullum omnino aestum in M observari debere. Quod si autem Luna habeat [declinationem] borealem, tum ad transitum Lunae per meridianum superiorem aquam attolli ad

44) Editio princeps et Manuscriptus: $2b^3$ loco $3b^3$. Correxit EJA.

spatium[44]) $=\frac{LQq}{3b^3P}(9P^2p+p^3-r^3)$; at dum Luna in alterutro circulo horario sexto versetur, tum aquam ad libellam naturalem fore constitutam; Luna autem infra horizontem ad meridianum appellente, aquam infra libellam depressum iri per spatium $=\frac{LQq}{3b^3P}(9P^2p+p^3-r^3)$; contrarium denique fore aestum, si Luna habeat declinationem australem. In tali igitur Maris tractu quotidie semel tantum aqua affluet, semelque refluet, si quidem Luna habeat declinationem; nam si Luna aequatorem occupat, aestus omnino erit nullus.

130. Ex hoc casu aptissime explicari posse videtur Phaenomenon illud aestus singularis, qui in portu Tunquini ad Batsham observatur, ubi omnino ut in praesente casu dum Luna in aequatore versatur, Mare nullum aestum sentit, at dum Luna removetur ab aequatore vel versus boream vel versus austrum, quotidie aqua semel tantum affluit semelque refluit, prorsus ut calculus monstravit; scilicet si Lunae declinatio fuerit borealis, aqua versus Lunae occasum, hoc est post transitum Lunae per meridianum super horizonte, affluit, versus ortum vero defluit, quae retardatio ab inertia aquae et motu ad littora provenire intelligitur ut supra. Contra vero si Lunae declinatio sit australis, aqua deprimitur Luna ad occasum inclinante, Luna autem oriente, attollitur: quae Phaenomena apprime conveniunt cum casu modo exposito. Est praeterea elevatio poli Tunquini 20° 50′, borealis, atque Mare utrinque cum peninsulis tum insulis ab utroque oceano Pacifico et Indico fere prorsus separatur, saltem ut libera communicatio non adsit: praeterea hic idem Maris tractus, qui versus boream ad littora regni Tunquini terminatur, extenditur ultra aequatorem ad gradus circiter 45, cuius latitudinis sinus circiter duplo maior est, quam sinus latitudinis borealis 20° 51′: Quocirca ex his circumstantiis per nostram Theoriam eadem ipsa singularia Phaenomena aestus Maris observari debent, quae actu observantur: atque hoc modo si ullum adhuc dubium circa nostram theoriam reliquum fuisset, id resolutione huius mirabilis Phaenomeni funditus sublatum iri confidimus.

Beantwortung
verschiedener
Fragen
über
die Beschaffenheit, Bewe-
gung und Würckung
der
Cometen.

Berlin, zu finden bey Ambrosius Haude.
1 7 4 4.

Commentatio 67 indicis ENESTROEMIANI

BEANTWORTUNG VERSCHIEDENER FRAGEN ÜBER DIE BESCHAFFENHEIT, BEWEGUNG UND WÜRCKUNG DER COMETEN

I

Wie die Cometen von andern Sternen zu unterscheiden?

Wenn ein Stern an dem Himmel gesehen wird, welcher einen langen Schweif mit sich führet; so ist wohl kein Zweifel, daß derselbe nicht ein Comet seyn sollte. Es werden aber öfters von den Astronomis solche Sterne für Cometen ausgegeben, welche entweder keinen mit blossen Augen sichtbaren Schweif haben, oder auch nicht einmahl durch die Ferngläser in einer solchen Gestalt erscheinen: und dahero wird die Frage nicht ohne Grund gethan, wie man die Cometen unter den übrigen Sternen allezeit richtig erkennen soll. Die Beantwortung dieser Frage wird aber nicht schwehr fallen, wenn man auf die zweyerley Arten der Sterne, welche gewöhnlich an dem Himmel erscheinen, Achtung giebt. Alle diese Sterne sind entweder Fixsterne, oder Planeten. Jene ob sie sich gleich täglich um die Polos herum zu drehen scheinen, so behalten sie doch beständig unter sich einerley Stellung; diese aber, nehmlich die Planeten, verändern immer ihre Stelle, so wohl unter sich, als in Ansehung der Fixsterne, und werden deswegen Irrsterne genennt.

Die Anzahl der erstern ist so groß, daß dieselbe fast nicht bestimmet werden kann. Denn je besserer Ferngläser man sich bedienet, je mehr entdecket man solche Sterne, welche mit blossen Augen nimmer gesehen werden können. Hingegen erstreckt sich die Zahl der Planeten nur auf 5 welchen schon von den ältesten Zeiten her die Nahmen, Saturnus, Jupiter, Mars, Venus und Mercurius, beygeleget worden[1]). Man pflegte zwar zu denselben noch die Sonne und den Mond zu zehlen; allein ausser dem, daß sich diese Cörper von den ordentlichen Sternen ihrer scheinbaren Grösse nach allzu deutlich unterscheiden, so ist zu mercken, daß, nachdem der Stillstand der Sonne bewiesen worden, dieselbe füglicher mit den

1) Im 18. Jahrhundert fehlen die finiten Formen der Hilfsverben besonders häufig. EJA.

Fixsternen in einerley Ordnung gesetzet, und an ihre Stelle die Erde selbst zu den Planeten gerechnet wird. Der Mond aber ist nur ein Trabant, oder Satelles der Erde, welcher deswegen nicht verdienet, den Haupt-Planeten beygezehlet zu werden. Derselbe erhält also in dem sämtlichen Welt-Gebäude keinen höhern Rang, als die 4 Satellites, welche um den Jupiter, und die 5 welche um den Saturnum, vermittelst der Ferngläser, wahrgenommen werden. Man könnte zu dieser Classe, dahin der Mond gehöret, auch noch den Satellitem zehlen, welchen man vor etlichen Jahren in Engelland um die Venus endecket zu haben vorgiebt. Weil aber diese Erfindung seit der Zeit noch nicht genugsam bekräftiget worden; so wird rathsamer seyn, sich inzwischen mit der bekannten Zahl zu begnügen.

Ausser diesen zweyen Arten der Sterne, welche ordentlicher Weise an dem Himmel gesehen werden, erscheinen bißweilen ausserordentliche Sterne, die aber gleichwohl zu einer von diesen beyden Gattungen gezehlet werden müssen. Denn behält ein solcher neuer Stern, in Ansehung der Fixsterne, beständig einerley Stelle; so muß er auch zu dieser Classe gehören, und kan nicht füglicher, als ein ausserordentlicher Fixstern genennet werden. Dergleichen sind von Zeit zu Zeit verschiedene wahrgenommen worden, welche sich entweder nur eine kurtze Zeit haben sehen lassen, oder mit einer besondern Abwechselung bald grösser, bald kleiner geschienen. Man siehet anjetzo einige Sterne, von welchen die Alten gar nichts gewust haben; und hinwiederum finden sich bey diesen einige beschrieben, welche jetzt gantz und gar nicht mehr sichtbar sind. Unter allen diesen veränderlichen und ausserordentlichen Fixsternen ist derjenige am merckwürdigsten, welchen man zu Ende des Jahrs 1572 in der Cassiopea entdecket, und 16 Monathe lang an eben demselben Platz des Himmels gesehen. Derselbe hatte weder einen Schweif, noch einen Bart, oder Comam, sondern funckelte, wie die Fixsterne, und übertraf so gar den Sirium weit an Grösse, dergestalt, daß man denselben am hellen Tage bemercken konnte. Von dergleichen ausserordentlichen Fixsternen könnte noch sehr viel merckwürdiges angeführet werden, um daraus die grossen Veränderungen, welche schon würcklich an solchen Welt-Cörpern, die die Erde an Grösse weit übertreffen, vorgegangen sind, darzuthun, und dadurch die Meynung derjenigen, welche eine unumstößliche Beständigkeit in der Welt behaupten, zu wiederlegen. Allein, um auf die vorgelegte Frage zu antworten; so müssen wir insonderheit die andere Art der ausserordentlichen Sterne in Bewegung ziehen, welche ihre Stelle am Himmel gleich den Planeten verändern, und deswegen mit Recht ausserordentliche Planeten genennet werden können. Diese Sterne sind nun die Cometen, und daher ist ein Comet nichts anders, als ein ausserordentlicher Stern, welcher in Ansehung der Fixsterne seine Stelle verändert. Wer also die obgedachten 5 Planeten wohl kennet, und ausser denselben an dem Himmel einen neuen Stern beobachtet, welcher in Ansehung der Fixsterne eine Bewegung zu haben scheint, derselbe kan versichert seyn, daß er einen Cometen entdeckt habe, der Stern mag mit einem Schweif versehen seyn, oder nicht. Ungeachtet aber dieses das sicherste Kennzeichen eines Cometen ist, so ist es doch bisweilen schwer, sich desselben zu bedienen, wenn es nehmlich die Zeit und andere Umstände nicht erlauben, einen solchen Stern so lange zu observiren, biß man eine Aenderung in seiner Stelle gewahr wird; oder wenn seine Bewegung so langsam ist, daß in geraumer Zeit keine merckliche Veränderung beobachtet werden kan. In diesen Fällen muß man also zu andern Kennzeichen seine Zuflucht nehmen. Die Natur des Lichts kan öfters allen Zweifel so gleich heben, indem das Funckeln für ein sicheres Merckmahl eines Fixsterns gehalten wird, und ein schwächeres Licht einen Planeten anzuzeigen pfleget; am sichersten aber ist es, einen guten astronomischen Tubum darnach zu richten. Denn da dadurch alle Fixsterne nur als Punckte erscheinen, die Planeten aber mercklich vergrössert werden, welche Eigenschaft auch den Cometen zukommt; so ist es leicht, bey dem ersten Anblick eines solchen neuen Sterns zu bestimmen, ob derselbe ein ausserordentlicher Fixstern, oder ein Comet sey? weil in jenem Fall die scheinbare Grösse durch den Tubum vermindert, in diesem aber vermehret wird. Hierzu kommt noch, daß die Cometen, wenn auch gleich kein Schweif vorhanden ist, dennoch mit einer Coma, oder Bart, gemeiniglich umgeben sind, welches sich durch einen guten Tubum deutlich zu erkennen giebt. Im vorigen Jahre hat man allhier einen solchen Cometen wahrgenommen, von welchem man, ungeacht er sehr klein und keine Spur eines Schweifs vorhanden war, dennoch gleich versichern konnte, daß derselbe würcklich ein Comet wäre; da man aber denselben etliche Tage sehen konnte, so wurde man durch seine merckliche Bewegung völlig davon bekräfftiget.

II

Ob die Cometen feurige Cörper wie die Fixsterne, oder dunckele, wie die Planeten seyn?

Man hat durch die Observationes, mit Zuziehung der gesunden Vernunfft, deutlich erweisen, daß die Fixsterne so wohl, als die Sonne, feurige Cörper seyn, und ihr Licht von sich selbsten haben. Dahingegen die Planeten nicht sichtbar sind, als insofern dieselben von der Sonne erleuchtet werden. Hieraus entstehet also ein anderer Haupt-Unterscheid zwischen den Fixsternen, und den Planeten, indem jene feurige und selbst leuchtende, diese aber dunckle und nur ein entlehntes Licht besitzende Welt-Cörper, und in diesem Stück unserer Erde völlig ähnlich sind. Die Planeten bekommen nehmlich ihr Licht von der Sonne, um welcher sie sich alle in gesetzten Zeiten herum bewegen. Die Fixsterne aber, welche in einer erstaunend grossen Entfernung, so wohl von uns, als der Sonne, abstehen, würden gantz und gar nicht gesehen werden können, wenn dieselben nicht ihr eigenes Licht besässen. Unter allen Planeten ist der Saturnus der weiteste von der Sonne, der nächste Fixstern aber mag wohl mehr, als 1000 mahl weiter von der Sonne entfernet seyn, und es ist wahrscheinlich, daß auch die Fixsterne unter sich eben so weit von einander abstehen. Hieraus folget, daß ein jeglicher Fixstern eben ein solcher feuriger und grosser Cörper, als die Sonne sey, und auch einen eben so grossen Raum um sich habe, zu dessen Erleuchtung und Erwärmung er bestimmt seyn mag. Weil nun die Sonne von der Erde und den übrigen Planeten umgeben, und zu derselben Erhaltung vermittelst des Lichts und der Wärme bestimmt ist; so ist zu vermuthen, daß ein jeglicher Fixstern gleichfalls einige Planeten um sich herum habe, und dieselben so wohl durch die Erleuchtung, als die Erwärmung, unterhalte. Da wir nun ferner die gröste Ursache zu glauben haben, daß die übrigen Planeten unserer Sonne eben so wohl, als unsere Erde, mit Einwohnern angefüllt sind; so haben wir keine Ursache zu zweifeln, daß nicht auch alle diejenigen Planeten, womit ein jeglicher Fixstern umgeben ist, ebenfalls mit lebendigen Creaturen angefüllt seyn solten, als wodurch die Allmacht und Weisheit des Schöpfers noch viel deutlicher verherrlichet werden muß.

Der Einwurf, welchen man gegen diese Planeten der Fixsterne machen könnte, daß man noch keinen derselben, auch durch die besten Telescopia, habe wahrnehmen können, ist sehr leicht aus dem Weg zu räumen. Weil der Mercurius der nächste Planet an der Sonne ist, so scheinet er bey derselben immer so nahe zu stehen, daß man ihn sehr selten entweder kurtz vor der Sonnen Aufgang, oder gleich nach derselben Untergang, und das nur auf eine kleine Zeit sehen kan, indem er sich niemahls über 28 Grad von der Sonne entfernet. Diese scheinbare Entfernung beruhet auf der Distantz der Erde von der Sonne. Wären wir im Saturno, so würde uns der Mercurius nimmer über 3 Grad, die Venus aber nimmer über 5 Grad von der Sonne abzustehen scheinen, und dahero können die Einwohner des Saturni weder den Mercurium noch die Venus jemahls sehen, indem ihnen diese Planeten beständig unter den Sonnen-Strahlen verborgen bleiben. Weil nun die Fixsterne etliche tausendmahl weiter von der Sonne entfernet sind, als der Saturnus; so ist klar, daß man aus einem Fixsterne keinen ein[z]igen von den Planeten, welche ihren Lauf um die Sonne haben, jemahls würde sehen können. Nun aber befinden wir uns in gantz gleichen Umständen, in Ansehung der Fixsterne, und dahero ist es nicht möglich, daß wir jemahls einen Planeten, welcher unter das Systema eines Fixsterns gehöret, zu Gesicht bekommen können.

Über dieses aber, daß ihre scheinbare Distantz von dem Fixstern allzu klein seyn würde, so stehet auch noch ihre dunckle Eigenschaft im Wege; denn es ist natürlich, daß ein Cörper, welcher kein eigenes Licht hat, nicht so weit gesehen werden könne, als ein feuriger. Wir können solches leicht an dem Jupiter, welcher der gröste Planet in unserm Systemate ist, abnehmen. Denn, wenn derselbe etliche 1000 mahl weiter, biß in den Bezirck des nechsten Fixsterns versetzet werden sollte; so würde gewiß nicht das geringste mehr davon zu sehen seyn. Hieraus folget nun von selbsten, daß die Cometen keine zu einem Fixstern gehörende Planeten seyn können, welches jedoch im folgenden ausführlicher gezeigt werden soll. Hier kommen wir nun hauptsächlich auf die vorgelegte Frage, ob die Cometen feurige oder dunckle Cörper seyn, oder welches gleich viel ist, ob dieselben ihr eigenes Licht, oder ein entlehntes zu uns werfen? Wir haben die wichtigsten Gründe, das letztere zu behaupten. Denn erstlich haben sie nicht das lebhafte und blitzende Licht, welches wir bey den von sich selbst leuchtenden Fixsternen wahrnehmen, sondern sie scheinen durch die Tubos noch blässer, als die Planeten, welches ein deutliches Zeichen ist, daß ihr Licht ihnen anders woher mitgetheilt sey. Hernach, weil ihre scheinbare Grösse durch gute Ferngläser vermehret wird; so ist gewiß, daß dieselben uns weit näher seyn müßten, als die Fixsterne, und folglich müste ihr Licht, wenn es ihr eigen wäre, noch stärcker seyn, als derselbe. Endlich aber wird unsere Meynung am kräftigsten dadurch bestätiget, daß sich alle Cometen völlig unsern Augen entziehen, ehe sie sich von uns so weit, als der Jupiter, entfernen. Woraus denn unwiedersprechlich folget, daß die Cometen nicht nur kein eigenes Licht besitzen, sondern daß auch das entlehnte Licht weit schwächer sey, als bey den Planeten. Wir können auch die Ursache von dieser Verdunckelung leicht anzeigen, wenn wir auf den innern Zustand der Cometen etwas genauer acht haben wollen, welches wir auf die Beantwortung der folgenden Frage versparen.

III

Was die Cometen eigentlich für Cörper seyn?

Es ist im vorigen schon dargethan worden, daß die Cometen gleich den Planeten dunckle Cörper seyn, und ihr Licht nicht von sich selbsten haben. Da nun auch schon angeführet worden, und im folgenden mit mehrerem gezeiget werden soll, daß die Cometen, so lange sie uns sichtbar sind, der Sonne weit näher stehen, als der Saturnus, der doch sein gantzes Licht von der Sonne erhält; so ist kein Zweifel, daß nicht alle Cometen ihr Licht von der Sonne haben sollten, und daß uns folglich dieselben nur in so fern sichtbar sind, als sie von der Sonne erleuchtet werden. Man nimt aber fast bey allen Cometen, und insonderheit bey denjenigen, welche mit einem Schweif versehen sind, eine Comam, oder einen Bart, wahr, welcher dieselben gleichsam als ein Nebel umgiebt, und von welchem man gemeiniglich den wahren und eigentlichen Cörper des Cometen, welcher der Kern, oder Nucleus genennet wird, durch gute Fern-Gläser deutlich unterscheiden kan. Da nun dieser Nebel den Cometen allenthalben umgibt, dennoch aber der Kern dadurch gesehen werden kan; so können wir diesen Nebel mit nichts besser, als mit der Luft, oder dem Dunstkreise unserer Erde, vergleichen. Da aber gleichwohl bey den übrigen Planeten, welche ohne Zweifel gleich der Erde mit einem Dunstkreiß versehen sind, dieser Umstand nicht wahrgenommen wird; so ist zu schliessen, daß dieser Dunstkreiß bey den Cometen weit dichter, als unsere Luft, und einem starcken Nebel ähnlich seyn müsse. Was die Grösse dieses Dunstkreises betrifft, so ist insonderheit merckwürdig, daß der Durchschnitt desselben gemeiniglich etliche mahl grösser ist, als der Durchschnitt des Cörpers der Cometen selbst. Wenn wir also einen Cometen so groß setzen, als die Erde, deren Durchmesser 1720 deutsche Meilen beträgt; so würde sich diese neblichte Luft um den Cometen auf eine Höhe von etliche 1000 deutschen Meilen erstrecken. Nun aber stehen die höchsten Wolcken in unserer Luft selten mehr, als eine halbe deutsche Meile von der Erde ab: laßt uns aber eine gantze Meile für die Höhe unsers Dunstkreises annehmen, so wird doch der Dunstkreiß eines Cometen etliche tausend grösser seyn. Aus der erstaunlichen Grösse dieser Athmosphaere der Cometen ist leicht abzunehmen, daß sich diese Cörper in einem gantz andern Zustande befinden müssen, als die Erde, oder ein anderer ordentlicher Planet. Die Witterung muß dahero in einem Cometen im höchsten Grad ungestüm und heftig seyn, daß kaum zu glauben ist, daß sich darinne lebendige Creaturen aufhalten können. Und wenn wir zeigen werden, daß ein solcher Comet zu einigen Zeiten der Sonne sehr nahe kömmt, zu andern aber sich sehr weit von derselben entfernet; so scheinet diese Unbeständigkeit noch weit mehr den Einwohnern entgegen zu seyn. Nach dem Systemate des WHISTONS[2]) muß sich die Erde bey der Schöpfung in einem ziemlich ähnlichen Zustande befunden haben. Denn die Heil[ige] Schrift berichtet uns, daß es auf der Erde nicht nur wüste und leer, sondern daß auch alles Gewässer mit der Luft vermischt gewesen, welches

2) WILLIAM WHISTON, *A new theory of the earth*, Zweite Ausgabe (London, 1708). EJA.

daraus zu schliessen, daß gleich am zweyten Tage die oberen Wasser von den untern geschieden worden, welches nicht füglicher erkläret werden kan, als daß damahls der so unordentliche Dunstkreiß in seine ordentliche Gräntzen gebracht worden. Hier ist nicht der Ort, dieses weitläufiger zu erzehlen. Es ist genug, gewiesen zu haben, daß der Zustand der Cometen zu Erhaltung lebendiger Geschöpfe gäntzlich unbequem scheinet. Wenn nun dieses bey den Planeten, als eine von den Haupt-Absichten des Schöpfers angesehen werden muß; so folget, daß die Cometen zu einem gantz andern Endzweck erschaffen und bestimmet worden, als die Planeten. Man könnte zwar einwenden, daß diese Cörper vielleicht mit einer andern Art Einwohner angefüllet wären, als unsere Erde, welche den verwirrten Zustand ertragen könnten, und diese Meynung dadurch bekräftigen, weil es nicht wahrscheinlich wäre, daß GOtt der HErr so viele grosse Cörper ohne Einwohner hervorgebracht haben würde. Allein, ausser dem, daß uns der wahre Endzweck GOttes gäntzlich unbekannt ist; so müssen wir doch zugeben, daß unsere Erde vor den sechs Schöpfungs-Tagen gleichfalls von allen Einwohnern entblösst gewesen, wodurch der andere Grund des gemachten Einwurfs zernichtet wird. Vielleicht sind viele Cometen zu einem gleichen Endzweck bestimmt, um daraus zu seiner Zeit bequeme Wohnungen für neue lebendige Geschöpfe zu bereiten. Vielleicht mögen dieselben aufbehalten werden, um gewisse Veränderungen auf den bewohnten Planeten hervorzubringen, wie denn obgedachter WHISTON der Meynung ist, als wenn die Sündfluth durch einen Cometen verursachet worden. Im übrigen aber ist der menschliche Verstand viel zu schwach, um etwas gewisses über diese Puncte zu bestimmen; wir müssen nur glauben, daß diese Welt-Cörper zu Ausführung gantz besonderer Absichten erschaffen worden, und noch erhalten werden. Nun aber wiederum auf den eigentlichen Zustand des Cörpers der Cometen zu kommen; weil die ungeheure Athmosphaere derselben mit flüßiger Materie angefüllt ist, so scheinet es, daß der Kern bloß allein aus einer festen Materie zusammen gesetzt sey. Ob sich derselbe aber wie die Erde um eine gewissen Axe herum drehe, solches kan aus den Observationen nicht bestimmet werden, weil man denselben auch durch die besten Tubus nicht so deutlich sehen kan, daß man einige Veränderung darauf bemercken könnte.

Der grosse Dunstkreiß, womit die Cometen umgeben sind, ist ohne allen Zweifel die vornehmste Ursache, warum an denselben ein so schwaches Licht bemercket wird. Denn da die Sonnen-Strahlen erst durch diesen dichten Nebel dringen müssen, ehe sie auf den Kern fallen; so geht dadurch ein guter Theil derselben verlohren, und wird daher der Kern um so viel weniger erleuchtet. Ferner, da die Strahlen, welche von dem Kern wiederum zurück kommen, die

Athmosphaere gleichfalls paßiren müssen, so wird der Verlust verdoppelt, und uns folglich ein weit schwächeres Licht mitgetheilet, als wenn der Comet ein ordentlicher Planet wäre. Hieraus wird nun begreiflich, warum uns die Cometen nicht sichtbar werden, ehe und bevor sie sich auf eine gewisse Distantz uns genähert haben, welche gemeiniglich die Distantz der Sonne kaum übertrift, da wir doch die Planeten in einer weit grösseren Distantz noch ziemlich deutlich sehen können. Wir können auch ferner aus diesem Grunde sicher schliessen, daß wenn uns ein Comet sehr helle scheinet, wie derjenige, welcher anjetzo ins besondere unsere Aufmercksamkeit verdienet, derselbe uns gewiß sehr nahe seyn müsse. Ungeachtet aber dieses auch durch die Parallaxin bestätiget wird, so kommt doch dagegen ein Einwurf vor, dessen Beantwortung die Beschaffenheit des wahren Zustandes der Cometen noch besser an den Tag legen wird.

Von den Planeten, und andern dunckeln Cörpern wird jederzeit von der Sonne nur ungefehr die Helfte erleuchtet. Wenn uns nun nicht eben diese erleuchtete Helfte entgegen gekehret wird; so sehen wir den Planeten nicht in seinem vollen Schein, welches bey der Venere, dem Mercurio, auch dem Marte, fürnehmlich aber bey dem Mond, bemercket wird, als von welchem wir bisweilen gar nichts, bißweilen nur die halbe, bißweilen die gantze erleuchtete Helfte zu Gesicht bekommen. Wenn aber ein solcher Cörper sehr weit von uns entfernet ist, als der Jupiter und der Saturnus; so ist die von der Sonne erleuchtete Helfte von derjenigen, welche uns entgegen gesetzt ist, fast nicht mercklich unterschieden, dahero wir diese Planeten beständig in vollem Licht erblicken. Dieses voraus gesetzt, so pflegt dieser Einwurf gemacht zu werden: Wir sehen die Cometen immer mit vollem Licht; wenn nun dieselben nicht sehr weit von uns entfernet wären, so müsten wir öfters von denselben nur die Helfte, oder noch weniger zu Gesicht bekommen. Da aber dieses nicht geschicht[3]), so folget, daß die Cometen von uns sehr weit, ja weiter als der Mars, entfernet seyn müssen.

So wichtig dieser Einwurf dem ersten Ansehen nach scheinet, und auch so gar verschiedene Astronomos bewogen, den Cometen eine sehr entfernte Stelle anzuweisen, so deutlich kan derselbe doch, wenn man die Würckung ihrer grossen Atmosphaere in Erwegung ziehet, gehoben werden. Denn es ist genugsam bekannt, daß wegen der Luft, und der darinne entstehenden Strahlenbrechung, jederzeit mehr, als die Helfte von der Oberfläche der Erde, durch die Sonne erleuchtet wird. Der Unterscheid ist zwar geringe, und beträgt nur etwas weniges über einen halben

3) Eine alte Form von «geschieht». EJA.

Grad, indem die Horizontal-Refraction nicht mehr, als 33 bis 34 Minuten ausmacht. Weil aber diese Würckung von der Höhe und Dichte unserer Atmosphaere herrühret; so ist leicht zu begreiffen, daß wenn dieselbe weit höher und dichter wäre, auch ein grösserer Theil der Erde über die Helfte erleuchtet werden würde. Da nun die Athmosphaere der Cometen nicht nur weit dichter ist, als unsere Luft, sondern auch etliche 1000 mahl weiter davon ausgedehnet ist; so muß die Brechung der Sonnen-Strahlen so groß seyn, daß bey nahe die gantze Oberfläche eines Cometen von der Sonne auf einmahl erleuchtet wird. Folglich mag der Comet, in Ansehung der Sonne und der Erde, eine Stelle haben, wie man immer will, so muß beständig die gantze Helfte desselben, welche gegen uns stehet, erleuchtet seyn, und uns deswegen allezeit mit vollem Lichte scheinen. Wenn sich also ja in einem Cometen Einwohner befinden sollten, so würden dieselben keiner Abwechselung von Tag und Nacht gewahr werden, sondern eines beständigen Tages geniessen, ungeachtet sie durch die dichte und grosse Atmosphaere das Bild der Sonne nicht wohl deutlich zu Gesicht bekommen würden.

IV
Was von dem Schweif der Cometen zu halten sey?

Das fürnehmste Kennzeichen, wodurch sich die Cometen in den Augen des gemeinen Volcks von andern Sternen unterscheiden, ist ohne Zweifel der Schwantz. Und wenn gleich die Astronomi auch andere ausserordentliche Planeten für Cometen ausgeben, welche ohne Schwantz erscheinen, so pflegen dieselben dennoch von dem gemeinen Mann nicht zu dieser Classe gezehlet zu werden. Hier kommt also erstlich die Frage vor, ob alle diejenigen Cometen, welche ohne Schweif erscheinen, auch in der That desselben beraubet sind, oder nicht? Gemeiniglich erscheinet dieser Anhang in einer sehr veränderlichen Gestalt, indem derselbe bald länger, bald kürtzer, gesehen wird, bisweilen auch gar verschwindet. Unter allen Cometen scheinet der Anno 1680 und 81 in Ansehung des Schwantzes am merckwürdigsten zu seyn, welcher sich, da er am grösten war, über 100 Grad erstreckte. Der Schwantz des Cometen von 1742 betrug nur etliche Grad, und da gegen das Ende seiner Erscheinung die Luft durch das Monden-Licht etwas erleuchtet wurde; so konnte man denselben nicht mehr gewahr werden. An dem sehr kleinen Cometen Anno 1743 wurde gar kein Schwantz bemercket, und der jetzige Comet von 1744 hatte anfänglich einen Schwantz nur von 2 oder 3 Graden, welcher aber in etlichen Wochen biß auf 20 Grad angewachsen. Aus allen dergleichen Observationen läßt sich aber fast gar nichts über die würckliche Länge des Schweifes schliessen; denn da das Licht desselben so sehr schwach ist, so kommt es hierbey sehr starck auf die Beschaffenheit unserer Luft an. Ist dieselbe klar, und von keinem andern Licht erleuchtet, so sieht man den Schweif nicht nur deutlicher, sondern auch länger, als wenn die Luft neblicht und von dem Mond erleuchtet ist. Dahero nicht zu vermuthen ist, daß eine solche Veränderung von einer in dem Cometen würcklich vorgegangenen Veränderung herrühre. Hierauf beruhet die sichtbare Länge des Schweifs hauptsächlich auf desselben Stellung und Lage, in Ansehung der Erde. Sollte derselbe gerade gegen uns zu, oder von uns abgekehret seyn; so würden wir nichts davon sehen können, je mehr aber derselbe seitwerts gewendet wird, je geringer wird die Verkürtzung, unter welcher er erscheinet. Aus diesem Grunde ist es also möglich, daß uns eben derselbe Schweif bald lang, bald kurtz, erscheinet, bald gantz und gar unsichtbar ist. Hieraus können wir aber gleichwohl die Folge nicht ziehen, daß in den Schweifen der Cometen gar keine Veränderung vor sich gehe, vielmehr scheinet die Natur derselben so beschaffen zu seyn, daß dieselbe nothwendig beständigen Veränderungen unterworfen seyn müssen.

Was nun ferner die Materie, woraus die Cometen-Schweife bestehen, anlanget; so ist gewiß, daß dieselbe über alle massen subtil seyn müsse. Dieses erhellet aus ihrer Durchsichtigkeit, indem sogar die kleinsten Sterne dadurch erkannt werden können. Wir wissen aber aus der Erfahrung, daß man auch durch die subtileste Wolcke nicht nur keine Sterne, sondern nicht einmahl die Sonne sehen könne. Nun aber ist eine Wolcke nichts anders, als ein Nebel in der Entfernung, und da man durch einen Nebel in der Nähe noch wohl etwas sehen kan; so ist klar, daß je weiter ein Nebel von uns entfernet ist, je mehr derselbe seine Durchsichtigkeit verliehren müsse. Da nun die Cometen-Schweife gewöhnlich viel tausend, ja Millionen mahl weiter von uns entfernet sind, als die Wolcken, und dennoch durchsichtig sind; so ist unstreitig, daß ihre Materie noch viel mehr mahl subtiler seyn müsse, als der geringste Nebel. Wenn also ja die Cometen-Schweife gleich unseren Wolcken aus Dünsten beständen, wie einige dafür halten; so müste doch die Quantität dieser Dünste so geringe seyn, daß sie kaum in einige Betrachtung gezogen werden könnte. Nach den Rechnungen des berühmten NEWTONS[4]) folget, daß alle wässerichte Dünste, aus welchen der Schweif eines Cometen bestehen könnte, zusammen genommen, nicht einmahl einen cubischen Zoll Wasser betragen. Es ist demnach

4) ISAAC NEWTON, *Philosophiae Naturalis Principia Mathematica* (Londini, 1687), Buch III, Satz 41, S. 503. EULER gibt NEWTONS Meinung nicht ganz richtig wieder. EJA.

schwehrlich zu glauben, daß diese Schweife aus wässerichten Dünsten zusammen gesetzt seyn sollten, vielmehr scheinet die Stellung derselben diese Meynung völlig umzustossen. Man hat nehmlich bey allen Cometen beständig wahrgenommen, daß ihre Schweife nach einer geraden Linie von der Sonne abgekehret sind. Nun aber ist kein Grund anzuzeigen, warum diese Dünste nur von der der Sonne entgegen gesetzten Seite hinaufsteigen, und sich nur allein in dieser Gegend aufhalten sollten, vielmehr scheinen alle Eigenschaften der Dünste dieser beständigen Erfahrung entgegen zu stehen. Andere Erklärungen von der Natur dieser Cometen-Schweife sind eben so viel Schwierigkeiten unterworfen, daß man sich also noch nicht im Stande befindet, etwas zuverläßiges davon zu sagen. Vielleicht ist ein Schweif eine nothwendige Folge des grossen und ungestümen Dunstkreises, womit die Cometen umgeben sind, und wenn wir die Eigenschaften solcher Cörper, nebst der Natur des Lichts, deutlicher einsehen sollten, so würde es vielleicht nicht schwer fallen, die wahre Ursache davon anzuzeigen. Es sind einige auf die Gedancken gefallen, daß die hellen Schweife der Cometen eine grosse Aehnlichkeit mit den Nord-Lichtern haben möchten, welche auf unserer Erde wahrgenommen werden. Dieser Einfall scheinet sehr glücklich ausgesonnen zu seyn, indem der Ort, wo die Nord-Lichter erscheinen, weit höher über der Erde steht, als die Athmosphaere hinreichet, und also könnte aus gleichem Grunde auch aus den Cometen eine ähnliche Materie weit über die Atmosphaere derselben hinaufsteigen, und da diese die Höhe unserer Athmosphaere viel 1000 mahl übertrift; so wäre es kein Wunder, wenn die Schweife der Cometen so unbegreifflich lang wären. Weil aber die wahre Ursache der Nord-Lichter noch nicht genugsam bekannt ist, so läst sich auch die angeführte Meinung noch zur Zeit weder genugsam bestätigen, noch wiederlegen. Dem ungeachtet, wäre es ohne Zweifel eine wichtige Entdeckung, wenn man nur zeigen könnte, daß die Schweife der Cometen, und die Nord-Lichter aus einerley Ursachen herkämen. Man könnte vielleicht auch mit gutem Grunde sagen, daß die Schwäntze der Cometen von den Sonnen-Strahlen herrühreten, weil dieselben beständig von der Sonne abgekehrt gesehen werden. Denn es ist aus der Erfahrung klar, daß die Sonnenstrahlen auf diejenigen Cörper, welche davon erleuchtet werden, eine kleine Kraft ausüben, und die subtilsten Theilchen derselben würcklich in Bewegung setzen. Hieraus folget, daß die Sonnenstrahlen die subtilsten Luftstäubchen gleichfalls von sich stossen müssen, auf diese Art aber werden dieselbe gegen die Erde getrieben, daß also daher keine sonderbare Würckung zu verspühren ist. Weil aber in den Cometen auch die von der Sonne abgekehrte Seite erleuchtet wird, so müssen dadurch die subtilsten Theilchen von dem Cörper des Cometen weggestossen, und folglich in die der Sonne gegen über liegende Gegend getrieben werden. Man könnte zwar einwenden, daß, da wahrscheinlicher Weise auch alle Cörper gegen den Cometen schwehr sind, eben diese Schwehre die Entfernung der subtilsten Theilchen verhindern müste. Allein, da die Schwehre dieser kleinen Theilchen kaum mercklich seyn kan, so wird dieselbe leicht von der Gewalt der Strahlen überwunden. Hierzu kommt noch, daß gegen das Ende des Dunstkreises die Schwehre weit geringer seyn müsse, als bey dem Kern, und dahero die Fortstossung der Sonnenstäubchen um so viel weniger aufgehalten werden kan. Sind dieselben aber einmahl so weit von dem Cörper des Cometen entfernet, daß ihre Schwehre verschwindet; so ist keine[5]) Hinderniß mehr übrig, wodurch eine noch grössere Entfernung gehemmet werden könnte. Nach dieser Erklärung begreifft man sogleich, warum die Schweife der Cometen immer von der Sonne abstehen, und wenn auch eine kleine Abweichung von dieser Direction wahrgenommen wird, so muß die Ursache davon in der krummlinichten Bewegung des Cometen selbst gesucht werden, indem diese Sonnenstäubchen dem veränderten Lauf des Cometen sogleich nicht folgen können. Laut dieser Meynung wären also die Schweife der Cometen nichts anders, als eine grosse Menge Sonnenstäubchen, welche durch die Lichtstrahlen von dem Cörper des Cometen weggestossen worden, und sich in der Gegend hinter dem Cometen versammelt haben; und die von diesen Stäubchen zurück geprellte Sonnenstrahlen stellen uns die Erscheinung des Schwantzes vor. Diese Erklärung scheinet auch noch durch einen sonderbaren Umstand, welcher bey dem jetzigen Cometen wahrgenommen wird, bestätiget zu werden. Nehmlich, die nach der Sonne zustehende Seite des Barts erscheinet etwas eingedruckt, welches eine nothwendige Folge der Krafft der Strahlen ist, indem die äussersten Theile der Athmosphaere in dieser Gegend nach dem Kern des Cometen gestossen werden. Sollte nun diese Ursache der Wahrheit gemäß seyn, so siehet man sogleich in was für Fällen die Cometen mit Schweifen begabt seyn müssen. Denn erstlich wird dazu eine solche grosse und dichte Athmosphaere erfordert, daß auch die von der Sonne abstehende Helfte beynahe gantz erleuchtet wird, welcher Umstand durch alle Observationes sehr schön bestätiget wird. Hernach muß auch das Licht, welches der Comet von der Sonne empfängt, sehr starck seyn. Dieses wird auch durch die Erfahrung bekräftiget, indem an den Cometen, welche von der Sonne sehr weit abstehen, fast gar kein, oder nur ein sehr kurtzer Schweif bemercket wird; hingegen aber, je mehr sich ein Comet der Sonne nähert, je länger erscheinet der Schweif, in sofern nehmlich die Stelle der Erde so beschaffen ist, daß man denselben [nicht] in einer allzustarcken Verkürtzung anschaut. Diese Erklä-

5) Am Anfang des 18. Jahrhunderts waren alle Substantive auf «nis» weiblich. EJA.

rung hat den sich um die Astronomie so hoch verdient gemachten KEPLER[6]) zum Urheber, und da dieselbe leicht so ausgeführet und befestiget werden könnte, daß fast kein Zweifel übrig bliebe; so gereichet dieser Einfall nicht wenig zu Vermehrung des Ruhms dieses grossen Mannes. Endlich ist zu mercken, daß die Schweife von verschiedenen Cometen der Länge nach sehr starck unterschieden seyn können, je nachdem der Dunstkreiß derselben grösser oder kleiner ist, und dieselben weiter oder näher von der Sonne abstehen. Der Schweif des jetzigen Cometen möchte sich an der Länge wohl auf 600 000 Deutsche Meilen belauffen; über welche, ob gleich sehr grosse Länge man sich nicht mehr so starck zu verwundern Ursache findet, wenn man die Krafft und die Geschwindigkeit der Lichtstrahlen in Erwegung ziehet. Wollte man aber die Meynung von den wässerichten Dünsten behaupten; so müste dieselbe durch diese so erstaunliche Länge gleich umgestossen werden.

V
Wie die Bewegung der Cometen beschaffen sey?

Weil sich die Erde beständig innerhalb 24 Stunden um ihre Axe vom Abend gegen Morgen herum drehet, wir aber diese Bewegung nicht gewahr werden; so scheinen uns alle ausser dem Erdwirbel befindliche Cörper eine gleiche Bewegung zu haben, und sich innerhalb 24 Stunden vom Morgen nach Abend um eben dieselbe Axe herum zu drehen. Diese scheinbare Bewegung ist also auch den Cometen, mit allen übrigen Sternen, gemein, und dahero sehen wir dieselbe bald auf- bald untergehen, wenn ihre Declination nicht grösser ist, als die Polus-Höhe, in welchem Fall sie beständig über dem Horizont bleiben. Ausser dieser allgemeinen Bewegung haben aber die Cometen einen sonderbaren Lauf, wodurch dieselben von den Fixsternen, als bey welchen nur die allgemeine Bewegung statt findet, unterschieden werden, und diese gleichsam den Cometen eigene Bewegung wird aus der Veränderung ihrer Stelle, in Ansehung der Fixsterne, beurtheilet. Wenn zum Exempel ein Comet heute dicht bey einem Fixstern stünde, morgen aber um 5 Grad von eben demselben Stern entfernet wäre; so sagt man, daß dieser Comet in dieser Zeit einen Weg von 5 Grad an dem Himmel durchgelauffen. Giebt man auf die Direction dieses Wegs, welchen ein Comet durch seine eigene Bewegung zurück legt, acht; so wird man leicht finden, ob dieselbe vom Morgen gegen Abend, oder vom Abend gegen Morgen, oder von Norden gegen Süden, oder von Süden gegen Norden, oder aber nach einer

6) JOHANNES KEPLER, *De cometis libelli tres* (1619) = *Gesammelte Werke,* Band 8 (München, 1963), S. 129–262, besonders S. 226. EJA.

zwischen diesen 4 Haupt-Gegenden liegenden Gegend, gerichtet sey. Die Ordnung der 12 Himmlischen Zeichen, welche die Sonne in einem Jahr, der Monde aber in einem Monath durch zu lauffen scheinet, gehet vom Abend gegen Morgen. Wenn also ein Comet seinen eigenen Lauf vom Abend gegen Morgen richtet, so sagt man, daß er nach der Ordnung der Zeichen fortgehe, und rechtläufig sey; geht aber der Comet vom Morgen gegen Abend, so läufft er wieder die Ordnung der Zeichen, und wird dahero rückläufig genannt, welche Benennungen auch bey den Planeten in den Calendern häufig vorkommen. Ungeachtet aber die Planeten uns bald rechtläufig, bald rückläufig, bald geschwinder, bald langsamer, fortzugehen scheinen, so bleiben dieselben doch beständig in dem Zodiaco, oder Thierkreise, und entfernen sich von der Ecliptic, oder dem Weg der Sonne, nicht über 6 Grad. Die Cometen aber sind nicht nur eben diesen Veränderungen, welche wir bey den Planeten in Ansehung ihrer Bewegungen wahrnehmen, unterworffen, sondern ihre Bahn ist auch gemeiniglich von dem Thierkreiß sehr unterschieden, und es ist nicht leicht ein Ort an dem Himmel zu bestimmen, wo man nicht schon einen Cometen beobachtet hätte. Diese so grosse Irregularität, welche bey dem Lauf der Cometen wahrgenommen wird, ist aber hauptsächlich der Bewegung der Erde zuzuschreiben. Denn da uns die Bewegung der Planeten sehr regulär und gleichförmig vorkommen würde, wenn wir dieselben aus der Sonne anschauen könnten; so würden wir auch, wenn wir uns in der Sonne befänden, keine so grosse Unrichtigkeit in dem Lauf der Cometen wahrnehmen. Man hat aber in der Astronomie Mittel ausgefunden, jederzeit den Ort am Himmel zu bestimmen, wo man aus der Sonne die Cometen sehen würde; und da findet sich in der That, daß die meisten Irregularitaeten bey dem Lauf dieser Cörper wegfallen. Denn da erstlich die Bahn der Cometen, wie wir dieselbe aus der Erde sehen, gemeiniglich nicht nach einer geraden Linie an dem Himmel, oder nach einem grossen Zirkul desselben gerichtet ist; so würden einem Zuschauer in der Sonne alle Cometen nach solchen grossen Zirkeln sich zu bewegen scheinen, woraus dann folget, daß ein Comet sich immer in eben demselben Plano befinde, und daß dieses Planum durch die Sonne gehe. Hernach würde ein Zuschauer in der Sonne den Cometen entweder immer rechtläufig, oder immer rückläufig sehen, und auch in seiner Geschwindigkeit keine so grosse Ungleichheit wahrnehmen. Er würde zwar darinne eine Veränderung beobachten, allein dieselbe würde sich dergestalt nach der Distantz des Cometen von der Sonne richten, daß die Bewegung um so viel geschwinder scheinen würde, je näher der Comet zu der Sonne käme. Aus allem diesen folget also unstreitig, daß sich der Lauf der Cometen nach der Sonne richte, und daß die Sonne gleichsam der Mittelpunckt der Bewegung sowohl der Cometen als der Planeten seyn müste. Um derohalben den Lauf eines Cometen richtig zu beurtheilen, so muß man nicht auf die Bewegung desselben, wie uns solche auf der Erde vorkömmt, sondern auf diejenige, welche man aus der Sonne wahrnehmen würde, acht haben. Auf diese Weise wird man aber eine grosse Aehnlichkeit zwischen dem Lauf der Cometen und der Planeten gewahr werden, woraus die wunderbare Ordnung in dem sämmtlichen Welt-Gebäude um so viel deutlicher erhellet. Die Planeten bewegen sich um die Sonne herum, aber nicht in Zirkullinien, sondern ihre Bahn stellet eine ablange Rundung vor, in welcher die Sonne also stehet, daß sie einer Spitze näher ist, als der andern. Der Ort nun, wo ein Planet der Sonne am nächsten kommt, heisset sein Perihelium, oder Sonnen-Nähe; wo er aber von der Sonne am weitesten entfernet ist, wird das Aphelium, oder Sonnenferne, genannt. Bey den Planeten ist aber der Unterschied zwischen der grösten und kleinsten Entfernung von der Sonne nicht groß, und dieses ist die Ursache, warum ein Planet bey nahe immer gleich weit von der Sonne entfernet bleibet, worauf ohne Zweifel auch der ruhige und in einem Gleichgewicht befindliche Zustand ihrer Athmosphaere beruhet. Die Bahnen der Cometen sind nun auch solche ablange Rundungen; allein der Unterschied zwischen der Sonnenferne und Sonnen-Nähe ist weit grösser, dahero dieselben zu einer Zeit der Sonne ziemlich nahe kommen, zu einer andern aber sich sehr weit von derselben entfernen. Weil also die Cometen zu der Zeit, da sie der Sonne am nächsten sind, eine grosse Hitze, hingegen aber, wenn sie am weitesten davon abstehen, eine sehr heftige Kälte auszustehen haben; so scheinet diese ungleiche Witterung die fürnehmste Ursache ihrer grossen und unordentlichen Athmosphaere zu seyn. Der jetzige Comet scheinet in seinem Perihelio der Sonne nicht viel näher zu kommen, als der Erde, in seinem Aphelio aber mag er sich wohl viel weiter, als der Saturnus davon entfernen, welches letztere fast von allen Cometen zu verstehen ist. Der Comet Anno 1742 war in seinem Perihelio der Sonne eben so nahe, als die Venus. Der Comet von 1680 aber kam der Sonne weit näher als der Mercurius; ja er berührete dieselbe fast, als er sich um dieselbe herum schlung. Diese ablangen krummen Linien, in welchen sich sowohl die Cometen, als Planeten, um die Sonne herum bewegen, werden in der Geometrie Ellipses genannt. In denselben sind insonderheit zwey Puncte merckwürdig, welche die Foci, oder die Brennpuncte, heissen, und sind deswegen allhier zu mercken, weil sich die Sonne immer in einem von diesen Puncten befindet. Die gerade Linie, welche durch eine solche Ellipsin der Länge nach mitten durchgezogen wird, pflegt die Axe, und der Mittelpunkt derselben das Centrum genannt zu werden. Die Foci oder Brennpuncte liegen nun immer in der Axe, und stehen beyderseits von dem Centro gleich weit ab, sie sind auch unter sich um so viel weiter von einander entfernt, je mehr die Ellipsis ablang ist, oder je mehr die Länge grösser ist, als die Breite. Wird aber die Länge der Breite vollkommen gleich, in welchem Fall die Ellipsis sich in einen Zirckul verwandelt; so vereinigen sich beyde Foci in dem Centro. Diese Anmerckungen können hinreichend seyn, um sich einen klaren Begriff von den Bahnen der Cometen zu machen. Wenn man aber die wahre Bahn eines Cometen aus seiner Bewegung in derselben bestimmen will, so werden dazu nachfolgende 6 Stücke erfordert. Erstlich muß man die Entfernung des Perihelii von der Sonnen wissen. Zweytens muß auch die Distanz des Aphelii von der Sonne bekannt seyn, ob sich gleich dieselbe fast nimmer genau bestimmen läßt. Es ist aber um die sichtbare Bewegung zu erklären genug, wenn man dieselbe nur sehr groß annimmt, um welcher Ursache willen NEWTON dieselbe unendlich groß zu setzen pflegt, in welchem Fall die Ellipsis in eine Parabel verwandelt wird. Drittens, weil diese Cometen-Bahn die Ecliptic, oder die Bahn der Erde nothwendig durchschneiden muß, und die beyden Durchschnitte die Nodi genennt werden, so müssen die Stellen derselben angezeigt werden. Aus denselben erkennt man folglich, wo der Comet aus der Sonne durch die Ecliptic zu gehen gesehen wird. Vierdtens muß auch der Winckel bestimmet werden, unter welchem die Bahn der Cometen die Ecliptic durchschneidet, welches die Inclination der Cometen-Bahn genennet wird. Fünftens muß man die Lage des Perihelii, in Ansehung der Ecliptic, anzeigen, und wissen, wie weit dasselbe von den Nodis entfernet ist. Sechstens muß endlich die Zeit bekannt seyn, zu welcher der Comet im Perihelio gewesen. Hat man diese sechs Puncte accurat bestimmt, so kan man alsdenn leicht auf eine jegliche vorgegebene Zeit ausrechnen, an was für einem Punct des Himmels der Comet erscheinen, und wie weit derselbe von der Erde entfernet seyn müsse; wenn auch derselbe wegen allzu grosser Entfernung nicht einmahl gesehen werden kan. Wenn man auch den zweyten Punct, nehmlich die Distanz des Aphelii von der Sonne, genau ausfindig machen könnte; so würde daraus leicht die Zeit des gantzen Umlaufes des Cometen gefunden, und auch die künftigen Zeiten angezeigt werden können, wenn eben derselbe Comete wiederum in unsern Gesichtskreiß kommen, und wo derselbe sichtbar seyn werde. Sowohl die Cometen, als die Planeten, bewegen sich in ihren Elliptischen Bahnen dergestalt um die Sonne, daß wenn zu verschiedenen Zeiten aus der Sonne nach dem Ort des Planeten oder Cometen gerade Linien gezogen werden, die dadurch abgeschnittenen Stücke ihrer Bahn sich unter einander verhalten, wie die inzwischen verflossenen Zeiten, und dahero folget, daß je näher der Comet der Sonne kömmt, je schneller seine Bewegung werden müsse. Und aus diesem Grundsatze, wenn die Zeit bekannt ist, da der Comet in seinem Perihelio gewesen, so kan die Zeit bestimmet werden, zu welcher er an eine jegliche Stelle seiner Bahn kommen muß. Wenn nun für dieselbe Zeit auch der Ort ausgerechnet wird, wo die Erde alsdenn stehet, so findet man durch die Geometrie leicht, an welchem Punckt des Himmels uns zu derselben Zeit der Comet erscheinen müsse. Aus allem diesem erkennet man also genugsam, daß der Lauf der Cometen sich nach allgemeinen Grundgesetzen der Natur richte, und daß man folglich mit Recht hoffen könne, mit der Zeit eine gründliche Erkänntniß von denselben zu erlangen.

VI

Ob man die Erscheinung der Cometen nicht vorher verkündigen könne?

Da die Cometen nichts anders sind, als ausserordentliche Planeten, und auch ihren Lauf nach eben denselben Grundgesetzen vollenden; so scheinet keine Hinderniß in dem Wege zu sein, daß man nicht die Erscheinungen der Cometen eben so richtig, als der Planeten, vorher verkündigen könnte. Wenn man die bey Beantwortung der vorhergehenden Frage angezeigten 6 Stücke, insonderheit aber das zweyte genau bestimmen könnte, so hätte die gantze Sache freylich nicht die geringste Schwierigkeit. Allein, eben dieser Umstand ist mit so viel Difficultaeten verknüpft, daß man nicht leicht hoffen kan, dieselben so bald zu überwinden. Zu genauer Bestimmung der Distantz des Aphelii eines Cometen von der Sonne, welche zu diesem Ende unumgänglich nöthig ist, werden nicht nur die accuratesten Observationes erfordert, wozu noch zur Zeit die Astronomischen Instrumente nicht hinlänglich sind, sondern man müßte auch einen Cometen eine weit längere Zeit beobachten können, als uns gemeiniglich vergönnet wird. Etliche Monathe, in welchen uns ein Comet gewöhnlicher Weise seine sichtbare Gegenwart ertheilet, sind in Betrachtung seines völligen Umlaufs um die Sonne weit zu kurtz, als daß man daraus seinen gantzen Periodum ausrechnen könnte. So weit man es anjetzo sowohl in der theoretischen als practischen Astronomie gebracht, so würde es doch sehr schwehr fallen, nur die Erscheinung eines Planeten, als des Saturni, vorher zu verkündigen, wenn man denselben nur etliche Monathe lang observirt hätte, und keine andern Observationen zu Hülffe nehmen könnte. Weil man aber die Planeten von so vielen Jahren, ja Seculis her, observiret, und dieselben nicht nur bey ihren Periheliis, sondern auch den Apheliis mehrmahlen gesehen; so ist kein Wunder, daß man auch zu einer weit vollkommeneren Erkenntniß ihres Laufs gelanget ist. Sollte sich also ein Comet, wenn er von der Sonne weg gegen seinem Aphelio fortrücket, nicht so bald unsern Augen entziehen, oder uns gar in seiner Sonnenferne noch sichtbar bleiben; so würde uns eine gleiche vollkommene Erkenntniß seines Laufs nicht schwehr fallen. Der jetzige Comet scheinet in diesem Stücke vor andern was voraus zu haben, und uns eine hinlänglichere und vielleicht vollständige Erkenntniß seines Laufs zu versprechen, indem uns derselbe, allen Ansehen nach, noch viele Monathe sichtbar bleiben wird, wenn sich nur die Herren Astronomi alle mögliche Mühe geben werden, accurate Observationes darüber anzustellen. Es ist aber hierbey wohl in Erwegung zu ziehen, daß der geringste Fehler, welcher bey den Observationen begangen worden, viele Jahre in der Zeit des gantzen Umlaufs betragen könne. Dahero man aus diesem Grunde nicht wohl hoffen kan, die künftige Erscheinung eines Cometen bey Tag und Stunden vorher zu bestimmen. Diese Schwierigkeit könnte nun vermieden werden, wenn man aus den alten Observationen bestimmen könnte, daß eben derselbe Comet schon vormahls wäre gesehen worden. Denn in diesem Fall würde man die wahre Zeit seines Umlaufs um die Sonne auf das genaueste anzeigen, und seine Erscheinungen auf alle künfftige Zeiten ausrechnen können. Weil es aber hier auf die wahre Bahn, welcher der Comet um die Sonne beschreibt, ankommt; so muß man nicht auf die Uebereinstimmung des aus der Erde beobachteten Laufs sehen. Denn wenn auch eben derselbe Comet in seiner Elliptischen Bahn um die Sonne wiederkommt, die Erde aber sich nicht wiederum auch an eben der Stelle, wo sie vormahls gestanden, befindet, so müssen wir nothwendig die Cometen in gantz andern Sternbildern erblicken. Weil auch ein Comet wegen seines schwachen Lichts vor unsern Augen verschwindet, wenn er nur von uns weiter, als die Sonne entfernet ist; so kan es auch leicht geschehen, daß eben derselbe Comet öfters wiederkommt, ohne daß wir ihn gewahr werden. Wäre zum Exempel der Comet, welcher uns anjetzo so groß erscheint, 6 Monathe früher oder später gekommen, so würden wir gewiß nichts davon zu Gesicht bekommen haben: weil er alsdenn von uns viel weiter, als die Sonne, entfernt gewesen seyn würde, da er uns doch anjetzo bey nahe 10 mahl näher ist, als die Sonne. Dahero kan es leicht seyn, daß eben dieser Comet schon vormahls mehrmahlen durch sein Perihelium passirt, und daß er entweder gar nicht, oder sehr klein, und dabey in gantz andern Himmels-Gegenden, gesehen worden. Hernach kan es auch geschehen, daß wir zwey gantz verschiedene Cometen an dem Himmel an einerley Ort erblicken, welche doch aus der Sonne an gantz verschiedenen Orten würden gesehen worden seyn. Um dieser Ursache willen kan man aus der Vergleichung und Gegeneinanderhaltung des observirten Laufs verschiedener Cometen nichts zuverläßiges bestimmen, ehe und bevor man nicht alle diese Cometen-Bahnen, so wie solche aus der Sonne gesehen seyn würden, genau ausgerechnet hat. Diese Arbeit hat der berühmte Astronomus HALLEY[7]) übernommen, und von sehr vielen, von einigen Seculis her observirten Cometen, ihre Bahnen bestimmet; es ist aber zu bedauren, daß die alten Astronomi so nachläßig in den Observationen der Cometen gewesen sind, und daß man sich dahero auf die HALLEYanische Bestimmung, oder welche aus den alten Observationen gezogen worden, nicht so fest verlassen kan. Sollte man aber in diesem Stück mit der Zeit glücklicher werden, und in einer solchen genauen Beschreibung der Cometen-Bahnen finden, daß zwey Cometen in ihren Periheliis gleich weit von der Sonne entfernt gewesen, daß ihre Bahnen die Ecliptic an einerley Orten durchschnitten, und eine gleiche Inclination gehabt haben, und daß auch endlich[8]) ihre Perihelia an eben demselben Orte des Himmels gestanden: so würde man sicher schliessen können, daß diese Cometen nicht von einander unterschieden wären, und also die Zeit ihres Umlaufs genau anzeigen können. Es könnte zwar seyn, daß dieser Comet inzwischen ein oder mehrmahl schon wiederum gekommen, und um obangeführter Ursache willen nicht gesehen worden wäre: allein aus dieser Ungewißheit würde man sich leicht helffen können, wenn man nur von beyden Erscheinungen richtige Observationen hätte, welche doch bey dieser gantzen Untersuchung voraus gesetzt werden.

Wenn man nun alle diese Umstände wohl betrachtet, so wird man leicht begreiffen, wie viel auf die Prophezeyungen der künftig kommenden Cometen zu halten sey. Die meisten sind bißher nicht eingetroffen, und wenn ja unter so viel fehl geschlagenen eine eintreffen sollte, so würde dieses einem blossen Glücke zuzuschreiben seyn. Einige, welche sich mit dergleichen Vorherverkündigung abgegeben haben, haben sich falsche Begriffe von dem Lauf der Cometen gemacht, dergleichen sich noch viele finden, und haben darauf ihre Muthmassungen gegründet. Andere haben sich Astrologischer Gründe bedienet, und aus gefährlichen Aspecten die Zukunfft der Cometen geschlossen; weil etwa in vorigen Zeiten etliche mahl auch in solchen Fällen Cometen mögen erschienen seyn. Alles dieses dienet aber nicht wenig zu Bestätigung der wahren Grund-Gesetze der Bewegung der Cometen, welche hier kürtzlich erklähret worden, und deren Erfindung wir dem grossen NEWTON zu dancken haben. Vorher hatten sich die Astronomi vielerley und solche ungereimte Meynungen hierüber ausgesonnen, daß es zu weitläufig seyn würde, derselben zu erwehnen. Erstlich meynte man, daß die Cometen nicht unter die

7) EDMOND HALLEY, *Astronomiae cometicae synopsis* (Oxford, 1705). *Philosophical Transactions* 24 (1704–1705), 1882–1899. EJA.

8) In einigen Exemplaren dieser Ausgabe steht «nehmlich» statt «endlich». EJA.

Welt-Cörper zu rechnen, sondern daß dieselben nur in unserer Luft befindliche entzündete Ausdünstungen wären. Nachdem diese Meynung wiederleget worden, so haben einige die Cometen beständig über den Saturnum hinaus versetzet, und denselben keinen nähern Zutritt zu uns verstatten wollen. Anjetzo aber findet man um so viel weniger Ursache an der Wahrheit des NEWTONianischen Systematis zu zweifeln, je mehr man Gelegenheit bekommt Cometen zu observiren, und die Observationen dagegen zu halten.

VII
Wie groß die Anzahl der Cometen sey?

Dieses ist eine Frage, welche wohl niemahls richtig wird beantwortet werden können. So viel ist gewiß, daß die Anzahl derselben sehr groß seyn müsse. Denn wenn unter allen, welche bißher erschienen, keine völlige Uebereinstimmung wahrgenommen würde; so müßten zum allerwenigsten sich würcklich so viel Cometen an dem Himmel befinden, als bißher gesehen worden. Wenn aber ja einige Cometen etliche mahl sollten beobachtet worden seyn, so würde doch die Anzahl nicht mercklich verringert. Wir müssen vielmehr glauben, daß sich dieselbe weit höher belauffe, als man sichs einbilden sollte. Denn, weil man erstlich in den alten Zeiten in Aufsuchung der Cometen nicht so sorgfältig gewesen ist, als jetzo, so ist sehr wahrscheinlich, daß viele kleine Cometen, dergleichen man anjetzo nicht leicht aus der Acht läßt, gantz und gar nicht sind angemercket worden: auch ist die Historie nicht so alt und vollständig, daß darinne sich nur alle grosse Cometen aufgezeichnet finden sollten. Man hat allhier vor einem Jahre einen kleinen Cometen gesehen, welcher auf den meisten übrigen Observatoriis nicht ist wahrgenommen worden, der also leicht aus dem Catalogo der Cometen hätte weggelassen werden können, und ein solches Schicksal mag schon vielen andern, so wohl in den letztern als absonderlich in den alten Zeiten, wiederfahren seyn. Zweytens kan es leicht seyn, daß viele Cometen, wenn sie uns so nahe kommen, daß sie gesehen werden könnten, dennoch am Himmel der Sonne so nahe stehen, daß sie nicht zum Vorschein kommen können. Wie man denn einmahl am Tage, bey einer totalen Sonnen-Finsterniß, einen grossen Cometen gesehen haben soll, welcher ohne diesen sonderbaren Zufall nimmer würde bemerckt worden seyn. Drittens können auch viele Cometen in der Südlichen Himmels-Helfte sichtbar seyn, welche entweder gar nimmer, oder nur des Tages über unsern Horizont zu stehen kommen; von welchen wir folglich aus Mangel tüchtiger Observatorum in den Südlichen Gegenden der

Erde nichts erfahren. Vierdtens kan es auch leicht geschehen, daß zwar eben derselbe Comet öffters in sein Perihelium kommt, die Erde aber sich alsdenn immer an einem solchen Ort ihrer Bahn aufhält, daß der Comet wegen allzugrosser Entfernung von uns nicht sichtbar seyn kan. Fünftens, da einige Cometen in ihrem Perihelio so weit, als die Erde von der Sonne, entfernet sind; so ist glaublich, daß es auch solche Cometen gebe, welche der Sonne nimmer so nahe als der Mars, oder der Jupiter, oder gar der Saturnus, kommen; dergleichen können folglich aus der Erde nimmer gesehen werden. Sechstens, wäre es auch möglich, daß der völlige Umlauf eines Cometen eine so lange Zeit erforderte, daß derselbe seit Erschaffung noch nicht hätte zu seinem Perihelio gelangen können. Weil ferner über dieses die Entfernung des Aphelii von der Sonnen nicht bekannt ist, so könnte es seyn, daß ehe ein Comet sein Aphelium erreicht, derselbe einem benachbarten Fixstern allzunahe käme, und sich mit desselben seinem Systemate vereinigte, folglich das Systema der Sonne völlig verliesse. Da nun die Fixsterne ohne Zweifel auch ihre besondern Cometen um sich haben, so könnte es hinwiederum geschehen, daß sich derselben auch einige von da weg begeben, und sich in unser Systema einschlichen. Die Möglichkeit hiervon kan nun nicht wiederleget werden, allein die Distantz des nächsten Fixsterns von der Sonne ist groß, daß ein solcher Comet lange vor Erschaffung der Welt hätte ausreisen müssen, um erst jetzt zu uns zu gelangen. Diese Meynung kann aber auch nicht wohl behauptet werden, seit dem man angefangen, accuratere Observationen von den Cometen zu machen. Denn ob man gleich daraus die wahre Weite ihres Aphelii nicht bestimmen kan; so kan man daraus doch zeigen, daß dieselbe bey weitem nicht so groß sey, als die Distantz des nächsten Fixsterns. Es ist im Gegentheil zu vermuthen, daß einige Cometen in ihrem Aphelio nicht einmahl so weit, als der Saturnus von der Sonne, abstehen. Man hat von einem schon so gar gefunden, daß sein Aphelium der Sonne näher sey, als der Jupiter, das Perihelium gleich wohl weiter als der Mars abstände, und daß dieser Comet ungefehr in 4 Jahren seinen ordentlichen Lauf um die Sonne vollende. Sollte sich nun die Sache also verhalten, so müßte dieser Comet vielmehr zu den Planeten gezehlet werden. Denn ausserdem, daß sein Lauf mit den Planeten übereinkäme, so kan sich um denselben auch nicht der unordentliche Dunstkreiß, noch ein Schweif befinden, welche Eigenschafft sonsten die Cometen von den Planeten hauptsächlich unterscheidet. Aus eben diesem Grunde wäre also zu vermuthen, daß noch mehrere Cometen unter die Anzahl der Planeten versetzt zu werden verdienten.

VIII
Ob die Cometen einige Würckung auf die Erde haben können?

Dieses ist die fürnehmste Frage, welche von den meisten Leuten über diese Materie gethan zu werden pfleget. Vor diesem hielte man die Cometen für gewisse Zeichen und Vorboten besonderer darauf folgender Landesplagen, und anjetzo finden sich noch viele, welche diese Himmels-Cörper nicht ohne Schrecken ansehen können. Um nun auf diese Frage gründlich zu antworten, so muß dieselbe in zwey gantz differente Fragen zertheilet werden. Die erstere ist, ob man die Erscheinung der Cometen nicht als ein besonderes Zeichen des göttl[ichen] Zorns, und daher zu besorgen stehender Strafen, anzusehen habe, oder nicht? Die andere ist von dieser gäntzlich unterschieden, ob die Cometen nicht durch eine natürliche Krafft Veränderungen auf der Erde hervor bringen können? Was nun die erstere Frage betrifft; so muß man in Beantwortung derselben alle Behutsamkeit gebrauchen, um darüber weder ein unrichtiges noch verwegenes Urtheil zu fällen. Daß die Menschen die Cometen für gefährliche Vorboten angesehen, daran mag wohl die damals so sehr im Schwange gewesene Astrologie die fürnehmste Ursache gewesen seyn, als worinne dem Gestirn ein sonderbarer Einfluß auf die Erde und so gar in die Handlungen der Menschen zugeeignet worden. Da nun aber diese ungereimte Meynung größtentheils ausgerottet worden; so bleibet gleichwohl die Frage noch übrig, ob die Cometen nicht als Zeichen bevorstehender Glücks- oder Unglücks-Fälle angesehen werden müsten, ob sie gleich dieselben nicht durch einen Einfluß unmittelbarer Weise würckten. Diejenigen, welche diese Meynung behaupten, führen an, daß da sich in allen Theilen der Welt die gröste Ordnung und Uebereinstimmung befinde, und alles in dem vollkommensten Grad mit einander verbunden sey, man auch nothwendig zugeben müste, daß die Erscheinung der Cometen, als eine ausserordentliche Begebenheit in der Welt, auch mit gantz besondern Veränderungen verknüpft sey. Es würde hingegen eine grosse Verwegenheit seyn, wenn man sich einbilden wolte, daß diese Erscheinungen bloß von ungefehr geschehen, und uns von der Vorsehung GOttes ohne einen besondern Endzweck zugeschickt würden. Und da uns die Betrachtung des Welt-Gebäudes zur Erkenntniß und Verherrlichung des Schöpfers leiten soll, dieser Würckung aber bey den meisten Menschen aus einer unverantwortlichen Nachläßigkeit kein Raum gelassen wird; so wird es sehr wahrscheinlich, daß solche ausserordentliche Erscheinungen an dem Himmel, als welche die Aufmercksamkeit der Menschen insonderheit zu erwecken pflegen, um so viel mehr zur Ausübung dieser grossen Pflicht antreiben sollen. Ja man kan sicher sagen, daß wenn die Erscheinungen der Cometen diese Würckung haben sollten, dieselbe gewiß den Absichten des Schöpffers gemäß wäre. Was aber die Bedeutung solcher sonderbaren Himmels-Begebenheiten anlangt, so bezeuget die heilige Schrifft, daß Gott zur Ankündigung besonderer Gerichte auch merckwürdige Vorbothen gesandt habe, und daß insonderheit zu den letzten Zeiten viel Zeichen am Himmel gesehen werden sollen. Wir ersehen auch gantz deutlich aus der Offenbarung Johannis, daß noch vor dem Ende der Welt erschreckliche Straff-Gerichte über das menschliche Geschlecht ergehen sollen, welche vielleicht auch durch gantz ausserordentliche Erscheinungen am Himmel werden angekündiget werden. Vielleicht können die Cometen auch natürlicher Weise so gar selbst zu Vollstreckung dieser bevorstehenden Gerichte als Werckzeuge dienen, welche Betrachtung zur andern Frage: Was die Cometen natürlicher Weise für eine Würckung auf der Erde haben können? gehöret. Um hierauf richtig zu antworten, müssen drey Fälle wohl in Erwegung gezogen werden. Der erste ist, wenn die Cometen sehr weit von uns entfernt sind; der andere, wenn uns dieselben ziemlich nahe kommen, und der dritte, wenn sie allzunahe anrücken. In diesen Fällen kan die Würckung von nichts anders herrühren, als von der Anziehungs-Krafft, welche bey allen himmlischen Cörpern wahrgenommen wird. Also ziehet die Sonne alle Planeten und Cometen an sich, der Mond wird sowohl gegen die Sonne, als die Erde gezogen, und die Erde selbst auch hinwiederum gegen den Mond, wodurch die Ebbe und Fluth des Meeres verursachet wird. Die Grösse dieser Anziehungs-Krafft beruhet theils auf der Grösse des anziehenden Cörpers, theils auf dessen Entfernung. Je grösser der anziehende Cörper ist, um so viel grösser ist auch gemeiniglich seine Krafft; hingegen je weiter derselbe entfernet ist, je kleiner wird dieselbe, und das nach der Verhältniß der Quadrate, dergestalt, daß wenn der anziehende Cörper zweymal weiter entfernet ist, seine Krafft viermahl kleiner, in einer dreyfachen Distanz neunmahl kleiner wird, und so fort. Aus diesem Grunde kan man also sagen, daß die Erde nicht nur gegen die Sonne und den Mond, sondern auch gegen alle übrigen Planeten, Fixsterne und Cometen, gezogen werde, daß aber die Krafft der letztern theils wegen ihrer geringen Grösse, theils wegen ihrer grossen Entfernung, nicht mercklich sey. Die Cometen sind gemeiniglich in Ansehung ihres Cörpers nicht grösser, als die Erde oder die Venus. Wenn also ein Comet so weit als die Venus von uns entfernt ist, wenn uns dieselbe auch am nächsten ist; so kan auch seine anziehende Krafft nicht grösser, und folglich nicht zu mercken seyn. Im ersten Fall kan also den Cometen gantz und gar keine Würckung auf die Erde zugeschrieben werden, so lange dieselben nehmlich noch so weit von uns entfernet sind, daß ihre anziehende Krafft nicht grösser ist, als diejenige, welche die Planeten auf uns ausüben. Wenn uns aber die Cometen näher, als die Venus kommen, dennoch aber noch viel weiter, als der Mond von uns abstehen; so ist zwar ihre Krafft noch sehr geringe, und nicht vermögend, den Lauf der Erde um die Sonne zu verwirren, inzwischen aber kann dieselbe doch hinlänglich seyn, die Fläche ihrer Bahn einigermassen zu verändern. Diese Würckung kan auf zweyerley Art mercklich werden, entweder in der Schiefe der Ecliptic, oder in der Fortrückung der Aequinoctial-Puncte, als welche schon durch die Krafft des Monds jährlich um 50 Secunden zurück geschoben werden. Diese Verändrungen werden durch die Stellung des Cometen, wenn er uns am nächsten ist, und durch den scheinbaren Ort der Sonne dergestalt bestimmt, daß wenn der Comet eine Nördliche Breite hat, und die Sonne im Widder steht, die Schiefe der Ecliptic vermehret, wenn aber die Sonne in der Waage stehet, vermindert werde. Ist aber die Sonne im Krebs, so werden die Aequinoctial-Puncte vorwärts, im Steinbock aber rückwärts gerücket. Hat aber der Comet eine Südliche Breite, so sind diese Würckungen contrair. Ueberhaupt aber müssen dieselben um so viel grösser seyn, je näher uns der Comet ist. Weil uns nun schon öfters Cometen so nahe gekommen, so müssen vermuthlich schon mehrmahl solche Verändrungen vorgegangen seyn. Dieses wird auch durch die Vergleichung der alten Astronomischen Observationen mit den heutigen sehr schön bekräfftiget, indem man zu allen Zeiten eine ziemliche Veränderung in der Schiefe der Ecliptic wahrnimmt, dahero man kein Bedencken tragen darf, diese Würckung den Cometen zuzuschreiben. Aus diesem Grunde müste der jetzige Comet, theils die Aequinoctial-Puncte mehr zurück schieben, theils die Schiefe der Ecliptic vermehren, welches die künftigen Observationen lehren werden. Weil nun die Abwechselung der 4 Jahres-Zeiten absonderlich von der Schiefe der Ecliptic herrühret, so können die Cometen eine empfindliche Würckung auf unserm Erdboden verursachen.

Am allermeisten aber muß diese Würckung zu verspüren seyn, wenn uns ein Comet noch weit näher kommen sollte, als im vorhergehenden Falle angenommen worden. Alsdenn würde es bey den gemeldeten Verändrungen nicht bleiben, sondern die gantze Bewegung der Erde um die Sonne müste so wohl, als die Bahn selbst, sehr mercklich verändert werden. Das Jahr könnte entweder länger oder kürtzer werden, und da anjetzo die Erde fast immer gleich weit von der Sonne entfernet bleibet, so könnte dieselbe alsdenn zu einer Zeit der Sonne weit näher kommen, zu einer andern aber viel weiter sich davon entfernen, welche Veränderung sehr betrübte Folgen nach sich ziehen würde. Sollte aber ein Comet sogar an die Erde stossen, so würden die schon gemeldeten Zufälle weit grösser, und erschrecklicher werden. Daß der Schweif eines Cometen, wenn derselbe die Erde berühren sollte, einige besondere Würckung haben könnte, ist nicht zu vermuthen, wenn nehmlich derselbe aus nichts anders, als den subtilsten Sonnenstäubchen bestehet. Da aber der Dunstkreiß eines Cometen ohne Zweifel aus ziemlich dichten Dünsten bestehet; so könnte durch die Annäherung desselben unsere Lufft dergestalt mit Wasser angefüllt werden, daß daher nothwendig eine Sündfluth entstehen müste, und deswegen ist die Meynung des WHISTONS nicht so leicht zu verwerfen, welcher die Sündfluth von einem Cometen herleitet. Es wollen zwar einige eine solche Annäherung der Cometen als eine unmögliche Sache ansehen, allein dieselben haben entweder die wahre Beschaffenheit dieser Cörper nicht recht eingesehen, oder sich von dem Lauf derselben verkehrte Begriffe gemacht. Der jetzige Comet ist kaum 40 mahl weiter von uns entfernt als der Mond, warum sollte es nun nicht möglich seyn, daß ein Comet einmahl der Erde noch viel näher gekommen wäre. Inzwischen ist doch nicht zu befürchten, daß uns ins künftige ein Comet so nahe kommen werde, weil durch die Annäherung seines Dunstkreises nothwendig eine neue Sündfluth entstehen müste, wir aber gleichwohl die Versicherung haben, daß die Erde nicht mehr durch Wasser verderbet werden soll.

Man könnte aber dagegen einwenden, daß ein Comet der Erde noch weit näher, als bey der Sündfluth geschehen, kommen, und also durch seine Anziehungs-Krafft nicht nur nichts von seinem Wasser verliehren, sondern uns noch dazu einen guten Theil unserer Athmosphaere rauben könnte, wodurch denn zwar keine Sündfluth, dennoch aber der völlige Untergang der Erde, und folglich der jüngste Tag, verursachet werden könnte. Allein, ungeachtet gegen die Möglichkeit dieses Untergangs aus der Natur-Wissenschafft nichts eingewendet werden kan; so hat uns doch die göttliche Offenbahrung von dieser Furcht befreyet. Denn da uns unser Heyland die Versicherung gegeben, daß die Zeit des Endes der Welt nicht nur den Engeln im Himmel, sondern auch ihm selbst verborgen sey; so kan dasselbe auch durch keine natürliche Ursache veranlasset werden. Ungeachtet uns aber der völlige Lauf der Cometen, ihre Anzahl und Wiedererscheinungen nicht bekannt ist; so müssen wir doch glauben, daß die Engel, als weit vollkommenere Geister, eine vollständige Erkenntniß dieser Himmels-Cörper besitzen, und also sogar den Tag und die Stunde des Endes vorher wissen würden, wenn dasselbe durch einen Cometen verursachet werden sollte.

Fortgesetzte Beantwortung

der

Fragen

über

die Beschaffenheit, Bewe-
gung und Würckung

der

Cometen.

Berlin, zu finden bey Ambrosius Haude.
1744.

Commentatio 68 indicis ENESTROEMIANI

FORTGESETZTE BEANTWORTUNG DER FRAGEN ÜBER DIE BESCHAFFENHEIT, BEWEGUNG UND WÜRCKUNG DER COMETEN

I

Warum erscheinet der Schweif der Cometen bald gerade, bald gebogen, bald gespalten?

Wenn man die Schweife der Cometen nur überhaupt betrachtet, so findet man, daß dieselben immer von der Sonne abgewandt sind, wie schon im vorigen bemercket worden.[1]) Inzwischen werden doch bisweilen einige besondere Umstände wahrgenommen, welche von dieser Regel abzuweichen scheinen, dergleichen diejenigen sind, deren in der vorgelegten Frage Erwehnung geschicht. Weilen aber so wohl die Beugung als Spaltung der Schweife die obgemeldte Direction sehr wenig verändert, so kann deswegen keine Ausnahme gemacht werden. Was nun diese verschiedene Erscheinungen, welche sich insonderheit bey dem jetzigen Cometen ereignen, betrift, so beruhet ohne allen Zweifel der Grund davon auf der wahren Ursache, aus welcher die Schweife der Cometen entspringen, und so lange uns diese unbekannt ist, so können auch die gemeldeten Umstände nicht erkläret werden. Wir haben oben verschiedene Meynungen der Gelehrten über den Ursprung der Schweife angeführet, unter denselben aber keine der Wahrheit gemässer befunden, als diejenigen, welche der um die Astronomie so hoch verdiente KEPLER[2]) angegeben. Diese Erklärung wird nun durch diese verschiedene Erscheinungen welche an den Schweifen der Cometen wahrgenommen werden, völlig ausser allem Zweifel gesetzet, indem alle diese besondern Umstände daraus gantz natürlich fliessen. Um aber dieses desto deutlicher darzuthun; so muß vor allen Dingen angemercket werden, daß wann die Schweife der Cometen nach einer geraden Linie von der

1) Im 18. Jahrhundert fehlen die finiten Formen der Hilfsverben besonders häufig. EJA.

2) J. KEPLER, *De cometis libelli tres* (1619) = *Gesammelte Werke*, Band 8 (München, 1963), S. 129–262, insbesondere S. 226. EJA.

Sonnen abgewandt seyn sollten, dieselben uns auch am Himmel gerade, oder nach einem grossen Zirkul ausgestreckt erscheinen müßten. Wofern also hierinne eine Abweichung beobachtet wird, so kann man auch sicher schliessen, daß der Schweif entweder in der That gebogen, oder nicht gäntzlich von der Sonne abgelehnet seyn müsse. Man muß aber hierbey nicht vergessen, die Refraction in Erwegung zu ziehen, kraft welcher uns alles am Himmel höher über dem Horizont zu stehen scheint. Nahe am Horizont ist die Würckung der Refraction am grösten, und beträgt ungefehr einen halben Grad, in einer Höhe von 20 Graden aber ist dieselbe nicht mehr merklich. Wenn also der Schweif eines Cometen mit einem Ende den Horizont fast berühret, und mit demselben einen schiefen Winkel macht, so sehen wir dieses untere Ende um einen halben Grad höher, als es in der That ist, den obern Theil aber sehen wir in seiner wahren Stelle. Folglich kan uns in diesen Umständen der Schweif eines Cometen gekrümmt erscheinen, wenn derselbe gleich in der That gerade ist. Weil nun der jetzige Comet einige Zeit her sehr nahe am Horizont gesehen worden, so kann es seyn, daß die wahrgenommene Krümmung des Schweifes meistentheils der Refraction zuzuschreiben ist. Dem ohngeachtet aber kann man nicht läugnen, daß der Schweif nicht in der That etwas gekrümmet seyn solte, und dieses läßt sich auch gantz deutlich erweisen aus der vorhergegebenen Erklärung des KEPPLERS. Denn wenn wir setzen, daß die Schweife aus denen von dem Dunstkreiß der Cometen durch die Kraft der Sonnen-Strahlen weggestossenen Theilchen bestehen, so können wir von allen diesen sonderbahren Erscheinungen die Ursache leicht anzeigen. Wir müssen zu diesem Ende drey verschiedene Fälle betrachten. Der erste ist, wann ein Comet fast stille steht oder in seiner Bahn sehr langsam fortgehet: der zweyte, wenn er sich geschwind aber nach einer geraden Linie beweget, und der dritte, wenn sein Lauf nicht nur geschwind, sondern auch merklich gekrümmet ist.

Im ersten Fall bleiben die von den Sonnen Strahlen ausgestossenen Theilchen in einer geraden Linie, welche rückwerts verlängert durch die Sonne gehen würde. Denn weilen der Comet keinen oder nur einen sehr langsamen Lauf hat, so behalten die folgenden ausgestossenen Theilchen mit den vorhergehenden einerley Direction. Daher muß in diesem Fall der Schweif am Himmel auch in einer geraden Linie von der Sonne abgewandt erscheinen. Dieses geschiehet nur, wenn ein Comet noch sehr weit von seinem Perihelio, oder Sonnen Nähe, entfernet ist: denn alsdann ist seine Bewegung nicht nur sehr langsam, sondern auch fast nach einer geraden Linie entweder gegen die Sonne, oder von derselben weggerichtet, welcher Lauf keine merkliche Veränderung in der Lage des Schweifs verursachen kann, wie aus dem folgenden mit mehrerem erhellen wird.

Im zweyten Fall, da der Comet geschwind nach einer geraden Linie fortzugehen gesetzet wird, ist klar, daß die Theilchen, welche jetzt ausgestossen werden, nach einer andern Gegend ihren Lauf richten, als diejenigen, welche schon vor einiger Zeit, da der Comet noch an einem andern Orte stund, sind weggestossen worden. Da nun auch diese noch einen Theil des sichtbaren Schweifes ausmachen, so folgt nothwendig, daß in diesem Fall der Schweif theils gebogen, theils nicht schnur gerad von der Sonne abgewandt erscheinen müsse. Diese Abweichung und Krümmung muß auch um so viel grösser seyn, je geschwinder theils die Bewegung ist, theils je mehr die Direction von der Sonne abweichet. Dieser Fall ereignet sich nun, wenn ein Comet sich nahe bey seinem Perihelio befindet, und es war eben um diese Zeit, als der Schweif des jetzigen Cometen gebogen erschienen ist.

Im dritten Fall muß aber die gröste Veränderung entstehen, wenn der Lauf des Cometen nicht nur schnell, sondern auch sehr starck gekrümmet ist. Alsdenn müssen die Theilchen, welche zum Exempel gestern sind ausgestossen worden, eine gantz andere Lage haben, als diejenigen, welche erst heute herausgestossen werden. Und weilen der sichtbare Schweif auch noch aus sehr viel Theilchen bestehet, welche schon vor einiger Zeit aus dem Cometen gekommen; so kann derselbe nicht anders als gebogen erscheinen. Hieraus wird auch klar, daß, da diese Theilchen einige Zeit in dem grossen Raum stehen bleiben, und inzwischen durch die nachfolgenden ein neuer Schweif hervor gebracht werden kann, nothwendig ein gespaltener oder in verschiedene Aeste zertheilter Schweif zum Vorschein kommen könne. Dieser Fall ereignete sich nun wenn der Comet durch sein Perihelium gehet, allwo sein Lauf nicht nur am schnelsten, sondern auch am meisten gekrümmet ist. Der gegenwärtige Comet ist den ersten Martii durch sein Perihelium gegangen, allwo er der Sonne ungefehr zweymahl näher gewesen, als der Mercurius. Folglich war sein Lauf sehr geschwind, und gekrümmt, massen[3]) er in einem Tage bey nahe 6 Grad um die Sonne gelaufen. Weilen nun um eben diese Zeit der Schweif nicht nur gespalten erschienen, sondern auch sehr viel Schweife zugleich gesehen worden; so wird dadurch nicht nur die vorgelegte Frage beantwortet, sondern auch zugleich die KEPPLERische Erklärung von dem Ursprung der Cometen Schweife auf das festeste bestätiget.

Weilen hier angeführet worden, daß der jetzige Comet in seinem Perihelio der Sonne näher, als der Mercurius, gestanden; im vorigen aber Erwehnung geschehen, daß derselbe der Sonne nicht einmahl so nahe, als die Venus kommen würde; so ist

---

3) Eine alte Form von «da», «weil». EJA.

nöthig allhier zu erinnern, daß alles dasjenige, was in der Beantwortung der vorigen Fragen von dem Lauf des jetzigen Cometen vorgebracht worden, auf Observationes gegründet gewesen, über welche man die nöthige Rechnung nicht angestellet hat. Seit der Zeit aber hat es sich gefüget, daß einige genauere Observationes zu Handen gekommen, aus welchen die wahre Bahn dieses Cometen sehr genau hat bestimmet werden können. Weilen nun dieser Comet beständig viel weiter von uns entfernet gewesen, als in den vorhergehenden Beantwortungen angegeben worden, so folget daraus, daß so wohl der Cörper desselben weit grösser, und sein Schweif viel länger gewesen sey, als vorher gemuthmasset worden.

II

Warum machen die Cometen bey dem Ende ihrer Erscheinung dem Ansehen nach einen Zickzack, welches dem Hevelio Gelegenheit gegeben zu glauben, daß sie von dem Saturnus-Himmel an eine Spiralem beschreiben?

Bey dieser Frage ist vor allen Dingen zu bemercken, daß diese schlangenförmige Bewegung nicht bey allen Cometen wahrgenommen werde, daher des HEVELII[4]) Muthmassung schon wegfällt. An den Cometen 1681 und 1742 wurde nichts dergleichen beobachtet. Wenn aber solches bey andern geschehen, so folget daraus keines weges, daß sich würcklich eine solche irregulaire Bewegung in dem Lauf derselben befinden sollte. Die Planeten scheinen uns auch bißweilen zurück, und denn wieder vorwerts zu gehen, und also einen Zickzack am Himmel zu beschreiben; es ist aber anjetzo genugsam bekannt, daß die Ursache davon der Bewegung der Erde zuzuschreiben sey. Eben dieser Umstand findet nun bey den Cometen gleichfalls statt, indem uns ihr Lauf gantz anders vorkommt, als er in der That ist. Der Lauf des Cometen Anno 1742 schien uns vorwerts oder directus zu seyn, ungeachtet derselbe würcklich rückgängig war, und der jetzige Comet ist in der That um die Sonne rechtläuffig, und scheinet uns doch am Himmel zurück zu gehen. Hiervon ist nun die Bewegung der Erde die wahre Ursache. Denn nach dem dieselbe geschwinder oder langsamer, der Comet näher oder weiter, und die Direction seiner Bewegung mehr oder weniger von der Direction der Erde unterschieden ist; so müssen daher auch gantz unterschiedliche Erscheinungen entstehen. Also kann es

4) Siehe J. HEVELIUS, *Cometographia* (Danzig, 1668), S. 670 und das Titelbild. EJA.

sehr wohl geschehen, daß ein Comet, welcher einige Zeit rechtläuffig gewesen, zurück zu gehen, und denn wiederum vorwerts zu lauffen anfängt, wodurch uns folglich eine schlangenförmige Bewegung vorgestellet wird. Wenn man also auch diese Erscheinungen nicht gründlich genug erklären könnte; so würde daraus doch nichts zur Bekräftigung der Meynung des HEVELII zu schliessen seyn. Nachdem aber der grosse NEWTON auf das deutlichste bewiesen, daß die Cometen in ihrer Bewegung eben die selbigen allgemeinen Grund-Gesetze der Natur beobachten, nach welchen sich alle Planeten richten, so müssen alle Zweifel, welche man etwan bey dem sehr wunderlich scheinenden Lauf der Cometen haben könnte, völlig wegfallen. Denn nachdem so wohl NEWTON selbst, als HALLEY, alle vorher erschienene Cometen, von welchen richtige Observationen vorhanden waren, nach dieser Theorie untersuchet hatten; so hat sich bey allen die schönste Uebereinstimmung gefunden, und die wunderbarsten Erscheinungen waren gantz natürliche Folgen derselben.

III

Wenn ein Comet in seinem Aphelio und also die Schwere gegen die Sonne sehr geringe ist, woher entstehet alsdenn die Krümmung in seinem Lauff, daß er wieder zurück kommt?

Alle Vernunft-Schlüsse, welche man aus den Observationen ziehen kann, stimmen darinne überein, daß alle Cometen so wohl als die Planeten beständig gegen die Sonne gezogen werden, und derselben krummlinichter Lauf bloß allein von dieser Kraft, welche die Attraction der Sonne genennt wird, herrühre. Ohne eine solche Kraft würden alle himmlische Cörper, welche einmahl in Bewegung gesetzt worden, mit einerley Geschwindigkeit nach geraden Linien beständig fortlauffen. Die Würckung dieser Kraft, oder Attraction auf die Cometen und Planeten ist demnach zweyfach. Erstlich wird dadurch ihre Geschwindigkeit entweder vermehret oder vermindert, und denn zweytens wird durch diese Kraft die krummlinichte Bewegung verursachet. Was die erstere Würckung anlangt, so ist leicht zu begreiffen, daß wenn sich ein Cörper von der Sonne entfernet, seine Geschwindigkeit verändert werden müßte, indem er beständig gegen die Sonne, das ist, rückwerts gezogen wird. Gehet aber der Cörper gegen die Sonne, oder kommt er der Sonne immer näher; so muß seine Bewegung immer geschwinder fort werden. So lange aber ein Cörper von der Sonne gleich weit entfernet bleibt, so lange behält er auch in seinem Lauf einerley Geschwindigkeit. Weilen nun die Planeten fast immer einerley Weite von der Sonne behalten, so ist klar, daß auch ihre Geschwindigkeit fast gleichförmig seyn müsse. Aus diesem Grunde ist nun die Geschwindigkeit der Cometen der grösten Veränderung unterworffen. Denn da sich dieselben im Aphelio sehr weit von der Sonne entfernen, in ihrem Perihelio aber derselben sehr nahe kommen; so wird ihre Bewegung vom Perihelio biß zu dem Aphelio langsamer, vom Aphelio aber bis zum Perihelio schneller, und folglich läuft ein Comet am geschwindesten, wenn er in seinem Perihelio ist, am langsamsten aber, wenn er sich im Aphelio befindet.

Die andere Würckung, welche die Attraction der Sonne in dem Lauf der Planeten und Cometen hervorbringt, bestehet in der Krümmung ihrer Bahn. Hierbey kommt es aber auf nachfolgende drey Stücke an. Erstlich auf die Grösse der Attraction, zweytens auf die Geschwindigkeit, und drittens auf die Direction des Laufs in Ansehung der Sonne. Je grösser die Attraction ist, je grösser ist auch die daher entstehende Krümmung, wenn die beyden übrigen Stücke einerley bleiben. Von der Geschwindigkeit dependiret die Krümmung dergestalt, daß je geschwinder der Cörper läuft, je kleiner die Krümmung ist, welche daher entstehet, wenn nehmlich die Attraction und die Schiefe einerley ist. Bey den vorgelegten Fragen ist nicht nöthig auf die Schiefe zu sehen: sondern man darf nur merken, daß auch eine sehr schwache Attraction fähig ist, eine grosse Krümmung hervorzubringen, wenn nur die Geschwindigkeit klein genug ist. Da nun bey einem Cometen, welcher in sein Aphelium geht, die Geschwindigkeit immer abnimmt, so kann dieselbe endlich so klein werden, daß die in dieser Gegend so sehr geschwächte Attraction doch hinreichend genug ist, die erforderte grosse Krümmung hervorzubringen. Und dieses geschiehet würcklich, wenn die Bahn eines Cometen eine Ellipsis ist, wie solches die Berechnung gantz deutlich vor Augen legt.

Man könnte hier noch fragen, warum ein Comet seinen Lauf nicht eher, als er sein Aphelium erreicht, solcher gestalt krümme, weilen vorher die Attraction der Sonne noch mächtiger ist, und solches leichter bewerckstelligen könnte. Hier muß man nun auch auf den dritten Umstand, nehmlich auf die Schiefe seines Laufes sehen, worauf die Krümmung zugleich mit beruhet. In diesem Stück verhält sich die Würckung solcher Gestalt. Wenn ein Cörper gerade zu der Sonne, oder gerade von derselben wegläufft, so leidet derselbe in seinem Lauf gar keine Krümmung, eben so wenig, als ein Stein, welcher entweder auf die Erde gerade hinab fält, oder senckelrecht hinauf geworfen wird, eine krumme Linie beschreibt. Je mehr aber die Direction des Laufs von der Linie, welche aus dem Cörper zu der Sonne gezogen wird, abweicht, je grösser wird die Krümmung. Und wenn die Direction mit dieser Linie einen rechten Winckel macht, so wird die Krümmung am grösten, wenn nemlich die Attraction und die Geschwindigkeit einerley bleibt. Wenn nun ein

Comet von der Sonne weg gegen sein Aphelium gehet, so ist anfänglich seine Schiefe sehr groß, oder seine Direction macht mit der nach der Sonne gezogenen Linie einen sehr kleinen Winckel, dahero entstehet fast gar keine Krümmung. Weil dieselbe aber doch nicht völlig nichts ist, so wird besagter Winckel nach und nach grösser, bis er endlich so groß wird, daß die Attraction, ob sie gleich beständig abnimmt, dennoch mächtig genug wird, den schon sehr langsam gewordenen Cometen völlig herum zu schwingen. Sollte aber gleichwohl noch über diesen Punct einiger Zweifel übrig seyn; so ist zu rathen, daß man sich die Gesetze der Bewegung besser bekannt mache, und die daraus entstehenden Würckungen durch die Rechnung genau bestimmen lerne.

IV

Man siehet durch die Athmosphaere der Cometen die Fixsterne,
(welches wir zwar selbst nicht wahrgenommen,
uns aber von andern versichert ist)
wie kann denn die Dunst-Kugel so dicke seyn,
als sie der Hr. Verfasser angiebt?

Was oben von der grossen Dichte der Athmosphaere der Cometen beygebracht worden, dagegen wurde eingewendet, daß man gleichwohl durch dieselbe bißweilen Sterne erblicket hätte, woraus denn folgen müßte, daß selbige bey weitem nicht so dichte seyn könnte, als angezeiget worden. Bey dieser Frage ist nun erstlich zu untersuchen, ob die angegebene Observation, daß Fixsterne durch die Athmosphaere der Cometen gesehen worden, ihre Richtigkeit habe, oder nicht? und denn zweytens, ob daraus der gemachte Schluß richtig folge? Was das erste betrift, so ist zu bemercken, daß ohngeachtet uns der Himmel so häuffig mit Sternen angefüllet zu seyn scheinet, es dennoch sehr selten zu geschehen pflegt, daß ein Comet nur sehr nahe bey einem Fixsterne vorbey gehet, will geschweigen denselben bedecken. Daher ereignet sich dieser Fall so selten, daß man von allen bißher observirten Cometen sehr wenige dergleichen Observationen anzuführen im Stande ist. Weder der Comet Anno 1680 und 81 noch einer von den folgenden biß jetzo, ist jemahls einem Fixstern so nahe gekommen, daß er denselben bedeckt hätte. Wenn also die in der vorgelegten Frage vorgegebene Observation sich bey einem von den letzteren Cometen zugetragen haben soll, so ist dieselbe für sehr verdächtig zu halten. Vielleicht haben diejenigen, welche solches gesehen zu haben vorgeben, den Schweif mit zu der Atmosphaere gerechnet; in welchem Fall es gar nichts ungewöhnliches seyn würde, wenn dadurch Sterne gesehen worden wären, als welches sehr öfters geschicht, und noch fast bey allen Cometen wahrgenommen worden.

Inzwischen kann man gleichwohl der würcklichen Athmosphaere der Cometen die Durchsichtigkeit nicht völlig absprechen, indem man dadurch ihren Kern oder wahren Cörper ziemlich deutlich erkennen kann, welches nicht geschehen würde, wenn die Atmosphaere nicht durchsichtig wäre. Die Art und Weise, wie oben der Ursprung der Schweife erkläret worden, setzet auch einen ziemlichen Grad der Durchsichtigkeit voraus, weil sonst die Sonnen Strahlen nicht durchgehen, und so wohl den gantzen Cörper erleuchten, als durch die Wegstossung kleiner Stäubchen den Schweif formiren könnten. Da also im obigen die Durchsichtigkeit nicht nur nicht geläugnet, sondern auch so gar unwiedersprechlich dargethan worden, so kann die angeregte Observation, daß bißweilen Fixsterne durch die Athmosphaere der Cometen gesehen worden, wenn dieselbe auch ihre völlige Richtigkeit hätte, keineswegs zu Bestreitung unsers Satzes angeführet werden. Die Zeit erlaubet inzwischen nicht, alle alte Observationen nach zu sehen, um zu forschen, ob jemahls eine solche Observation von einem tüchtigen Astronomo gemacht worden; denn auf das Vorgeben solcher Leute, welche in dieser Wissenschaft nicht genugsam geübet, kann man sehr wenig trauen. Es ist unterdessen doch so viel gewiß, daß die Athmosphaere der Cometen nur einen gewissen Grad der Durchsichtigkeit habe, welcher bald grösser bald geringer seyn kann, wegen der grossen Veränderungen, welchen diese Cörper ausgesetzet sind, daher wir auch zu einer Zeit den Kern weit deutlicher unterscheiden können, als zu einer andern. Weilen aber derselbe unmittelbar hinter der Athmosphaere stehet, so wird auch kein so grosser Grad der Durchsichtigkeit erfordert, um denselben zu sehen, als andere Cörper welche sehr weit davon entfernet sind. Wir können zum Exempel durch einen nicht allzu dichten Nebel Cörper, welche sehr nahe hinter demselben stehen, noch wohl sehen, andere aber, welche sich sehr weit hinter demselben befinden, werden unserm Gesichte völlig entzogen. Und also wäre es wohl möglich, daß die Athmosphaere der Cometen uns die Sterne völlig verbergen könnte, ohngeachtet der Kern sichtbar bleibt.

Wenn aber auch der Dunstkreis eines Cometen ja dergestalt wäre aufgekläret worden, daß man einen Fixstern dadurch gesehen hätte, so könnte daraus doch kein gründlicher Einwurf gegen die angegebene Dichte hergeleitet werden. Die Durchsichtigkeit, und die Dichte der Cörper, sind zwey so verschiedene Eigenschaften, daß beyde fast nichts mit einander gemein haben. Denn es giebt sehr dichte Cörper, als Wasser, Glaß, Crystall, und der Diamant, welche doch vollkommen durchsichtig sind: dahingegen andere, welche weit leichter und dünner sind, keine

Lichtstrahlen durchlassen. Deswegen könten auch bey dem Dunstkreiß eines Cometen diese zwey Eigenschaften, nehmlich die Dichte, und die Durchsichtigkeit, sehr wohl zusammen statt finden, indem es zur Erklärung der Erscheinungen nur auf die Refraction ankömmt, welche, ohne der Durchsichtigkeit Abbruch zu thun, so groß seyn kann, als es die Hervorbringung des Schweifs erfordert.

V
Warum erscheinet der Comet durch ein kleines Sehe-Rohr heller und deutlicher, als durch ein grosses?

Dieses ist ein Umstand, welcher nicht nur bey den Cometen, sondern auch bey den Planeten, und überhaupt bey allen Cörpern, welche durch Vergrösserungs-Gläser angesehen werden, Statt findet. Um die Ursache davon anzuzeigen, muß man auf nachfolgende drey Stücke wohl Acht haben.

Erstlich auf die Natur der Gläser. Denn ob uns diese gleich vollkommen durchsichtig scheinen; so gehet doch ein guter Theil der Lichtstrahlen dadurch verlohren. Weilen nun die sichtbare Helle eines Cörpers von der Menge der Strahlen, welche in unsere Augen fallen, herrühret; so können die Gläser überhaupt nicht anders, als die Objecta verdunckeln. Diese Verduncklung wird um so viel grösser, je mehr Gläser in ein Sehe-Rohr gesetzet werden, und daher pflegt man in die Astronomischen Tubos nicht mehr als 2 Gläser zu machen, damit die Verduncklung nicht ohne Noth vermehret werde. Es ist also eine ausgemachte Sache, daß von allen Objectis, welche wir durch Gläser ansehen, weniger Strahlen in unsere Augen fallen, und daß uns dieselben folglich dunckler scheinen müssen, als wenn wir dieselben mit blossen Augen besehen. Jedoch wird dieser Abgang öfters durch die Vergrösserung ersetzet, indem eine deutliche Vorstellung nicht allein von der Helle, sondern auch von der scheinbaren Grösse herkömmt, wovon im folgenden zweyten Stück gehandelt werden soll.

Man hat nehmlich die Kunst gefunden, zwey oder mehr geschliffene Gläser dergestalt in eine Röhre zu setzen, daß uns dadurch die Objecta weit grösser erscheinen. Und hierinnen bestehet der grosse Vortheil, welchen wir den Sehe-Röhren, und den Vergrösserungs Gläsern zu dancken haben. Ein dadurch gesehener Cörper scheinet uns also einen grösseren Raum einzunehmen, als wenn man denselben mit blossen Augen betrachtet. Inzwischen kommen daher nicht nur nicht mehr, sondern so gar weniger Strahlen in unser Auge, wie im vorigen Punckt dargethan worden ist: und also scheinen uns die Strahlen, welche mit blossen Augen in einem kleinern Raum vereinbaret sind, durch ein Vergrösserungs Glaß in einem grössern Raum zerstreuet, und folglich muß das Objectum dunckler gesehen werden. Dahero sind alle Fern- und Vergrösserungs-Gläser diesem Fehler unterworfen, daß je mehr dadurch die scheinbare Grösse vermehret wird, die Helle um eben so viel verdunckelt werde. Dergleichen Instrumente können also nicht anders mit Vortheil gebraucht werden, als bey solchen Cörpern, welche von Natur ein sehr helles Licht haben, dessen Verringerung der Deutlichkeit keinen sonderlichen Abbruch thut. Die Planeten und andere entfernte Cörper, deren Licht sehr helle ist, können demnach durch Fern-Gläser mit sehr grossem Vortheil betrachtet werden. Weil aber die Cometen ein weit schwächeres Licht haben, so gewinnt man durch die Fern-Gläser um so viel weniger, je mehr dieselben vergrössert werden. Dieser Umstand ereignet sich deutlich bey den Microscopiis, wodurch die Objecta kaum gesehen werden können, wenn dieselben nicht durch ein besonderes starckes Licht erleuchtet werden.

Drittens ist aber auch fürnehmlich auf die Güte der Fern-Gläser zu sehen, welche erstlich in der Reinigkeit der Gläser, und denn in der accuraten Schleiffung und Figur derselben bestehet. Die erstere Vorsichtigkeit ist nöthig, damit so wenig Strahlen, als möglich, verlohren gehen. Die andere ist so beschaffen, daß ein geringer Fehler, welcher in der Figur der Gläser begangen worden, das gantze Instrument unbrauchbar macht. Es kommt aber hauptsächlich auf das Objectiv Glaß, oder auf dasjenige, welches gegen die Sterne gekehret wird, an. Diese Gläser erfordern einen solchen Fleiß, und eine so grosse Geschicklichkeit des Künstlers, daß öfters unter 100 kaum eines geräth, welches mit Vortheil gebrauchet werden kann. Und diese Schwierigkeit wird um so viel grösser, je länger die Röhre ist, zu welcher das Glaß gehöret. Da es also um ein langes Fernglaß, welches zugleich ohne Mangel ist, eine sehr seltene Sache ist; so hat man sich nicht zu verwundern, daß die meisten, deren man sich bedienet, die erwünschte Würckung nicht haben. Dieses mag auch mit eine von den fürnehmsten Ursachen seyn, warum man die Cometen durch sehr lange Sehe-Röhren gemeiniglich weit dunckler, und undeutlicher erblicket, als durch kürtzere. Sollte die Röhre in Ansehung dieses dritten Punckts ohne Fehler seyn, so würde man sich über diesen Mangel nicht so sehr zu beschweren Ursache haben. Denn ich bin von einigen Freunden, welche mit sehr herrlichen Tubis versehen sind, versichert worden, daß sie durch dieselben den jetzigen Cometen sehr deutlich, und auch noch ziemlich helle erblicket haben, und das noch bey 100 mahl grösser, als er mit blossen Augen gesehen wird.

VI

Bewegen sich die Cometen auch um ihre Axe, wie bey den meisten Planeten wahrgenommen wird?

Dieses ist eine Frage, welche nicht wohl anders, als durch die Erfahrung ausgemacht werden kann. Dieses ist aber eine solche Sache, welche den grösten Schwierigkeiten unterworfen ist. Wir sind über diesen Punckt noch nicht einmahl von allen Planeten gewiß, welche man doch fast täglich betrachten kann. Man hat aus der Veränderung der Flecken, welche in dem Jupiter wahrgenommen worden, geschlossen, daß er sich ungefehr in 10 Stunden um seine Axe herumdrehe, und aus gleichen Beobachtungen hat man gefunden, daß sich der Mars in ungefehr 24 Stunden herumdrehe. Wegen der Venus ist man sehr lange Zeit in Ungewißheit gewesen, bis endlich BIANCHINI 5) durch sehr fleißige Observationen dargethan, daß ihre Herumdrehung in 23 Tagen 6) geschehe. Von dem Mercurio und Saturno läßt sich noch diese Stunde nichts bestimmen. Da wir also in Ansehung der Planeten noch in einer so grossen Ungewißheit stehen; so ist wenig Hoffnung vorhanden, daß man bey den Cometen zu einer vollkommeneren Erkenntniß gelangen sollte. Was hierüber bey den Planeten endecket worden, hat man den sehr langen und guten Fern-Gläsern zu dancken, durch welche man darinne nicht nur einige Flecken, sondern auch eine Veränderung an denselben hat wahrnehmen können. Weilen nun aber durch diese Instrumente bey den Cometen nicht viel auszurichten ist, wie im vorigen angemerket worden, und man auch gemeiniglich einen Cometen nicht lange genug betrachten kann; so ist aus der blossen Erfahrung über diese Frage nichts zuverläßiges zu melden. Die Theoretische Erkenntniß scheinet auch in diesem Stücke nicht mehr Licht zu geben. Denn ungeachtet wir aus den nunmehr erkannten Bewegungen, welche der Jupiter, Mars und Venus, nebst der Erde um ihre Axe haben, sicher schliessen können, daß sich auch eine ähnliche Bewegung bey dem Saturno und Mercurio befinden müsse; so kann man doch hieraus keine Folge auf die ausserordentlichen Planeten, dergleichen die Cometen sind, ziehen. Denn dieselben sind in so viel andern Stücken von den Planeten so sehr unterschieden, daß man keine genugsame Ursache hat, in diesem Stück eine Aehnlichkeit zu behaupten. Den Planeten scheinet diese Bewegung hauptsächlich zum Vortheil ihrer Einwohner mitgetheilet zu seyn. Weilen nun schwerlich zu glauben, daß die Cometen mit

5) FRANCESCO BIANCHINI (1662–1729), *Hesperi et Phosphori nova phaenomena, sive observationes circa planetam Veneris* (Romae 1728). EJA.

6) Eigentlich 24 Tage und einige Stunden, loc. cit., S. 58–67. TS.

Einwohnern begabet sind; so fält auch die Ursache einer solchen Bewegung weg. Und wenn auch ja diese Cörper bewohnet seyn sollten, so würden sie doch einen schlechten Vortheil von dieser Bewegung zu geniessen haben: indem der so irregulaire Lauf allen Vortheil zernichten würde. Über dieses, da wegen des grossen Dunstkreises der gantze Cörper auf einmahl erleuchtet wird, so scheinet auch kein Grund vorhanden zu seyn, warum sich derselbe um eine Axe herum drehen sollte, indem diese Bewegung in den Planeten nur zu einer Abwechselung zwischen Tag und Nacht dienet.

Einige sind der Meynung, daß die Cometen vormals ordentliche Planeten gewesen, und erst durch den Anstoß eines anderen Cometen aus ihrer runden Bahn in eine ablange versetzet worden. In diesem Falle müßten also die Cometen vormahls eine Bewegung um ihre Axe gehabt haben, und weilen diese durch einen Stoß nicht getilget werden kann, so müßte dieselbe noch fort dauren. WHISTON[7]) aber glaubt, daß die Planeten vormals Cometen gewesen, und daß insonderheit die Erde in den von Mose erwehnten sechs Tagen zu einem ordentlichen runden Lauf gebracht und wohnbar gemacht worden. Er behauptet ferner, daß die Erde vor der Sündfluth keine tägliche Bewegung gehabt, sondern daß dieselbe ihr erst bey der Sündfluth eingedrücket worden. Nach dieser Meynung müßte also die Erde, da sie noch ein Comet war, auch keine tägliche Bewegung um ihre Axe gehabt haben, weil sie solche ohne Zweifel behalten haben würde, woraus denn folget, daß die Cometen ohne eine solche Bewegung in ihren Bahnen fortgiengen. Diese beyden Meynungen sind aber noch weit mehreren Schwierigkeiten unterworfen, als die vorgelegte Frage selbst, und wird daher rathsamer seyn, unsere Unwissenheit über diese Frage frey zu gestehen, als einer ungegründeten Antwort beyzupflichten.

7) WILLIAM WHISTON, *A new theory of the earth*, Zweite Ausgabe (London, 1708). EJA.

VII

Newton leitet die Praecessionem aequinoctiorum von der abplatten Figur der Erde sowohl, als aus dem Mond her, die abplatte Figur der Erde aber von ihrer Bewegung um die Axe. Wenn also die 6te Frage mit ja zu beantworten wäre; so würden die loca Perihelii und nodi in einem Cometen veränderlich seyn. Wenn dieses ist, woran kann man einen Cometen, der ein tempus periodicum von etlichen 100 Jahren hat, wieder erkennen?

Wenn wir die Punckte am Himmel, wo der Aequator die Ecliptic durchschneidet, als fest und unbeweglich ansehen; so scheinen nicht nur die Aphelia und die Nodi der Planeten, sondern auch so gar alle Fixsterne, eine Bewegung zu haben. Denn da vor ungefehr 2000 Jahren das Frühlings Aequinoctium mit dem ersten Stern des Widders übereingekommen; so ist anjetzo dieser Stern bey nahe 30 Grad weit davon entfernet. Entweder müsten also die Fixsterne, oder die gedachten Aequinoctial-Punckte, beweglich seyn. Das letztere ist nun der Wahrheit weit mehr gemäß, als das erstere; indem es weit natürlicher ist, den Durchschnitt des Aequators mit der Ecliptic, als welche bloß allein auf der Stellung der Axe der Erde beruhet, veränderlich zu setzen, als eine solche Bewegung allen Fixsternen zuzuschreiben. Wenn wir also die Fixsterne als unbewegliche Stellen am Himmel ansehen; so werden die Aequinoctial-Punckte dergestalt veränderlich, daß sie jährlich um 50 Secunden gegen die Ordnung der Himmlischen Zeichen zurück rücken. Diese Veränderlichkeit rühret nun einzig und allein von der Stellung der Axe der Erde her, als welche wandelbar und nicht beständig gegen einen Punckt an dem Himmel gerichtet ist. Ist nun diese Axe veränderlich, so ist der Aequator der Erde eben dieser Veränderung unterworfen, und folglich auch der Durchschnitt des Aequators mit der Ecliptic. Der Aequator ist ein grosser Zirkul der Erde, welcher allenthalben von den beyden Polis 90 Grad weit entfernet ist, und die Ecliptic ist diejenige Bahn, in welcher sich der Mittel-Punckt der Erde um die Sonne herum beweget. Weilen nun die Axe der Erde auf der Ecliptic schief auf stehet, so durchschneidet der Aequator die Ecliptic, und der Durchschnitt ist eine gerade Linie, welche, so sie beyderseits bis zu den fast unendlich weit entfernten Fixsternen verlängert wird, daselbst die sogenannten Aequinoctial-Puncte vorstellet, welche folglich alle Jahr um 50 Secunden zurückgehen. Von dieser Bewegung ist die Ursache nach dem Newton in der anziehenden Kraft, theils aller Theile der Erde, theils des Monds zu suchen, als daher er die Ebbe und Fluth des Meers hergeleitet. Denn gleichwie die Nodi der Mond-Bahn zurückgehen, also zeiget dieser grosse Mann, daß auch aus gleichen Gründen die Aequinoctial-Punckte zurück gehen müssen. Um dieser Ursache willen müssen wir glauben, daß aller Planeten Axen gleichfals beweglich seyn müssen. Weil wir aber kaum derselben Bewegung um ihre Axe wahrnehmen können; so ist nicht zu vermuthen, daß wir jemals die Veränderung, welche in der Stellung derselben vorgehet, werden bemercken können. Eben so wenig werden auch die Einwohner anderer Planeten des Rückgangs unserer Aequinoctial-Punckte gewahr werden, als welches Phaenomenon nur uns betrifft, und auf die übrigen Planeten keinen Einfluß hat.

Nachdem nun dieses von der Praecessione Aequinoctiorum angemercket worden, so wollen wir die Frage in Erwegung ziehen, ob die Aphelia und Nodi der Planeten würklich beweglich sind oder nicht, und das in Ansehung der Fixsterne, weil, wie schon erwiesen, die Aequinoctial-Punckte keine festen Stellen am Himmel darstellen. Viele Astronomi sind der Meynung, daß die Aphelia und Nodi der Planeten beständig gegen einerley Fixsterne gerichtet bleiben, und folglich in Ansehung unserer Aequinoctial-Punckte mit den Fixsternen jährlich um 50″ fortrücken. Andere halten diese Stellen am Himmel auch in Ansehung der Fixsterne für veränderlich, die Bewegung aber, welche sie ihnen zuschreiben, ist so gering, daß dieselbe in einigen Seculis kaum bemercket werden kann. Und deswegen sind die Observationen nicht zureichend, diesen Streit zu entscheiden. Denn so weit man es anjetzo in der Accuratesse zu observiren gebracht hat, so ist man doch nicht vermögend, die Aphelia und Nodos der Planeten näher, als auf etliche Minuten zu bestimmen. Will man aber für die alten Zeiten aus den damahls gemachten Observationen diese Stellen berechnen; so kann man leicht um einen oder gar etliche Grade fehlen, und folglich die Bewegung derselben bey weitem nicht so genau erkennen, als zu diesem Ende nöthig ist. Wir müssen derohalben[8]) unsere Zuflucht zu der Theorie nehmen. Wenn wir nun setzen, daß alle Planeten gegen die Sonne gezogen werden nach den Gesetzen, welche NEWTON entdecket, so folget gantz deutlich, daß sowohl die Aphelia, als die Nodi der Hauptplaneten, in Ansehung der Fixsterne unveränderlich seyn müssen. Wenn wir aber ferner nach der NEWTONianischen Philosophie annehmen, daß die Planeten nicht nur von der Sonne angezogen werden, sondern daß dieselben auch gegen einander eine anziehende Krafft

8) Eine alte Form von «deshalb». EJA.

ausüben, welche aber in Ansehung der Sonnen-Krafft fast für nichts zu rechnen: so wird man leicht abnehmen, daß gleichwohl durch die Länge der Zeit endlich von diesen kleinen Kräfften eine merkliche Würkung entstehen müsse, welche sich in der Lage der Apheliorum und Nodorum äussern würde, ungeachtet man Ursache zu vermuthen hat, daß diese Punckte eben so offt rückwerts als vorwerts verrücket werden. Man hat auch noch bey keinem andern Planeten einiges Merckmal von dieser Würckung wahrnehmen können, als bey dem Saturno, wann ihm der Jupiter am nächsten kommt. Denn da der Jupiter der größte Planet ist, und folglich eine sehr starcke Anziehungs-Krafft hat, bey dem Saturno aber die Anziehungs-Krafft der Sonne schon ziemlich schwach wird; so ist jene Krafft schon vermögend, den Lauf des Saturni dergestalt zu verwirren, daß seine Abweichung von den Astronomischen Tabellen beobachtet werden kann.

Um nun endlich auf die Cometen zu kommen, so ist aus obigem klar, daß wenn dieselben nur allein von der Sonne angezogen würden, weder ihre Nodi, noch ihre Perihelia, in Ansehung der Fixsterne veränderlich seyn könnten. Sollte aber ein Comet nahe bey einem Planeten vorbey gehen, so müßte freylich sein Lauf um so viel mehr verrücket werden, je geringer seine Entfernung von dem Planeten gewesen. Da aber dergleichen Umstände sich sehr selten ereignen, und woferne der Comet einem Planeten nicht sehr nahe kommt, die Würckung nicht groß seyn kann; so kann auch keine so grosse Verwirrung in seinem Lauf entstehen, daß derselbe bey seiner Wiederkunft unerkenntlich seyn sollte. Hieher kommt noch, daß man absonderlich der alten Cometen Perihelia und Nodos bey weitem nicht so accurat weiß, als nöthig wäre, eine solche Veränderung in den künftigen Zeiten wahrzunehmen. Inzwischen aber bleibet dennoch gewiß, daß wenn ein Comet einem Planeten sehr nahe kommen sollte, seine Bahn in allen Stücken dergestalt verändert werden könnte, daß es unmöglich wäre, denselben ins künfftige wieder zu erkennen.

VIII

Wenn die Verrückung der Nodorum und Apheliorum eines Planeten nur von der praecessione aequinoctiorum, wie Streetius davor hält, herkommt, so wäre der Zweifel Num. 7 leicht zu heben. Doch wäre hernach die Frage, warum die andern Planeten und Cometen ihre loca Apheliorum nicht ändern solten, und nur die Erde allein, da sie sich doch so wohl als die Erde um ihre Axen bewegen?

Der Zweifel, welcher in dieser Frage enthalten, ist schon grösten Theils in der vorigen Beantwortung gehoben worden. Denn wenn nach dem STREETIO[9]) die Aphelia der Planeten still stehen, so verstehet er solches in Ansehung der Fixsterne, und also hat das Aphelium der Erde nichts vor den andern voraus, indem es gleichfalls als stillstehend angenommen wird. Also kann man nicht sagen, daß die Bewegung des Aphelii der Erde von ihrer Umwälzung um die Axe herrühre, indem dadurch nur die Aequinoctial-Punckte verändert werden. Wenn nun ein jeglicher Planet auch seine besondern Aequinoctial-Puncte hat, und dieselben gleichergestalt beweglich sind, wie aus ihrer Umdrehung um ihre Axen zu vermuthen ist: so hat auch die Erde in diesem Stücke nichts vor den andern Planeten voraus, daß man fragen könnte, warum solches in der Erde allein statt finde. Vielleicht hat man sich in dieser Frage an die Unbeweglichkeit der Nodorum der Planeten gestossen, und weil von der Erden-Bahn keine Nodi angegeben werden, ihre Aequinoctial-Punckte für ihre Nodos gehalten, und da diese beweglich sind jene aber nicht, so mag daher die gemachte Schwierigkeit entstanden seyn. Hierauf dienet zur Antwort, daß die Aequinoctial-Punckte keinesweges mit den sogenannten Nodis in Vergleichung gesetzt werden können. Um dieses deutlicher zu machen, so hat man zu mercken, daß ein jeglicher Planet seine besondere Ecliptic, und auch seinen besondern Aequatorem hat. Die Ecliptic ist die Fläche oder das Planum, worinne sich der Planet um die Sonne beweget, welches, so es bis zu den Fixsternen verlängert wird, daselbst einen grossen Zirckul vorstellt. Der Aequator aber eines Planeten ist das Planum, welches durch sein Mittelpunckt perpendicular auf die Axe, um welche sich

9) THOMAS STREETE (1622–1689), *Astronomia Carolina. A new theorie of the coelestial motions* (London, 1661), S. 23; *An appendix to the Astronomia Carolina* (London, 1664), S. 29–30. Über den Streit mit WING, siehe VINCENT WING (1619–1668), *Examen Astronomiae Carolinae* (London, 1665), S. 28–32 und THOMAS STREETE, *Examen examinatum* (London, 1666), S. 17–24. EJA.

der Planet herumwälzet, gezogen wird, und welches folglich am Himmel auch einen grossen Zirckul vorstellet. Die Nodi eines jeglichen Planeten sind nun die Punckte, wo sich die Ecliptic desselben mit der Ecliptic der Erde durchschneidet, und daraus ist klar, daß in diesem Verstande die Erde selbst keine Nodos haben könne. Ein jeglicher Planet hat ferner auch seine Aequinoctial-Punckte, in welchen sich desselben Ecliptic mit seinem eignen Aequatore durchschneidet. Jene Punckte, nehmlich die Nodi, werden nun bey dem STREETIO als unbeweglich angenommen; diese aber, nehmlich die Aequinoctial-Punckte, sind vermutlich bey allen Planeten, eben wie bey der Erde veränderlich, jedoch bey einem mehr, als bey dem andern, welche Veränderung, wie schon gemeldet, durch die Observationen nicht so leicht bemercket werden kann.

IX

Wenn das Planum orbitae cometae gegen das Planum der Ecliptic unter einem kleinern Winckel als 90° inclinirt ist, so gehet der Comet von Abend gegen Morgen, wie die Planeten: ist aber dieser Winckel grösser, so geht er von Morgen gegen Abend.

Dieses ist die richtigste Art, sich die rückgängigen Cometen vorzustellen, und ihren Lauf mit der Welt-Ordnung zu vergleichen. Man muß aber hierbey auf ihren wahren Lauf, welchen sie aus der Sonne zu haben scheinen, würden sehen, und nicht auf denjenigen, welcher uns auf der Erde erscheinet, indem es öfters geschehen kann, daß ein Comet, welcher uns rückgängig vorkommt, in der That rechtläuffig ist, und hinwiederum ein rückläuffiger uns rechtläuffig scheint. Ein Comet ist aber in der That rechtläuffig, wenn er aus der Sonne betrachtet nach der Ordnung der Himmlischen Zeichen fort zu gehen scheinet, hingegen aber rückläuffig, wenn er sich gegen diese Ordnung beweget. Man beurtheilet nun diesen Lauf nach der Ecliptic, als wo sich die 12 Himmlischen Zeichen, deren Ordnung einmahl fest gesetzt ist, befinden, und ungeachtet die Bahnen der Cometen gemeiniglich ziemlich schief auf der Ecliptic aufstehen, so werden doch ihre Stellen dergestalt mit der Ecliptic verglichen, daß man ihre Länge und Breite zugleich anzuzeigen pflegt. Aus der Veränderung der Länge, welche aus der Sonne erscheint, wird nun leicht gefunden, ob der Comet recht- oder rückläuffig sey. Wenn aber die Bahn eines Cometen mit der Ecliptic einen rechten Winckel macht, so kann man nicht sagen, ob derselbe rechtläuffig oder rückläuffig ist, indem er immer einerley Länge behält, und nur allein seine Breite verändert. Dieses voraus gesetzt, so betrachte man einen Cometen, wenn er in seinem Nodo die Ecliptic durchschneidet, und messe den Winckel, welchen seine Bahn, darinne er fortgehet, mit der Ecliptic, nach der Ordnung der Zeichen, formiret, welche die Inclination der Cometen Bahn geben wird. Nun ist klar, daß wenn dieser Winckel kleiner ist, als 90 Grad, der Comet nach der Ordnung der Zeichen, fortgehe, und also rechtläuffig sey: wenn aber besagter Winckel grösser ist, als 90°, so wird der Comet zurück gehen. Wenn man derowegen den Inclinationis Winckel der Cometen Bahnen auf diese Art ausdrücken sollte, so hätte man nicht nöthig zu melden, ob der Comet recht- oder rückläuffig wäre, indem sich dieses aus dem vorigen von selbsten verstünde.

X

Wenn sich die Sonne um ihre eigene Axe vom Abend gegen Morgen bewegt, so scheinet es, als ob die Planeten motu vertiginis contrario sich bewegen müssen.

Die hier erwehnte Schwierigkeit kommt hauptsächlich von der nicht genugsam bestimmten Redens Art her, wodurch man dergleichen Kreißförmige Bewegungen zu beschreiben pflegt. Man pflegt zu sagen, daß die Sonne nach ihrer täglichen sichtbaren Bewegung vom Morgen gegen Abend fortgehe. Wenn man aber betrachtet, daß die Sonne, welche heute bey ihrem Untergang gegen Abend gestanden, morgen wiederum gegen Morgen aufgeht; so könnte man mit eben dem Recht sagen, daß die Sonne unterdessen vom Abend gegen Morgen gelauffen, wodurch, der Redens Art nach, eine gantz wiederwärtige Bewegung ausgedrückt wird. Wenn man die Herumdrehung einer Schraube andeuten will, so pflegt man zu sagen, daß dieselbe entweder von der Rechten zur Lincken, oder von der Lincken zur Rechten, geschehe. Allein keine von diesen Redens Arten determinirt die Bewegung völlig, indem beyde so wohl von einerley, als von wiederwärtigen Bewegungen gebraucht werden können. Denn wenn vorwärts die Bewegung von der Rechten zur Lincken geschehen, so geschiehet dieselbe hinterwerts von der Lincken zur Rechten. Dahero solte, um alle Verwirrung zu vermeiden, dieser Umstand auch bemercket werden. Man könnte also von einer Schraube sagen, daß sie von der Rechten vorwärts herum gegen die Lincke gedrehet werde, und dadurch würde eben diejenige Bewegung angedeutet werden, als wenn man sagte, die Schraube würde von der Lincken hinterwerts herum gegen die rechte gedrehet. Eine gleiche

Bewandtniß hat es auch mit den Redens-Arten wenn man sagt, daß sich die Sonne und die Planeten vom Abend gegen Morgen herum drehen, als welche gleichfals die Sache nicht völlig bestimmen, und zu dem in der Frage enthaltenen Zweifel mögen Anlaß gegeben haben. In der Astronomie wird aber diese Verwirrung vermieden, wenn man sagt: daß diese Bewegungen der Sonne und der Planeten um ihre Axen, nach der Ordnung der Himmlischen Zeichen geschehe, und wenn man sich nur dieser Redens-Arten bedienet, so wird man bald sehen, daß die gemachten Schwierigkeiten von selbsten verschwinden.

XI

Die Sonne und ein Comet können in ihrer Bewegung nicht erhalten werden, ausser sie bewegen sich um einen Ruhe-Punckt, dessen Entfernung von dem Mittel-Punct der Sonne sich verhält zu der Weite von dem Mittel-Punckt des Cometen, wie die Quantität der Materie des Cometen zu der Quantität der Materie der Sonne. Wenn also ein Comet eine sehr ablange Ellipsin beschreibet, und er sich im Aphelio befindet, so müste das Centrum gravitatis commune des gantzen Systematis Planetarii verrückt werden, und extra Centrum Solis fallen. Man könte zwar darauf antworten daß dieses deswegen nicht geschehen, weil nicht alle Planeten auf einer Seite der Sonne in einer geraden Linie auf einmahl stehen.

Es ist in der NEWTONianischen Philosophie eine ausgemachte Sache, daß sich die Planeten nicht um den Mittelpunckt der Sonne, sondern um das allgemeine Centrum gravitatis der Sonne, und der Planeten, wie auch der Cometen, bewegen, und daß dieses Centrum gravitatis würcklich still stehe, dahero sich die Sonne selbst um dasselbe herum bewege. Die Sonne übertrifft aber alle Planeten und Cometen an ihrer Grösse so sehr, daß erwehntes Centrum gravitatis selten nur ausser der Sonne fällt. Unterdessen ist nicht zu läugnen, daß wenn sich alle Planeten zugleich auf einer Seite, und auch alle Cometen auf eben derselben Seite in ihren Apheliis befinden sollten, der Mittelpunckt der Sonne sehr mercklich von dem Centro gravitatis aller dieser Cörper entfernet seyn müßte, und wenn dieser Fall zur Würcklichkeit kommen sollte, man auch den Effect beobachten würde.

XII

Wenn ein Astronomus in der Sonne stünde,
und der Comet ihm einen Planeten bedeckte,
was würde diese Zusammenkunft vor eine Würckung haben?
die Astronomi haben angemerckt, daß Saturnus seinen Weg ändere,
wenn zwischen ihm und der Sonne Jupiter
in einer geraden Linie stehet.

Was für eine Würckung in diesem Fall entstehen müßte, ist schon oben erkläret worden, da wir gezeiget, was die Annäherung eines Cometen gegen einen Planeten verursachen könne. Es kommt aber hier nicht sowohl auf die aus der Sonne scheinbare Coniunction, als auf die würckliche Nähe an, wie im vorigen schon zur Gnüge dargethan worden.

XIII

Wenn ein Comet eine Bewegung um seine Axe hat,
so muß dieselbe immer langsamer werden,
je näher er zu der Sonne kommt.
Wir schliessen dieses aus der Analogie mit den Planeten,
unter welchen sich diejenigen am geschwindesten
um ihre Axen drehen, die von der Sonne
am weitesten entfernt sind.

Daß der Schluß, welcher hier aus einer vermeinten Analogie gemacht wird, nicht Stich hält, ist aus demjenigen, was schon bey den vorigen Fragen angemercket worden, deutlich zu ersehen. Denn daß die Planeten, welche näher bey der Sonne sind, sich langsamer um ihre Axe herum drehen, ist erstlich sehr ungewiß, indem uns noch diese Bewegung an dem Saturno und Mercurio unbekannt ist: hernach da der Mars sich etwas langsamer als die Erde, nehmlich in 24 Stunden und 40 Minuten herum drehet, ungeachtet er weiter von der Sonne ist, als diese, so kann die vermeynte Regel schon nicht bestehen. Man könte zwar einwenden, daß der Unterschied zwischen der Entfernung des Martis u[nd] der Erde nicht groß genug sey, allein da zwischen der Erde und der Venus, in Ansehung der Entfernung von der Sonne, kein grösserer Unterscheid, und sich die Venus dennoch erst in 23 Tagen einmahl herum drehet; so kann man hieraus nicht anders schliessen, als daß die Herumdrehung der Planeten gantz und gar nicht nach einer festen Regel aus der

Distanz von der Sonne bestimmet werden könne. Daß aber eben derselbe Planet, wenn er in seiner Bahn der Sonne näher kommt, sich langsamer um seine Axe herumdrehen sollte, ist weder der Theorie noch der Erfahrung gemäß. Denn die erstere lehret uns, daß ein Cörper, welcher einmahl eine wälzende Bewegung bekommen, dieselbe beständig unverändert erhalte, woferne dieselbe nicht durch äusserliche Kräfte verwirret wird. Nun aber ist die Kraft der Sonne nicht vermögend, eine solche Würckung hervorzubringen, und kann dahero nicht behauptet werden, daß die wälzende Bewegung der Planeten durch die Herannäherung zu der Sonne gehemmet worden könnte. Und wenn auch ja dieses geschehen sollte, welches doch den Principiis der Mechanic völlig zuwider ist; so würde doch die darauf folgende Entfernung den erlittenen Abgang unmöglich wieder ersetzen können. Aus der Erfahrung aber ist zur Gnüge bekannt, daß die tägliche Bewegung der Erde im geringsten nicht langsamer werde, wenn sich dieselbe in ihrem Perihelio befindet, da sie doch alsdenn der Sonne um etliche 1000 Deutsche Meilen näher ist, als in ihrem Aphelio, und man auch den allergeringsten Unterschied in der Zeit sehr leicht würde bemercken können.

XIV
Solte man nicht hieraus einen erträglichern Zustand vor die Einwohner der Cometen schliessen, wenn welche vorhanden wären.

Worinne der Vortheil, welchen die Einwohner eines Cometen von der langsamen Bewegung um die Axe im Perihelio geniessen könnten, bestehe, ist nicht leicht zu ergründen. Wenn ein Comet in seinem Perihelio der Sonne noch näher, als der Mercurius kommt; so muß die Hitze gewiß gantz ausserordentlich groß, und vielleicht der Comet selbst in Feuer und Flammen gesetzt werden. Wenn sich alsdenn nun der Comet sehr langsam herumdrehete; so müßte die Hitze an einem Ort allzusehr überhand nehmen. Man hätte also vielmehr Ursache zu behaupten, daß alsdenn die Herumdrehung weit geschwinder wäre, damit die so grosse Hitze an einem Ort nur eine sehr kurtze Zeit dauren möchte. Im übrigen aber ist so viel gewiß, daß sich in einem Cometen unmöglich dergleichen Einwohner befinden, als auf unserer Erde, oder in einem andern Planeten, indem keine von diesen Arten eine weit grössere Kälte, als im Saturno, und darauf eine noch grössere Hitze, als im Mercurio, ausstehen könnte. Man ist demnach genöthiget zu sagen, daß wenn ja die Cometen mit Einwohnern angefüllet sind, dieselben von einer gantz verschiedenen und uns unbegreiflichen Natur seyn müßten, und in diesem Fall können wir auch nicht sagen, was denselben, in Ansehung der herumwelzenden Bewegung um die Axe, zuträglicher seyn würde oder nicht.

XV

Wie ist es möglich, daß ein Comet sich aus einem Systemate in das andere bewege, wie der Herr Autor behauptet. Solte er seine vim centripetam, die ihn gegen einen Fixstern anziehet, verliehren, oder solten in beyden Brenn-Punckten seiner Elliptischen Bahn Sonnen liegen?

Ungeachtet im vorhergehenden nirgens behauptet worden, daß ein Comet würcklich aus einem Systemate in ein anderes herüber gehe; so kann doch die Möglichkeit nicht geleugnet werden. Die Bewegung eines Planeten oder Cometen rühret von der ihm anfänglich eingedruckten Bewegung her. Dieselbe kann nun so beschaffen seyn, daß der Cörper entweder in einem Zirckul, oder in einer Ellipsi, oder in einer Parabel, oder auch sogar in einer Hyperbel seinen Lauf vollführe. Wenn wir nun den Fall betrachten, daß sich ein Planet oder Comet in einer Ellipsi bewege, so kann dieselbe mehr oder weniger ablang seyn, und bey den meisten Cometen ist dieselbe so ablang, daß sie in ihrem Aphelio sich viel weiter, als der Saturnus von der Sonne, entfernen. Nichts stehet also der Möglichkeit im Wege, daß sich ein Comet viel 1000 mahl weiter, als der Saturnus von der Sonne entfernte, und in diesem Fall könnte es geschehen, daß ehe er seinen Lauf wiederum nach der Sonne umwendete, [er] einem andern Fixstern so nahe käme, daß er von der Kraft desselben zu seinem Systemate hingerissen würde, und alsdenn eine gantz andere krumme Linie zu beschreiben anfienge. Solte aber dem Cometen anfänglich eine solche Bewegung eingedruckt worden seyn, nach welcher er eine Parabel oder Hyperbel beschreiben müßte, so ist gewiß, daß derselbe nimmer wieder gegen die Sonne kommen, sondern seine Bewegung beständig von der Sonne hinweg fortsetzen würde. Weilen er nun auf diese Art endlich in ein anderes Systema einrücken müsste, so würde er uns völlig geraubet werden. Es ist zu diesem Ende nicht nöthig, daß man noch eine neue Kraft in dem andern Foco annehme: man darf nur die Rechnungen, welche der grosse NEWTON über diese Materie angestellt, nachsehen, so werden alle Zweifel, welche durch blosse Worte nicht so leicht gehoben werden können, wegfallen.

XVI

*Durch die Annäherung eines Cometen an die Sonne, Planeten,
und andere Cometen, kann seine Bahn so verrückt werden,
daß er bey seiner Rückkehr zu uns gantz unerkenntlich wird.*

Auf diese Frage ist in einer von den vorhergehenden Beantwortungen schon genugsame Erläuterung gegeben worden. Denn da wir keine andern Mittel haben, eben denselben wiederum zu erkennen, als durch die Uebereinstimmung der Bahnen, welche aus den Observationen hergeleitet worden; so ist freylich nicht möglich, eben denselben Cometen, wenn er auch, nachdem er eine so merckliche Veränderung in seinem Lauf erlitten wieder sichtbar würde, zu erkennen, wenn wir nicht die Veränderung selbst beobachtet hätten. Diese Frage trift aber die Planeten so wohl, als die Cometen. Denn laßt uns setzen, daß zum Exempel der Mars, wenn er in seiner Coniunction mit der Sonne, und uns folglich nicht sichtbar ist, einen solchen Stoß erlitten, wodurch sein Lauf völlig geändert würde; wie würden wir alsdenn, wenn wir ihn hernach wieder zu Gesichte bekämen, gewiß seyn können, daß es der sogenannte Planet Mars wäre. Wir würden zwar den vorigen Martem nicht mehr sehen, aber doch keine Merckmahle übrig behalten, woraus wir sicher schliessen könnten, daß der nachgehends uns erscheinende neue Himmels-Cörper eben der Mars wäre. Da nun eine solche Veränderung bey den Planeten möglich ist, so hat man sich nicht zu verwundern, wenn bey den Cometen ein gleiches geschehen solte. Unterdessen haben wir doch keine gegründete Ursache, zu glauben, daß bißher mit einem Cometen eine solche Veränderung vorgegangen.

XVII

*Da die Cometen sich gemeiniglich in solchen Flächen bewegen,
welche mit der Ecliptic sehr grosse Winckel machen;
so scheinet es nicht wohl möglich zu seyn,
daß ein Comet der Erde sehr nahe kommen,
oder gar an dieselbe stossen sollte?*

Wenn ein Comet würcklich an die Erde stossen soll, so werden dazu drey Umstände erfordert. Erstlich muß sich der Comet in der Fläche der Ecliptic oder Erden-Bahn befinden. Zweytens muß er auch alsdenn durch die Bahn der Erde selbst gehen, und drittens muß sich eben um diese Zeit auch die Erde in demselben Punckt ihrer Bahn befinden, wo der Comet durchgehet. Was den ersten Umstand anlangt, so treffen wir denselben bey allen Cometen an. Denn da die Bahn eines jeglichen Cometen die Ecliptic in zweyen Punckten durchschneidet, welche die Nodi der Cometen-Bahn genennet werden, so gehen auch alle Cometen zweymahl durch die Ecliptic, indem sie ihren Lauf einmahl vollenden. Wenn uns nehmlich ein Comet beständig, seiner grossen Entfernung ungeachtet sichtbar bleiben sollte, so würden wir denselben bey einem jeglichen Periodo durch seine elliptische Bahn zweymahl in der Ecliptic erblicken. Der andere Umstand trifft nur bey den wenigsten Cometen ein. Denn da die Cometen in ihrem Perihelio gemeiniglich der Sonne näher kommen, als die Erde, im Aphelio aber sich weiter von der Sonne entfernen, als der Saturnus; so kommt es darauf an, wie weit ein Comet alsdenn von der Sonne entfernet ist, wann er durch die Ecliptic, oder die Fläche, darinne sich die Erde beweget, gehet. Sollte nun ein Comet in einem von seinen Nodis eben so weit von der Sonne entfernet seyn, als die Erde, wenn sie sich in eben demselben Punct der Ecliptic befindet, so würde der Comet in diesem Nodo nicht nur durch die Fläche der Erden-Bahn, sondern so gar durch die Bahn oder den Weg der Erde selbst gehen. Folglich findet dieser zweyte Umstand bey den wenigsten Cometen statt. Denn gemeiniglich ist ihre Entfernung von der Sonne alsdenn, wenn sie sich in einem Nodo befinden, entweder viel grösser, oder viel kleiner, als die Distantz der Erde von der Sonne, und da bey einem jeglichen Umlauf sowohl die Nodi als ihre Entfernung von der Sonnen keine Veränderung leiden, es wäre denn, daß ihr Lauf durch die allzugrosse Nähe eines Planeten oder andern Cometen verwirret würde; so kann man auch von den meisten Cometen versichert seyn, daß dieselben die Erde nicht berühren werden. Der letzt erschienene Comet war in seinem heraufsteigenden Nodo weiter von der Sonne entfernet, als der Mars, im hinabsteigenden aber derselben näher, als der Mercurius, dahero die Erde von diesem Cometen, wenn er auch tausendmahl wieder kommen sollte, nimmer das geringste zu befürchten haben wird. Es ist noch ungewiß, ob sich dieser Umstand bey einem einigen[10]) Cometen befindet, unterdessen ist es aber nicht unmöglich, daß ein Comet von dieser Beschaffenheit würcklich da sey. Sollte nun dieses geschehen, so wäre es freylich auch möglich, daß sich die Erde zu eben der Zeit, da der Comet durch den Weg der Erde durchgehet, sich zugleich in eben dem Punckt ihrer Bahn befände, ungeachtet der Comet auch viel tausendmahl kommen könnte, ohne die Erde zu berühren.

Da nun unter 1000 Cometen vielleicht kaum einer befindlich, bey welchem der zweyte Umstand statt hat, und dieser auch seinen Periodum wohl 1000 mahl

10) Eine alte Form von «einzigen». HCI.

absolviren könnte, ehe es sich solchergestalt fügete, daß derselbe und die Erde zugleich in einem Punckt einträfen; so wird die Wahrscheinlichkeit, daß jemahls ein Comet die Erde würcklich berühren sollte, sehr geringe, und verschwindet fast gäntzlich. Daß aber niemahls ein Comet der Erde ziemlich nahe kommen sollte, ist kaum zu vermuthen, indem schon würcklich solche Cometen erschienen, welche von der Erde nicht so sehr weit entfernet gewesen. Was in solchen Fällen für Veränderungen entstehen können, ist im vorigen schon ausführlich erkläret worden.

XVIII
Warum ein Nebel in der Ferne undurchsichtiger sey, als in der Nähe?

Die Ursache hiervon ist leicht zu begreifen, wenn man nur auf die Natur der Strahlen Acht hat. In der Luft entstehet ein Nebel, wenn dieselbe mit solchen Theilchen angefüllet ist, welche nicht durchsichtig sind, und uns folglich für sich selbst sichtbar[11]) werden, da die Theilchen eines völlig durchsichtigen Cörpers, an sich selbst nicht sichtbar sind. Hernach beliebe man zu erwegen, daß je näher sich das Auge befindet, je kleiner der Raum sey, welchen man auf einmahl übersehen kann, im Gegentheil aber übersiehet man einen um so viel grössern Raum, je weiter man sich entfernet. Wenn wir uns also nahe bey einem Nebel befinden, so fallen uns nicht so viel undurchsichtige Theile in die Augen, als wenn wir weit davon sind, und dieses ist die Ursache, warum uns ein Nebel in der Entfernung weit dichter vorkommt, als in der Nähe. Und daher kommt es folglich auch, daß die hinter einem Nebel befindlichen Cörper um so viel eher unsichtbar werden, je weiter wir uns davon entfernen. Hierbey hat man aber noch zu betrachten, daß je weiter ein Cörper von uns absteht, je weniger Lichtstrahlen von demselben in unsere Augen fallen, wenn auch die Luft vollkommen rein ist. Ist nun die Luft neblicht, so werden viele von den Strahlen aufgehalten, daß sie nicht zu uns kommen können. Wenn also ein Cörper weit hinter einen Nebel von uns entfernet ist, so wird noch die kleine Anzahl der Strahlen, welche zu uns kommen würden, durch die unreine Luft um so viel mehr vermindert, so daß uns daher der Cörper selbst leicht völlig unsichtbar wird.

11) Erste Ausgabe: «nicht sichtbar» anstatt «sichtbar». TS.

XIX

Ob man bey den Observationen der Cometen nicht auch auf derselben Parallaxen, und auf die Variation, welche von der Bewegung der Lichtstrahlen herrühret, zu sehen habe? und ob die Athmosphaere der Cometen selbst nicht die Observationen unrichtig machen könne?

Wann sich ein Comet so nahe bey uns befindet, daß seine Parallaxis mercklich wird, so muß man freylich bey der Correction der Observationen ausser der Refraction auch mit auf die Parallaxen sehen, und dahero pflegen sich die Astronomi alle Mühe zu geben, um dieselbe zu entdecken. Gemeiniglich aber ist dieselbe fast gar nicht zu mercken. Denn wenn ein Comet uns auch zweymahl näher ist, als die Sonne, so beträgt doch die Parallaxis nicht mehr, als eine halbe Minute, und dieses nur auf dem Horizont; dann je höher ein Gestirn über den Horizont kommt, je kleiner wird seine Parallaxis. Da man nun in Bestimmung des wahren Orts eines Cometen gemeiniglich nicht wohl näher als auf eine Minute versichert seyn kann, so ist die Correction, welche aus der Parallaxi gezogen wird, von keiner Erheblichkeit.

Was die andere Correction anlangt, welche daher rühret, daß die Lichtstrahlen nicht plötzlich, sondern eben wie der Thon, erst nach Verfliessung einiger Zeit zu uns kommen, so kann dieselbe in gewissen Fällen so mercklich werden, daß man sich in Bestimmung des Orts eines Cometen betrügen würde, wenn man darauf nicht Acht haben wollte. So lange sich der Comet noch so weit von der Sonne befindet, so kann sich diese Variation nicht leicht auf eine halbe Minute erstrecken, und dieses nur in gewissen Stellungen in Ansehung der Sonne, darinne Cometen selten erscheinen, in andern Fällen wird diese Variation noch weit kleiner. Wenn aber der Comet sich der Sonne sehr nähert, und einen sehr schnellen Lauf bekommt, so kann die Variation auch weit grösser werden. Bey dem Mercurio beträgt dieselbe in gewissen Fällen mehr, als eine Minute, wenn nun ein Comet der Sonne noch näher als der Mercurius steht, so kann sich die Variation bißweilen auf 2 und mehr Minuten belauffen, welche Correction ein habiler Astronomus nicht aus der Acht lassen muß.

Die Atmosphaere eines Cometen selbst aber kann uns seine scheinbare Stelle nicht verrücken. Denn ungeachtet die aus dem Kern des Cometen kommenden Strahlen in der Athmosphaere eine Brechung leiden; so kommen doch alle endlich aus der Athmosphaere selbst, und stellen uns folglich den wahren Ort derselben vor.

Beschluss von dem wahren Lauf des jüngst erschienenen Cometen.

Da in den vorhergehenden Beantwortungen verschiedene Muthmassungen über den wahren Lauf des letzten Cometen beygebracht worden, welche aber auf solche Observationen gegründet gewesen sind, worauf die zu diesem Ende erforderte weitläufige Rechnung nicht füglich angestellt werden konnte; so wird es den geneigten Lesern, und insonderheit den Liebhabern der Astronomie, nicht unangenehm seyn, wenn wir zum Beschluß dieser Blätter dasjenige anführen, welches seit der Zeit aus Observationen, die mit richtigen Instrumenten gemacht sind, durch die Rechnung über den Lauf dieses Cometen heraus gebracht worden.

Nach allen Nachrichten, welche wir von der Erscheinung dieses Cometen erhalten, ist derselbe zuerst in der Schweitz zu Lausanne[12]) wahrgenommen worden, nehmlich schon den 13ten December des vorigen Jahres, da des Abends um 8 Uhr seine Länge im Widder 28° 26′ mit einer Nördlichen Breite von 15° 11′ gefunden worden.

In Paris hat man ihn zu erst den 3ten Januarii gesehen, und des Abends um 5 Uhr seine Länge im Widder 14° 11′ mit der nördlichen Breite von 17° 33′ befunden: den 18ten Januarii aber Abends um 7 Uhr wurde daselbst die Länge im Widder 6° 57′ die Breite 18° 37′ observirt. Allhier in Berlin hatte man den 18ten Feb[ruarii] um 6 Uhr 43′ da der Comet bey dem Stern Marchab vorbey gieng, Gelegenheit, seinen Platz am Himmel sehr genau zu bestimmen, und fand seine Länge in den Fischen 19° 57′ mit einer nördlichen Breite von 19° 11′. Nach dieser Zeit ist der Comet ziemlich geschwind gegen die Sonne gegangen, und seine Breite hat auch sehr mercklich abgenommen, daß man ihn des Abends nur sehr kurtz nach der Sonnen Untergang, des Morgens aber vor der Sonnen Aufgang hat erblicken, und also keine accuraten Observationes anstellen können. Zu Ende des Februarii aber kam er mit der Sonne in Coniunction, und verschwand bald darauf völlig in der Morgenröthe, daß man nichts mehr als seinen Schweif erblicken konnte.

Nach diesen Observationen schien uns der Comet rückläuffig, weilen er gegen die Ordnung der himmlischen Zeichen aus dem Widder in die Fische getreten. Wenn man denselben aber aus der Sonne hätte betrachten können, so würde man ihn rechtläuffig gesehen haben, und so war auch sein wahrer Lauf beschaffen. Er kam von seinem Aphelio und eilte zu seinem Perihelio, welches er den zweyten Martii

12) Diese Observationen wurden von Jean Philippe Loys de Chéseaux gemacht. Siehe J. P. Loys de Chéseaux, *Traité de la comète qui a paru en Decembre 1743 & en Janvier, Fevrier & Mars 1744* (Lausanne et Genève, 1744). EJA.

Morgens früh erreichte. Zu dieser Zeit war er der Sonne näher, als der Mercurius, und seine Entfernung von derselben war 4379 halbe Erd-Durchmesser, deren jeder 860 deutsche Meilen beträgt, folglich war er zu dieser Zeit weiter von uns entfernet, als der Mercurius, wenn er uns am nächsten ist. Uebrigens war dieser Comet beständig weiter von uns, als anfänglich gemuthmasset worden. Denn biß gegen das Ende des Januarii war derselbe etwas weiter von uns, als die Sonne, und wenn er nicht so ausserordentlich groß gewesen wäre, würde man ihn auch anfänglich nicht gesehen haben. Die Cometen von Anno 1680 und 1742 waren nicht einmahl so groß, als die Erde, dieser aber mag nach seinem Diametro wohl dreymahl so groß gewesen seyn. Da derselbe nun so viel weiter gewesen, als wir anfangs vermuthet hatten; so war auch sein Schweif weit länger, als er angegeben worden. Den 18ten Februarii muß derselbe also wenigstens 8000 halbe Diameter der Erde, das ist ohngefehr 6 Millionen deutsche Meilen, lang gewesen seyn, welche erstaunliche Länge wohl nicht erkläret werden könnte, wenn dieselbe nicht von den Sonnen Strahlen erzeuget würde, deren unbegreifliche große Geschwindigkeit allein hinlänglich ist, eine solche Würckung hervorzubringen.

Die sichtbare Breite des Cometen nahm anfangs zu, hernach als er nahe zur Sonne kam, wiederum ab: die Länge aber, weil er rückgängig war, wurde vermindert. Weilen nun der wahre Lauf nicht aus dem sichtbaren beurtheilet werden kann, sondern aus demjenigen, welchen er aus der Sonne betrachtet zu haben scheinen würde; so wollen wir hier die wahren Longitudines und Latitud[ines] Heliocentricas für die obangeführten Zeiten, da richtige Observationes gemacht worden, hersetzen,

		Long. helioc.	Lat. helioc.
A. 1743	d. 13 Dec.	♉ 24° 14′	8° 50′
A. 1744	d. 13 Jan.	♉ 27° 44′	12° 37′
	d. 18 Jan.	♊ 1° 46′	16° 45′
	d. 18 Feb.	♋ 2° 55′	39° 24′

woraus deutlich erhellet, daß sein wahrer Lauf nach der Ordnung der Zeichen gegangen. Den 13ten December war der Comet ungefehr um die Hälfte weiter von der Sonne, als die Erde, dahero sein Lauf noch ziemlich langsam gewesen. Woraus abzunehmen, daß er schon lange vorher durch die Ecliptic gegangen, und sich in seinem Nodo ascendente befunden. Dieses ist geschehen den 7ten Augusti Anno 1743, ungefehr um Mittag, da seine Länge aus der Sonne gesehen, war ♉16° 20′. Von der Zeit an hat seine wahre Breite oder Entfernung von der Ecliptic immer zu genommen, bis den 25ten Februarii um 1 Uhr Nachmittags, da seine gröste Breite war 48° 30′. Seine wahre Länge war damahls im Löwen 16° 20′, die Erde aber wurde aus der Sonne im 6ten Grad der Jungfrau gesehen, und also folgte der Comet der Erde noch nach, oder der Comet gieng uns nach der Sonne durch den Meridianum, und die Coniunctio mit der Sonne folgte etwa 3 Tage darauf, nehmlich d[en] 28ten Feb[ruarii] da seine wahre Breite sehr geschwind wieder abnahm, indem er schon den 4ten Martii, morgens um 8 Uhr, durch die Ecliptic gieng, und seinen Lauf nach Süden richtete. Sein Nodus descendens war aus der Sonne gesehen im Scorpion 16° 20′ und damahls war er von der Sonne ungefehr so weit, als der Mercurius entfernet.

Er gieng also um diese Zeit fast durch die Bahn des Mercurii, der Mercurius aber war zu eben der Zeit nicht weit von diesem Punct in seiner Bahn, nehmlich im 27ten Grad des Scorpions, daß folglich der Comet dem Mercurio sehr nahe gekommen seyn muß. Dahero, wenn dieser Comet auch eine seiner Grösse gemäße Anziehungs-Krafft gehabt, so muß dadurch der Mercurius viel gelitten haben, und sein Lauf sehr mercklich verwirret worden seyn. Inskünfftige müßte also der Lauf des Mercurii nicht mehr mit den Astronomischen Tabellen übereinstimmen, welches die Zeit lehren muß. Es wird demnach der Mühe werth seyn, daß die Herrn Astronomi den Mercurium fleißig observiren, und wie weit er von den Ephemeridibus abweicht, wohl bemercken. Sollte es aber gar geschehen seyn, daß der Comet den Mercurium mit sich fortgeschlept hätte, so würden wir gar einen Planeten aus unserm Sonnen-Kreiß verlohren haben, und würde also dieses Jahr für die Astronomie eine sehr bedenckliche Zeit seyn.

Jedoch scheinet der Comet dem Mercurio nicht den gäntzlichen Untergang gebracht zu haben, indem er demselben nicht näher, als auf 5000 halbe Erd-Durchmesser, gekommen. Es ist aber zu mercken, daß wenn in den Observationen, worauf sich diese Bestimmungen gründen, nur ein sehr geringer Fehler vorgegangen, diese Distantz leicht um die Helfte kleiner seyn, ja gar verschwinden könnte: dahero die künftigen Observationen uns hierüber genauer unterrichten müssen.

Nach dem 4ten Martii wurde also sowohl die Breite, als die Declination des Cometen südlich: derowegen ungeachtet derselbe von der Sonne durch unsern Meridianum gieng, so erhob er sich doch erst nach der Sonne über unsern Horizont, und gieng auch lang vor derselben unter, welches die Ursache ist, daß wir denselben seit der Zeit nicht wieder zu Gesicht bekommen haben. In den Südlichen Gegenden der Erde aber muß dieser Comet noch in seinem vollen Glantz einige Zeit vor der Sonnen Aufgang sichtbar gewesen seyn, indem uns derselbe nach dem 4ten Martii fast eben so nahe gewesen, als vorher jemahls: wozu noch kommt, daß die Cometen, nach dem sie von ihrem Perihelio zurück kommen, viel heller scheinen, als vorher.

Es wäre deswegen zu wünschen, daß man auch in der Südlichen Helfte der Erde eben so aufmercksam darauf seyn möchte, als man in der Nördlichen gewesen, und daß sich daselbst tüchtige Astronomi finden möchten, welche den Lauf desselben fleißig bemerckten, um aus der Vergleichung ihrer Observationen mit den unsrigen die wahre Bahn dieses Cometen noch genauer zu bestimmen. Man könnte sich so gar versprechen, daß man durch Hülffe dergleichen Observationen die Zeit seines Umlaufs ziemlich genau bestimmen, und die Zeit seiner Wiederkunft vorher zu sagen im Stande seyn würde. Denn aus den schon bekannten Observationen erhellet zur Gnüge, daß dieser Comet sich in der That in einer Ellipsi beweget, welche aber so ablang ist, daß derselbe in seinem Aphelio weit über den Saturnum hinauf steigt. Die Zeit seines Umlauffs dürfte auch wohl zum wenigsten 4 bis 500 Jahr betragen. Dahero wir unter den schon mit Fleiß observirten Cometen keine Spur finden, daß eben dieser Comet seit A[nn]o 1300 wäre gesehen worden. Weswegen die Prophezeyungen derjenigen, welche diesen Cometen vorher angekündigt haben wollen, gäntzlich ohne Grund sind, und die Erfüllung einem blossen Hazard zuzuschreiben ist.

In unsern Nördlichen Gegenden werden wir um angeführter Ursachen willen, diesen Cometen, weil er seinen Lauf je länger je mehr gegen Süden lencket, nicht mehr zu Gesicht bekommen: in den Südlichen Theilen der Erde aber dürfte derselbe noch wohl bis in den Augustum, oder gar in den September hin, sichtbar bleiben, wenn man ihn mit Fleiß durch gute Tubos verfolgen wollte. Den 14. April dieses Jahres wird dieser Comet den Einwohnern der Südlichen Länder im 8ten Grad des Widders erscheinen, mit einer Südlichen Breite von 29 Graden, und seine Distantz von der Erde wird bey nahe um die Helfte grösser seyn, als die Distantz der Sonne von der Erde. Den 1sten Julii wird er im 27sten Grad des Widders mit einer Südlichen Breite von 47°, den 6ten September aber im 3ten Grad des Widders mit einer Südlichen Breite von 53 Grad gesehen werden, alsdann aber wird er schon über 2 und $\frac{1}{2}$ mahl weiter von uns entfernet seyn, als die Sonne, dahero er nur durch gute Tubos sichtbar seyn wird. Dieser Comet wird also sich in einem Zickzack zu bewegen scheinen, indem er erstlich aus dem Widder zurück in die Fische, denn wieder vorwerts bis zu Ende des Widders, und endlich wieder zurück bis zum Anfang desselben gehen wird; welcher wunderliche Gang doch gantz natürlich aus seinem ordentlichen Lauf in seiner elliptischen Bahn folget.

Einige Observationen von denen, die auf dem Königlichen Observatorio allhier von denen Hrn. Astronomis sind gemacht worden.

Anno 1744 war des Cometen	Länge	nördl. Breite
15. Jan. 7 Uhr 35′ 55″	♈ 8° 28′	18° 19′
17. Jan. 6 Uhr 50′ 13″	♈ 7° 25′	18° 33′
2. Feb. 9 Uhr 5′ 40″	♈ 0° 54′	19° 5′
5. Feb. 8 Uhr 50′ 8″	♓ 29° 35′	19° 21′
9. Feb. 7 Uhr 57′ 44″	♓ 27° 2′	19° 54′
17. Feb. 6 Uhr 31′ 18″	♓ 20° 54′	19° 22′
18. Feb. 6 Uhr 42′ 47″	♓ 19° 57′	19° 11′

Folgende cometische Observationes sind von einem geschickten Frauenzimmer[13]) gemacht, welche dem Verleger ohngefähr zu Händen gekommen und die er dem Publico anbey mittheilen wollte.

Freytags den 3ten Jan. war der Himmel Abends recht klar und schön gestirnt. Um 8 Uhr ward ein Comet wahrgenommen, welcher gegen West-Süd-West stand, und einem Stern zweyter Grösse ziemlich gliche, man konte ihn also gantz deutlich sehen. Wenn man ihn aber gegen die 3 hellen Sterne im Pegaso und den Kopf der Andromeda (welches Sterne zweyter Grösse sind) genau betrachtete, so war er zwar wohl so groß, doch aber etwas blasser anzusehen, darneben hatte er auch einen schwachen Strahl, der nicht gar deutlich zu erkennen war. Als man seine rechte Stelle am Himmel suchte, fand er sich bey dem nördlichen Fische neben dem Arm der Andromeda, hernach als der Mond nach 8 Uhr aufgieng, wurden die kleinen Sterne am Himmel sehr blaß, der Comet aber blieb gut zu sehen.

Den 4. Januar war der Himmel bewölckt und trübe.

Sonntags den 5. Jan. Abends um 7 Uhr, fieng sichs an aufzuklären, und um 9 Uhr war der Himmel helle. Der Comet hatte an Grösse und Deutlichkeit etwas zugenommen. Man bemerckte, daß sein Lauf gegen die Ordnung der Zeichen von

13) Margarethe Kirch. Siehe «Gewitter Observationes 1744. M. K.». Handschrift in der Crawford Bibliothek, Königliche Sternwarte, Edinburgh. Signatur 2.67. EJA.

Die Bahn des Cometen Anno 1744. vom 3 Januarii an.

Osten nach Westen gerichtet, und daß derselbe sehr langsam, weil seine Stelle wenig verändert war. Man bemühete sich dieselbe genauer zu finden, und befand vermittelst eines Micrometers, daß der Comet diesen Abend um 9 Uhr von dem Stern im südlichen Arm der Andromeda, welcher im Bayero mit ζ bezeichnet ist, 3 Grad 51 Minuten abstand, sein Schweif rührete fast an diesen Stern. Durch einen 6 schuhigen Tubum zeigte sich der Comet schön und groß. Er hatte einen hellen Kern; der Dunstkreiß um denselben breitete sich weit herum aus, so daß der Kern nur wie ein sehr kleines rundes Küchlein darinnen zu sehen war. Der Dunstkreiß schien ziemlich dicht, und endigte sich sehr kenntlich, daß man den Diameter desselben messen konnte, solcher wurde also durch einen 2 Schuhigen Tubum auf 18 Minuten geschätzt.

Montags den 6. Jan. gegen Abend ward der Himmel etwas überzogen. Um ein viertel auf 6 Uhr sahe man den Cometen zwischen den Wolcken hervor blicken. Es wurde hierauf heller, und gegen 6 Uhr maß man den Cometen wieder von dem Stern ζ Bayeri und fand, daß er nur 22 Minuten seit gestern fortgerücket war. Mit freyem Gesichte konnte man bemercken, daß der Comet mit dem Kopf der Andromeda, und dem Sterne Algenib, ein fast recht wincklicht Dreyeck bezeichnete.

Den 7. Jan. war es bewölckt und trübe.

Mittwochs den 8. Jan. Abends um 6 Uhr war zwar ein heller Blick, daß man den Cometen sehen konte, es bezog sich der Himmel aber gar bald, daß nichts gewisses observiret werden konnte, und wurd gantz trübe.

Donnerstags den 9. Jan. gegen Abend bezog sich der Himmel mit Wolcken, um 7 Uhr aber blickten die Sterne hervor, und nach 9 Uhr war der Himmel gantz klar. Der Comet stund mehrentheils mitten über den beyden hellen Sternen, nehmlich dem Kopf der Andromeda und dem Algenib, so daß er noch wie ein, (wie wohl sehr stumpfes) Dreyeck mit diesen Sternen machte. Man konnte daraus leicht abnehmen, daß der Comet in etlichen Tagen mit ihnen in gerader Linie stehen, und also zwischen denenselben durchgehen würde. Diesen Abend um 7 Uhr stund der Comet von dem Stern im Arm der Andromeda, welchen Bayer ζ benennet, 5 Grad 48 Minuten, und sein Strahl rührete vollkommen an denselben Stern.

Den 10. Jan. war des Abends die Luft gar dicke und unrein, darum konte man den Cometen nicht gut sehen.

Den 11., 12. und 13. war es trübe.

Dienstags den 14. Jan. hatte sich der Himmel nun wieder aufgekläret. Der Comet war diesen Abend schön hell zu sehen, er stund zwischen dem Kopfe der Andromeda und dem Algenib, und machte bey nahe mit diesen Sternen eine gerade Linie, jedoch noch nicht völlig. Es stund noch ein Sternlein zwischen dem Cometen und dem Algenib in einer Linie, welches aber, weil es nur ein Stern sechster Grösse, von wenigen mit blossen Augen bemercket wurde. Dieses Sternlein stehet in der Kette der Andromeda und wird in der DOPPELMAYRischen Charte mit Y, im BAYERO aber ist es im Pegaso befindlich und mit χ bezeichnet.

Mittewochs den 15. Jan. war der Himmel auch klar. Diesen Abend stund der Comet dem genanten Sternlein in der Kette der Andromeda am nächsten. Um 7 Uhr stund er von diesem Sternlein 1 Grad 28 Minuten nördlich. Durch einen guten 6 schuhigen Tubum sahe man ein kleines unbenantes Sternlein dichte am Kopfe des Cometen stehen, welches er auch mit seiner Athmosphaere noch etwas bedeckte. Etwa einen halben Grad von demselben Sternlein, und dicht neben dem Schweif des Cometen, war noch ein grösseres Sternlein, welches man auch durch den 2 Schuhigen Tubum sehen konnte. Diese beyden Sternlein sahe man durch verkehrt stellende Tubos zur rechten. Diese Umstände werden nur darum bemercket, weil man hernach den Cometen noch etliche Tage bey diesen Sternlein wahrgenommen, dergestalt, daß er dicht neben denselben langsam fort gerücket und mit seinem Schweif darneben weggestreiffet.

Donnerstags den 16. Jan. war der Himmel Abends erstlich bewölckt, nach 7 Uhr fingen die Sterne an zu blicken, und wurd bald hernach gantz klar. Der Comet stund diesen Abend völlig in gerader Linie mitten zwischen den beyden hellen Sternen, nehmlich dem Kopf der Andromeda und dem Algenib. Das Sternlein aber in der Kette Andromedä war schon etwas ausgewichen.

Freytags den 17. Jan. Abends ließ sich der Comet wieder gut sehen. Sein Lauf war noch langsam, er stund noch mehrentheils in gerader Linie mit den beyden vorgemeldten hellen Sternen. Der Diameter des Cometen mit seiner Athmosphaere betrug 20 Minuten, und sein Schweif erstreckte sich damahls der Länge nach auf 7 Grad.

Sonnabends den 18. Jan. war es des Abends sehr helle. Der Comet sahe gar ansehnlich aus, daß man ihn wohl als einen Stern erster Grösse schätzen konte. Der Strahl hatte sich bisher immer nach dem Stern im Arm der Andromeda, welchen BAYER ζ benennet, gerichtet. Diesen Abend aber fing er an davon abzuweichen, und wandte sich von der Zeit an je länger je mehr gegen den Kopf Andromedä.

Sonntags den 19. Jan. war wiederum des Abends ein heller Himmel. Nun konte man wohl erkennen, daß der Comet mercklich aus der Linie der beyden hellen Sterne, (nemlich des Kopfs Andromedä und Algenib), heraus getreten, und daß er seinen Lauf auf den Stern Marcab richtete. Obgleich der Mond helle schien, so war doch der Comet gut zu sehen.

Montags den 20. Jan. war die Luft des Abends sehr dicke, so daß man wenige

Sterne sehen konte, doch blickte der Mond, der Comet, und die grösten Sterne, durch die dicke Luft.

Den 21. Jan. war es gewölckt und trübe.

Mittewochs den 22. Jan. schien der Mond des Abends sehr helle, welches auch verursachete, daß die Sterne sehr blaß aussahen. Der Comet hatte noch einen sehr langsamen Lauf, und änderte seinen Stand in etlichen Tagen nicht mercklich. Er war auch ohngeachtet des Mondscheins noch gut zu sehen.

Donnerstags den 23. Jan. war wieder heller Mondschein. Der helle Stern im Flügel des Pegasi, Algenib genandt, sahe viel kleiner aus, als der Comet, welcher noch immer gegen den Stern Marcab zurückete.

Den 24. und 25. Jan. war es klar, Abends trüber Himmel.

Sontags den 26. Jan. war der Himmel Abends gantz klar, der Mond schien sehr hell. Mit blossen Augen schien es, als wenn der Comet fast eine gerade Linie mit den Sternen Algenib und Scheat machte, aber doch war er noch nicht recht in der Linie, sondern stund noch ein wenig höher als diese Sterne.

Den 27. Jan. war der Himmel völlig mit Wolcken überzogen.

Dienstags den 28. Jan. war der Himmel klar. Der Mond welcher fast voll, schien sehr hell, doch war der Kopf des Cometen deutlich und helle zu sehen. Er war grösser als die grössesten Sterne, nur ein wenig blasser, sein Schweif aber war wegen des hellen Mondscheines ziemlich schwach. Der Comet machte nun fast eine gerade Linie mit denen Sternen Algenib und Scheat, doch aber nicht vollkommen, er war also seit dem 26. Jan. wenig gerückt.

Mittewochs den 29. Jan. war wiederum heller Himmel. So bald man nur einige Sterne mercken konnte, so sahe man den Cometen mit am ersten. Er stund nun in gerader Linie mit Algenib und Scheat, dem erstern etwas näher, als dem letztern.

Den 30. und 31. Jan. war der Himmel Abends gewölckt, desgleichen auch den 1. Feb.

Sonntags den 2. Feb. war der erste helle Abend nach dem Mondscheine. Der Comet zeigte sich Abends gar bald, ehe sonst noch ein Stern zu sehen war, als ein grosser Stern. Wie es hernach immer dunckler wurde, und der Mond untergegangen war, so zeigte sich der Comet nebst seinem Strahl so groß und hell gläntzend, daß sich jederman verwunderte. Denn bißher hatte der Strahl wegen des Mondscheins gar blaß ausgesehen, jetzo aber war er sehr helle, es wurde die Länge desselben auf 15 Grad geschätzet.

Den 3., 4. Feb. Abends gewölckt.

Mittewochs den 5. Feb. ward der Himmel bald hell, bald mit Wolcken überzogen. Die Luft schien nicht gantz rein zu seyn, daher der Comet fast nicht so ansehnlich aussahe, als Sontags den 2. Febr. Daher einige glaubten, der Comet wäre schon im abnehmen. Sein Diameter betrug noch vollkommen 20 Minuten.

Donnerstags den 6. Febr. Abends, da die Luft gantz rein war, erschien der Comet wieder gar hell. Er gab einen solchen Glantz von sich, daß man mit freyem Gesichte keinen Nebel um ihn bemerckte, und durch den Tubum war auch nicht viel vom Nebel um ihn zu sehen. Nahe an dem Kopfe oder Stern des Cometen gläntzte der Strahl oder Schweif auch sehr, hernach aber wurde er oberwärts, da er sich ausbreitete, immer blasser. Merckwürdig war es, daß der Strahl diesen Abend wie zertheilet in zwey Strahlen erschien, da das südliche Theil viel kürtzer, als das nördliche war. Eben dieses wurde hernach noch etliche Mahl bemercket, wenn die Luft recht rein und klar war.

Den 7. Februar war es trübe.

Den 8., 9., 11. und 12. Febr. war heller Himmel, und darneben mehrentheils strenge Kälte. Der Comet hat sich diese Abende allemahl hell und gläntzend gezeiget. Er übertraf an Grösse alle andre Sterne, die sich damahls am Himmel zeigten, und gieng noch immer gegen den Stern Marcab zu.

Nachher war der Himmel bewölckt bis den 16. Februar.

Sontags den 16. Febr. ward der Himmel Abends hell. Der Comet zeigte sich auch gar bald. Er war dem Stern Marcab ziemlich nahe kommen, man sahe also, daß seine Geschwindigkeit zugenommen. Um 6 Uhr stund er nur noch 2 Grad von ihm ab, man sahe itzt deutlich, daß der Comet südlich vorbey gehen würde.

Montags den 17. Febr. Weil wir vermutheten, daß der Comet wohl des Morgens schon etwas zum Vorschein kommen würde, so sahe man sich diesen Morgen nach ihm um. Etwas vor 6 Uhr sahe man ihn gegen Ost-Nord-Ost durch die Dünste wie einen Stern aufgehen, wie er etwa 3 bis 4 Grad über dem Horizont war, so konnte man seinen Strahl oder Schweif erst recht mercken. Derselbe gieng von Osten nach Norden, dem Horizont fast parallel. Je höher nun der Comet kam, desto deutlicher wurde er.

Diesen Abend nehmlich den 17. Febr. um 6 Uhr stund der Comet nur noch 1 Grad von dem Stern Marcab. Dieser Stern sahe gegen den Cometen so wohl im Tubo, als mit blossen Augen, überaus klein, daher man die Grösse des Cometen gegen diesen Stern, welcher doch ein Stern zweyter Grösse, wohl beurtheilen konnte.

Dienstags den 18. Febr. Abends schien der Mond gar helle, dennoch war der Comet groß zu sehen. Wegen der hellen Demmerung konnte man vor 6 Uhr nichts observiren. Um 6 Uhr Abends war die nächste Coniunction des Cometen mit dem Stern Marcab schon vorbey. Es war der Comet zwar noch gar nahe an dem Stern, der Länge nach war er 8 Minuten vorbey oder nach Westen zu, der Breite nach stund er $13\frac{1}{2}$ Min[uten] von diesem Stern südlich. Der Comet verdunckelte mit seinem Glantz und Strahlen den Marcab, so daß ihn auch einige mit blossen Augen darbey nicht sehen konnten. Um 8 Uhr sahe man den Cometen gegen West-Nord-West untergehen, sein Strahl war gerade vom Horizont in die Höhe gerichtet.

Den 19., 20. und 21. waren die Abende bewölckt und trübe.

Sonnabends den 22. Febr. hatte der Himmel sich ziemlich aufgeklähret, nur gegen Westen war es noch etwas bewölckt. Gegen 6 Uhr Abends sahe man den Cometen durch die Wolcken blicken. Ob nun gleich die Abend-Demmerung noch gar helle war, und der Mond auch helle schien, so war doch der Comet so hell gläntzend, fast wie der Jupiter zu sehen, daß man ihn mit Verwunderung betrachtete. Seinen Strahl konnte man gar lang sehen, weil ihn die Wolcken mehrentheils verdeckten, wo er aber durch die Wolcken-Spalten hindurch schien (welches mehrentheils nahe am Kopfe geschahe) war er auch so hell, gelb und gläntzend, als wenn lauter Gold-Strahlen aus dem Cometen quelleten. Man konnte erstlich den Stern Marcab nicht finden, weil es noch so helle war, endlich erblickte man ihn, und sahe, daß der Comet sich schon ziemlich von ihm entfernet hatte. Der Stern und der Comet liessen sich nicht mehr völlig mit dem 2 schuhigen Tubo zusammen fassen. Da nun dieser Tubus 5 volle Grad und ein wenig drüber fasset, so muste der Comet etwas mehr als 5 Grad vom Marcab abstehen, mehrere Sterne konnte man bey ihm nicht sehen, er verbarg sich auch bald gäntzlich hinter den Wolcken. Sein Strahl sahe etwas krumm aus.

Sonntags den 23. Febr. Abends war helle und klarer Himmel. Der Comet zeigte sich deutlich, aber die andern Sterne waren wegen des hellen Mondenscheines so blaß, daß man kaum den Marcab erblicken konnte. Man sahe wohl, daß der Comet etwas weiter von ihm weggerücket, und suchte, ob man nicht die Sterne im Halse des Pegasi finden könnte, denn allem Vermuthen nach, muste der Comet nicht weit davon stehen; aber es war keine Möglichkeit etwas zu sehen, daher hat man nur mit blossen Augen geurtheilet, daß der Comet mit dem Marcab und dem Kopf der Andromeda eine gerade Linie machte. Der Schweif war eben so wie gestern, etwas krumm und Säbelförmig.

Montags den 24. Febr. Früh Morgens suchte man den Cometen wieder. Der Himmel was recht heiter und klar. Der Comet zeigte sich viel besser, als vor 8 Tagen des Morgens. Diesen Morgen sahe man den Strahl erst in die Höhe kommen, welcher gar schief von Osten nach Norden lag. Der Kopf des Cometen folgte gar

14) Erste Ausgabe: 6 anstatt 5. Verbessert nach der Handschrift «Gewitter Observationes 1744. M.K.» EJA.

bald hernach, um 5 Uhr 20 Minuten[14]). Man hoffte etwa einen Stern im Halse Pegasi zu erblicken, aber es war vergebens, denn des Tages Anbruch verursachte durch seine zunehmende Helle, daß wenig Sterne zu sehen waren und wie alle andre Sterne unsichtbar worden, sahe man doch noch den Cometen nebst der Venus. Der Comet blieb so gar bis 1 viertel Stunde vor der Sonnen Aufgang sichtbar; die Venus aber noch einige Minuten länger.

Diesen Abend, nemlich den 24. Febr. schien der Mond sehr helle, man konnte also ebenfalls kein Sternlein bey dem Cometen sehen. Den Stern Marcab erblickte man ein wenig, der Comet war auch gut zu sehen, und sein Schweif wie vorigen Tages. Er gieng um 6 Uhr 43 Minuten unter.

Dienstags den 25. Febr. Abends war die Luft sehr dicke und neblicht. Der meistvolle Mond schien noch ziemlich, der Comet aber war sehr schlecht wegen der hellen Abenddemmerung zu sehen, doch konnte man noch seinen Untergang bemercken, welches geschahe um 6 Uhr 24 Minuten. Dieses ist das letzte Mahl, daß man ihn des Abends hat erblicken können. Denn hierauf ist der Himmel über 8 Tage lang immer bewölckt und trübe gewesen.

Über den Schweif des Cometen sind über dieses noch folgende Betrachtungen gemacht worden.

Dieser Comet erschien anfänglich zu Berlin, nehmlich den 3. Jan. als ein Stern, welcher einen blassen Schimmer oder Nebel um sich hatte, und einen kleinen Schweif von sich strahlete, etwa 2 bis 3 Grad lang. Nach Verfließung weniger Tage war der Strahl schon etwas länger worden, doch war er gar blaß, und eine kleine Bewegung darinne zu mercken, als wenn er vom Winde beweget würde. Wenn man ihn genau betrachten wollte, so schien der Strahl etwas kürtzer zu seyn, als wenn man ihn nur obenhin ansahe. Es war zuweilen, als wenn er gar unsichtbar werden und verschwinden wollte, so daß einige die Schuld ihren Augen gaben, daß sie den Strahl nicht recht sehen könnten, ingleichen daß der Strahl bald länger bald kürtzer erschien.

In der Mitte des Januarii war der Comet heller, und sein Strahl oder Schweif deutlicher. Um diese Zeit war derselbe 7 bis 8 Grad lang. Er ist auch nach und nach immer länger worden, der Strahl war nahe am Kopf des Cometen schmahl, hell und deutlich. Von da an gieng er etwas ausgebreitet von ihm ab, und wo er am breitesten war, da vergieng auch seine Deutlichkeit; es gieng aber ein schmahler spitziger Strahl mitten durch über das breite Theil lang herüber, welches insonderheit von denen gesehen ward, die gute Augen hatten. Diese Spitze konnte man auch bey dem Mondschein sehen, und also das Zunehmen des Strahls wohl bemercken.

Es wurde also dieser Comet dem Ansehen nach immer heller und grösser. Im Febr. sahe man fast keinen Nebel um ihn, sondern er war wie ein sehr grosser Stern von der ersten Grösse, jedoch funckelte er nicht, als ein Fixstern. Man konnte ihn eher dem Jupiter vergleichen, denn er glänzte und war schön von Farbe. Der Strahl schimmerte auch gelblich. Seine Länge war den 2. Febr. etwa 15 Grad, und die Breite desselben, oben wo er am breitesten war, etwas über 1 Grad, und nahe am Kopfe aber etwa einen halben Grad. Der Kopf oder der Comet selbst hielt 20 bis 21 Minuten. Die Länge des Strahls nahm hernach noch sehr zu; denn man konnte ihn den 6. Febr. auf 25 Grad lang, und noch wohl drüber schätzen, und die Breite, wo er am breitesten, war 3 bis 4 Grad. Es wurde bemercket, daß der Strahl, wo er anfieng blaß zu werden, sich getheilet. Der Haupt-Strahl, wo die lange Spitze war, gieng nördlich dicht am Haupt der Andromeda vorbey, und reichete bis an deren Gürtel: der andre Strahl welcher südlich stund, war etwa halb so lang. Diese Theilung des Schweifes wurde hernach allezeit, wenn der Himmel gantz rein war, bemercket, wenn aber die Luft nicht gantz klar war, so konnte diese Zertheilung nicht genau gesehen werden. Man hat weiter keine sonderliche Veränderung an ihm wahrgenommen, bis den 23. und 24. Febr. Abends. Denn nachdem es etliche Tage zuvor trübe gewesen, und der Himmel sich erst den 23. und 24. des Abends aufgeklähret; so sahe man den Strahl bey ziemlich hellen Mondschein dennoch sehr hell und gläntzend. Er breitete sich nur etwas aus, und zwar ziemlich nahe an dem Cometen, übrigens war er gar schmahl und darbey etwas in die Höhe gekrümmet, etwa in der Form eines Säbels. Der runde Theil war nördlich und war gar dicht und gelb, die hole Seite südlich und etwas dünner und blasser, am äusersten Ende aber wurde der Strahl gantz spitz. Den 25. Febr. sahe man ihn noch eben so, aber durch die sehr dicke Luft gantz blaß, hernach war der Himmel bey 8 Tage trübe, und da indessen die Sonne immer später, der Comet aber früher untergegangen; so konnte man des Abends ferner nichts mehr von ihm sehen.

Es war auch nicht möglich, sich wegen des trüben und gewölckten Himmels eher wieder nach dem Cometen umzusehen, bis den 5. Martii des Morgens. Der Himmel war zwar anfänglich noch ziemlich bewölckt, es blickten aber doch früh um 4 Uhr einige Sterne, und gegen Osten wurde man durch die ziemlich dicke Luft einige Strahlen gewahr, welche etwas schief nach Norden zu etwa 10 Grad hoch in die Höhe giengen. Es schien, als wenn etliche Cometen neben einander stünden, welche am Horizont von Dünsten bedecket würden, daher es kein Wunder, daß einige Leute gesagt, sie hätten 3 oder auch wohl 4 Cometen des Morgens gesehen;

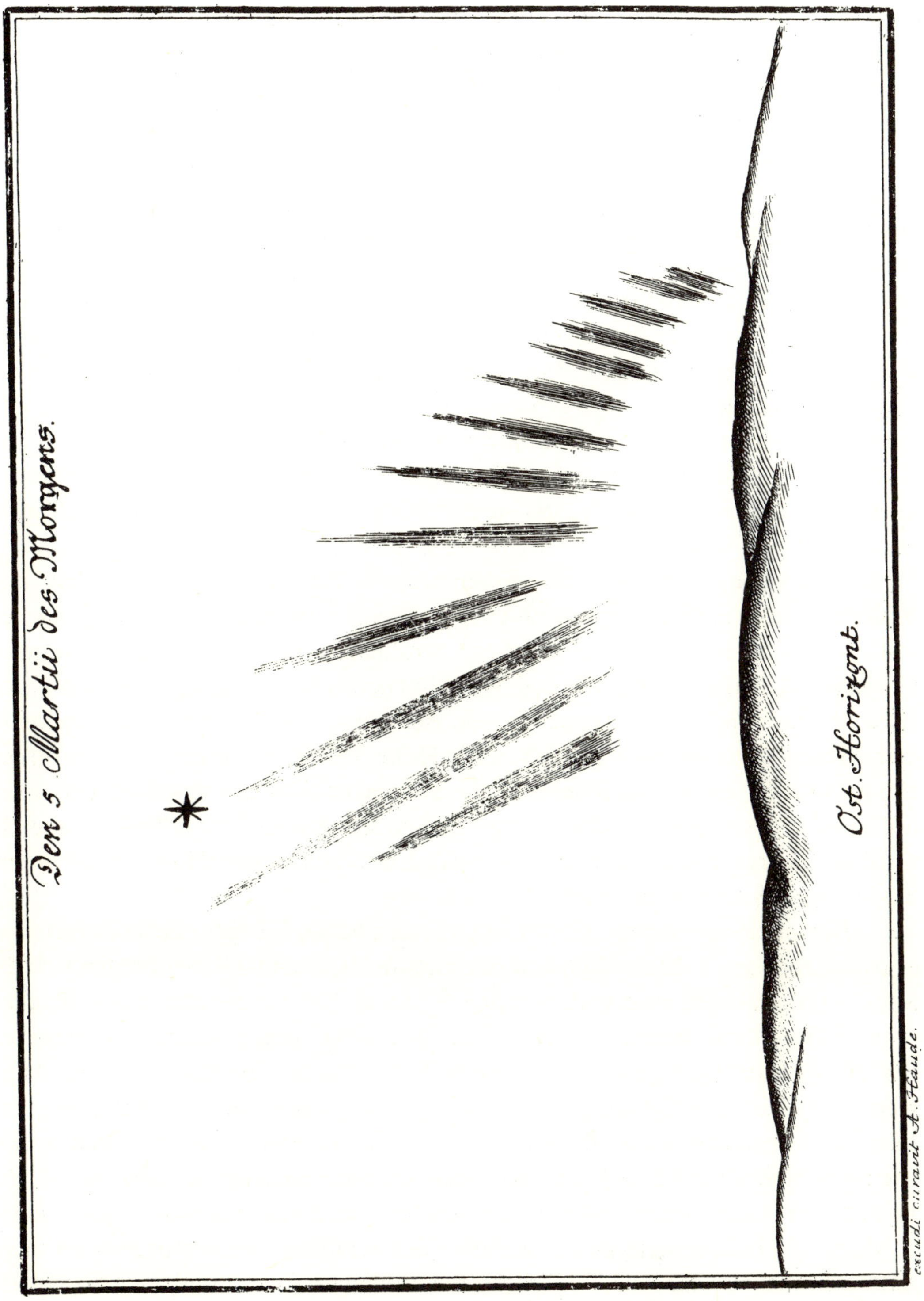
Den 5 Martii des Morgens.
Ost. Horizont.
excudi curavit A. Haude.

denn diese Strahlen sahen nicht viel anders aus. Man hatte aber Ursache zu vermuthen, daß der einige Comet seinen Schweif so ausgebreitet, und daß der Cörper selbst hinter dem Horizont seyn, und bald hervor kommen würde. Als man sich hernach auch an dem gantzen Himmel umsahe, so erblickte man gegen Nord-Ost zwischen zertheilten Wolcken einen Nordschein, welcher gelb und helle war, auch liessen sich Strahlen darinnen sehen, welche sich mit einer mercklichen Geschwindigkeit gegen die Cometen Strahlen zu bewegten, aber auch schnell wieder zurück giengen, dergleichen schnelle Bewegung bey denen Nordscheinen öfters geschiehet. Jedoch sahe dieser Anblick gar fürchterlich aus, denn der Cometen Schweif war so ungemein ausgebreitet, und der Nordschein schien den gantzen Himmel in Bewegung zu setzen. Man kam deswegen fast auf die Gedancken, daß die Oestlichen Strahlen auch vom Nordschein herrühreten, und also gar nichts mehr vom Cometen vorhanden wäre; allein nach einiger Betrachtung vergieng der Nordschein, und die Strahlen gegen Osten blieben beständig stehen, sie erhoben sich aber mit den Sternen, also daß, da sie erst etwa 10 Grad hoch über den Horizont hervor giengen, sie eine Stunde hernach auf 20 Grad lang zu sehen waren.

Sie sahen auch blasser und trüber aus, als die Nordscheins-Strahlen, und behielten immer einerley Gestalt und Figur. Wenn man eine Vergleichung machen sollte, so könnte man sich die Figur dieser Strahlen als einen grossen ausgebreiteten Flügel vorstellen, welcher hinter dem Horizont herauf stieg, dessen Länge um 5 Uhr etwa 20 Grad, und die Breite auch fast so viel betrug. Denn je höher er kam, desto mehr Strahlen kamen auch südlich in die Höhe, bis sie durch des Tages Anbruch blasser wurden, und hernach nebst denen Sternen endlich gar verschwunden. Man hoffte zwar den Cometen selber noch vor Sonnen-Aufgang zu erblicken, aber so wohl die Morgenröthe, als auch die dicke Luft, welche den Horizont einnahm, hinderten, daß man weiter nichts sehen konnte.

Hierauf erwartete man mit Verlangen einen hellen Morgen. Den 6. Mart. war es trübe, aber den 7. Mart. früh war der Himmel hell. Um 4 [Uhr] zeigten sich die Strahlen gegen Osten wiederum, wie den 5. Martii. Sie stunden wie grosse Balcken etwa 15 Grad vom Horizont in die Höhe, etwas schief gegen Norden zu gelencket. Der nördlichste Strahl gieng nahe vor dem Stern Scheat südlich vorbey, von da an giengen die übrigen Strahlen vom Ost-Horizont neben einander bis unter den Delphin, sie erreichten aber denselben nicht völlig. Mitten in den Strahlen blickte der Stern, welcher auf dem Maule des Pegasi zu finden, der im BAYERO mit ε in der DOPPELMAYERschen Charte aber mit a bezeichnet ist. Zwey Tage zuvor, nemlich den 5. Mart. stund dieser Stern gantz nahe über den Strahlen, diesen Morgen aber meist mitten drinne: also sahe man, daß sich die Strahlen verlängert und auch noch mehr

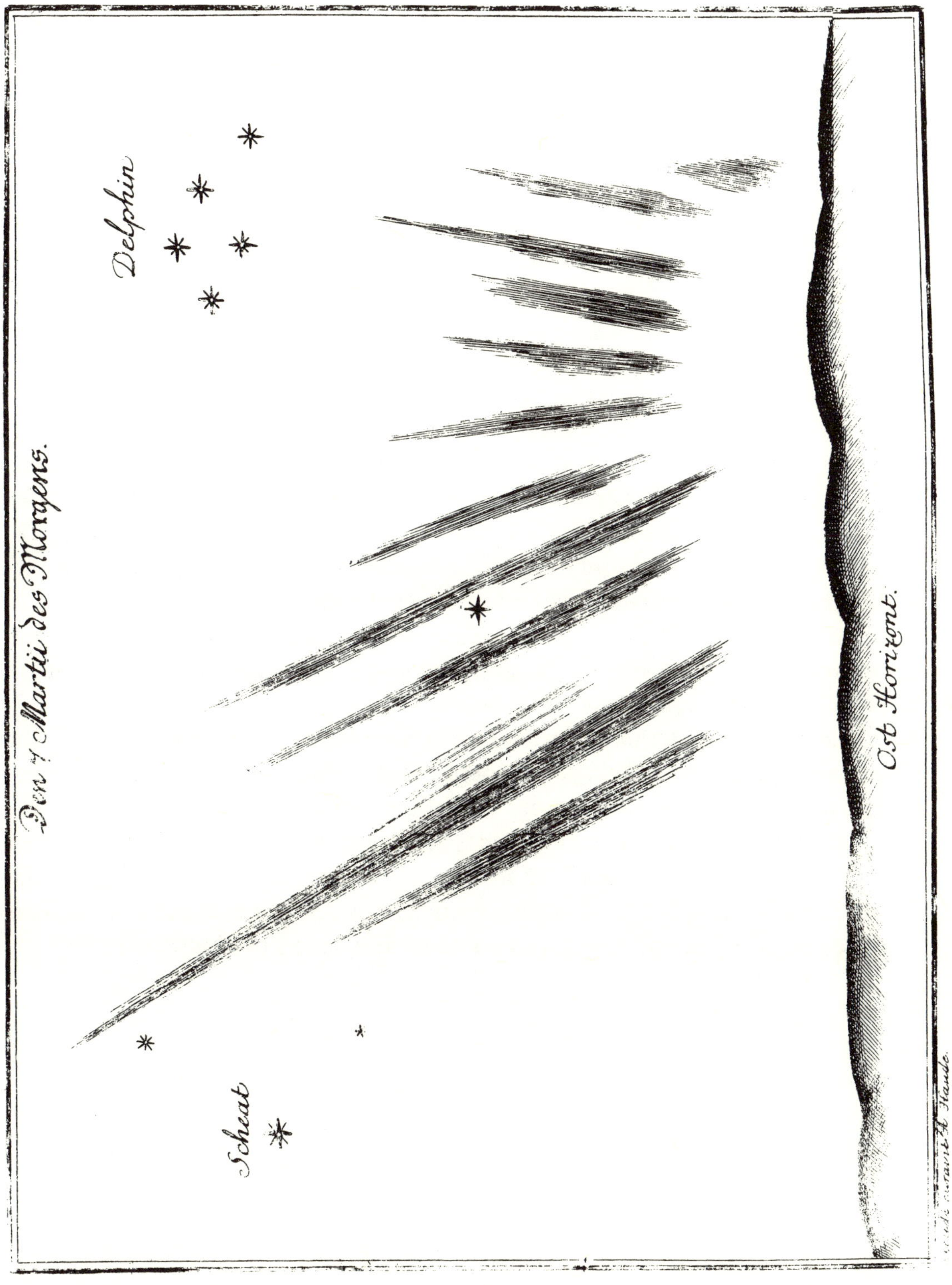
Den 7 Martii des Morgens.
Delphin
Scheat
Ost Horizont.

ausgebreitet hatten. Sie stiessen nicht gantz bis an den Horizont, denn es war daselbst eine blasse Helle, welche durch des Tages Anbruch noch mehr zunahm, bis endlich die Strahlen wegen Annäherung des Tages nicht mehr konnten gesehen werden. Man gab Achtung, bis die Sonne aufgieng, ob der Comet selbst noch zum Vorschein kommen möchte, aber es war vergebens.

Den 8. Martii früh war trüber Himmel.

Den 9. Martii früh um 4 Uhr war der Himmel meist hell, die Strahlen zeigten sich wieder in Osten, wie den 7. Martii, erstreckten sich aber noch etwas höher. Die südlichen Strahlen, welche allezeit etwas kürtzer, als die Nördlichen, hatten nunmehr mit ihrer Länge den Delphin vollkommen erreichet.

Den 10. Martii früh sahen die Strahlen noch so aus, wie vorigen Morgen, nur etwas blasser.

Den 11. Martii war die Nacht trübe und ein hefftiger Sturm-Wind.

Den 12. Mart. war der Himmel des Morgens klar. Gegen 5 Uhr stunden die Strahlen ziemlich hoch, sie sahen aber gar blaß aus, nicht heller als die Milchstrasse.

Der scheinbare Lauf dieses Cometen, wird nun in der beygefügten Kupffer-Platte No. 1 vorgestellet, worinne auch die Länge des Schweifes, wie solche nach und nach zu genommen, bemercket ist. Um sich aber von den vielen Strahlen, welche im Merz Monath des Morgens vor der Sonnen Aufgang gesehen worden, einen deutlichern Begrif zu machen; so sind davon die in den Platten No. 2 und 3 befindlichen Abbildungen gemacht worden.

Weil man hier in Berlin wegen des trüben Himmels den Lauf des Cometen nicht länger, als biß auf den 25. Februar hat bestimmen können; so hat man aus fremden Observationen den Weg, welchen der Comet biß den letzten Februar beschrieben, noch mit in der Tabelle bemerckt, welches deßwegen Anmerckungswürdig ist, weil der Comet in dieser Zeit seinen Lauf so sehr gekrümmet, und so plötzlich gegen die Ecliptic gerichtet hat. Wodurch denn dasjenige, welches von dem wahren Lauf des Cometen aus der Theorie angeführet worden, sehr schön bestätiget wird.

DE RELAXATIONE MOTUS PLANETARUM

Commentatio 89 indicis ENESTROEMIANI
Opuscula varii argumenti 1 (1746), p. 245–276

1. A verae physicae principiis nimium abhorret eorum sententia, qui planetas in spatio ab omni materia vacuo moveri statuunt. Non attingam hic controversiam famosissimam de vacuo, utrum in mundo locum inveniat, nec ne? quicquid enim de hac quaestione sentiatur, spatium, in quo planetae et cometae circa solem volvuntur, omnino vacuum esse nequit. Cum enim ad haec corpora in orbibus suis continenda vi opus sit centripeta, huiusmodi vis in vacuo absoluto concipi non potest, nisi cum plerisque philosophis Anglis, qui se magnum NEUTONUM sequi profitentur, attractionem mutuam inter ipsas omnium corporum proprietates referre velimus. Interim tamen etiamsi hanc opinionem amplecteremur, tamen universi spatium ab omni materia vacuum servare non possemus. Nullum enim est dubium, quin radii lucis cum materia quadam subtili sint coniuncti; quoniam in vacuo quidem radios lucis statuere absurdissimum foret. Cum igitur in coelo nullum quasi punctum assignari queat, ad quod radii lucis tam solis et stellarum non continuo penetrent, totum coeli spatium necessario illa materia subtili, qua radii constant, perpetuo erit repletum. Sive enim radii tanquam effluvia ex sole eiaculentur, sive instar soni per medium quoddam elasticum propagentur, utroque casu totum spatium, in quo planetae moventur, materia quadam plenum habebimus. Quae cum negari nullo modo queat, si eum aetherem appellemus, huius certe aetheris existentia solidissimis argumentis evicta est censenda.

2. Cum igitur aether universum coeli spatium impleat, quaestio nascitur difficillima, utrum planetae ab eo resistentiam patiantur nec ne? nisi enim aether cum planetis communem habeat motum, ob inertiam ipsis necessario resistere debet; sin autem aether eadem celeritate qua planetae circa solem revolvatur, cometae, qui cursum secundum directiones oppositas habent directum, eo maiorem resistentiam sentire deberent. De cometis quidem, cum eorum apparentia non sit

perennis, minus certum iudicium petitur; neque enim effectus resitentiae percipitur, neque temporis spatium ad hoc sufficit, nisi resistentia esset valde magna, quod tamen non probabile videtur. Verumtamen ex pluribus eiusdem cometae reversionibus, tempora periodica conferendo aliquid concludi posset; si enim perpetuo eodem tempore elapso ad perihelium reverteretur, nullum inde resistentiae indicium haberetur; sin autem eius tempus periodicum sensim decrescere deprehenderetur, tum in hoc ipso effectus resistentiae esset collocandus. Sic cometa, qui Anno 1682 apparuit, idem esse censetur, qui Anno 1607 et Anno 1531 est observatus; et secundum calculos HALLEII tempus a prima apparentia ad secundam anno circiter excedit tempus a secunda ad ultimam elapsum: hincque ideo indubitatum haberemus testimonium resistentiae aetheris in motu cometarum. Infra enim demonstrabitur, a resistentia tempora periodica continuo decrescere debere. Tum vero etiam ob constantem caudarum cometicarum directionem aether in quiete versari videtur: si enim materia aetheris simul cum planetis circa solem pari motu gyraretur, caudas cometarum secum abripere sicque efficere deberet, ut caudae perpetuo in eam plagam, in quam motus aetheris dirigitur, deflecterent; quod cum non observetur, non immerito suspicamur aetherem in coelo quiescere, vel saltem non eiusmodi habere motum constantem et velocem, quo caudarum directio immutari posset.

3. Quoniam itaque aether non solum revera adest, sed etiam nullo motu notabili cieri videtur, fieri omnino nequit, ut planetae ab illo nullam resistentiam sustineant; a qua partim tempora periodica, partim orbitarum excentricitates minui debeant, uti in sequentibus demonstrabitur. Interim tamen nulla fere huiusmodi mutatio ab Astronomis, etiamsi observationes intervallo 4000 annorum factae inter se conferantur, deprehenditur. Hinc autem plus non sequitur, quam mutationes has tam esse exiguas, ut ad eas percipiendas vel ne hoc quidem tempus sufficiat, vel observationum praesertim antiquissimarum accuratio non suppetat. Quanquam enim tam remotae observationes praesto sunt, tamen difficile est ex iis eiusdem planetae tempora periodica, quae diversis seculis tenuerit, colligere, atque inter se comparare. Tutissime enim cuiusque planetae tempus periodicum inveniri censetur, si eius observationes novissimae cum antiquissimis conferantur; hoc autem modo neque id tempus periodicum, quod planetae nunc convenit, reperitur, neque id, quod ipsi in antiquitate obtigerat; sed medium quoddam tempus prodit, quod sine dubio praesenti tanto est maius, quantum ab antiquo deficit. Neque etiam observationes medii aevi tuto in subsidium vocare licet, cum eae plerumque nimis sint crassae, quam ut quicquam certi inde concludi posset. Motus tamen solis vel potius terrae hunc resistentiae effectum non obscure declarare videtur: quo longius enim

in antiquitatem regrediamur, anni spatium longius creditum fuisse, quam nunc quidem comperimus. Namque ante nostra tempora annus tropicus nunquam minor est habitus quam 365 d. 5 h. 49′, cum tamen is nunc ab astronomis aliquot scrupulis secundis minor statuatur. Quod autem ad excentricitatis decrementum attinet, etsi terrae excentricitas olim maior est putata quam nunc, tamen observationes veteres minus sunt exactae, quam ut vera excentricitas illorum temporum inde concludi possit. Quoniam tamen omnes astronomi antiqui orbitae terrae unanimiter maiorem excentricitatem tribuerunt, quam hodierni, hinc non infirmum argumentum pro resistentia aetheris evincenda, suppeditari videtur.

4. Quidquid autem sit, tantum certe hinc tuto colligitur, resistentiam aetheris esse minimam, quod quidem summa eius raritas aperte declarat. Nullum enim est dubium, quin aether sit materia aere longe rarior, etiamsi eius elasticitas aeris elaterem maxime superet. Resistentia autem pendet a densitate medii, quae si fuerit minima, ipsam resistentiam insensibilem esse oportet. Ex celeritate quidem radiorum solis, si cum celeritate soni conferatur, liquido colligitur, si densitas aetheris ad densitatem aeris naturalis ponatur ut 1 ad m, elasticitas autem aetheris ad elasticitatem aeris ut n ad 1, fore, sicut in dissertatione de luce[1]) ostendi, $mn = 387\,467\,100\,000$, pro quo numero utamur isto rotundo $mn = 400\,000\,000\,000$. Si igitur elasticitas aetheris millies maior assumatur, quam aeris, ita ut sit $n = 1000$, fiet $m = 400\,000\,000$, totiesque densitas aetheris minor foret densitate aeris; in tali autem medio rarissimo corpus admodum densum, cuiusmodi sunt planetae, diutissime moveri potest, antequam portionem sensibilem celeritatis suae amittat. Si numerus n adhuc foret minor, alter numerus m tanto maior prodiret: at quia durities corporum a vi aetheris elastica proficiscitur, siquidem eam a causa mechanica derivare velimus, numerus n millenario vix minor statui poterit; unde alteri numero m maiorem valorem quam $400\,000\,000$ tribuere non licebit. Deductus quidem est valor iste mn ex parallaxi solis horizontali 13″, et tempore 8′, quo radii a sole ad terram pertingere putantur; quorum utrumque, cum nondum certissime sit evictum, fieri omnino potest, ut valor producti mn adhuc multo sit maior; hincque ergo resistentia aetheris multo adhuc minor admitti posset.

5. Antequam autem hanc resistentiam ad planetarum motus applicemus, videamus quantum detrimentum corpus, quod in aethere libere in directum promo-

1) E 88, «Nova theoria lucis et colorum», *Opuscula varii argumenti*, tom. 1, (Berolini, 1746), pp. 169–244 = *Opera omnia* III 5, pp. 1–45. EJA.

veatur, neque ab ullis viribus sollicitetur, ab eius resistentia perpeti debeat. Sit corpus hoc sphaericum, eiusque radius ponatur $= a$: eius autem densitas se habeat ad densitatem aeris naturalis ut k ad 1, ita ut istius corporis densitas ad densitatem aetheris, in quo movetur, futura sit ut km ad 1. Habuerit hoc corpus initio celeritatem altitudini b debitam, nunc autem percurso spatio $= x$, sit eius celeritas debita altitudini v. Hoc ergo statu si resistentia aetheris similis sit resistentiae aeris, motus corporis retardabitur a vi, quae se habet ad ipsius massam ut $\frac{3v}{8kma}$ ad 1, unde obtinebimus hanc aequationem:

$$dv = -\frac{3v\,dx}{8kma}, \quad \text{seu} \quad -\frac{dv}{v} = \frac{3\,dx}{8kma},$$

cuius integrale est[2])

$$l\frac{b}{v} = \frac{3x}{8kma};$$

ex qua aequatione cognoscitur, quomodo spatio quocunque emenso celeritas corporis se habitura sit ad celeritatem initialem. Si scilicet quaeramus, quantum spatium huic corpori percurrendum sit, donec celeritatis suae partem $\frac{1}{1000}$ amittat, poni debet $\sqrt{v} = \frac{999}{1000}\sqrt{b}$, seu $\frac{\sqrt{b}}{\sqrt{v}} = 1 + \frac{1}{999}$, et $\frac{b}{v} = 1 + \frac{2}{999}$ quam proxime. Hinc erit $l\frac{b}{v} = \frac{2}{999}$ seu $\frac{2}{1000}$ proxime, ideoque $\frac{3x}{8kma} = \frac{2}{1000}$ et $x = \frac{16kma}{3000}$; spatium ergo percurret, quod se habebit ad eius semidiametrum a ut $16km$ ad 3000. Cum igitur sit $m = 400\,000\,000$, si corpus sit ferreum, erit $k = 7000$, eaque proportio prodibit ut 14 900 000 000 ad 1; ad quod certe spatium percurrendum longissimum tempus requiretur.

6. Collocemus iam terram in locum huis globi ita ut a denotet semidiametrum terrae. Quodsi iam tota terra ex aqua constaret, foret $k = 850$, sin autem tota ex auro, quod est corpus ponderosissimum, esset conflata, foret $k = 16\,000$; unde cum terra universa gravior aqua, levior tamen auro sit putanda, litera k medium quemdam valorum inter 850 et 16 000 sortietur; et cum in imis visceribus multo

2) l denotat logarithmum. EJA.

gravior videatur, quam circa superficiem, fortasse non multum a veritate aberrabimus, si literae k valorem 8000 tribuamus; ita ut eius resistentia foret, si celeritas eius altitudini v debita ponatur, $= \frac{3v}{8 \cdot 8000 \cdot 400\,000\,000 \cdot a}$. Hoc autem modo resistentia foret comparata, si terra nullis poris, per quos aether libere transire queat, esset pervia: cum autem aether per corpora etiam densissima satis libere permeet, terra in motu suo in exiguam tantum partem aetheris impinget, dum reliqua pars transitum inveniens nullam resistentiam parit. Hinc ergo resistentia quoque multo minor erit statuenda, quam assumsimus, et cum fortasse vix decima pars motui resistat, resistentia, quam ante assumseramus $= \frac{3v}{8kma}$, vix decimae eius parti aequabitur, eritque fere $= \frac{v}{27kma}$. Cum autem hic nihil certi definire liceat, ne nostras speculationes ad casum a natura nimis abhorrentem restringamus, pro numero $27km$ adhibeamus literam μ, cuius loco deinceps quivis valor substitui queat. Foret autem si praecedentes coniecturae veritati essent consentaneae circiter $\mu = 100\,000\,000\,000\,000$; fieri autem potest, ut iste numerus multo maior minorve existat, praecipue pro variis planetis seu cometis.

7. Quemcunque igitur haec litera μ valorem habeat, in hac dissertatione investigare constitui, quantum motus tam planetarum quam cometarum ab ista resistentia perturbari debeat. Manifestum autem est, quando celeritas planetae ob hanc resistentiam diminuitur, eum a vi centripeta propius ad solem admoveri, hincque vicissim eius motum accelerari debere. Quocirca ab hac resistentia motus planetarum accelerabitur potius quam diminuetur, et dum eorum distantiae a sole sensim decrescent, tempora periodica contrahentur, quae mutatio quanta esse debeat hic perscrutabor. Sequens scilicet problema resolvere constitui, quo quaeritur motus corporis ad quoddam centrum fixum in ratione reciproca duplicata distantiarum attracti, dum perpetuo resistentiam patitur quadratis celeritatum proportionalem. Tractavi equidem iam hoc ipsum problema in mechanica, eiusque solutionem ad aequationes perduxi. Verum cum resolutio harum aequationum nimis difficilis videretur, neque etiam institutum operis perfectam evolutionem requireret, ex iis, quae ibi sunt tradita, parum ad plenam huius motus cognitionem colligere licet. Inde tamen aequationes principales, quas ibi inveneram, desumam, easque ad praesens institutum accommodabo, ne iis ex primis principiis eliciendis nimium sit immorandum.

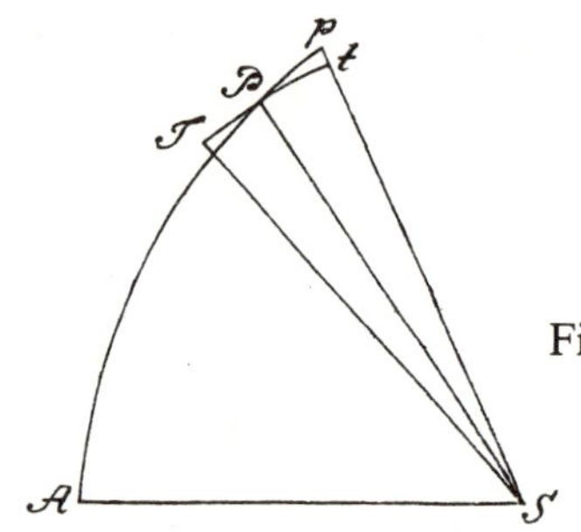

Fig. 1

8. Sit igitur S [fig. 1] centrum solis, ad quos omnes tam planetae quam cometae pellantur viribus quadratis distantiarum reciproce proportionalibus. Sit praeterea f ea a sole distantia, in qua corpus ad solem tanta vi urgetur, quanta, si in superficie terrae versaretur, deorsum niteretur, seu sit f ea distantia, in qua vis centripeta aequalis sit vi gravitatis, quam unitate designemus. Quam distantiam, ut inveniamus, sit T tempus periodicum terrae et A eius axis transversus[3]), erit $T = \frac{\pi A\sqrt{A}}{f}$, ubi si $\sqrt{A}$ in partibus millesimis pedis rhenani exprimatur, atque expressio $\frac{\pi A\sqrt{A}}{f}$ per 250 dividatur, tempus in minutis secundis expressum prodibit[4]). Si ergo tempus periodicum T in minutis secundis exhibeatur erit $\frac{f}{A} = \frac{\pi\sqrt{A}}{250\,T}$. Cum autem semidiameter terrae sit $a = 19\,615\,791$ ped. Par. posita solis parallaxi horizontali 13″ fiet distantia media terrae a sole $= 15866\,a$, ideoque axis transversus orbitae $A = 622\,468\,600\,000$ ped. paris. seu in pedibus Rhenanis $A = 644\,211\,428\,198$. Annus autem sidereus est 365 d. 6 h. 9′36″: ita ut in minutis secundis sit $T = 31\,558\,176$. Ex his itaque valoribus calculo subducto reperitur $\frac{f}{A} = 0{,}01010677$, hincque ob $A = 31732\,a$ erit $f = 320{,}708\,a$, et $ff = 102854\,a^2$. Cognita ergo hac distantia f pro omnibus planetis et cometis eadem, in quavis distantia a sole y vis attractiva solis erit $= \frac{ff}{yy}$, siquidem vis gravitatis unitate exprimatur. Ponamus ergo primum planetam in perihelio suo A esse versatum, ita ut hoc loco eius directio motus fuerit normalis ad rectam SA. Sit ista distantia perihelii a sole AS $= h$; celeritas vero, quam in A habuerit debita sit altidudini b.

3) Id est, axis maior. EJA.

4) Id est, $T = \frac{\sqrt{1000}}{250}\,\frac{\pi A^{3/2}}{f}$, ubi A in pedibus Rhenanis exprimitur. EJA.

Deinde pervenerit planeta in locum P, et ducta recta SP, item perpendiculo ex S in tangentem[5]) ST, vocentur: SP $= y$, ST $= p$, et arcus AP $= s$, celeritas autem planetae hoc loco debita sit altitudini v. Designet nunc α semidiametrum corporis planetae, ac ponatur brevitatis gratia $\mu\alpha = c$. His ita praemissis in Mech. Tom I. pag. 433[6]) perveni ad has aequationes

$$v = \frac{bhh}{pp}\, e^{-s:c} \quad \text{et} \quad \frac{ff}{yy} = \frac{2bhh\, dp}{p^3\, dy}\, e^{-s:c}$$

ubi e denotat numerum cuius logarithmus hyperbolicus $= 1$.

9. Ponamus brevitatis causa $2bhh = ffg$, et posterior aequatio transibit in hanc formam:

$$\frac{e^{s:c}\, dy}{yy} = \frac{g\, dp}{p^3},$$

quae si resistentia evanesceret, quod eveniret si μ ideoque c esset quantitas infinita, abiret in

$$\frac{dy}{yy} = \frac{g\, dp}{p^3},$$

unde sectio conica, in qua planeta esset progressurus, si nulla adesset resistentia, facile eruitur. Hunc ergo casum, quo resistentia est nulla, prius expediamus, quam resistentiae rationem habeamus. Aequatio autem

$$\frac{dy}{yy} = \frac{g\, dp}{p^3}$$

integrata dat

$$\frac{1}{h} - \frac{1}{y} = \frac{g}{2hh} - \frac{g}{2pp}$$

quoniam in A uti fit $y = h$, simul esse debet $p = h$. Hinc erit

5) Linea PT est tangens ex P. In hac commentatione non opus est punctis p et t, quae non definiti sunt. EJA.

6) E15, *Opera omnia* II 1, p. 364. EJA.

$$\frac{g}{2pp} = \frac{gy - 2hy + 2hh}{2hhy} \quad \text{et} \quad pp = \frac{ghhy}{2hh + (g - 2h)\,y}.$$

Vocemus nunc angulum ASP $= t$, eritque

$$dt = \frac{p\,dy}{y\sqrt{(yy - pp)}},$$

quae si, postquam loco p valor ante inventus fuerit substitutus, recte tractetur, deducetur ad hanc integralem:

$$y = \frac{gh}{h + (g - h)\cos t},$$

ex qua oritur

$$p = \frac{gh}{\sqrt{(hh + 2h(g - h)\cos t + (g - h)^2)}}.$$

Quemadmodum illa aequatio, si ponatur $t = 0$, seu $\cos t = 1$ praebet distantiam perihelii a sole AS $= h$, ita ponendo $t = 180°$ seu $\cos t = -1$, invenietur distantia aphelii a sole $= \frac{gh}{2h - g}$, unde axis transversus erit $= \frac{2hh}{2h - g}$, et focorum distantia $= \frac{2h(g - h)}{2h - g}$. Ex quibus excentricitas orbitae, seu focorum distantia ad axem transversum applicata erit $= \frac{g - h}{h}$; sic si sit $g = h$, erit ubique $y = h$, et corpus sine excentricitate in circulo moveretur, sin autem sit $\frac{g - h}{h} = 1$ seu $g = 2h$, erit $y = \frac{2h}{1 + \cos t}$, corpusque movebitur in parabola; quorum duorum casuum prior ad motum planetarum, posterior vero ad motum cometarum propius accedit.

10. Quoniam vero pro resistentia arcus curvae s in computum ingreditur, quo difficultates hinc nascendas facilius removeamus, in hypothesi vacui quantitatem arcus AP $= s$ definiri conveniet. Cum autem sit $ds = \sqrt{(dy^2 + y^2 dt^2)}$, ob

$$dy = \frac{gh(g - h)\,dt \sin t}{(h + (g - h)\cos t)^2}$$

reperietur:

$$ds = \frac{gh\,dt\,\sqrt{(hh + 2h(g-h)\cos t + (g-h)^2)}}{(h + (g-h)\cos t)^2}$$

quae expressio, cum generaliter ad formam integralem perduci nequeat, duos casus, quorum alter ad circulum proxime accedat, alter ad parabolam, perpendamus. Sit ergo primo $g - h$ quantitas minima, ac ponatur $g - h = \beta$, fiet proxime[7]):

$$ds = \frac{gh\,dt}{(h + \beta\cos t)^2}\left(h + \beta\cos t + \frac{\beta\beta}{2h}\sin t^2\right)$$

seu

$$ds = g\,dt\left(1 - \frac{\beta}{h}\cos t + \frac{3\beta\beta}{4hh} + \frac{\beta\beta}{4hh}\cos 2t\right)$$

cuius integrale est:

$$s = gt - \frac{\beta g}{h}\sin t + \frac{3\beta\beta gt}{4hh} + \frac{\beta\beta g}{8hh}\sin 2t.$$

11. Pro parabola autem ubi est $g = 2h$, erit[8])

$$ds = \frac{2h\,dt\,\sqrt{2}\,(1 + \cos t)}{(1 + \cos t)^2} = \frac{2h\,dt\,\sqrt{2}}{(1 + \cos t)^{3/2}}.$$

At est $1 + \cos t = 2(\cos\frac{1}{2}t)^2$; ideoque

$$ds = \frac{h\,dt}{(\cos\frac{1}{2}t)^3};$$

hinc fit

$$s = \frac{h\sin\frac{1}{2}t}{(\cos\frac{1}{2}t)^2} + \frac{h}{2}\int\frac{dt}{\cos\frac{1}{2}t},$$

porroque integrando[9]):

$$s = h\left(\frac{\operatorname{tag}\frac{1}{2}t}{\cos\frac{1}{2}t} + l\operatorname{tag}(45^\circ + \tfrac{1}{4}t)\right).$$

7) Intellegimus $\sin^2 t$ loco $\sin t^2$, idem $\cos^2 t$ loco $\cos t^2$. EJA.

8) Intellegitur $\sqrt{(2(1+\cos t))}$ loco $\sqrt{2}(1+\cos t)$. EJA.

9) tag denotat tangentem. EJA.

At si orbita tantum proxime ad parabolam accedat, ponatur $g = 2h - \gamma h$; fietque

$$ds = \frac{g\,dt\sqrt{(2(1-\gamma)(1+\cos t)+\gamma\gamma)}}{(1+\cos t-\gamma\cos t)^2}.$$

Cum igitur sit $1 + \cos t = 2\cos\frac{1}{2}t^2$, erit

$$ds = \frac{g\,dt\sqrt{(4(1-\gamma)\cos\frac{1}{2}t^2+\gamma\gamma)}}{(2\cos\frac{1}{2}t^2-\gamma\cos t)^2}.$$

Haec aequatio perducitur porro ad hanc formam:

$$ds = \frac{g\,dt\sqrt{((2-\gamma)^2\cos\frac{1}{2}t^2+\gamma\gamma\sin\frac{1}{2}t^2)}}{(2(1-\gamma)\cos\frac{1}{2}t^2+\gamma)^2}$$

seu

$$ds = \frac{g\,dt\sqrt{((2-\gamma)^2\cos\frac{1}{2}t^2+\gamma\gamma\sin\frac{1}{2}t^2)}}{((2-\gamma)\cos\frac{1}{2}t^2+\gamma\sin\frac{1}{2}t^2)^2}.$$

Si iam γ tam sit parvum, ut $\gamma\gamma$ prae reliquis terminis negligi queat, erit:

$$ds = \frac{g(2-\gamma)\,dt\cos\frac{1}{2}t}{((2-\gamma)\cos\frac{1}{2}t^2+\gamma\sin\frac{1}{2}t^2)^2} = \frac{ggh\,dt\cos\frac{1}{2}t}{(g\cos\frac{1}{2}t^2+\gamma h\sin\frac{1}{2}t^2)^2}$$

quae reducitur ad hanc formam:

$$s = \frac{gh\sin\frac{1}{2}t}{g\cos\frac{1}{2}t^2+\gamma h\sin\frac{1}{2}t^2} + \frac{1}{2}\int\frac{gh\,dt\cos\frac{1}{2}t}{g\cos\frac{1}{2}t^2+\gamma h\sin\frac{1}{2}t^2}$$

cuius integrale debito modo sumtum fit:

$$s = \frac{gh\sin\frac{1}{2}t}{g\cos\frac{1}{2}t^2+\gamma h\sin\frac{1}{2}t^2} + \frac{h\sqrt{g}}{2\sqrt{(g-\gamma h)}}\,l\,\frac{\sqrt{g}+\sin\frac{1}{2}t\cdot\sqrt{(g-\gamma h)}}{\sqrt{g}-\sin\frac{1}{2}t\cdot\sqrt{(g-\gamma h)}}.$$

Hinc si ponatur $t = 180°$, ut sit $\sin\frac{1}{2}t = 1$ et $\cos\frac{1}{2}t = 0$, prodibit semissis perimetri huius ellipsis maxime oblongae

10) $\sqrt{2h}$ denotat $\sqrt{(2h)}$. EJA.

$$= \frac{g}{\gamma} + \frac{h\sqrt{g}}{2\sqrt{(g-\gamma h)}}\, l\, \frac{\sqrt{g} + \sqrt{(g-\gamma h)}}{\sqrt{g} - \sqrt{(g-\gamma h)}}$$

axis transversus autem erit $= \dfrac{2h}{\gamma}$, hincque tempus periodicum[10]) $= \dfrac{2\pi h\sqrt{2h}}{\gamma f\sqrt{\gamma}}$.

12. Introducamus nunc rationem resistentiae; et cum in vacuo invenissemus

$$y = \frac{gh}{h + (g-h)\cos t}$$

sive

$$\frac{1}{y} = \frac{1}{g} + \frac{(g-h)\cos t}{gh},$$

assumamus pro aethere hanc aequationem

$$\frac{1}{y} = \frac{1}{g} + \frac{(g-h)\cos t}{gh} + P,$$

ubi quidem ob resistentiam minimam P erit quantitas quoque minima, et quia P est functio ipsius t, ponatur $dP = Q\,dt$, et $dQ = R\,dt$. Hinc ergo erit

$$\frac{dy}{yy} = \frac{(g-h)\,dt \sin t}{gh} - Q\,dt.$$

At ex aequatione

$$dt = \frac{p\,dy}{y\sqrt{(yy-pp)}},$$

unde obtinetur

$$\frac{1}{pp} = \frac{1}{yy} + \frac{dy^2}{y^4\,dt^2},$$

et loco y valore suo substituto

$$\frac{1}{pp} = \frac{1}{gg} + \frac{2(g-h)\cos t}{ggh} + \frac{(g-h)^2}{gghh} + \frac{2P}{g} + \frac{2(g-h)P\cos t}{gh}$$
$$+ PP - \frac{2(g-h)Q\sin t}{gh} + QQ,$$

ubi quidem quadrata terminorum minimorum PP et QQ tuto negliguntur. Hinc autem per differentiationem prodit:

$$\frac{dp}{p^3} = \frac{(g-h)\,dt\sin t}{ggh} - \frac{Q\,dt}{g} + \frac{(g-h)\,P\,dt\sin t}{gh} + \frac{(g-h)\,R\,dt\sin t}{gh}.$$

Resumamus nunc aequationem principalem

$$\frac{e^{s:c}\,dy}{yy} = \frac{g\,dp}{p^3}$$

et quia c est quantitas maxima, erit $\frac{s}{c}$ fractio valde parva, nisi quidem corpus iam plurimas revolutiones absolverit. Quare si calculum non ad plures revolutiones simul extendere velimus pro $e^{s:c}$ tuto scribere poterimus $1+\frac{s}{c}$; hincque haec nascetur aequatio:

$$\frac{dy}{yy} + \frac{s\,dy}{cyy} = \frac{g\,dp}{p^3}$$

ubi si pro $\frac{dy}{yy}$ et $\frac{dp}{p^3}$ valores ante exhibitos substituamus, terminis utrinque aequalibus deletis, remanebit[11]:

$$\frac{(g-h)\,s\,dt\sin t}{cgh} = \frac{(g-h)\,dt\sin t}{h}(P+R)$$

seu ob $R = \frac{ddP}{dt^2}$, erit $\frac{s\,dt^2}{cg} = ddP + P\,dt^2$ quae si methodo alibi exposita integretur dabit:

$$P = \frac{\sin t}{cg}\int s\,dt\cos t - \frac{\cos t}{cg}\int s\,dt\sin t,$$

erit enim

$$Q = \frac{\cos t}{cg}\int s\,dt\cos t + \frac{\sin t}{cg}\int s\,dt\sin t$$

et

11) Editio princeps: $s\sin t$ loco $s\,dt\sin t$. Correxit EJA.

$$R = -\frac{\sin t}{cg}\int s\,dt\cos t + \frac{\cos t}{cg}\int s\,dt\sin t + \frac{s}{cg}$$

ex quibus valoribus veritas aequationis $\frac{s}{cg} = P + R$ comprobatur.

13. Ut autem constantes, quas haec integralia secum gerunt, determinemus, ad motus initium est spectandum, ubi fit $t = 0$, $y = h$, et $dy = 0$. Ideoque hoc casu tam P, quam Q evanescere debet, quod eveniet, si integralia $\int s\,dt\cos t$ et $\int s\,dt\sin t$ ita sumantur, ut evanescant posito $t = 0$. Loca autem sequentium apheliorum et periheliorum reperientur ponendo $dy = 0$, unde fit $Q = \frac{(g-h)\sin t}{gh}$, qui valor si substituatur, aequatio haec[12])

$$\frac{c}{h}(g-h)\sin t = \cos t\int s\,dt\cos t + \sin t\int s\,dt\sin t$$

definiet omnia aphelia et perihelia, in quae planeta seu cometa successive pertinget. Cognitis autem locis apheliorum et periheliorum seu valoribus angulorum t, hinc quoque distantiae maximae et minimae determinabuntur ope aequationis:

$$\frac{1}{y} = \frac{1}{g} + \frac{(g-h)\cos t}{gh} + P.$$

Superest ergo ut integralia $\int s\,dt\cos t$ et $\int s\,dt\sin t$ investigemus; cum autem valor arcus s per formulam finitam exhiberi nequeat, istae expressiones ita transformentur:

$$\int s\,dt\cos t = s\sin t - \int ds\sin t,$$

$$\int s\,dt\sin t = -s\cos t + \int ds\cos t,$$

qui valores substituti dabunt:

$$P = \frac{s}{cg} - \frac{\sin t}{cg}\int ds\sin t - \frac{\cos t}{cg}\int ds\cos t$$

$$Q = -\frac{\cos t}{cg}\int ds\sin t + \frac{\sin t}{cg}\int ds\cos t$$

12) Editio princeps: $(g-h)$ loco $\frac{c}{h}(g-h)$. Correxit EJA.

ubi haec integralia iterum ita accipi debent, ut evanescant posito angulo ASP $= t = 0$. Cum igitur sit

$$ds = \frac{gh\,dt\sqrt{(hh + 2h(g-h)\cos t + (g-h)^2)}}{(h + (g-h)\cos t)^2}$$

videndum est, utrum hoc differentiale, cum per $\sin t$ tum etiam per $\cos t$ multiplicatum, integrabile existat nec ne?

14. Formula quidem altera $\int ds \sin t$, ad hanc formam perducitur ut sit posito $g - h = \zeta h$

$$\int ds \sin t = \frac{g\sqrt{(1 + 2\zeta\cos t + \zeta\zeta)}}{\zeta(1 + \zeta\cos t)} + g\int \frac{dt \sin t}{(1 + \zeta\cos t)\sqrt{(1 + 2\zeta\cos t + \zeta\zeta)}}.$$

Quin etiam posterioris membrum integrale assignari potest per quadraturam circuli, ita ut debita addita constante sit[13])

$$\int ds \sin t = \frac{g\sqrt{(1 + 2\zeta\cos t + \zeta\zeta)}}{\zeta(1 + \zeta\cos t)} - \frac{2g}{\zeta\sqrt{(1-\zeta\zeta)}}\,\mathrm{A\,tag}\,\frac{\sqrt{(1 + 2\zeta\cos t + \zeta\zeta)}}{\sqrt{(1-\zeta\zeta)}}$$
$$-\frac{g}{\zeta} + \frac{2g}{\zeta\sqrt{(1-\zeta\zeta)}}\,\mathrm{A\,tag}\,\frac{1+\zeta}{\sqrt{(1-\zeta\zeta)}}.$$

Altera autem formula non aeque tractari potest; fit enim

$$\int ds \cos t = \int \frac{g\,dt \cos t\sqrt{(1 + 2\zeta\cos t + \zeta\zeta)}}{(1 + \zeta\cos t)^2}$$

quae ad sequentia integralia simpliciora reducitur,

$$\int ds \cos t = \frac{g\cos t\sqrt{(1 + 2\zeta\cos t + \zeta\zeta)}}{\zeta\sin t(1 + \zeta\cos t)}$$
$$+ g\int \frac{dt \cos t}{(1 + \zeta\cos t)\sqrt{(1 + 2\zeta\cos t + \zeta\zeta)}} + \frac{g}{\zeta}\int \frac{dt\sqrt{(1 + 2\zeta\cos t + \zeta\zeta)}}{(1 + \zeta\cos t)\sin t^2},$$

13) A tag denotat arcum tangentis. EJA.

vel hoc modo:

$$\int ds \cos t = \frac{g \sin t \sqrt{(1+2\zeta \cos t+\zeta\zeta)}}{(1-\zeta\zeta)(1+\zeta \cos t)}$$
$$-\frac{g}{1-\zeta\zeta}\int \frac{\zeta\, dt\,(\zeta^2+2\zeta \cos t+\cos t^2)}{(1+\zeta \cos t)\sqrt{(1+2\zeta \cos t+\zeta\zeta)}}.$$

Cuius integrale cum etiamnum exhiberi nequeat, si fortasse ab ipso arcu s pendeat, conveniet notasse:

$$s = -\frac{\zeta g \sin t \sqrt{(1+2\zeta \cos t+\zeta\zeta)}}{(1-\zeta\zeta)(1+\zeta \cos t)} + \frac{g}{1-\zeta\zeta}\int \frac{dt\,(1+2\zeta \cos t+\zeta\zeta \cos t^2)}{(1+\zeta \cos t)\sqrt{(1+2\zeta \cos t+\zeta\zeta)}}.$$

Potest etiam formula $ds \cos t$ ad quantitates consuetas reduci ponendo $\cos t = x$, ut sit $dt = \frac{-dx}{\sqrt{(1-xx)}}$. Hinc autem fiet

$$\int ds \cos t = -g\int \frac{x\,dx \sqrt{(1+2\zeta x+\zeta\zeta)}}{(1+\zeta x)^2\sqrt{(1-xx)}}$$

estque

$$s = -g\int \frac{dx \sqrt{(1+2\zeta x+\zeta\zeta)}}{(1+\zeta x)^2\sqrt{(1-xx)}}.$$

Quomodocunque autem hae formulae tractentur, earum integralia in quantitatibus finitis exhiberi nequeunt.

15. Quae difficultas cum impediat, quominus hoc problema generaliter pertractare queamus, eos saltem casus evolvamus, qui in mundo locum habent, et pro quibus ante valorem arcus s proxime eruimus. Sit igitur excentricitas $\frac{g-h}{h}$ valde parva, maneatque ut ante $g-h=\zeta h$, ita ut litera supra adhibita β sit $=\zeta h$; eritque pro motu planetarum in orbitis parum excentricis:

$$ds = g\,dt\,(1-\zeta \cos t+\tfrac{3}{4}\zeta\zeta+\tfrac{1}{4}\zeta\zeta \cos 2t)$$

et

$$s = gt-\zeta g \sin t+\tfrac{3}{4}\zeta\zeta\, gt+\tfrac{1}{8}\zeta\zeta g \sin 2t.$$

Hinc erit pro formulis superioribus:

$$\int ds \sin t = g \int dt\,(\sin t - \tfrac{1}{2}\zeta \sin 2t + \tfrac{5}{8}\zeta\zeta \sin t + \tfrac{1}{8}\zeta\zeta \sin 3t)$$

et integrando

$$\int ds \sin t = g\,(1 - \cos t + \tfrac{1}{4}\zeta \cos 2t - \tfrac{5}{8}\zeta\zeta \cos t - \tfrac{1}{24}\zeta\zeta \cos 3t - \tfrac{1}{4}\zeta + \tfrac{2}{3}\zeta\zeta).$$

Simili modo erit

$$ds \cos t = g\,dt\,(\cos t - \tfrac{1}{2}\zeta - \tfrac{1}{2}\zeta \cos 2t + \tfrac{7}{8}\zeta\zeta \cos t + \tfrac{1}{8}\zeta\zeta \cos 3t)$$

et integrando

$$\int ds \cos t = g\,(\sin t - \tfrac{1}{2}\zeta t - \tfrac{1}{4}\zeta \sin 2t + \tfrac{7}{8}\zeta\zeta \sin t + \tfrac{1}{24}\zeta\zeta \sin 3t).$$

Ex his nunc conficietur:

$$P = \frac{1}{c}(t - \sin t - \tfrac{1}{2}\zeta \sin t + \tfrac{1}{2}\zeta t \cos t + \tfrac{3}{4}\zeta\zeta t - \tfrac{2}{3}\zeta\zeta \sin t - \tfrac{1}{24}\zeta\zeta \sin 2t)$$

$$Q = \frac{1}{c}(1 - \cos t - \tfrac{1}{2}\zeta t \sin t + \tfrac{3}{4}\zeta\zeta - \tfrac{2}{3}\zeta\zeta \cos t - \tfrac{1}{12}\zeta\zeta \cos 2t).$$

16. Quoniam nunc in periheliis et apheliis est $Q = \dfrac{\zeta \sin t}{g}$, haec planetae loca pro pluribus sequentibus revolutionibus reperientur ex hac aequatione[14]):

$$\frac{\zeta c}{g} \sin t = 1 - \cos t - \tfrac{1}{2}\zeta t \sin t + \tfrac{3}{4}\zeta\zeta - \tfrac{2}{3}\zeta\zeta \cos t - \tfrac{1}{12}\zeta\zeta \cos 2t$$

cui aequationi satisfit, quoties $\sin t = 0$ et $\cos t = 1$; quare valores ipsius t erunt sequentes arcus: $0,\ 2\pi,\ 4\pi,\ 6\pi,\ 8\pi$, etc. denotante π angulum duobus rectis aequalem, quod indicium est omnes has radices praebere omnium periheliorum loca successiva. Locus ergo perihelii A manet fixus ita ut quoties planeta per rectam SA transit, tum simul soli sit proximus. Est ergo perihelium perinde atque in vacuo immobile, unde si forte absides orbitarum mobiles deprehendantur, causa

14) Editio princeps: c loco g. Correxit EJA.

frustra in resistentia aetheris quaeritur. Simili autem modo, quia corpus planetae a resistentia non de suo plano deflectitur, linea quoque nodorum cuiusque planetae a resistentia non mutabitur. Atque cum resistentia aetheris aphelia et perihelia non de suo loco moveat, idem quoque valebit, si resistentia concipiatur negativa, hoc est si planetae ab aethere in gyrum acto propellantur. Hancobrem promotio absidum, siquidem ulla datur, per vorticem aetheris planetas deferentem explicari quoque non poterit.

17. Quia perihelia perpetuo ad idem coeli punctum referuntur, dubitari omnio nequit, quin etiam aphelia quiescant; eo quod aphelia et perihelia inter se commutari possunt. Attamen videmus aphelium perihelio non e diametro opponi; si enim ponamus angulum $t = 180°$, aequationi inventae non satisfit, nisi et c simul sit infinitum, seu resistentia prorsus evanescat. Quo igitur aphelii locum facilius inveniamus, aequationem superiorem dividamus per $\sin t$, fietque

$$\frac{\zeta c}{g} = \frac{1 - \cos t}{\sin t} - \tfrac{1}{2}\zeta t + \frac{\zeta\zeta(9 - 8\cos t - \cos 2t)}{12 \sin t}.$$

Est vero $\dfrac{1 - \cos t}{\sin t} = \operatorname{tag} \tfrac{1}{2}t$, et ob $\cos 2t = 1 - 2\sin t^2$ erit

$$\frac{9 - 8\cos t - \cos 2t}{\sin t} = 8 \operatorname{tag} \tfrac{1}{2}t + 2\sin t,$$

hincque fiet:

$$\frac{\zeta c}{g} = \operatorname{tag} \tfrac{1}{2}t - \tfrac{1}{2}\zeta t + \tfrac{2}{3}\zeta\zeta \operatorname{tag} \tfrac{1}{2}t + \tfrac{1}{6}\zeta\zeta \sin t.$$

Ponatur nunc $t = \pi - z$, denotante π angulum $180°$, eritque z angulus valde parvus, et $\tfrac{1}{2}t = 90° - \tfrac{1}{2}z$; unde habebitur

$$\frac{\zeta c}{g} = \cot \tfrac{1}{2}z - \tfrac{1}{2}\zeta\pi + \tfrac{1}{2}\zeta z + \tfrac{2}{3}\zeta\zeta \cot \tfrac{1}{2}z + \tfrac{1}{6}\zeta\zeta \sin z.$$

Hic autem ob parvitatem primo termini z et $\sin z$ evanescunt; deinde quia novimus locum aphelii aeque ac perihelii esse fixum, necesse est ut quoque terminus $\tfrac{1}{2}\zeta\pi$ excedat, quippe quod re ipsa fieret, si supra quantitatem $e^{s:c}$ accuratius expressissemus. Hancobrem relinquetur:

$$\cot\tfrac{1}{2}z = \frac{\zeta c}{g(1+\frac{2}{3}\zeta\zeta)} \quad \text{seu} \quad \operatorname{tag}\tfrac{1}{2}z = \frac{g(1+\frac{2}{3}\zeta\zeta)}{\zeta c}.$$

Verum quia ζ est quantitas valde parva, erit proxime $\operatorname{tag}\frac{1}{2}z = \frac{g}{\zeta c}$ et ob $\frac{g}{\zeta c}$ fractionem minimam erit angulus $z = \frac{2g}{\zeta c}$. Hinc igitur aphelium [fig. 2], quod sit in A perihelio in P non directe opponitur, sed planeta e perihelio P egressus in aphelium A perveniet, antequam locum perihelio oppositum B attingat; huiusque differentiae seu anguli ASB tangens erit $= \frac{2g}{\zeta c}$ qui quidem angulus nunquam ultro aliquot minuta secunda exsurgit, ideoque sensibilis esse nequit.

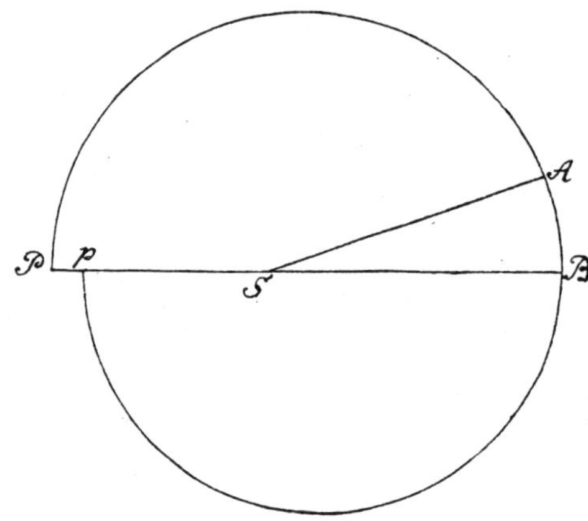

Fig. 2

18. Ad reliqua phaenomena motus planetarum in medio resistente investiganda, consideremus aequationem

$$\frac{1}{y} = \frac{1}{g} + \frac{\zeta}{g}\cos t + P,$$

existente

$$P = \frac{1}{c}\left(t - \sin t - \tfrac{1}{2}\zeta\sin t + \tfrac{1}{2}\zeta t\cos t + \tfrac{3}{4}\zeta\zeta t - \tfrac{2}{3}\zeta\zeta\sin t - \tfrac{1}{24}\zeta\zeta\sin 2t\right).$$

Hinc iam quoties planeta in perihelium revertitur, eius distantia a sole invenietur, si loco anguli t successive ponantur $0,\ 2\pi,\ 4\pi,\ 6\pi,\ 8\pi$, etc. In apheliis autem distantiae planetae prodibunt, si θ sumatur pro angulo, cuius tangens vel sinus est

$=\frac{2g}{\zeta c}$, loco t successive substituendo $\pi-\theta$, $3\pi-\theta$, $5\pi-\theta$, etc. quibus casibus est $\sin t=\frac{2g}{\zeta c}$, $\cos t=-1$; $\sin 2t=\frac{4g}{\zeta c}$ atque $\theta=\frac{2g}{\zeta c}$. Ex his sequentes orientur valores:

$$
\begin{array}{lll}
\text{Si} & t=0 & \text{erit}\quad P=0 \\
& t=\pi-\theta & P=\frac{1}{c}\left(\pi-\frac{1}{2}\zeta\pi-\frac{4g}{\zeta c}\right) \\
& t=2\pi & P=\frac{1}{c}(2\pi+\zeta\pi) \\
& t=3\pi-\theta & P=\frac{1}{c}\left(3\pi-\frac{3}{2}\zeta\pi-\frac{4g}{\zeta c}\right) \\
& t=4\pi & P=\frac{1}{c}(4\pi+2\zeta\pi) \\
& t=5\pi-\theta & P=\frac{1}{c}\left(5\pi-\frac{5}{2}\zeta\pi-\frac{4g}{\zeta c}\right).
\end{array}
$$

Cum igitur pro perihelio sit $\frac{1}{y}=\frac{1+\zeta}{g}+P$, et pro aphelio $\frac{1}{y}=\frac{1-\zeta}{g}+P$,

$$
\begin{array}{lll}
\text{Si} & t=0 & \text{erit}\quad \frac{1}{y}=\frac{1+\zeta}{g}+0 \\
& t=\pi-\theta & \frac{1}{y}=\frac{1-\zeta}{g}+\frac{\pi}{c}(1-\frac{1}{2}\zeta) \\
& t=2\pi & \frac{1}{y}=\frac{1+\zeta}{g}+\frac{2\pi}{c}(1+\frac{1}{2}\zeta) \\
& t=3\pi-\theta & \frac{1}{y}=\frac{1-\zeta}{g}+\frac{3\pi}{c}(1-\frac{1}{2}\zeta) \\
& t=4\pi & \frac{1}{y}=\frac{1+\zeta}{g}+\frac{4\pi}{c}(1+\frac{1}{2}\zeta) \\
& t=5\pi-\theta & \frac{1}{y}=\frac{1-\zeta}{g}+\frac{5\pi}{c}(1-\frac{1}{2}\zeta)
\end{array}
$$

etc. neglectis scilicet terminis $\frac{4g}{\zeta c}$ prae reliquis incomparabiliter parvis.

19. Ponamus, postquam planeta primum ex perihelio P est egressus, iam i integras revolutiones esse absolutas, ita ut nunc planeta iterum in perihelio versetur, sitque $t = 2i\pi$, erit ergo

$$\frac{1}{y} = \frac{1+\zeta}{g} + \frac{2i\pi}{c}(1+\tfrac{1}{2}\zeta)$$

et cum terminus posterior prae priori sit valde parvus erit:

$$y = \frac{g}{1+\zeta} - \frac{2i\pi(1+\frac{1}{2}\zeta)gg}{c(1+\zeta)^2}.$$

Pro sequente autem aphelio, ponendo $t = (2i+1)\pi - \theta$, erit

$$\frac{1}{y} = \frac{1-\zeta}{g} + \frac{(2i+1)\pi}{c}(1-\tfrac{1}{2}\zeta),$$

hincque

$$y = \frac{g}{1-\zeta} - \frac{(2i+1)\pi(1-\frac{1}{2}\zeta)gg}{c(1-\zeta)^2}$$

successivae ergo planetae distantiae a sole in perihelio et aphelio sequenti modo diminuentur.

	Distantia planetae a sole
I. Perihelium	$\frac{g}{1+\zeta} - 0$
Aphelium	$\frac{g}{1-\zeta} - \frac{\pi(1-\frac{1}{2}\zeta)gg}{c(1-\zeta)^2}$
II. Perihelium	$\frac{g}{1+\zeta} - \frac{2\pi(1+\frac{1}{2}\zeta)gg}{c(1+\zeta)^2}$
Aphelium	$\frac{g}{1-\zeta} - \frac{3\pi(1-\frac{1}{2}\zeta)gg}{c(1-\zeta)^2}$
III. Perihelium	$\frac{g}{1+\zeta} - \frac{4\pi(1+\frac{1}{2}\zeta)gg}{c(1+\zeta)^2}$
Aphelium	$\frac{g}{1-\zeta} - \frac{5\pi(1-\frac{1}{2}\zeta)gg}{c(1-\zeta)^2}$ etc.

In qualibet ergo revolutione perihelium ad solem propius accedit per intervallum

$$\frac{2\pi(1+\frac{1}{2}\zeta)gg}{c(1+\zeta)^2};$$

aphelium vero admovetur intervallo

$$\frac{2\pi(1-\frac{1}{2}\zeta)gg}{c(1-\zeta)^2};$$

media ergo distantia decrescet intervallo circiter $=\frac{2\pi gg}{c}$; et post i revolutiones hoc decrementum distantiae mediae erit $=\frac{2i\pi gg}{c}$.

20. Cum igitur post i planetae revolutiones sit distantia perihelii a sole:

$$\frac{g}{1+\zeta}-\frac{2i\pi(1+\frac{1}{2}\zeta)gg}{c\,(1+\zeta)^2}$$

atque sequens aphelii distantia a sole:

$$\frac{g}{1-\zeta}-\frac{(2i+1)\pi(1-\frac{1}{2}\zeta)gg}{c(1-\zeta)^2}$$

his addendis prodibit post i revolutiones orbitae axis transversus:

$$\frac{2g}{1-\zeta\zeta}-\frac{4i\pi gg}{c(1-\zeta\zeta)^2}-\frac{\pi(1-\frac{1}{2}\zeta)gg}{c(1-\zeta)^2},$$

ubi quidem, quia de tempore definito est quaestio, tempus a perihelio ad aphelium omitti potest, ita ut post i revolutiones habeatur orbitae axis transversus

$$=\frac{2g}{1-\zeta\zeta}-\frac{4i\pi gg}{c(1-\zeta\zeta)^2},$$

quemadmodum etiam initio axis transversus assumtus est $=\frac{2g}{1-\zeta\zeta}$. Deinde si distantia perihelii a sole post i revolutiones, quae est

$$=\frac{g}{1+\zeta}-\frac{2i\pi(1+\frac{1}{2}\zeta)gg}{c\,(1+\zeta)^2},$$

a distantia aphelii a sole, quae eidem tempori conveniret, atque inventa est

$$= \frac{g}{1-\zeta} - \frac{2i\pi(1-\frac{1}{2}\zeta)gg}{c\,(1-\zeta)^2},$$

subtrahatur, prodibit distantia focorum post i revolutiones:

$$\frac{2\zeta g}{1-\zeta\zeta} - \frac{2i\pi\zeta gg(3-\zeta\zeta)}{c\,(1-\zeta\zeta)^2}$$

quae si per axem transversum ante inventum dividatur obtinebitur excentricitas orbitae pro hoc tempore:

$$= \zeta - \frac{i\pi\zeta g}{c},$$

quae initio erat $= \zeta$. A resitentia ergo excentricitas continuo minuitur, orbitaeque planetarum propius ad figuram circularem reducuntur.

21. Ex diminutione axis transversi sequitur diminutio temporis periodici. Sit enim T tempus periodicum planetae initio, Θ vero sit eius tempus periodicum, postquam iam i revolutiones absolvit. Quia ergo in una revolutione effectus resistentiae sensibilis esse nequit, erit T ad Θ in ratione sesquiplicata axis transversi initio, qui erat $= \frac{2g}{1-\zeta\zeta}$, ad axem transversum post i revolutiones, qui inventus est

$$= \frac{2g}{1-\zeta\zeta} - \frac{4i\pi gg}{c(1-\zeta\zeta)^2}.$$

Cum igitur horum axium transversorum ratio sit ut 1 ad $1 - \frac{2i\pi g}{c(1-\zeta\zeta)}$, erit:

$$T : \Theta = 1 : \left(1 - \frac{2i\pi g}{c(1-\zeta\zeta)}\right)^{3/2} = 1 : 1 - \frac{3i\pi g}{c(1-\zeta\zeta)}.$$

Hinc ergo fit

$$\Theta = T\left(1 - \frac{3i\pi g}{c(1-\zeta\zeta)}\right).$$

Post centum ergo revolutiones tempus periodicum cuiusque planetae, quod ante erat $= T$, nunc diminutum erit tempore

$$= \frac{300\pi g T}{c(1-\zeta\zeta)} = \frac{942{,}4779\, g T}{c(1-\zeta\zeta)}.$$

Hoc igitur decrementum temporis periodici primum ipsi tempori periodico erit proportionale. Quocirca decrementa temporum periodicorum, quae diversi planetae eodem temporis tractu sunt perpessa, erunt inter se ut valores formulae $\frac{g}{c(1-\zeta\zeta)}$; unde haec decrementa primo erunt directe ut distantiae mediae a sole, tum vero ob $c = \mu\alpha$ reciproce ut diametri planetarum, si quidem eorum corpora sint similia. Tertio vero decrementa erunt maiora, quo excentricitates fuerint maiores. Ex quibus concluditur Martem inter omnes planetas maximum temporis periodici sui detrimentum pati debere.

22. Applicemus haec, quantum circumstantiae permittunt, ad motum terrae, ac primo quidem eius excentricitatem spectemus, unde aequatio maxima pendet. Quodsi autem omnis temporis Astronomos inter se conferamus, ingens quidem discrimen deprehendemus, sed eiusmodi tamen, ut aequatio maxima plerumque a veteribus maior sit habita quam a recentioribus. Namque sequentes

Autores	aequationem	maximam	statuerunt
Cl. Ptolemaeus	2°	23′	0″
Rex Alphonsus	2	10	0
Ioh. Newtonus	2	4	47
Keplerus in Tab.	2	3	46
Tycho Brahe	2	3	15
Ioh. Maginus	2	3	12
Bullialdus	2	2	41
T. Street	1	59	6
Horroxius	1	59	0
W. Leyburn	1	57	0
Isaac Newton	1	56	20
De la Hire	1	55	42
Cassini	1	55	51
Chev. de Louville	1	54	45
Wurzelbaur	1	57	14

Hic fere cum decursu temporis aequatio decrescere cernitur, praecipue si cogitemus non omnes hos autores quantitatem istius aequationis ex propriis observationibus, sed saepe longe anterioribus conclusisse. Parva quidem plerisque his numeris fides haberi potest, praesertim antiquissimis; verumtamen si TYCHONIS BRAHEI et NEUTONI determinationes reliquis veriores putentur, [diminutio] aequatio[nis] terrae intervallo 100 annorum fere 7 minuta censenda esset. Cum igitur excentricitas ζ sit ut 2° 3′ 15″, foret $\frac{100\pi g}{c} = \frac{1}{17}$ fere. Hinc autem tempus periodicum uno seculo sexta sui parte diminueretur, quod cum sit absurdissimum, diminutio ista aequationis maximae, siquidem tanta locum haberet, alii causae adscribi deberet.

23. Perpendamus ergo tempus periodicum terrae, annique longitudinem praesentem cum praeteritis, quantum fieri potest, conferamus. Ac primo quidem modus, quo ab Astronomis tempus anni determinari solet, ita est comparatus, ut inde numquam praesentis cuiusvis anni quantitas cognosci possit: nisi enim duo aequinoctia, quorum momenta comparantur, plurium annorum spatio a se invicem distent, ob errores, quibus huiusmodi observationes exui nequeunt, nihil certi inde concludi potest. Quando igitur Astronomi ex collatione duorum aequinoctiorum aliquot seculorum intervallo a se invicem remotorum quantitatem anni definiunt, haec determinatio neutri tempestati, sed mediae cuidam conveniet, siquidem errores observationum non sint notabiles. Hinc ergo difficillimum est nosse, quanto tempore nunc terra periodum absolvat; longe difficilius autem, quanto tempore terra ante aliquot secula periodum absolverit: quoniam nunc quidem ob sollertiam Astronomorum et summam subsidiorum copiam recentiores observationes veteribus longe sunt anteferendae. Nunc quidem terra ab aequinoctio ad idem recurrere statuitur tempore 365d. 5h. 48′ 55″; sed quia ad hoc observationes vetustiores sunt usurpatae, verisimile est annum iam aliquanto breviorem esse. Antiquissimos certe Astronomos non constat defectum anni tropici ab anno Juliano 365d. 6h. animadvertisse; hincque satis probabile videtur, tum temporis anni longitutinem propius ad istam mensuram accessisse quam nunc. Seculis autem proxime elapsis duratio anni tropici maior quam 365d. 5h. 49′ est credita; unde fortasse a veritate non multum recedemus, si sumamus tempus periodicum terrae singulis seculis quinis minutis secundis minui; neque etiam observationibus repugnaret, si hoc decrementum adhuc maius extaret.

24. Ponamus ergo decrementum anni seculare esse 5″; fietque

$$\frac{942{,}4779\,gT}{c(1-\zeta\zeta)} = 5'';$$

ubi T denotat annum sidereum seu 31 558 176″. Cum itigur excentricitas ζ sit fere $= \frac{1}{60}$, erit

$$\frac{c}{g} = \frac{31\,558\,176 \cdot 942{,}4779}{5 \cdot 0{,}999722},$$

hincque subducto calculo reperietur:

$$\frac{c}{g} = 5\,950\,231\,000.$$

Est vero, si a denotet semidiametrum terrae, $c = \mu a$, et distantia terrae media a sole $g = 15866a$; unde fit

$$\frac{c}{g} = \frac{\mu}{15866};$$

atque $\mu = 94\,406\,320\,000\,000$, qui valor satis prope accedit ad illum, quem supra per coniecturam constituimus, nempe $\mu = 100\,000\,000\,000\,000$, qui valor adhuc minus decrementum produxisset. Atque adeo hinc patet resistentiam ab aethere oriundam maiorem non esse, quam ut cum observationibus conciliari queat. Quare cum ex motu planetarum non sensibiliter imminuto defectus materiae coelum implentis minime sequatur, hinc gravissimum argumentum, quo Angli attractionem suam tueri conantur, corruit: si enim coelum materia subtili est repletum, nullam causam amplius invenimus, cur gravitatis universae causam mechanicam negare velimus. Ceterum si terrae tempus periodicum intervallo mille annorum 50″ minuitur, reliquorum planetarum tempora periodica eodem intervallo, quantum ex eorum diametris apparentibus colligi licet, minuentur, ut sequitur:

♄ 33″; ♃ 15″; ♂ 62″; ♀ 25″ et ☿ 33″.

25. Cum igitur tam pro terra quam pro reliquis planetis numerus $\frac{c}{g}$ sit tantopere magnus, diminutio excentricitatis plurimorum seculorum spatio plane erit imperceptibilis. Namque si ζ ponatur pro aequatione maxima: etiamsi aequa-

tio maxima non sit excentricitati simpliciter proportionalis, tamen in parvis differentiis haec ratio satis exacte locum habet; spatio 100 annorum aequationis maximae decrementum fit $= 314\zeta \cdot \frac{g}{c} = \frac{\zeta}{19\,000\,000}$ quam proxime. Quare cum sit $\zeta = 1^\circ\ 56'\ 10'' = 6970''$, integro seculo elapso aequatio maxima minuetur $\frac{697''}{1\,900\,000} = \frac{1''}{2720}$, ita ut demum post 2720 secula hoc decrementum ad unum minutum secundum exurgeret. Deinde etiam ob eandem rationem angulus ASB [fig. 2] quo aphelium oppositionem perihelii antevertit, omnino erit insensibilis. Tangens enim istius anguli est $= \frac{2g}{\zeta c}$, ideoque pro terra erit: $\operatorname{tag} \mathrm{ASB} = \frac{1}{2\,975\,115\,500\,\zeta}$. Est autem excentricitas $\zeta = 0{,}01692$, unde erit $\operatorname{tag} \mathrm{ASB} = \frac{1}{50\,338\,900} = 0{,}000000019865$, ideoque ipse angulus $\mathrm{ASB} = \frac{1'}{14643} = \frac{1''}{244} = \frac{1}{4}''' = 15^{\mathrm{iv}}$.

26. His igitur expeditis restaret, ut motum cometarum quoque in medio resistente expenderemus: at formula pro arcu s supra inventa tam est complicata, ut inde ad nullas certas conclusiones pervenire liceat. Quia autem pro planetis vidimus, eorum tempora periodica eo magis ceteris paribus diminui, quo eorum orbitae magis sint excentricae, hinc satis tuto colligimus, tempora periodica cometarum satis notabiliter in quavis revolutione diminui debere, ita ut mirum non sit, cometam 1682 primum post intervallum 76 annorum, postea vero post 75 annos ad perihelium rediisse: simul vero patet ob hanc perturbationem locum perihelii non mutari debuisse.

RECHERCHES PHYSIQUES SUR LA CAUSE DE LA QUEÜE DES COMETES, DE LA LUMIERE BOREALE, ET DE LA LUMIERE ZODIACALE

Commentatio 103 indicis ENESTROEMIANI
Mémoires de l'académie des sciences de Berlin [2] (1746), 1748, p. 117–140

1. Il y a beaucoup d'affinité entre les queües des Cometes et la Lumiere Boreale. En effet la queüe d'une Comete doit offrir à un spectateur placé sur sa surface dans l'hemisphere opposé au soleil, un Phénoméne presque semblable à celui de la Lumiere Boreale sur la Terre; avec cette difference que la Lumiere de la queüe des Cometes est perpetuelle, beaucoup plus forte, et environne ces Corps de toutes parts, au lieu que la Lumiere Boreale ne paroit qu'en certains tems, et pour l'ordinaire dans une certaine région seulement du Ciel. Depuis lorsque le Ciel brille à nos yeux de l'eclat d'une forte lumiere Boreale, on ne sauroit douter, que la Terre ne paroisse comme ornée d'une petite queüe du coté Septentrional, aux spectateurs qui sont placés hors d'elle, par exemple, dans la Lune. Ainsi quoiqu'outre cette direction, il y ait encore plusieurs autres differences, qui distinguent l'Aurore Boreale des Queües des Cometes, il reste pourtant une ressemblance si considerable entre ces deux Phénoménes, que nous sommes tout à fait fondés à dériver leur origine de la même cause; de sorte que si l'on sait bien la veritable cause de l'un, on ne sauroit être dans l'ignorance à l'égard de l'autre. Il est constant que le célébre Mr. DE MAIRAN[1]), qui prétend avoir trouvé la cause de la Lumiere Boréale dans la Lumiere Zodiacale, se propose d'expliquer aussi les queües des Cometes par le même principe[2]). Mais comme plusieurs Cometes paroissent avec des queües, avant que d'avoir atteint la Lumiere Zodiacale, il en naît une objection importante

1) Voir JEAN JACQUES D'ORTOUS DE MAIRAN, *Traité physique et historique de l'aurore boréale* (Paris, 1733), p. 3. EJA.

2) Loc. cit., pp. 274–275. TS.

contre cette explication même de l'Aurore Boreale; et cette difficulté jointe à plusieurs autres, qu'on peut former contre cette Hypothese, d'ailleurs extrémement ingénieuse, lui ôte beaucoup de sa vraisemblance. Je me persuade d'être en etat d'assigner une cause qui puisse satisfaire à l'explication de l'un et de l'autre de ces deux Phénoménes, et qui soit en même tems si bien liée avec les autres verités fondamentales de la Physique, qu'il ne sera presque plus permis de la revoquer en doute.

2. On se sert ordinairement de plusieurs Argumens pour prouver que les queües des Cometes, aussi bien que l'Aurore Boreale, ne sont pas de purs phantomes et des illusions de la vüe, comme l'Arc en Ciel, les Halos etc. mais qu'il existe effectivement dans les endroits où on les apperçoit des Corpuscules, d'où part cette Lumiere: quoiqu'on convienne que plusieurs Phénoménes particuliers de l'Aurore Boreale se produisent par la réflexion et par la réfraction des rayons. Etant donc certain que dans les régions où existent les Cometes, et dans le voisinage de la Terre, où les Aurores Boreales paroissent de tems en tems, il y a des Corpuscules propres à la représentation de ces Phénoménes, il naît de là une double Question; savoir, quels sont ces Corpuscules? et d'où ils ont été apportés dans ces régions? Mr. de Mairan y répond, en disant que ces Corpuscules viennent de l'Atmosphére Solaire, où il place la Lumiere Zodiacale, et sont transportés de cette Atmosphere dans les endroits, où ils produisent les Phenomenes susdits. Ce sentiment est sujet à bien des difficultés, puisqu'on ne sauroit comprendre, comment cette Atmosphere Solaire peut s'etendre à une si grande distance du Soleil, pourquoi elle ne s'etend aussi qu'autour de l'Equateur du Soleil, et comment plusieurs Cometes ont des queües, quoiqu'elles n'atteignent jamais cette Atmosphere. Il est donc beaucoup plus probable que la Lumiere Zodiacale meme procede d'une pareille cause que les Queües des Cometes et l'Aurore Boreale, en sorte que les particules qui forment la Lumiere Zodiacale émanent du Soleil, de la même maniere que les particules, auxquelles nous attribuons les Queües des Cometes et la Lumiere Boreale, procédent de ces Corps mêmes. En effet le Soleil est environné autour de son Equateur d'une lumiere qui s'etend à une distance prodigieuse, et qu'on appelle Zodiacale, les Cometes ont une lumiere sous l'apparence de Queüe dans la région à peu prés opposée au Soleil, et la Terre paroit quelquefois revetüe d'une lumiere qui se manifeste au Septentrion. Quelque difference qu'il y ait à divers egards entre ces Phénoménes, elle n'est pourtant pas telle qu'ils ne puissent etre produits par la même cause. Car la diversité des Corps du Soleil, de la Terre et des Cometes peut tellement modifier l'action d'une même cause, qu'il paroisse en resulter des effets

extrémement dissemblables. C'est pourquoi je ne balance point à déduire ces trois Phénoménes de la même cause.

3. Et d'abord pour ce qui regarde la nature des Corpuscules qui nous représentent ces Phénoménes, il faut, puisqu'ils sont visibles et qu'ils nous envoyent des rayons, qu'ils soient lumineux, ou du moins eclairés. Mais, quoique la Lumiere Zodiacale, comme venant du Soleil, semble prouver que ces particules ont une lumiere propre, on ne sauroit d'un autre coté supposer rien d'embrasé ou d'ardent dans les Queües des Cometes, puisque les Cometes elles mêmes sont des Corps opaques, et que, ce nonobstant, elles jettent plus de lumiere que leurs queües. Mais comme les Planetes et les autres Corps opaques, qui sont à une grande distance de nous, paroissent, lorsque le Soleil les éclaire, briller d'une lumiere à peu prés egale à celle des Etoiles fixes, qui sont des Corps naturellement lumineux, il semble suffisant pour expliquer les Phénoménes des Queües des Cometes, de la Lumiere Zodiacale, et de l'Aurore Boreale, de supposer qu'il y a dans ces endroits là des particules opaques, qui reçoivent la lumiere du Soleil. Ce n'est pas que je voulusse depouiller entierement ces particules de toute lumiere propre, puisqu'il peut arriver, malgré leur opacité naturelle, qu'en passant d'une Atmosphére plus épaisse dans un air plus libre, leur etat d'equilibre change de maniere à leur faire acquerir les vibrations requises pour former des rayons lumineux. Il se peut même que ces particules, detachées de l'Atmosphére par une cause que j'indiquerai dans la suite, soient d'une nature ignée, et qu'elles s'embrasent facilement d'elles mémes. De plus, comme les Etoiles sont pour l'ordinaire visible à travers ces Phénoménes, il faut que la matiere formée de l'assemblage de ces particules soit trés rare, mille fois plus et au dela que le nuage le plus delié. Car un nuage qui est transparent de prés, perd, dés qu'il s'eloigne toute sa pellucidité, parce que les Corps placés à certains intervalles les uns des autres, paroissent au Spectateur d'autant plus prés entr'eux, que leur eloignement est grand. Si donc les Queües des Cometes conservent leur transparence à une aussi enorme distance, il faut nécessairement que les particules dont elles sont composées, soient à de très grands intervalles les unes des autres. C'est de là que NEWTON à tiré cette conclusion fort vraisemblable, que toute la matiere qui forme la plus longue queüe d'une Comete, etant comprimée, ne rempliroit peut-etre pas l'espace d'un pouce cubique[3]).

3) ISAAC NEWTON, *Philosophiae Naturalis Principia Mathematica* (Londini, 1687), Liber III, propositio 41. Cette citation d'EULER est inexacte. NEWTON écrit: «Ideoque globus aeris nostri digitum unum latus, ea cum raritate quam haberet in altitudine semidiametri unius terrestris, impleret omnes Planetarum regiones ad usque sphaeram Saturni et longe ultra.» EJA.

4. Pour expliquer ces phénoménes, je dis que les rayons du Soleil peuvent chasser des Atmosphéres des Planetes les particules les plus subtiles, sur lesquelles ils agissent. Car si les rayons de lumiere partoient effectivement du Soleil, comme NEWTON le prétend, avec une vîtesse aussi grande que l'est celle que les Observations leur attribuent, il n'y auroit aucun lieu de douter qu'ils n'enlevent avec une extrême force les corpuscules contre lesquels ils heurtent. Mais si l'on établit, au lieu du mouvement veritable des rayons, une propagation de flots de lumiere à travers l'ether, que je crois avoir demontrée dans ma Theorie de la lumiere et des Couleurs[4]), de maniere que cette propagation de lumiere dans l'ether se fasse comme celle du son dans l'air, il semble plus difficile d'expliquer, comment de semblables flots peuvent enlever les particules qui voltigent dans l'Atmosphére. Cependant comme un son vehement excite non seulement un mouvement vibratoire dans les particules de l'air, mais qu'on observe encore un mouvement réel dans les petites poussieres trés legéres qui voltigent dans l'air, on ne sauroit douter que le mouvement vibratoire causé par la lumiere ne produise un semblable effet. Nous voyons en effet que les rayons du Soleil rassemblés par le miroir ardent écartent et dissipent avec une grande force les plus petits corpuscules qui sont placés au foyer; et de cette force que l'Experience nous montre dans les rayons reünis, nous sommes en droit de conclurre, que chaque rayon en a une semblable, quoique beaucoup moindre. La même conséquence découle de la nature même des rayons. Car, quoique les particules, dont le mouvement vibratoire fait la lumiere, ne s'ecartent pas sensiblement des lieux qu'elles occupent, cependant il y a quelque espace trés petit dans lequel elles se meuvent, et ce mouvement suffit pour ebranler un peu les corpuscules les plus légers, contre lesquels elles heurtent; lequel ébranlement etant continuellement repeté, il faut qu'à la fin ces corpuscules s'avancent d'un espace sensible. Il est évident que cela demande un tems considerable; et cela à proportion de l'epaisseur de ces particules, et de la résistance de la pesanteur, qui peut même etre telle qu'elle détruise entiérement cet effet. A moins donc que ces particules ou petites poussieres qui voltigent dans l'Atmosphere de quelque Planete, ne soient de la derniere tenuité, que leur force de pesanteur ne puisse etre surmontée, et qu'elles ne demeurent assez longtems exposées à l'action des rayons du Soleil, il sera difficile qu'elles puissent etre chassées à une distance considerable. Toutes ces circonstances doivent etre soigneusement remarquées, comme essentielles à l'intelligence de l'explication suivante.

---

4) E 88, «Nova theoria lucis et colorum», *Opuscula varii argumenti* 1 (1746), pp. 169–244 = *Opera omnia* III 5, pp. 1–45. EJA.

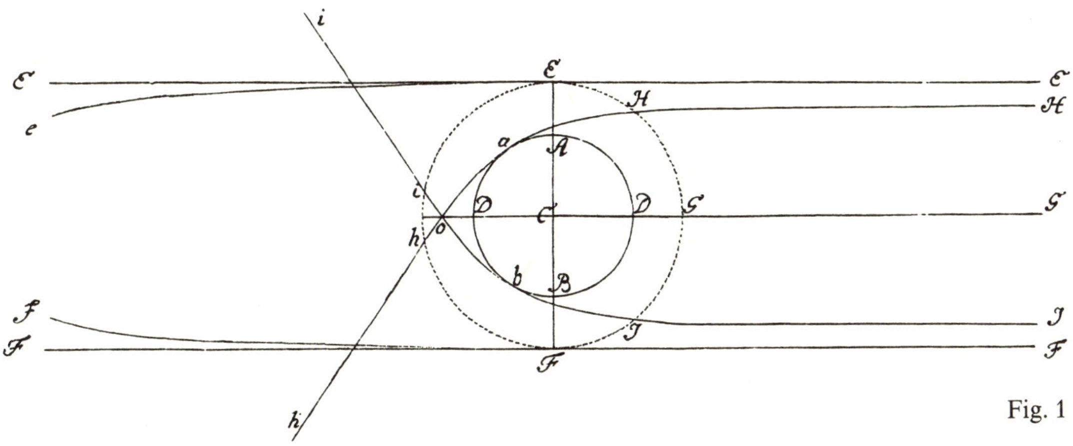

Fig. 1

5. Ayant donc établi cette force des rayons, je considererai d'abord une Comete (Fig. 1), dont le noyau ou le veritable Corps soit sphérique *aADBb*, et environné d'une Atmosphére pareillement sphérique *iEHGIFh*. Que les rayons du Soleil viennent du coté *EHGIF* suivant les directions paralléles *EEE*, *FFF*. Alors les rayons solaires *GGD*, qui tombent directement, ou qui atteignent le Corps de la Comete, en vertu de la force ci dessus prouvée, chasseront les particules subtiles, ou les especes de petites poussieres vers le Corps même de la Comete, et ne les pousseront par conséquent pas hors de son Atmosphére. Mais les rayons qui ne font que friser l'Atmosphere de la Cométe, comme *EEE*, *FFF*, enleveront avec eux les particules *E* et *F*, situées aux extremités de l'Atmosphere, et les emporteront suivant les directions *EE* et *FF*. Et comme ces particules pesent vers le Corps de la Comete, aussitot qu'elles seront chassées hors de l'Atmosphére, la double force, savoir celle de pesanteur, et celle que leur impriment les rayons du Soleil, leur fera suivre les lignes Courbes, *Ee*, *Ff*, et elles continüeront à s'eloigner ainsi toujours plus de la Comete même, puisqu'elles restent perpetuellement exposées à l'action des rayons. Il résultera un semblable effet des rayons les plus voisins de *EE*, *FF*, qui passent bien par l'Atmosphere de la Comete, mais qui ne pénetrent pas jusqu'au noyau. Les derniers rayons, qui produiront cet effet, seront *HHa*, *IIb*, qui aprés la réfraction qu'ils souffrent dans leur passage par l'Atmosphere, touchent le Corps de la Comete dans les points *a* et *b*, et souffrant alors une nouvelle réfraction, s'en vont suivant les directions *ahh*, *bii*. Elles emporteront donc avec elles suivant ces directions les particules les plus subtiles de l'Atmosphére, autant néanmoins que leur mouvement ne sera pas empeché et leur direction renduë courbe par la pesanteur. Mais on s'appercevra principalement de l'effet de la pesanteur, lorsque ces particules seront une fois poussées hors de l'Atmosphére,

parce que, tandis qu'elles y sont, leur pesanteur specifique presque égale à celle de l'Atmosphére, fait qu'elles n'y gravitent point. Cette pesanteur ne courbe pas seulement le mouvement de ces particules suivant la direction *hh*, *ii*, des rayons, mais parce qu'elles se rencontrent exposées à l'action des rayons *EE*, *FF*, et des autres rayons exterieurs qui suivent la même direction, leur mouvement se courbe continuellement d'autant plus vers ces directions *EE* et *FF*. De cette maniere donc, pendant un espace de tems suffisant, il sort une quantité considerable de ces particules hors de l'Atmosphére de la Comete, qui remplissent l'espace *EiefhF* derriere la Comete, diametralement opposé au Soleil; et cet espace à cause des rayons divergens *ahh*, *bii*, ne sera pas cylindrique, mais il représentera un Cone divergent. Cependant cette divergence, comme nous l'avons déja remarqué, est considerablement reprimée par l'action des rayons exterieurs. Lors donc que les particules repanduës dans cet espace nous sont visibles, elles offrent à nos yeux le Phénoméne, que nous avons coutume d'appeller la Queüe d'une Comete.

6. Il paroit donc par ce que nous venons de dire, que si la Comete etoit en repos et que son Corps fut rond, sa Queüe paroîtroit dans la Region directement opposée au Soleil, et qu'elle occuperoit dans le Ciel un espace un peu divergent. Pour la longueur de la Queüe, elle sera plus grande, à proportion que les rayons du Soleil auront exercé plus longtems leur action, ou que la Comete sera plus voisine du Soleil, parce qu'en ce dernier cas la force des rayons augmente considerablement. L'eclat de la Queüe sera aussi proportionné à la grandeur de l'Atmosphére de la Comete, et au nombre des particules qui auront cedé à l'action des rayons. Car plus l'Atmosphére est vaste, plus il y a de rayons qui peuvent passer à travers, sans toucher le Corps de la Comete. Or les Observations témoignent evidemment, que les Cometes sont le plus souvent environnées de trés amples Atmosphéres; ce qui confirme non seulement beaucoup l'explication de la Queüe des Cometes que nous donnons ici, mais sert encore à rendre raison d'un Phénoméne non moins remarquable. C'est que quand la Comete est placée dans une region, où l'on ne devroit voir que cette partie de son milieu, qui est opposée au Soleil, elle continüe cependant à paroître toute lumineuse, et ne revêt point les apparences des phases, qu'offrent les Planetes dans de semblables situations. Or comme on sait qu'à cause de la réfraction, il y a continuellement plus de la moitié de la Terre qui est illuminée, et que d'un autre coté la réfraction est d'autant plus grande qu'une Atmosphére est plus vaste et plus épaisse, il est clair que le Soleil doit illuminer beaucoup au delà de la moitié des Cometes. Ainsi dans notre figure la portion illuminée est representée par l'arc *aADBb*, qui est considerablement plus grand que la demi-

circonférence du Cercle; et il paroit probable que des portions encore beaucoup plus grandes des Cometes sont illuminées. Et même l'Atmosphére pourroit etre si etendüe, et douée d'une si grande force de réfraction, que toute la moitié opposée au soleil fut eclairée, de sorte que la Comete paroitroit pleine en toute situation. Mais quand même quelque partie, comme *ab*, ne recevroit point de rayons, cependant à cause de l'extreme convergence des rayons des extremités *ao*, *bo*, le cone d'ombre placé derriere la Comete devient si petit, que si par hazard il est tourné vers nous, ce qui ne peut arriver que rarement, on peut à peine l'appercevoir. Car ce cone *aob* etant environné de toutes parts d'une matiere eclairée, les tenebres qui y sont renfermées ne sauroient se montrer. La partie *ab* reçoit elle-même quelque clarté de la lumiere de l'Atmosphére, de sorte qu'il n'y a aucun endroit de la surface de la Comete, qui soit entierement destitué de lumiere.

7. Voyons à présent quel changement le mouvement de la Comete peut apporter dans la formation de la queüe; et d'abord supposons une Comete arretée dans la meme place du Ciel, où elle n'ait qu'un mouvement de rotation par lequel elle tourne autour de quelque Axe. On comprend aisément, que si l'Axe, autour duquel la Comete tourne, est dirigé vers le soleil, en sorte que la droite *DCD* represente cet Axe, ce mouvement ne sauroit troubler sensiblement la génération de la queüe. Car, lorsque la Comete tourne autour de son axe *DD*, les particules *E* et *F*, qui sont exposée à l'action des rayons, n'y sont point soustraites par le mouvement gyratoire, mais elles conservent perpetuellement à l'egard des rayons la même situation que si ce mouvement gyratoire n'existoit point; d'où s'ensuit que les rayons du soleil produiront le même effet, et enleveront avec eux les particules les plus subtiles de l'Atmosphére, suivant la même direction que dans le cas précedent. Toute la difference qui peut se rencontrer, c'est que la matiere, dont la queüe est formée, acquiere une sorte de mouvement gyratoire, qui en partie sera tout a fait lent, et en partie ne changera rien à l'aspect de la queüe. Mais si l'axe, autour duquel la Comete tourne est perpendiculaire à la droite *CG*[5]) tirée au soleil, comme est la ligne *AB*; en sorte que le Soleil soit perpetuellement visible dans l'Equateur de cette comete, les particules que les rayons du Soleil avoient mises en mouvement se soustrairont bientot à cause du mouvement de rotation à l'action de ces rayons, et cela d'autant plus promtement qu'elles seront plus voisines de l'Equateur. A moins donc que le mouvement de rotation ne soit extraordinairement lent, en sorte que le tems, pendant lequel ces particules soutiennent l'action des rayons, suffise pour

5) Edition originale: *CI* au lieu de *CG*. Corrigé par EJA.

les ecarter à une distance assez considerable, avant qu'elles s'ecartent, la formation de la queüe sera fortement empechée dans cette région. Mais plus nous approcherons des Poles de la Comete A et B, plus longtems ces particules demeureront soumises à l'action des rayons, tant à cause de la lenteur du mouvement de rotation, que de la grande réfraction des rayons; et ainsi les particules placées aux Poles mêmes soutiendront perpetuellement cette action. Dans ce cas par conséquent les rayons du soleil produiront autour des Poles de la Comete le même effet que s'il n'y avoit point de mouvement de rotation; mais il se détachera beaucoup moins de particules des autres régions de l'Atmosphére, et elles s'ecarteront à un beaucoup moindre intervalle; ce qui fera paroitre la queüe de la Comete fort foible. De ces deux cas qui sont les extremes opposés, il est aisé de conclurre, comment et combien le mouvement de rotation préjudicie à la formation de la queüe, en supposant l'axe incliné à la droite CG[6]) sous un angle oblique quelconque.

8. Comme le mouvement gyratoire de la Comete change en partie la grandeur, et en partie la figure de la queüe, de même son mouvement vrai dans l'orbite qu'elle décrit autour du soleil courbe un peu la direction de la queüe. En effet, soit (Fig. 2) $ABCDE$ l'orbite de la comete, qui ait son foyer dans le soleil S. Si la

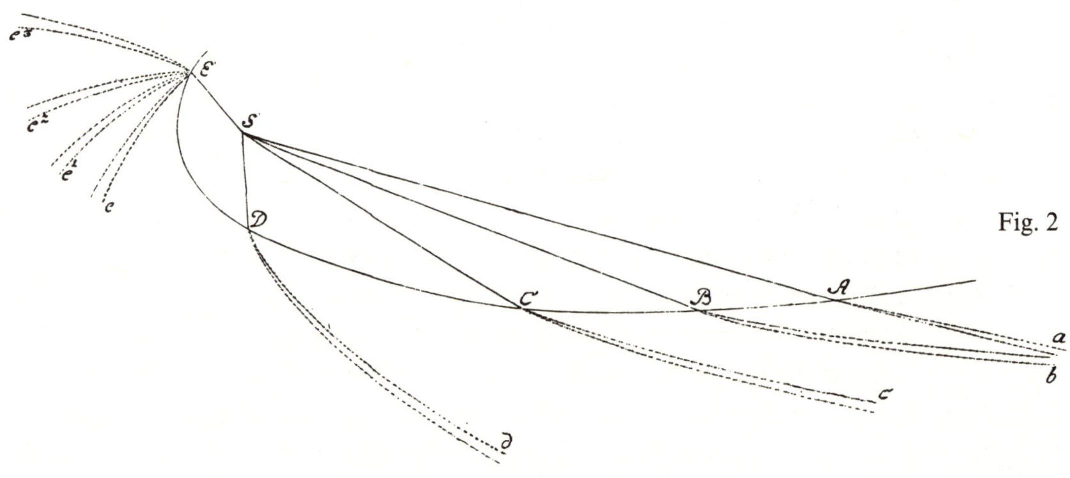

Fig. 2

Comete s'arretoit, sa queüe formée par les rayons du soleil Aa auroit une direction opposée à la droite SA. Si nous supposons à présent que la Comete avance dans son orbite suivant AB; pendant ce tems là de nouvelles particules seront continuellement chassées de son Atmosphére, suivant la direction des rayons du Soleil; laquelle changeant continuellement, et l'action par laquelle ces particules sont

6) Edition originale: CI au lieu de CG. Corrigé par EJA.

chassées n'etant pas trop rapide, les parties de la queüe les plus eloignées de la Comete ne suivront pas aussi bien son mouvement que celles qui en sont les plus proches. Ainsi dans la situation de la Comete B la queüe Bb ne sera pas placée suivant la droite SB prolongée, mais elle s'en écartera vers a, et se courbera tant soit peu; de maniere que ses extremités a et b se reüniront, ou du moins ne seront pas fort eloignées l'une de l'autre. Alors la Comete s'avançant toujours vers C, toute la queüe, qui est continuellement réparée et augmentée par les nouvelles particules qui y sont chassées, suivra à la verité le mouvement de la Comete, mais de maniere cependant que les parties les plus eloignées s'y conformeront moins; d'où il arrivera que la queüe de la Comete aura perpetuellement un peu de courbure, et s'ecartera de la direction du Soleil SC ou SD; déclinaison qui sera d'autant plus grande, que la Comete sera müe avec plus de rapidité, et que son orbite sera plus courbe. Cela est non seulement si conforme aux Loix du mouvement, que cette déclinaison et cette courbure de la Queüe pourroit etre exactement déterminée par le calcul, si l'on connoissoit exactement la vîtesse, avec laquelle les particules qui forment la queüe sont chassées; mais encore toutes les Observations confirment la même chose au sujet de ce Phénoméne. Mais quand la Comete passe par le Perihélie, où non seulement son mouvement est le plus rapide, mais où son orbite a aussi le plus de courbure, la direction de la queüe doit alors s'ecarter beaucoup de l'opposition du Soleil. Il peut même arriver, si la Comete parcourt avec trop de vitesse cette région, que l'extreme force des rayons du Soleil engendre une nouvelle queüe, avant que la premiere puisse suivre; et dans ce cas la Comete paroitra avoir en E deux ou plusieurs queües Ee, Ee^1, Ee^2, Ee^3. C'est ce Phénoméne meme qui a été observé dans la Comete de 1744 à laquelle on apperçut plusieurs queües, tant qu'elle fut autour de son Perihélie. Cette Comete fut pendant quelque tems si voisine du Soleil, que pendant une partie de ce tems il auroit pu s'engendrer une nouvelle queüe, quand même elle n'en auroit point eu auparavant; mais comme elle en avoit déja une, et que cette nouvelle production fut quelquefois reïterée, la cause de cette pluralité de queües qu'on y observa, est assez manifeste.

9. Jusqu'à présent nous avons supposé le corps de la Comete parfaitement rond; mais on comprend aisément que la formation de la queüe dépend beaucoup de la figure de la Comete. Supposons donc (Fig. 3) que le corps de la Comete $aADB$ soit oblong par rapport au Soleil; ce qui fait que les rayons tombent suivant la direction GC dans la situation quelconque exprimée par la figure; car la diversité de la situation ne contribuë pas peu à la figure de la queüe. Mais, pour ne pas me jetter dans de trop grandes longueurs, il me suffira de considerer une situation

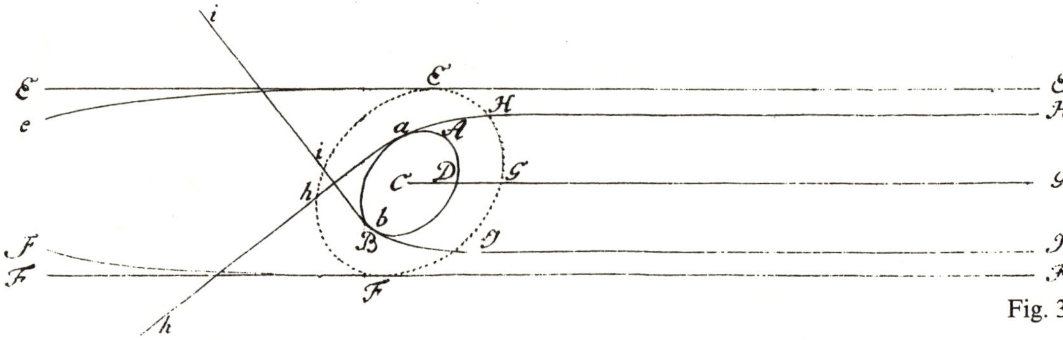

Fig. 3

unique, de laquelle on pourra aisément déduire les variations qui peuvent naître de toute autre. Et d'abord les rayons extremes *EE*, *FF*, en rasant l'Atmosphére de la Comete, en emportent des particules suivant la direction *Ee*, *Ff*, si l'on a en même tems egard à la pesanteur. Considerons ensuite les rayons *HH* et *II*, qui etant entrés dans l'Atmosphére y souffrent une réfraction, par laquelle ils touchent le corps même de la Comete en *a* et *b*. Donc ces rayons chasseront des particules de l'Atmosphére suivant les directions *ahh*, *bii*, dont l'une sera plus, l'autre moins inclinée à la direction *GG*[7]); et dans l'etat que la Figure exprime, la gravité fera encore plus décliner de la direction *GG* les particules *ii* que les particules *hh*. Ainsi quoique la force des rayons externes, dont les particules *hh* éprouvent l'action, soit peut-etre suffisante pour les pousser suivant la direction *FF*, il pourroit cependant arriver que la même force ne fut pas capable de contraindre les particules *ii* à garder la direction *EE*. Dans ce cas ces particules *ii* formeront une queüe particuliere moindre, dont la direction differera tantot plus, tantot moins, de celle de la queüe principale *EFef*, suivant que la situation de la figure ovale *AB* changera par rapport au Soleil. Nous découvrons par ce moyen la cause de ces queües fourchuës, qui ont été souvent observées, et en particulier dans la Comete de l'année 1744 à l'egard de laquelle les Observations ont fait voir que son Corps s'eloignoit extrêmement de la figure sphérique, puisque son plus grand axe etoit double de son moindre. Que s'il arrive une si grande déclinaison aux particules *hh*, que l'action des rayons exterieurs ne puisse les ramener à la direction, alors la queüe paroîtra fenduë en trois, ou bien la queüe principale aura deux branches, qui seront divergentes de part et d'autre. La même chose peut aussi arriver quand la Comete a une figure sphérique; car si (Fig. 1) l'obliquité des rayons *hh* et *ii* est si grande, que la direction des particules qu'ils chassent, ne puisse etre corrigée par le reste des

7) Edition originale: *Gg* au lieu de *GG*. Corrigé par EJA.

rayons, la queüe aura une forte divergence, tout prés du Corps de la Comete, et suivant la situation du Spectateur, pourra paroître separée en trois. Au moins en général la queüe de la Comete doit montrer auprés du Corps tant soit peu de divergence.

10. Aprés avoir ainsi exposé la cause, qui produit la queüe des Cometes, et avoir montré qu'elle s'accorde parfaitement avec tous les Phénoménes, de sorte qu'il n'est presque plus permis de douter de sa verité, voyons ce que la force des rayons du Soleil est capable de produire dans l'Atmosphére des Planetes. Et d'abord plusieurs s'etonneront peut-etre que les Planetes n'ayent point de queües, vû que leurs Atmosphéres sont constamment exposées à l'action du Soleil, et que les Cometes pour l'ordinaire ne conservent la leur que pendant un court espace de tems. Mais il y a une grande difference entre l'Atmosphére des Cometes et celle des Planetes. Les Cometes ont des Atmosphéres tres vastes, dont le diametre est souvent six fois, et au delà, plus grand que le diametre du noyau ou du vrai Corps de la Comete, au lieu qu'au contraire les Atmosphéres des Planetes excedent à peine les Corps de ces Astres de la centieme, ou même de la millieme partie. Car la plupart des Observations semblent confirmer que l'Atmosphére de la Terre ne s'etend presque pas au delà d'un mille d'Allemagne, quoique son diametre ait environ deux mille de ces milles. Dans les autres Planetes on remarque à peine la moindre trace d'Atmosphére. Ainsi les Atmosphéres des Planetes etant si subtiles et si petites qu'on peut les compter pour nulles au prix des Atmosphéres des Cometes, il n'est pas surprenant qu'il ne s'y engendre aucune queüe. L'extreme petitesse de l'Atmosphére des Planetes est encore cause, qu'aussi-tot que quelque particule en a été chassée, la pesanteur à cause du voisinage de la Planete est assez forte pour la ramener. Au contraire dans les Cometes, dont les Atmosphéres s'etendent si loin, la pesanteur des particules, lorsqu'elles commencent à s'en détacher, est déja si diminuée qu'elle ne sauroit empêcher l'effet des rayons, en supposant que la pesanteur vers le Corps des Cometes décroit aussi en raison doublée des distances de leur centre. Ajoutons qu'il passe fort peu de rayons par les Atmosphéres des Planetes, au lieu qu'il y en a beaucoup qui traversent les Atmosphéres des Cometes. Enfin ce qui met le plus d'obstacle à la génération des queües par rapport aux Planetes, c'est leur mouvement de rotation, qui empêche que la même particule puisse demeurer assez longtems exposée à l'action des rayons, à laquelle le mouvement en question la soustrait aussi tot. Toutes ces circonstances étant contraires à la production des queües, devoient nécessairement en priver les Planetes; et c'est par cet endroit principalement qu'elles different des Cometes.

11. Neanmoins, lorsque les rayons du Soleil passent par l'Atmosphére de quelque Planete, la force que nous avons démontré résider en eux, ne doit pas etre censée oisive et sans effet. Si elle ne sauroit y produire de ces Phénoménes qui s'appercoivent de loin, elle sera peut-etre propre à en faire naître de visibles dans une plus grande proximité. Voyons de quelle nature ils pourroient etre. Les particules qui voltigent dans l'air qui nous environne sont soumises à cette action des rayons du Soleil, quand nous voyons cet Astre à l'Horizon. Au lever du Soleil elles sont poussées vers l'Occident, et à son coucher vers l'Orient; et cette impulsion se faisant suivant la direction des rayons, elles doivent dans l'un et dans l'autre cas etre chassées de l'Atmosphére, et s'eloigner d'autant plus de la terre, qu'elles auront été plus longtems exposées à cette action des rayons; mais bientot, au moins dans nos contrées, ces particules doivent se soustraire à la force qui agit sur elles, à cause du mouvement diurne de la Terre, et elles retombent dans l'Atmosphére. Mais quand, vers le lever du Soleil, plusieurs de ces particules seront chassées du coté de l'Occident, et qu'elles s'elevent au dessus de la région de l'air épais, elles deviendront visibles aux habitans de la terre, qui sont Occidentaux à notre egard, et pour qui le Soleil n'est pas encore levé, et elles leur apporteront une espece de point du jour. De même au coucher du Soleil les particules seront chassées de nôtre air vers les contrées, pour lesquelles le Soleil est déja couché, et y produiront le crepuscule. Je ne crains point même d'affirmer que c'est là la veritable cause du crepuscule tant du matin que du soir. Car le terme du crepuscule arrivant, lorsque le soleil est plongé de 18° sous l'horizon, la réfraction des rayons ni la hauteur de l'Atmosphére déterminée par d'autres Phénoménes, ne sauroient expliquer cette clarté, car il faudroit pour cet effet augmenter la hauteur de l'Atmosphere jusqu'à 30 milles, élévation qui répugne à tout le reste des Phénoménes. Au lieu qu'en admettant l'action des rayons du Soleil qui a deja été assez démontrée par les queües des Cometes, et que la Lumiere Boreale prouvera bientot plus abondamment par rapport à la Terre, il est manifeste que cette lumiere dans la région du Ciel qui est située au dessus de nôtre air épais, doit préceder le lever du soleil, et suivre son coucher, et qu'à cause de sa grande distance de la terre, elle doit paroitre avec assez de force pour produire les Phénoménes du Crépuscule. Mais je ne m'arrête pas plus longtems à cet effet, et je passe à la cause de l'Aurore Boreale, qui paroit beaucoup plus cachée.

12. Tout-ce que nous avons dit jusqu'ici montre evidemment que les particules subtiles sont chassées de l'Atmosphére d'autant plus loin, que le Soleil demeure plus longtems prés de l'horizon, c'est à dire, qu'il s'eleve moins subitement au

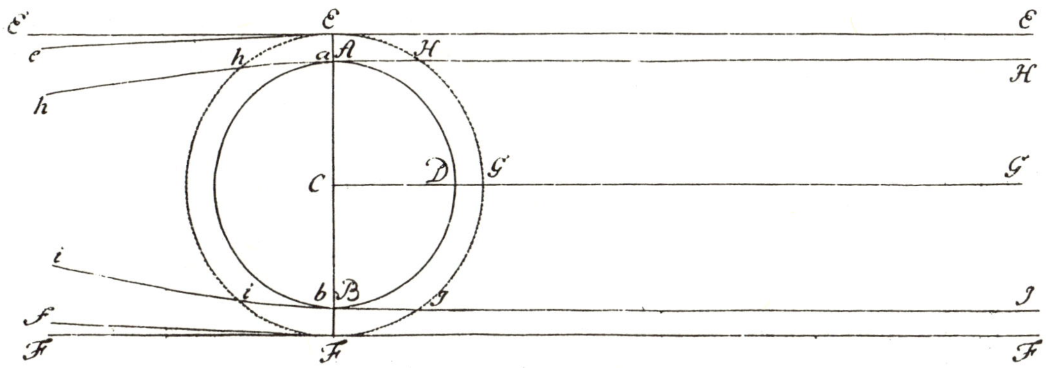

Fig. 4

dessus, ou s'enfonce au dessous. Par cette raison l'effet dont il s'agit doit etre beaucoup moindre dans les lieux de la Terre situés prés de l'Equateur que dans les contrées qui en sont plus eloignées. Autour des Poles donc de la Terre, où le Soleil pendant plusieurs jours consécutifs, est visible prés de l'horizon, cet effet doit etre trés grand, et chasser les particules subtiles à une grande distance de la Terre. Soit, par exemple, (Fig. 4) AB l'axe de la Terre, et A et B ses Poles; dans le tems des Equinoxes, où les rayons du Soleil GG tombent perpendiculairement sur l'Equateur D, le Soleil sera visible pendant quelque tems sous les Poles mêmes à l'Horizon, ou les rayons du Soleil EE et FF toucheront l'Atmosphére de la Terre dans les points E et F. Ils en chasseront donc les particules les plus subtiles suivant les directions EE, FF[8]), et à cause de la pesanteur elles s'ecarteront tant soit peu vers Ee et Ff[9]). Les rayons plus voisins de la Terre produiront un effet semblable, jusqu'à ce qu'ils touchent la Terre même dans les points a et b; tels sont les rayons $HHahh$, $IIbii$. Par conséquent à l'un et à l'autre une region fort etenduë au dessus de l'Atmosphére $Eehh$ et $Ffii$, se remplira de ces particules chassées, qui etant illuminées par le Soleil, seront visibles la nuit assez loin des Poles, et nous représenteront le Phénoméne, que nous appellons Aurore Boreale. En effet cette clarté du Ciel se montre d'abord vers les Poles, et ensuite elle est fort fréquente autour des Equinoxes; circonstances qui conviennent parfaitement à nôtre explication. Cela n'exclut pourtant pas entierement les tems plus eloignés des Equinoxes, puisque le Soleil, lorsqu'il éclaire l'autre Zone froide, exerce assez longtems son action sur l'Atmosphére, quoique le mouvement diurne rende cet effet beaucoup moindre.

8) Edition originale: Ee, Ff au lieu de EE, FF. Corrigé par EJA.
9) Edition originale: EE, FF au lieu de Ee, Ff. Corrigé par EJA.

L'expulsion des particules hors de l'Atmosphére de la Terre dépend principalement de l'etat de l'air dans les Zones froides; car quand le Ciel est serain dans ces régions, et que les rayons du Soleil ont un libre passage à travers l'Atmosphére, un beaucoup plus grand nombre de particules sont emportées, et à une distance plus considerable de la Terre, que si le Ciel etoit couvert, et que la plupart des rayons fussent absorbés par les vapeurs. L'on comprend par là que ce Phénoméne est fort variable, et qu'il n'est astreint à aucunes Loix, par lesquelles on puisse assigner sa quantité ou prédire son apparition.

13. Toutes les Observations qu'on a faites sur les Aurores Boreales, nous enseignent incontestablement, que la matiere, dont la lumiére produit ces Phénoménes, n'existe point dans nôtre Atmosphére; mais qu'elle est extremement eloignée de nous. Car quoique sa vraye distance ne puisse pas etre déterminée par les Observations, cependant l'eloignement des lieux, dans lesquels on voit souvent le même Phénoméne à la fois, met en droit de conclurre qu'il est placé à une trés grande distance de la surface de la Terre. Mais l'explication que nous donnons ici confirme manifestement la même chose; car les particules les plus subtiles etant, comme nous l'avons vû, poussées à une distance de la Terre d'environ 30 milles, dans le tems du point du jour et du crépuscule, quoiqu'elles ne demeurent pas à peine exposées une heure à l'action des rayons du Soleil; il est aisé de s'appercevoir que dans le voisinage des Poles, où cette action dure plusieurs jours de suite, de semblables particules doivent etre emportées à quelques milliers de milles de la Terre, en sorte que la hauteur de la Colomne illuminée *AEeh* ou *BFfi* peut quelque fois surpasser le diametre entier de la Terre. Que si la chose arrive, l'Aurore Boreale devient visible dans les contrées de la Terre, qui sont de part et d'autre à 60 degrés du Pole. Mais comme un aussi grand effet est trés rare à cause des raisons ci dessus alleguées, le spectacle de l'Aurore Boreale est peu familier aux païs qui sont à une distance considerable des Poles; mais dans ceux, qui en sont plus voisins, ce Phénoméne doit etre beaucoup plus fréquent, puisque les particules y sont visibles, sans avoir besoin d'etre poussées à un trop grand éloignement de la terre, et que l'action du Soleil pendant un jour, ou même pendant quelques heures, est suffisante pour les elever à la hauteur nécessaire. C'est pour cela qu'en Laponie et dans les autres païs de la Zone froide, on voit si souvent des Aurores Boreales, mais quand elles se manifestent dans nos contrées, nous pouvons en conclurre avec assurance que le Ciel a été sérein pendant un tems asses considerable au Pole, et que les rayons du Soleil ont pu y déployer leur force sans aucun obstacle. Par rapport à la situation de ces Aurores dans le Ciel, ce que nous avons dit montre clairement,

que la plus grande force de l'Aurore Boreale, ou son milieu, devroit etre vuë dans la région opposée au Soleil, si la Terre etoit privée du mouvement diurne. Mais quoique la matiere chassée hors de l'Atmosphére suive l'impression de ce mouvement diurne, c'est pourtant avec un peu plus de lenteur, d'où il arrive que le milieu de chaque Aurore Boreale devance un peu le lieu opposé au Soleil; ce qui s'accorde assez exactement avec les Observations qu'on a faites sur la déclinaison de ce Phénoméne par rapport au vrai septentrion.

14. Au reste, comme les particules chassées de l'Atmosphére de la Terre, dont la lumiere produit l'Aurore Boreale, ne sont pas [en] repos, et qu'au contraire elles sont agitées d'un mouvement perpetuel, le spectacle qui en résulte n'est pas tranquille, et l'on y remarque un ébranlement universel. De plus comme les rayons qui en viennent jusqu'a nous traversent l'Atmosphére, où les particules les plus grossieres qui voltigent dans nôtre air leur font souffrir plusieurs refractions et refléxions, il est manifeste que les apparences des Aurores Boreales doivent etre sujettes à de grands dérangemens. On comprend donc par là la cause générale de l'extreme varieté qui accompagne ces Phénoménes, quoique nous ne soyions pas en etat de rendre raison des singularités de chacun d'eux pris à part. Mais il se présente une Objection, qui n'est pas peu importante. Puisque de pareilles clartés dans le Ciel ne sont pas plus propres au Pole Septentrional qu'au Pole Austral, d'où vient que nous n'apprenons pas qu'on ait jamais observé d'Aurores Australes dans l'Hémisphere austral de la Terre? Je n'aurai pourtant pas de peine à lever ce doute. Car premierement, le froid excessif qui régne autour du Pole Austral, n'a pas encore permis d'en approcher au delà du 60me degré, et encore ne peut-on parvenir jusques là que presque au milieu de l'Eté, où il n'y a point d'Aurores Boreales. Ensuite dans les regions plus distantes du Pole Austral, qui sont perpetuellement habitées, de semblables Aurores Australes deviennent déja beaucoup plus rares, et au cas qu'il en paroisse quelquefois, peut-etre ne les remarque-t-on pas, ou du moins ne nous en communique-t-on pas les Observations. Une autre Objection contre cette Theorie semble pouvoir etre tirée du reste des Planetes, autour desquelles on ne voit jamais cette lumiere vers les Poles, que les particules chassées de leur Atmosphére devroient y montrer. Mais si nous reflêchissons seulement sur la distance, nous la trouverons d'une telle grandeur, qu'elle ne nous permet pas d'appercevoir une lumiere aussi foible. La Lune au moins, dira-t-on, qui est si voisine de nous, devroit nous présenter ce Phénoméne, et même il devroit y etre produit avec beaucoup de plus de force que sur la Terre à cause de l'extreme lenteur du mouvement de vertige de la Lune. Mais comme la Lune est ou entierement destituée d'Atmosphére, ou du

moins qu'elle n'en a qu'une trés mince, il n'est pas surprenant que ce Phénoméne n'existe pas autour de la Lune, ou que sa subtilité le rende imperceptible à notre vüe.

15. S'il y a autour du Soleil de semblables particules, sur lesquelles les rayons puissent exercer leur force, il en devra résulter un Phénoméne tout à fait constant, vu que l'action des rayons seroit perpetuelle, et sans aucun changement. Or les taches du Soleil, qu'on remarque etre placées à une distance assez considérable de sa surface, donnent lieu de conclurre que le Corps du Soleil est environné d'une Atmosphére, et je ne vois aucun sujet de douter qu'elle ne soit abondamment remplie de particules trés subtiles. Mais comme la pesanteur est trés grande dans le voisinage du Soleil, et qu'elle surpasse sans doute de beaucoup la force que les rayons exercent sur ces particules, leur expulsion seroit entierement arrêtée par là, sans le mouvement gyratoire du Soleil. C'est pourquoi prés des poles du Soleil les particules ne cederont point à l'impulsion, parce que la force centrifuge y evanouït entierement. Mais autour de l'Equateur du Soleil, où la force centrifuge est la plus grande, quoique la force des rayons soit encore bien moindre que la pesanteur, elle suffira neanmoins pour dilater considerablement la figure de l'Atmosphére Solaire. Car, sans la force des rayons, la pesanteur et la force centrifuge reünies donneroient une figure determinée et constante à l'Atmosphére du Soleil, et cette figure s'eloigneroit d'autant plus de la Sphérique, que la raison de la force centrifuge à celle de pesanteur seroit plus grande; au lieu que cette raison etant trés petite, la figure de l'Atmosphére seroit la plus prochaine de la sphérique. Mais si l'on a egard à l'action des rayons, dont la force est contraire à la pesanteur, et décroit comme elle en raison doublée des distances du Soleil, il en résulte une diminution de la pesanteur, et l'Atmosphére prend la même figure qu'elle auroit, si la pesanteur etoit beaucoup moindre, la force centrifuge ne souffrant aucun changement. Tout cela montre clairement que l'amplitude de l'Atmosphére doit s'augmenter considérablement autour de l'Equateur du Soleil; mais qu'autour des Poles elle doit à peine être plus grande, que si la force des rayons etoit nulle. Le Corps du Soleil sera donc environné d'une Atmosphére, dont la figure sphéroïdique sera fort applatie vers les Poles, et fort etendüe autour de l'Equateur; précisement comme Mrs Cassini[10]) et de Mairan[11]) représentent l'Atmosphére Solaire, dans laquelle ils placent la

10) Gian Domenico Cassini, *Mémoires de l'Académie Royale des Sciences depuis 1666 jusqu'à 1699* (Paris, 1730), tome 8, pp. 121–209. EJA.

11) J. J. Dortous de Mairan, *Traité physique et historique de l'aurore boréale* (Paris, 1733). EJA.

Lumiere Zodiacale. Ainsi il est extrémement vraisemblable que cette Lumiere Zodiacale n'est autre chose que le Phénoméne offert par la vüe de l'Atmosphére Solaire fort etendüe autour de l'Equateur; et cela est egalement confirmé par la figure et par la situation de ce Phénoméne.

16. Mais pour mettre dans un plus grand jour, combien la diminution de la pesanteur peut augmenter l'etendüe de l'Atmosphére Solaire autour de l'Equateur, faisons un calcul fondé sur les principes de l'Hydrostatique. Soit donc (Fig. 5) C

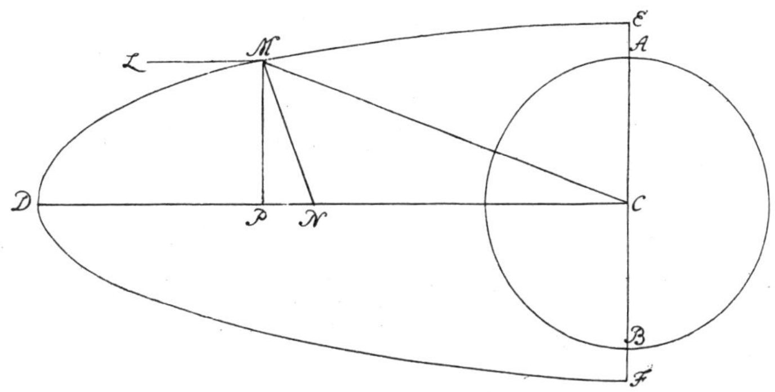

Fig. 5

le centre du Soleil, et AB son axe, autour duquel il tourne environ en 27 jours. Que EDF represente la section de l'Atmosphére faite par l'Axe du Soleil, dont nous recherchons la figure; laquelle etant arrivée à un etat permanent, il faut nécessairement que la direction moyenne des forces par lesquelles chaque particule extreme M est sollicitée, soit perpendiculaire à la surface de l'Atmosphére MN. Qu'on tire de M à l'axe CD la normale MP, et qu'on appelle $CP = x$, $PM = y$, et $CM = \sqrt{(xx + yy)} = z$. Que $\frac{ff}{zz}$ exprime la pesanteur de la particule M, par laquelle elle est pressée vers C; et que la force des rayons, qui écarte cette même particule du Soleil soit $= \frac{kk}{zz}$, en sorte qu'à présent la particule M dans la direction MC soit sollicitée par une force $= \frac{ff - kk}{zz}$. Ensuite, à cause du mouvement gyratoire, la particule M aura une force centrifuge proportionnelle à sa distance de l'Axe AB, laquelle soit $= \frac{x}{g}$ suivant la direction ML paralléle à CD. Or la normale MN etant la direction moyenne des forces MC et ML, il sera

$$CM : CN = \frac{ff - kk}{zz} : \frac{x}{g}.$$

Mais à cause de

$$PN = \frac{-y\,dy}{dx},$$

CN sera

$$= \frac{x\,dx + y\,dy}{dx} = \frac{z\,dz}{dx};$$

d'où résulte

$$z : \frac{z\,dz}{dx} = dx : dz = \frac{ff - kk}{zz} : \frac{x}{g},$$

ou bien

$$\frac{x\,dx}{g} = \frac{(ff - kk)\,dz}{zz};$$

laquelle equation etant integrée donne

$$\frac{xx}{2g} = c - \frac{ff - kk}{z}.$$

Mais si $x = 0$, CM deviendra $= CE$. Soit donc $CE = b$ et c sera

$$= \frac{ff - kk}{b},$$

d'où

$$xx = \frac{2g(ff - kk)(z - b)}{bz}.$$

A present pour trouver la plus grande amplitude CD, qu'on fasse $z = x$, et l'on aura $bx^3 = 2g\,(ff - kk)(x - b)$; equation cubique, dont la racine x donnera l'amplitude CD. Mais si cette équation a une racine affirmative, comme cela doit arriver dans le cas actuel, elle aura aussi nécessairement trois racines réelles, et alors il pourroit arriver que l'Atmosphére se changeât en Anneau, et environnât le Soleil, comme l'Anneau de Saturne entoure cette Planete. Les Observations ne permettent pas de décider, si la Lumiere Zodiacale est contigüe au Soleil, ou placée à quelque distance de cet Astre en forme d'anneau. C'est pourquoi il suffira d'avoir proposé ces conjectures jusque'à ce que l'experience permette de déterminer quelque chose de plus certain.

SUR L'ATMOSPHERE DE LA LUNE PROUVÉE PAR LA DERNIERE ECLIPSE ANNULAIRE DU SOLEIL

Commentatio 142 indicis ENESTROEMIANI
Mémoires de l'académie des sciences de Berlin [4] (1748), 1750, p. 103–121

1. En observant les momens de l'Eclipse du Soleil que nous eûmes ici le 25 Juillet 1748 et en rendant compte de mes observations[1]), je n'avois eu pour but que d'arriver à une détermination plus exacte du veritable mouvement de la Lune et de sa parallaxe. Mais il n'a pas laissé de se présenter dans le cours de l'Observation de cette Eclipse quelques autres Phénomenes remarquables, qui ne dépendoient, ni du mouvement de la Lune, ni de sa parallaxe, mais qui sembloient donner à connoitre la réalité de la réfraction des rayons qui rasent les bords de la Lune, et décider cette question agitée depuis long tems parmi les Astronomes; Si la Lune est environnée d'une Atmosphere, ou non? C'est ce qui m'engage à éxaminer ici plus exactement les phénomenes de cette espece, que j'ai observés pendant cette Eclipse, et à en rechercher les causes.

2. Pour observer cette Eclipse avec plus de succés, et soumettre à la mesure tout ce qu'elle offriroit de remarquable, j'avois préparé dans ma maison une chambre obscure, qui regardoit le Midi, et ayant dirigé vers le Soleil, par un trou pratiqué à la fenêtre, un tube Astronomique de 9 pieds, je reçus l'image de cet astre sur un papier blanc. J'affermis ce papier perpendiculairement à l'axe de la Lunette à une distance, telle que l'image du Soleil remplit exactement un cercle qui y etoit tracé, et je tirai ce tube jusqu'à ce que l'image fut representée de la maniere la plus distincte sur le papier, et qu'on put discerner clairement toutes les taches du Soleil, dont plusieurs etoient visibles sur son disque. La Machine etoit construite de telle

1) E 117, «Réflexions sur la derniere eclipse du Soleil du 25 juillet a. 1748», *Mémoires de l'Académie Royale des Sciences de Berlin* [3] (1747), 1749, 250–273 = *Opera omnia* II 30, pp. 51–72. EJA.

sorte, que tandis que le tube suivoit continuellement le mouvement du Soleil, le papier par un mouvement semblable, conservoit toujours la même distance à l'egard du tube, de sorte que l'image du Soleil demeuroit constamment dans le cercle tracé sur le papier.

3. La Machine étant ainsi montée, et l'image du Soleil s'y montrant tout de suite aux yeux, j'attendis l'arrivée de l'Eclipse, dont de petits nuages qui couvroient fréquemment le Soleil ne permirent pas d'observer le commencement. Il restoit même peu d'esperance d'observer les phases suivantes de l'Eclipse, le Ciel se couvrant de plus en plus de nuages. Cependant contre notre attente on put fort bien observer les principales phases, et sur tout l'anneau pendant toute sa durée. Comme les momens en ont été déterminés avec beaucoup de précision par M. KIES[2]), et qu'il en a fait son rapport à l'Académie, je ne les repeterai pas ici, me bornant uniquement aux choses qui se rapportent au but de ce Mémoire.

4. La Lune étant déjà entrée dans le disque du Soleil au delà de la moitié, de sorte que la figure du Soleil paroissoit déjà semblable à la Lune vers ses quadratures[3]), et que l'angle qui fermoit ses cornes, devenoit fort aigu, je remarquai

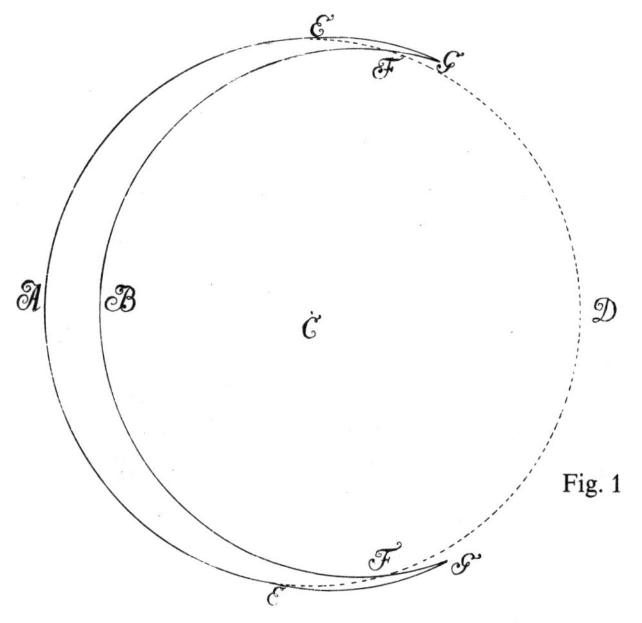

Fig. 1

2) *Mémoires de l'Académie Royale des Sciences de Berlin* [4] (1748), 1750, 99–102. EJA.

3) Edition latine: solisque figura iam admodum falcata (entièrement sous la forme d'une faucille) *au lieu de* semblable à la Lune vers ses quadratures. EJA.

premierement que le disque du Soleil ne demeuroit plus compris dans le cercle tracé sur notre papier, mais que les pointes des cornes en sortoient, quoique le bord du Soleil le plus eloigné de ces pointes demeurât cependant toujours dans les limites exactes du cercle. Ce Phénomene se présentoit tel qu'on le voit Fig. 1 où *AEDEA* est le cercle tracé sur le papier, et *GAGBG* la figure du Soleil éclipsé, dont les pointes *G*, *G*, s'etendoient de part et d'autre au delà du cercle, en sorte que les petites portions *EFG* débordoient, tandis que le reste du bord *EAE* étoit encore dans une exacte congruence.

5. Ces pointes *G*, *G*, continuoient à déborder de plus en plus hors du cercle, à mesure que les angles *G* et *G* des pointes devenoient plus aigus, jusqu'à ce qu'enfin ces pointes se reünirent, et le Soleil se montra sous la forme annulaire, où son disque forma sur le papier un cercle beaucoup plus grand, que celui dans lequel il avoit été d'abord exactement renfermé. Vers le milieu de la durée de cet anneau, on voyoit sa figure peinte sur le papier, telle qu'elle est représentée Fig. 2 où *AZBN* est le disque du Soleil, dont le point du sommet est en *Z*, et le point opposé en *N*, par lequel est menée la droite horizontale *AB*, dont l'extremité *A* regarde l'Orient, et l'autre extremité *B* l'Occident: *azbn* est le disque de la Lune et la droite *EF* qui passe par les centres *C* et *c* du Soleil et de la Lune, paroit etre distante de la verticale *ZN* d'un angle d'environ 40°, car je n'ai pas pris la mesure exacte de cet angle, étant plus attentif à d'autres phenomenes. La plus grande largeur de l'anneau étoit *Ff*, et la moindre *Ee*, qui etoit estimée égale à peu prés à la quatrieme partie de *Ff*.

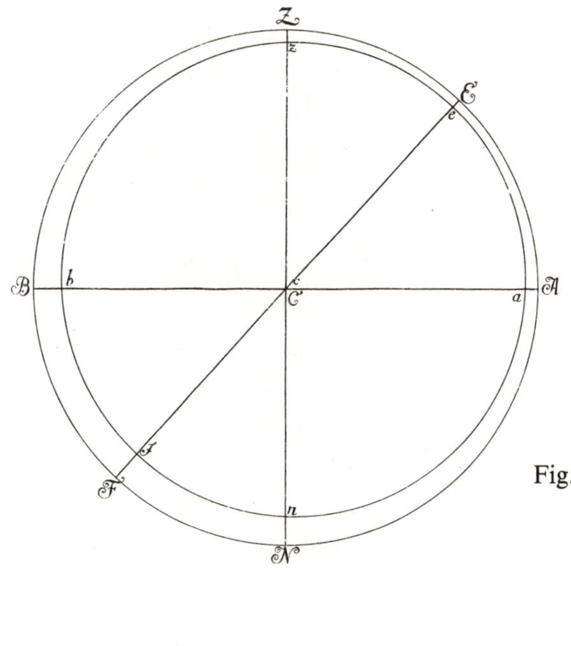

Fig. 2

6. Par le calcul Astronomique j'ai trouvé pour ce tems le demi-diametre apparent du Soleil $= 952''$ et le demi-diametre de la Lune $= 898''$, lesquelles mesures, (avant que le disque du Soleil eut souffert la dilatation qui lui arriva vers le tems, où l'anneau commença à se former,) s'accordent assez exactement avec l'observation, en sorte qu'ici la Theorie n'a aucun besoin d'étre corrigée. Je conclus outre cela de la durée de l'anneau, que la distance moindre des centres du Soleil et de la Lune devoit avoir été de $53''$ environ; d'où, si l'on suppose que le disque du Soleil n'ait souffert aucun accroissement dans la région Ff, où etoit la plus grande largeur de l'anneau, puisque nous avons vu que cet accroissement est seulement arrivé dans les endroits, où les bords du Soleil et de la Lune se touchoient réciproquement de plus prés, CF etoit $= 952''$, $cf = 898''$ et $Cc = 53''$, et par conséquent $Cf = 845''$, et à cause de cela la plus grande largeur de l'anneau $Ff = 107''$, ce qui s'accorde assez exactement avec la figure de l'anneau que nous avons tracée.

7. Il paroit de là, si le disque du Soleil ne s'etoit pas elargi vers la moindre largeur de l'anneau en Ee, quelle auroit dû etre cette largeur Ee. Car si on pose $CE = 952''$, à cause de $ce = 898''$ et $Ce = 951''$, la moindre largeur de l'anneau seroit $= 1''$; et cependant elle m'a manifestement paru égale à environ la quatrieme partie de la plus grande largeur Ff. La largeur Ee étoit donc à peu prés $= 26''$ quoiqu'elle n'eut cependant pas du avoir plus d'une seconde, si le disque du Soleil n'avoit souffert aucune dilatation. A Francfort sur l'Oder M. POLACK[4]) a observé le même phénomene, et la distance des centres du Soleil et de la Lune y etoit encore moindre qu'ici. Car l'anneau y ayant duré $4'$, j'en conclus que la moindre distance des centres a été d'environ $35''$, la plus grande largeur de l'anneau ayant été $89''$, et la moindre $19''$. Mais la moindre largeur etoit estimée comme sousdouble de la plus grande, et par conséquent de $44''$, en sorte qu'à Francfort de même qu'ici, le disque du Soleil etoit dilaté de $25''$, vers les endroits où la largeur de l'anneau étoit la moindre.

8. De là nait donc cette Question: *Quelle a été la cause de la dilatation du disque du Soleil, dans les endroits où le bord de la Lune le touchoit presque interieurement?* Pour la quantité de cette dilatation, que j'ai estimée $25''$, je ne la crois pas si certaine, qu'il ne puisse s'y etre glissé une erreur de plusieurs secondes: car, premierement la largeur de l'anneau n'a pu etre si exactement mesurée; et ensuite les

4) JOHANN FRIEDRICH POLACK (1700–1771), Professeur de mathématiques. EJA.

couleurs dont le bord du Soleil et celui de la Lune, étoient constamment revêtus, le premier de violet, et l'autre de rouge, empêchoient de discerner les extremités de ces bords. Malgré tout cela il ne sauroit demeurer aucun doute, que la moindre largeur de l'anneau n'ait surpassé ici 10″, et il est à propos d'examiner quelle peut avoir été la cause de sa dilatation.

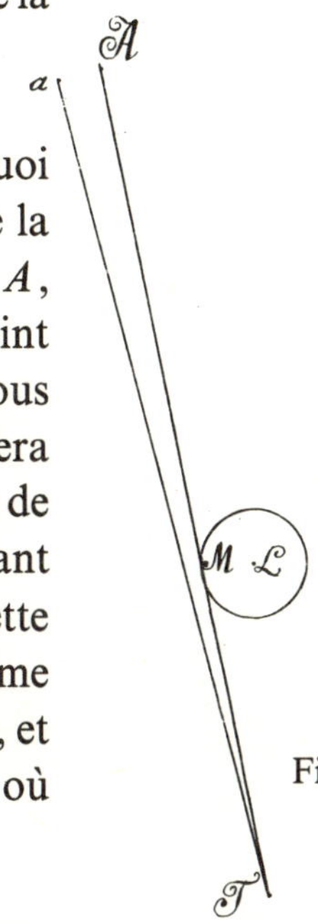

Fig. 3

9. La Question revient donc à ceci, c'est que nous expliquions, pourquoi l'extremité du bord du Soleil A [Fig. 3], qui etant vu par le rayon, AT, qui frise la Lune au point M devroit paroitre contigue de la Lune, n'est pourtant pas vu en A, mais en a, le spectateur etant supposé en T; de sorte que le lieu apparent de ce point differe de son vrai lieu de l'angle ATa, que nous avons trouvé de 25″, quoique nous l'estimions un peu plus petit? C'est en rendant raison de cela, qu'on fera comprendre, comment il peut arriver, que lorsque l'anneau ne devroit plus avoir de largeur, ou presque plus, cette largeur reçoit cependant un accroissement d'autant de secondes, qu'en contient l'angle ATa. Car si le bord du Soleil s'eloigne de cette maniere de celui de la Lune, dont il est tout à fait prés, il est clair que c'est la même cause, en vertu de laquelle les cornes du Soleil s'etendent avant et aprés l'anneau, et qu'il faut que la largeur de l'anneau même souffre une dilatation, dans l'endroit où elle est la moindre.

10. Avant que d'entreprendre l'explication de ce phenomene, il convient d'ecarter d'abord une opinion, qui au premier coup d'oeil n'est pas dénuée de probabilité. Ceux qui ont appris par la description des autres Eclipses annulaires du Soleil, que les deux largeurs de l'anneau, la plus grande et la plus petite, font ensemble une somme, plus grande que la difference entre les diametres apparens du Soleil et de la Lune; ceux, dis-je, auxquels cette observation est connüe, ont soupçonné pour l'ordinaire que, dans l'Eclipse annulaire de Soleil, le diametre de la Lune paroit un peu moindre, qu'il ne feroit, si on le voyoit hors du Soleil. Mais de la maniere, dont nous avons procedé à l'observation, il étoit parfaitement clair que la Lune n'avoit point souffert de diminution, mais que c'etoit le Soleil qui avoit reçu quelque accroissement. Car ceux qui pendant l'Eclipse même ont mesuré le diametre de la Lune, l'ont trouvé parfaitement conforme au calcul.

11. Pour revenir donc à cet aggrandissement du disque du Soleil, observé pendant l'Eclipse annulaire, on s'apperçoit d'abord, et par une legere attention à ce phenomene, que la cause doit en être cherchée dans la réfraction des rayons qui rasent le corps de la Lune. En effet on comprend aisément, que si la Lune etoit

environnée de quelque Atmosphere semblable à la notre, il devroit à cause de la réfraction des rayons qui la traversent obliquement, en résulter pleinement le même phénomene, qui consiste dans la déflexion des rayons. Non seulement donc il ne paroit rester aucun doute, qu'on doive attribuër à la Lune une certaine Atmosphere, mais on peut aussi déterminer la densité même de cette Atmosphere par la quantité de la réfraction.

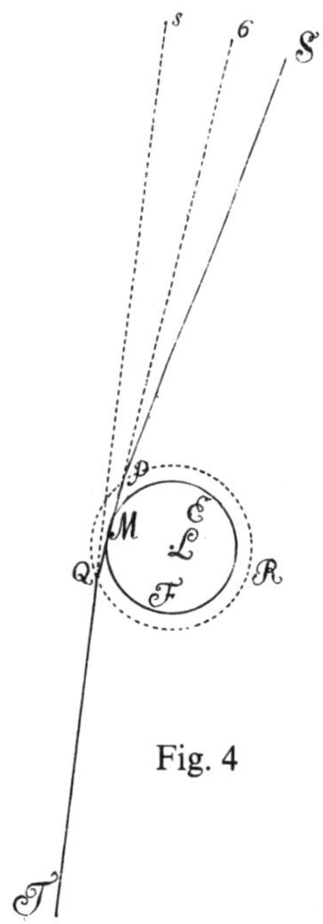

Fig. 4

12. Soit donc le corps de la Lune EMF [Fig. 4] ceint d'une Atmosphere déliée PQR, que nous supposerons être de la même densité dans toute sa hauteur, quoique sans doute elle devienne, comme celle de la Terre, toujours plus rare, à mesure qu'elle s'eloigne de la surface, jusqu'à ce qu'enfin elle se confond insensiblement avec l'ether, qui remplit tout l'espace celeste. Car comme nous savons par d'autres phenomenes, que l'Atmosphere de la Lune est trés subtile, en comparaison de celle dont la Terre est environnée, nous pourrons négliger sans erreur la diverse rareté qui peut y avoir lieu, suivant les diverses distances de la surface de la Lune. Cela posé, chaque rayon de lumiere qui entre dans l'Atmosphere de la Lune, souffrira une certaine réfraction, aprés laquelle il traversera l'Atmosphere en ligne droite, et lorsqu'il en sortira de nouveau pour rentrer dans l'ether, il éprouvera une nouvelle réfraction semblable à la premiere.

13. Considerons à présent le rayon de lumiere SP, partant du point S, soit que ce soit une Etoile, ou une particule appartenante au Soleil, lequel rayon tombe dans l'Atmosphere de la Lune en P, de maniere qu'aprés sa réfraction il rase la surface de la Lune en M. Le spectateur donc placé au point M de la Lune, n'apperçoit pas à cause de la réfraction le point S dans son vrai lieu, mais en σ, et ce point lui sera representé dans le Ciel par le rayon MP prolongé aprés sa réfraction. Or comme cette droite MP, parce qu'elle touche la surface de la Lune, représente l'horizon Lunaire, le spectateur placé en M appercevra un point lumineux dans l'horizon, lorsque ce point est encore effectivement caché sous l'horizon à l'angle $SP\sigma$; cet angle $SP\sigma$ sera donc égal à la réfraction horizontale, que les habitans de la Lune doivent sentir. Appellons donc cet angle $SP\sigma$, ou la réfraction horizontale vüe dans la Lune $= \alpha$, jusqu'à ce que nous puissions définir sa valeur avec plus de précision.

14. De plus, que ce rayon PM traverse toute l'Atmosphere de la Lune, et qu'il souffre une nouvelle réfraction en Q, où il entre dans l'ether, d'où suivant la ligne droite QT il parvienne jusqu'à la Terre, et frappe l'oeil de l'observateur en T. Cet

observateur jugera donc que le point lumineux, d'où ce rayon partoit, est dans la direction prolongée TQ, et par conséquent situé en s. Mais, si la Lune n'avoit pas été placée entre deux, l'observateur auroit apperçu ce même point dans son vrai lieu S. Que si la distance de ce point S est extrémement grande, eu égard à la distance de la Lune à la Terre, sans cet effet de la Lune, l'observateur T devroit l'appercevoir dans une direction parallele à la droite PS, laquelle à cause de l'angle PQs égal à l'angle $SP\sigma = \alpha$, sera inclinée à la direction TQ de l'angle double $= 2\alpha$.

15. Si donc la Lune n'avoit point d'Atmosphere, le point S seroit entierement invisible pour l'observateur placé en T, et demeureroit caché derriere la Lune à une distance de son bord $= \alpha$. C'est donc l'Atmosphere de la Lune qui rendra ce point S visible, et le montrera à l'observateur T, comme s'il etoit en s; et comme le bord de la Lune M, à cause du rayon MQT, se rapporte au même point du Ciel s, l'observateur placé en T appercevra un point lumineux S, contigu au bord de la Lune M; en sorte que la réfraction de l'Atmosphere Lunaire transportera dans le Ciel le point S de son vrai lieu S par l'espace $Ss = 2\alpha$. Donc la vüe d'une Etoile, ou d'un autre point lumineux quelconque dans le Ciel, ne nous est dérobée par la Lune, que quand ce point s'est déjà caché derriere elle au delà de l'espace $= 2\alpha$. Tant que cet espace est moindre, le point demeure visible hors du bord de la Lune.

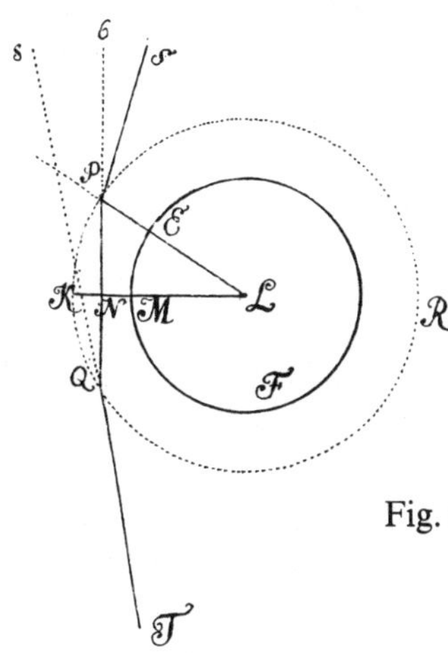

Fig. 5

16. Mais considerons aussi le cas, où le rayon SP [Fig. 5] pénétrant par l'Atmosphere de la Lune, ne rase plus son bord, mais passe à une distance donnée MN. Ce rayon souffrira donc une moindre réfraction, tant en P qu'en Q, puisqu'on doit concevoir l'Atmosphere plus rare, à mesure qu'elle s'eloigne de

la Lune; néanmoins on pourra également dans cette région plus rare considerer comme une ligne droite la route que le rayon PQ suit en traversant l'Atmosphere. Car un spectateur étant placé à la hauteur MN au dessus de la surface de la Lune, appercevra sans doute une moindre réfraction horizontale des Etoiles, et c'est à cette réfraction que sera égale l'angle $SP\sigma$, par lequel l'astre S paroitra elevé; et parce que le rayon PQ, qui parvient jusqu'à la Terre T, souffre une pareille réfraction en Q, l'observateur placé sur la Terre en T verra le point lumineux S situé en s, en sorte que l'intervalle Ss sera égal dans le Ciel au double angle $SP\sigma$.

17. Soit la distance MN de la surface de la Lune telle, que l'observateur placé sur la Terre en T la voye sous l'angle x, et comme toute la hauteur MK de l'Atmosphere de la Lune est fort petite, et la réfraction même trés petite en Q, le lieu apparent s de l'Etoile S lui paroitra distant du même intervalle $= x$ du bord M de la Lune. Mais le vrai lieu de cette Etoile qu'il appercevroit, si la Lune n'avoit point d'Atmosphere, sera plus proche du centre de la Lune de l'intervalle sS, qui se mesure par le double angle $SP\sigma$. Si donc cet angle $SP\sigma$, où la réfraction horizontale, qui répond à la hauteur $MN = x$ au dessus de la surface de la Lune, est posée $= \zeta$, et qu'on voye l'Etoile éloignée du bord de la Lune de l'intervalle $= x$, il faut, pour en inferer le vrai lieu de l'Etoile, approcher le lieu apparent vers le centre de la Lune par l'intervalle $sS = 2\zeta$. D'où, si pour chaque distance x du bord de la Lune on étoit assuré de la réfraction horizontale ζ qui y répond, il seroit aisé de déterminer le vrai lieu de chaque Etoile par son lieu apparent.

18. La dilatation de l'anneau Solaire observée à Berlin, donne lieu de conclurre, que comme l'anneau, lorsqu'il étoit le plus étroit, ne devoit avoir qu'une seconde, cette largeur qui étoit prête à s'evanouïr a reçu un accroissement de 25″, environ. Par conséquent si le bord du Soleil, ou quelque Etoile, a paru à 25″ de distance du bord de la Lune, la vraye distance doit etre estimée tout à fait nulle, à moins que peut être, à cause des raisons déjà indiquées, au lieu de 25″ il ne faille choisir un moindre nombre, comme 20 ou 15. Alors nous savons par les Observations trés abondantes, dont on se sert communément pour combattre l'Atmosphere de la Lune, que dés que la distance d'une Etoile au bord de la Lune surpasse seulement une minute, le changement que la réfraction Lunaire apporte à son lieu, devient tout à fait imperceptible.

19. Comme il paroit extremement difficile de déterminer par la theorie de la réfraction, l'effet de cette réfraction pour chaque distance du bord de la Lune,

parce que la diminution de la densité de l'Atmosphere nous est inconnüe, il ne paroit y avoir de meilleur moyen à employer que celui de l'estimation, en cherchant une formule, qui satisfasse le mieux aux phenomenes. Soit donc la distance apparente d'une Etoile quelconque à l'egard du bord de la Lune $= x''$, et l'effet de la réfraction qui répond à cette distance $= z''$, en sorte qu'on obtienne par ce moyen le vrai lieu de l'Etoile, en approchant davantage du centre de la Lune le lieu apparent de z''. Il faudra donc définir cette correction z par la distance x, de maniere qu'en posant $x = 20$, il en résulte aussi $z = 20$; mais qu'au cas que x soit $= 60$, alors la valeur de z devienne si petite qu'on ne puisse presque l'appercevoir, comme si elle etoit de $5''$.

20. Pour cet effet je prendrai une formule plus étenduë, et j'etablirai

$$z = \frac{A}{1 + Bx^n},$$

parce que je vois qu'une telle formule est extremément commode, lors que la distance x devient considerable, pour faire que la valeur z soit la plus petite possible, pourvu que l'exposant n ne soit que médiocrement grand. Que si l'on prend $n = 2$, et qu'on satisfasse aux deux conditions précedentes, on trouvera cette formule,

$$z = \frac{32}{1 + 0{,}0015\,xx},$$

ou

$$z = \frac{32}{1 + \frac{1}{666}xx};$$

mais si, au lieu de $5''$ que nous attribuons à la distance $x = 60''$, nous posons $4''$, cela donnera

$$z = \frac{40}{1 + \frac{1}{400}xx}.$$

Nous nous servirons donc un peu de cette formule, comme de la plus simple, jusqu'à ce que nous puissions en employer une plus certaine; car, quand même il y auroit de l'erreur, les conclusions qui en seront tirées, ne s'ecarteront pourtant pas de la verité, d'une maniere qui soit sensible.

21. C'est donc sur cette formule que j'ai construit la Table suivante, qui pour chaque distance apparente, où un Astre se trouve au bord de la Lune, fournit la correction suivant laquelle cette distance doit être diminuée, pour découvrir la veritable. La distance apparente du bord de la Lune est marquée par x, et la correction, ou l'effet de la refraction par z.

x	z	x	z	x	z	x	z	x	z	x	z
0″	40″	10″	32″	20″	20″	30″	12″	40″	8″	90″	2″
1	40	11	31	21	19	31	12	45	7	100	2
2	40	12	29	22	18	32	11	50	6	110	1
3	39	13	28	23	17	33	11	55	5	120	1
4	38	14	27	24	16	34	10	60	4	130	1
5	37	15	26	25	15	35	10	65	3[6])	140	1
6	36	16	24	26	15	36	9[5])	70	3	150	1
7	35	17	23	27	14	37	9	75	3	160	1
8	34	18	22	28	13	38	9	80	2	170	$\frac{1}{2}$
9	33	19	21	29	13	39	8	85	2	180	$\frac{1}{2}$
10	32	20	20	30	12	40	8	90	2		

22. En suivant donc cette Table, on satisfait non seulement au phénomene de la derniere Eclipse de Soleil, en sorte que le bord du Soleil est effectivement contigu au bord de la Lune, lorsqu'il paroit en être distant de 20″, mais encore on reconnoit par là que l'effet de la réfraction Lunaire a été imperceptible, avant que la proximité du bord de la Lune ait été au dessous d'une minute. En effet nous avons vu qu'à la distance de 180″, ou 3′, l'effet n'alloit pas même à une seconde; et pouvoit par conséquent etre compté pour rien; ce qui s'accorde parfaitement avec les Observations. Or comme cette Table indique, que si la distance apparente x evanouït tout à fait, l'effet de la refraction est 40″, c'est à dire, que le point lumineux est effectivement caché à cet intervalle derriere le disque de la Lune: si ce nombre étoit exactement juste, il s'ensuivroit que la réfraction horizontale pour les habitans de la Lune est de 20″. On comprend donc en même tems, que quand même cette Table renfermeroit quelque erreur, cependant la réfraction horizontale ne s'ecarteroit pas beaucoup du vrai.

5) Edition originale: 10 au lieu de 9. Corrigé par EJA.
6) Edition originale: 4 au lieu de 3. Corrigé par EJA.

23. Aussi-tot donc que le bord du Soleil est caché moins de 40″, derriere la Lune, il doit nous être visible; d'où il résulte, que si nous voulons définir une Eclipse annulaire du Soleil par le calcul, le commencement de l'anneau arrivera, avant que la distance des centres devienne égale à l'excés du demi-diametre du Soleil sur le demi-diametre de la Lune: et de la même maniere, l'anneau disparoitra un peu aprés que la distance des centres est devenue égale à la difference des demi-diametres. C'est à dire, que si l'on prend la parallaxe horizontale dans la Lune de 20″, l'anneau devroit paroitre aussi-tot que la distance des centres surpasseroit de 40″ la difference des demi-diametres; et par conséquent l'Eclipse seroit annulaire, quand même la distance des centres ne deviendroit jamais moindre, que la difference des demi-diametres, pourvu que cette distance ne surpasse pas cette difference de plus de 40″. L'Atmosphere de la Lune fait donc qu'une Eclipse, qui sans elle ne seroit pas annulaire, paroit cependant telle; et que l'anneau dure plus longtems, qu'il ne devroit durer suivant le calcul, en négligeant l'effet de l'Atmosphere de la Lune.

24. L'anneau ne sauroit pourtant paroitre durer aussi longtems, que cette réfraction de 40″ le demanderoit; car, de même que l'eclat du Soleil empêche de voir les Etoiles dans le Ciel, il est fort vraisemblable que l'anneau ne sauroit être apperçu du coté où il est trop étroit. En effet, tant que la largeur de l'anneau est extrémement petite, et ne s'accroit pas au delà de quelques secondes, elle sera invisible, et l'anneau ne se montrera, qu'aprés avoir aquis une largeur considerable. J'ai aussi observé dans la derniere Eclipse que l'anneau avoit paru tout à coup avec une largeur remarquable, et qu'il ne s'etoit point formé peu à peu et par degrés; ce qui prouve manifestement, qu'à cause de la lumiere du reste du Soleil, cet anneau n'a pu devenir visible, qu'aprés avoir aquis une largeur suffisante pour faire impression sur notre vüe.

25. Il convient aussi de remarquer ici, que ceux qui ont regardé directement le Soleil par un tube Astronomique, ont vu plus longtems l'anneau, que ceux qui ont consideré l'Image du Soleil representée sur le papier. Les premiers l'ont apperçû pendant l'espace de 82″, au lieu qu'il s'est à peine montré aux autres pendant une minute entiere. La raison de cette difference vient sans doute de ce que ceux qui regardoient le Soleil, recevoient une plus forte impression de ses rayons, et pouvoient par conséquent remarquer l'anneau, lorsque sa largeur étoit encore fort petite, au lieu que l'image du Soleil se peignant beaucoup plus foiblement sur le papier, il n'est pas étonnant que l'anneau ait commencé à y paroitre plus tard, et ait fini plutot; puisqu'ayant encore une largeur assez considerable, l'impression qu'en recevoit le papier étoit pourtant trop foible pour être apperçüe.

26. C'est aussi pour cela que tant avant qu'aprés l'anneau, lors que le Soleil paroissoit en croissant, les pointes de ses cornes ne s'exprimoient pas parfaitement sur le papier, mais elles paroissoient obtuses, et comme coupées vers les extremités; phénomene, dont la cause est manifeste par tout ce que nous avons dit. En effet les extremités des pointes G et G (Fig. 1) étoient trop étroites, pour que les rayons qui en partoient, produisissent leur image sur le papier, ou dans l'oeil. Car, dés là que la largeur de cette pointe ne surpassoit pas le diametre d'une Etoile, qui auroit été invisible dans cet endroit, elle ne pouvoit arriver à l'oeil de l'observateur; et dans l'endroit où l'on voyoit le bout des cornes, il y avoit déjà une largeur assez considerable. Ce phénomene est donc semblable à celui que nous avons rapporté ci-dessus, par rapport à la largeur que l'anneau doit avoir, avant que d'affecter le sens de la vüe.

27. Pour cette cause donc, il ne faut pas placer le commencement de l'anneau à l'instant, où la distance des centres du Soleil et de la Lune excede de 40″ la difference des demi-diametres, si tant est que nous estimions juste la réfraction horizontale dans la Lune en la faisant de 20″; mais le commencement de l'anneau doit être placé où sa largeur devient assez grande, pour surpasser le diametre des Etoiles invisibles dans cette région. Quant à moi, la largeur de l'anneau, au moment qu'il a paru, ne m'a pas semblé moindre que de 20″; d'où l'on peut inferer que l'anneau ne devient visible, que quand sa largeur s'est accrüe au delà de 20″, et qu'alors il commence à se montrer tout à coup avec cette largeur. On ne sauroit pourtant définir ce terme avec certitude, puisqu'à proportion de la bonté du tube par lequel on regarde le Soleil, on apperçoit l'anneau se fermer plutot, ou plus tard; et lorsqu'on reçoit l'image du Soleil sur le papier, cette difference s'etend encore plus loin.

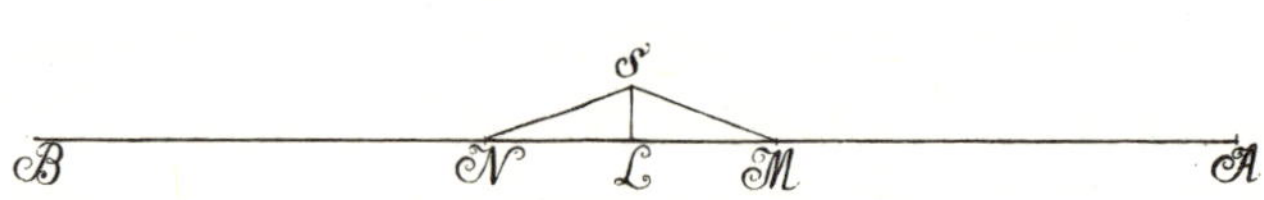

Fig. 6

28. Pour définir par le calcul les Eclipses annulaires de Soleil, soit [Fig. 6][7]) le demi-diametre apparent du Soleil $= a$, le demi-diametre apparent de la Lune $= b$, et la difference des demi-diametres $a - b = d$. Soit de plus la droite AB la route apparente du centre de la Lune à l'egard du centre du Soleil, qui soit consideré comme immobile en S. Alors la perpendiculaire SL

7) Edition originale: Fig. 2 au lieu de Fig. 6. Corrigé par EJA.

tombant sur AB fournira la moindre distance des centres, qui soit $SL = c$, et cette distance SL sera fort petite, si l'Eclipse est annulaire. Que l'Eclipse commence, lorsque le centre de la Lune parviendra en A, et que la fin tombe en B; la droite SA aussi bien que SB sera donc égale à la somme des demi-diametres $a + b$; et parceque SL est si petite, la droite même AB, ou la route apparente, que le centre de la Lune décrit depuis le commencement jusqu'à la fin, pourra être censée égale à la somme $SA + SB$, en sorte que AB soit $= 2a + 2b$.

29. Soit de plus t le tems de toute l'Eclipse, et le centre de la Lune peut être supposé parcourir l'espace $AB = 2a + 2b$ d'une maniere uniforme pendant le tems t. Ces choses etant posées, l'anneau commencera à paroître, lorsque le centre de la Lune sera parvenu en M, en sorte que SM soit égal à la difference des demi-diametres $a - b = d$, augmentée d'abord de $40''$, et ensuite diminuée de la quantité que doit avoir la largeur de l'anneau, avant que de devenir visible. Si donc nous supposons que l'anneau ne puisse être apperçu, à moins que sa largeur ne surpasse $20''$, le commencement de l'anneau tombera sur M, en sorte que $SM = d + 40'' - 20'' = d + 20''$; et de la même maniere, en prenant $SN = d + 20''$, la fin de l'anneau s'accordera avec le lieu du centre de la Lune N. Pourvû donc que $d + 20''$ soit plus grand que $SL = c$, l'Eclipse sera annulaire, c'est à dire, pourvû que la moindre distance des centres SL soit moindre que $d + 20''$.

30. On trouvera donc la durée de l'anneau, en cherchant la quatrieme proportionelle aux distances AB, MN et de tems t. Or ML sera $= \sqrt{((d + 20'')^2 - cc)}$, et par conséquent $MN = 2\sqrt{((d + 20'')^2 - cc)}$, d'où si l'on nomme le tems, pendant lequel l'anneau est visible, $= \theta$, on aura

$$2a + 2b : 2\sqrt{((d + 20'')^2 - cc)} = t : \theta$$

et par cons[équent]

$$(d + 20'')^2\, tt - cctt = (a + b)^2\, \theta\theta.$$

De là, si la durée de l'anneau θ est connüe, ainsi bien que la durée de l'Eclipse entiere t, et les demi-diametres apparens du Soleil et de la Lune a et b, d'où l'on a $d = a - b$, on conclurra réciproquement de l'observation de l'Eclipse annulaire la moindre distance des centres du Soleil et de la Lune $SL = c$. Car on aura

$$c = \sqrt{\left((d+20'')^2 - (a+b)^2 \frac{\theta\theta}{tt}\right)}.$$

31. À présent, si nous voulons appliquer cette formule à la derniere Eclipse annulaire de Soleil, nous avions le demi-diametre apparent du Soleil $a = 952''$, le demi-diametre apparent de la Lune $b = 898''$, et par cons[équent]

$$a - b = d = 54'', \quad \text{et} \quad d + 20'' = 74''.$$

Ensuite le tems de l'Eclipse entiere, recueilli des observations, est $3^{\mathrm{h}}\ 6'$, et le tems de la durée de l'anneau $1'\ 22''$; d'où résultera $t = 11\,160''$ et $\theta = 82''$, et à cause de $a + b = 1850''$, on trouvera la moindre distance des centres $SL = c$

$$= \sqrt{\left(74^2 - \frac{1850^2 \cdot 82^2}{11\,160^2}\right)} = \sqrt{5291} = 72''\ 44'''.$$

La moindre distance des centres auroit donc été $72\frac{3}{4}$ secondes, que par le calcul j'avois néanmoins trouvée seulement de $51''$.

32. Or j'ai fait voir dans ma Dissertation precedente [8]), où j'ai ramené cette Eclipse au calcul, que la parallaxe de la Lune, telle qu'on la suppose ordinairement dans les Tables, doit être considerablement diminuée, afin que la moindre distance des centres aille à $51''$, et il faudroit qu'elle le fut encore beaucoup davantage pour que cette augmentation de la moindre distance des centres parvint jusqu'à $72''$. Or, comme les élémens de cette Eclipse ne paroissent pas differer si considerablement du vrai, il est plus vraisemblable que la difference des demi-diametres d ne doit pas être augmentée de $20''$, mais d'un moindre nombre; ce qui arriveroit, si la réfraction horizontale dans la Lune n'etoit pas mise à $20''$, mais seulement à $15''$; car alors, au lieu de $d + 20''$, il faudroit écrire $d + 10''$, ou $64''$, d'où résulteroit $c = 62\frac{1}{2}''$, en sorte que la distance des centres trouvée par le calcul de $51''$ devroit seulement être augmentée de $11\frac{1}{2}''$; or il est assez probable que les élémens de cette Eclipse demandent cette petite correction.

33. Ces argumens établissent donc d'une maniere indubitable, que la Lune a aussi une Atmosphere, quoiqu'elle soit peut-être encore plus déliée, que ne

8) Voir note 1, p. 239 de ce tome. EJA.

l'estiment nos observations. Car le disque du Soleil ayant reçu un accroissement remarquable vers le milieu de l'Eclipse, quoiqu'on n'ait pas pu le mesurer avec exactitude, ce phénomene ne peut être attribué à aucune autre cause qu'à l'Atmosphere de la Lune. Je ne demande pourtant pas, qu'on fasse trop de fonds sur les déterminations que j'ai données pour la réfraction des rayons qui passent par l'Atmosphere de la Lune; et il est même fort vraisemblable par les observations comparées à la Theorie, que la réfraction horizontale de la Lune, que j'avois mis à 20″, va à peine au delà de 10″; d'où le diametre du Soleil peut recevoir un accroissement de 20″. Mais il est encore moins possible d'etablir quelque chose de certain sur la diminution de cette réfraction, d'une maniere qui convienne aux rayons qui passent plus loin de la Lune; quoique cette diminution paroisse si grande, que la réfraction devient tout à fait imperceptible, pour les rayons qui traversent l'Atmosphere de la Lune à quelques minutes de distance de son bord.

34. S'il nous arrivoit d'observer encore une fois une semblable Eclipse annulaire de Soleil, il faudroit soumettre à une mesure exacte tous les phénomenes que nous fourniroit l'accroissement du disque du Soleil, afin de définir par là non seulement la réfraction des rayons qui rasent le bord de la Lune, mais encore la Loi que suit la réfraction, à l'egard des rayons qui passent plus loin de la Lune. Mais comme il n'est presque pas permis de nous attendre encore à une pareille Eclipse, il faudra recourir à d'autres phénomenes, qui s'offrent plus souvent à nos regards. De ce genre ceux qui nous paroissent les plus propres au but dont il s'agit, ce sont les occultations des Etoiles fixes par la Lune; car, comme on peut appercevoir une Etoile fixe jusqu'à ce qu-elle devienne contiguë à la Lune, il faudra qu'on remarque dans son lieu quelque changement, qui proviendra de la réfraction Lunaire, pourvû que les instrumens se trouvent assez exacts pour indiquer ces petits détails.

35. Qu'on fasse donc choix d'une Etoile fixe, si voisine de celle qui souffrira l'occultation de la Lune, qu'on puisse les voir toutes deux à la fois par le tube; et que peu de tems avant que l'occultation arrive, on mesure par le moyen d'un excellent Micrometre la distance de ces Etoiles, qui paroit pouvoir se trouver à une seconde prés, pourvu que la longueur du tube ne soit pas moindre de dix pieds, et que la distance des Etoiles n'excede pas 15″. De cette maniere, parce que les rayons qui partent de l'Etoile qui va entrer dans l'occultation, ne souffrent encore aucune réfraction, on trouvera la vraye distance de ces Etoiles assez exactement exprimée en secondes. Alors qu'on attende l'occultation, et qu'au moment même où elle arrive, on mesure de nouveau la distance de ces Etoiles, en remarquant la difference entre

cette distance actuelle, et celle qui avoit été trouvée auparavant; et de la comparaison de ces distances on conclurra aisément la réfraction horizontale de la Lune, pourvû que le diametre apparent de la Lune ait été connu, aussi bien que sa situation à l'egard des deux Etoiles au moment de l'occultation; et si l'on ne peut déterminer ces choses en même tems par l'observation, il n'y a qu'à les tirer de la Theorie du mouvement de la Lune.

36. En effet, soit la distance vraye des Etoiles $= m''$, que le Micrometre aura indiquée longtems avant l'occultation, et qu'au moment même de l'occultation, l'Etoile qui va la souffrir, paroisse au bord de la Lune au point S [Fig. 7][9]); que

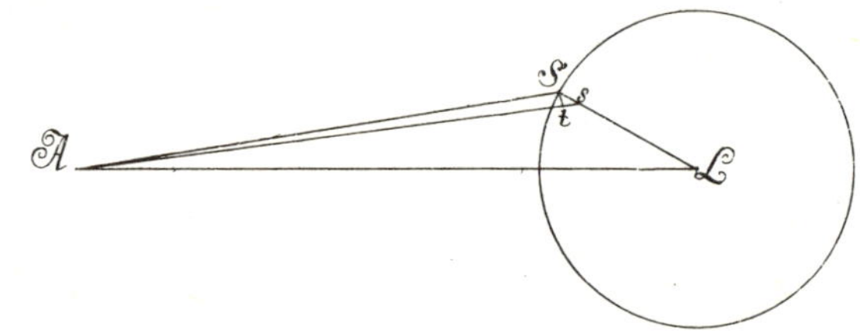

Fig. 7

l'autre Etoile soit en A, et alors supposons que la distance AS de ces Etoiles soit trouvée moindre, en sorte que AS soit $= m'' - \mu''$. De plus, comme par la Theorie le demi-diametre apparent de la Lune est assez exactement donné, et qu'on peut connoître l'angle ASL, ou par la Theorie, ou par la durée de l'occultation, ou de quelque autre maniere; soit cet angle $ASL = \varphi$, et parce qu'à cause de la réfraction le lieu de l'Etoile s'eloigne du centre de la Lune, que le vrai lieu de l'Etoile soit dans ce moment en s, on aura $As = m''$, et si[10]) $At = AS$ on retranche[11]) $st = \mu''$, d'où à cause de l'angle $Sst = 180° - \varphi$, on conclurra assez exactement l'effet de la réfraction

$$Ss = \frac{\mu''}{\cos(180° - \varphi)} = -\frac{\mu''}{\cos\varphi},$$

à la moitié duquel la réfraction horizontale de la Lune doit etre censée égale.

37. On fera bien de choisir, pour parvenir à cette fin, les occultations dans lesquelles l'Etoile fixe S entre sous le bord obscur de la Lune, ce qui arrive vers le

9) Edition originale: Fig. 2 au lieu de Fig. 7. Corrigé par EJA.
10) Edition originale: As au lieu de AS. Corrigé par EJA.
11) Edition originale: ST au lieu de st. Corrigé par EJA.

premier quartier. Car si l'Etoile arrive au bord éclairé de la Lune, sa lumiere est déjà offusquée avant l'occultation par l'eclat de la Lune, de maniere qu'elle disparoit avant le moment de l'occultation, à moins qu'elle ne soit de la premiere grandeur. En effet, quoique dans ce cas le bord de la Lune ne paroisse pas, cependant le moment même de l'occultation indique cet instant, où l'Etoile étoit contiguë au bord de la Lune. Cependant il ne sera pas inutile d'examiner aussi quelquefois l'arrivée des Etoiles au bord éclairé de la Lune, pour connoître par là, s'il y a quelque difference entre la réfraction auprés du bord éclairé, et la réfraction auprés du bord obscur? Car l'Atmosphere de la Lune étant continuellement exposée aux rayons du Soleil du coté de la région éclairée, on a lieu de soupçonner qu'elle est tellement attenuée dans ce tems-là, que la réfraction en devient beaucoup moindre. Et c'est peut-être là la cause, qui a empêché de remarquer jusqu'ici cet effet de la réfraction dans les occultations.

38. Ensuite une des choses auxquelles il faut apporter le plus de soin, c'est, avant que l'Etoile se cache, de rechercher perpetuellement, et à chaque moment, avec une extrême diligence, sa distance de l'autre Etoile fixe, parce que nous avons prouvé que cette distance doit diminuer peu à peu, avant son arrivée au bord de la Lune. Car, par cette suite d'observations, et en appellant la Theorie au secours, on pourra conclurre pour chaque Observation la distance apparente de l'Etoile au bord de la Lune, et l'on pourra déterminer la réfraction qui lui convient, d'une maniere semblable à celle que j'ai recommandée ci-devant pour le moment même de l'occultation. Et aprés que plusieurs Observations de cette nature auront été faites avec exactitude, on pourra s'en servir pour corriger la Table donnée au § 21 ou plutot pour en dresser sans peine une nouvelle, à l'aide de laquelle tous les phénomenes, qui naissent de l'Atmosphere de la Lune, pourront dans la suite être assignés avec beaucoup d'exactitude.

39. On n'a donc pas le moindre sujet de s'etonner que cette célébre Question, agitée déjà depuis longtems parmi les Astronomes; *Si la Lune a une Atmosphere, ou non?* n'ait pas encore été decidée. Car, quoique les occultations des Etoiles fixes par la Lune arrivent trés fréquemment, cependant l'effet de la réfraction est si petit dans les Etoiles fixes elles-mêmes, qui souffrent l'occultation, qu'on ne peut l'appercevoir en aucune maniere, à moins que leurs distances n'ayent déjà été trés soigneusement examinées par d'autres, et qu'on ne les mesure de nouveau avec toute l'exactitude possible, vers le tems même de l'occultation. Or il n'y a peut-être eu encore aucun Astronome, auquel il soit venu dans l'esprit de suivre cette voye pour faire des recherches sur l'Atmosphere de la Lune; ou s'il y en a eu qui soient par hazard

tombés sur cette methode, ils ont été forcés d'abandonner leur travail, faute d'instrumens assez exacts. Je ne sais, si ceux qui voudroient présentement prendre cette peine, ne feroient pas bien de préferer le Micrometre de feu Mr. KIRCH[12]), pourvû que les vis se terminent interieurement en pointe, à tous ceux qu'on fait à present avec tant de soins et à grands fraix. Car pourvu que les pointes des vis soyent une fois appliquées aux deux Etoiles, on s'appercevra facilement, si la distance des etoiles diminuë, et la diminution même se définira avec tout aussi peu de peine par la révolution des vis.

40. Les principes d'une saine Physique mettent eux mêmes l'Atmosphere de la Lune hors de doute; mais ce que les Observations nous ont découvert, n'en est pas moins admirable, c'est que cette Atmosphere de la Lune soit d'une si grande tenuïté, que son effet est presque imperceptible. Car la réfraction horizontale sur la Terre étant de plus d'un demi degré, si la Lune avoit une pareille Atmosphere, les Astres, en approchant du bord de la Lune, seroient transportes de leur place de plus d'un degré. Mais cet effet surpassant à peine 20″, il faut que l'air de la Lune soit presque 200 fois plus rare que le notre: d'où l'on peut conclurre, ou qu'il ne monte point du tout de vapeurs de la surface de la Lune, ou que la matiere de la Lune est si solide et si séche, qu'elle n'est presque sujette à aucune évaporation. Aussi par l'usage des longs tubes les Astronomes ont déjà appris, que ces taches obscures dans la Lune, qui sont prises vulgairement pour des eaux et des lacs, doivent plutot être des lieux arides, des cavernes, ou des forets, que des contrées humides.

12) GOTTFRIED KIRCH, «De simplicissima quadam specie micrometri», *Miscellanea Berolinensia* 1, 1710, 202–205. EJA.

PART OF A LETTER FROM LEONARD EULER, PROF. MATH. AT BERLIN, AND F.R.S. TO THE REV. MR. CASPAR WETSTEIN, CHAPLAIN TO HIS ROYAL HIGHNESS THE PRINCE OF WALES, CONCERNING THE GRADUAL APPROACH OF THE EARTH TO THE SUN.

Commentatio 183 indicis ENESTROEMIANI
Philosophical Transactions 46 (1749–50), 1752, p. 203–205

Berlin, June 28, 1749.

Monsieur LE MONNIER[1]) writes to me, that there is, at Leyden[2]), an Arabic Manuscript of IBN JOUNIS[3]) (if I am not mistaken in the Name, for it is not distinctly wrote in the Letter), which contains a History of Astronomical Observations. M. LE MONNIER says, That he insisted strongly on publishing a good translation of that Book. And as such a Work would contribute much to the Improvement of Astronomy, I should be glad to see it publish'd. I am very impatient to see such a Work which contains Observations, that are not so old as those recorded by PTOLEMY. For having carefully examined the modern Observations of the Sun with those of some Centuries past, although I have not gone farther back than the

1) PIERRE CHARLES LEMMONNIER (1715–1799). EJA.

2) EULER had originally located the manuscript at Oxford but communicated this correction before publication of the letter. In fact, manuscripts of only parts of the work are now extant at Leiden and Oxford. For details of these and the partial publication of the Leiden manuscript in the nineteenth century by A. P. CAUSSIN DE PERCEVAL, see DAVID A. KING, "Ibn Yūnus" in *Dictionary of scientific biography* (New York, 1970–1989), vol. 14, pp. 574–580. EJA.

3) IBN YŪNUS (d. 1009). EJA.

fifteenth Century, in which I have found WALTHER's[4]) Observations made at Nuremberg; yet I have observed that the Motion of the Sun (or of the Earth) is sensibly accelerated since that Time; so that the Years are shorter at present than formerly: The Reason of which is very natural; for if the Earth, in its Motion, suffers some little Resistance (which cannot be doubted, since the Space through which the Planets move, is necessarily full of some subtile Matter, were it no other than that of Light) the Effect of this Resistance will gradually bring the Planets nearer and nearer the Sun; and as their Orbits thereby become less, their periodical Times will also be diminish'd. Thus in Time the Earth ought to come within the Region of Venus, and in fine into that of Mercury, where it would necessarily be burnt. Hence it is manifest, that the System of the Planets cannot last for ever in its (present) State. It also incontestably follows, that this System must have had a Beginning: For whoever denies it, must grant me, that there was a Time, when the Earth was at the Distance of Saturn, and even farther; and consequently that no living Creature could subsist there. Nay there must have been a Time, when the Planets were nearer to some fixt Stars than to the Sun; and in this Case they could never come into the Solar System. This then is a Proof, purely physical, that the World, in its present State, must have had a Beginning, and must have an End. In order to improve this Notion, and to find with Exactitude, how much the Years become shorter in each Century; I am in Hopes that a great Number of older Observations will afford me the necessary Succours.

I beg you will present my Respects to the Royal Society; and am,

Yours, etc.

L. Euler.

4) BERNHARD WALTHER (1430–1504). There is a facsimile reprint of JOHANNES SCHÖNER's 1544 edition of the observations in *Joannis Regiomontani opera collectanea*, ed. by FELIX SCHMEIDLER (Osnabrück, 1972), pp. 627–641. EJA.

PART OF A LETTER FROM MR. PROFESSOR EULER TO THE REVEREND MR. WETSTEIN, CHAPLAIN TO HIS ROYAL HIGHNESS THE PRINCE, CONCERNING THE CONTRACTION OF THE ORBITS OF THE PLANETS.

Commentatio 184 indicis ENESTROEMIANI
Philosophical Transactions 46 (1749–50), 1752, p. 356–359

Berlin, Dec. 20, 1749.

You have done me much Honour in communicating an Extract of my last Letter[1]) to the illustrious Royal Society, November 2, 1749. I am still thoroughly convinced of the Truth of what I advanced therein, that the Orbs of the Planets continue to be contracted, and consequently their periodical Times grow shorter. But in order to put this Fact out of Doubt, we ought to be furnished with good ancient Observations, and also to be very sure of the Time elapsed, since those Observations, to this Day: Which we are not, with regard to the Observations that PTOLEMY has left us. For Chronologists, in fixing the Moments of those Observations, run into a Mistake, by supposing the Sun's mean Motion to be known; which ought rather itself to be determined by these same Observations. Now, if we reduce the Days marked by PTOLEMY to the Julian Kalendar, we run the Risque of committing an Error of a Day or two, in the whole Number of Days elapsed, from that to our Time; because the Course of the Julian Years, according to which every fourth ought to have been Bissextile, has been frequently interrupted by the Pontifices; of which we find some sure Marks in CENSORINUS and DION CASSIUS[2]).

---

1) E183, pp. 257–258 of this volume. EJA.

2) See CENSORINUS, *De die natali* (Leipzig, 1867), 20,8–21,6 and *Cassii Dionis Cocceiani Historiarum Romanarum quae supersunt*, ed. by URSUL PHILIP BOISSEVAIN (Berlin, 1895–1931), Book 48, ch. 33. EJA.

Wherefore it might well happen, since the Times mark'd by PTOLEMY, that there has really been a Day or two more than we reckon, and consequently, that PTOLEMY's Equinoxes, ought to be put a Day or two back; which would lengthen the Years of those Times. I was in hopes, that the Arabian Observations would not be liable to this Inconvenience; because the Julian Kalendar has not been interrupted for these last past twelve hundred Years[3]). The late Dr. HALLEY had also remark'd[4]), that the Revolutions of the Moon are quicker at present than they were in the Time of the ancient Chaldeans, who have left us some Observations of Eclipses. But as we measure the Length of Years by the Number of Days, and Parts of a Day, which are contained in each of them; it is a new Question, Whether the Days, or the Revolutions of the Earth round its Axis, have always been of the same Length. This is unanimously supposed, without our being able to produce the least Proof of it: Nor indeed do I see, how it could be possible to perceive such an Inequality, in case it had really existed. At present we measure the Duration of a Day by the Number of Oscillations, which a Pendulum of a given Length makes in this Space of Time: But the Ancients were not acquainted with these Experiments, whereby we might have been informed, whether a Pendulum of the same Length made as many Vibrations in a Day formerly as now. But even tho' the Ancients had actually made such Experiments, we could draw no Inferences from them, without supposing, that Gravity, on which the Time of an Oscillation depends, has always been of the same Force: But who will ever be in a Condition to prove this Invariability in Gravity? Thus, even supposing that the Days had suffered considerable changes; and that Gravity had been altered suitably thereto, so that the same Pendulum had always completed the same Number of Vibrations in a Day; it would nevertheless be still impossible for us to perceive this Inequality, were it ever so great. And yet I have some Reasons, deduced from Jupiter's Action on the Earth, to think, that the Earth's Revolution round its Axis continually becomes more and more rapid. For the Force of Jupiter so accelerates the Earth's Motion in its Orbit round the Sun, that the Diminution of the Years would be too sensible, if the diurnal Motion had not been accelerated nearly in the same Proportion. Wherefore, since we hardly at all remark this considerable Diminution in the Years, from thence I conclude, that the Days suffer much the same Diminution; so that the same Number will answer nearly to a Year.

---

3) At this point the original continues: "Or je viens d'apprendre que l'étoit à Leide et non pas à Oxford, que M. LE MONNIER a découvert un tel manuscript arabe." EJA.

4) *Philosophical Transactions* 19 (1695), 174–175. EJA.

LETTRE DU 11 MAI 1754 A ERICH PONTOPPIDAN

Commentatio 218 indicis ENESTROEMIANI
E. PONTOPPIDAN, Afhandling om Verdens Nyehed (Kiøbenhavn, 1757), p. 210–220.

Monsieur!

Tout sensible que je sois à l'honneur de votre lettre[1]), je m'estimeray bien heureux, si mes recherches peuvent jamais contribuer quelque chose à confondre la malice des *esprits forts*. Je croyois que l'approche successive des planetes vers le soleil fournissoit une preuve convainquante, tant que le systeme du monde, tel qu'il est aujourd'huy, n'ayt existé depuis toute l'eternité, ni qu'il sauroit subsister eternellement. Or, pour prouver ce sentiment *a priori*, et pour developper en meme temps, les questions, dont vous avez bien voulu m'honnorer, je prendray la liberté de vous proposer, *Monsieur*, les remarques suivantes.

1. Je dis donc, que l'espace du monde ne sauroit etre vuide de toute matiere, comme il seroit necessaire, si les planetes devoient decrire leurs orbites conformement aux regles etablies par KEPLER, et soutenues par le grand NEUTON, qui a eté obligé de supposer l'espace du monde entierement vuide, pour que les planetes, dans leur mouvement, n'y rencontrent aucune resistence, ayant bien reconnu, qu'une resistence devoit necessairement troubler leur mouvement. Or, quoique NEUTON ait soutenu vuide l'espace du ciel, je dis que ce sentiment ne sauroit etre admis, et qu'il est toujours detruit par la theorie de la lumiere, de quelque maniere qu'on l'envisage. Car, si, comme le veut NEUTON luy meme, la lumiere est dardée du soleil, avec cette terrible vitesse, qui luy fait parcourrir l'intervalle entre le soleil et la terre, en moins de huit minutes, tout l'espace doit etre absolument remplis de la matiere de la lumiere. Mais croyant avoir suffisamment refuté cette hypothese

1) Lettre de PONTOPPIDAN à EULER du 26 Mars 1754, *Afhandling om Verdens Nyehed* (Kiøbenhavn, 1757), pp. 203–209. EJA.

de l'emanation actuelle de la lumiere des corps lumineux, on est reduit par des raisons les plus fortes à soutenir, que la lumiere se repand par l'espace du ciel, de la maniere que le son est repandu par l'ouïe: Il faut donc, que l'espace du ciel soit rempli d'un fluide extremement elastique et rare, pour transmettre les rayons de la lumiere. Voila donc, dans l'un et l'autre cas, l'univers rempli d'une matiere subtile, qui ne sauroit manquer d'influer sur le mouvement des planetes. Cet effet peut etre double, car d'abord en tant que ce fluide est matiere, il faut que les corps qui s'y meuvent, rencontrent quelque resistence, qui tend à diminuer leur vitesse, et, en tant que cette matiere est en mouvement, elle choquera les planetes, selon la direction, et c'est sur quoi roule votre premiere question, que par le mouvement de la lumiere, les planetes devroient etre de plus en plus eloignées du soleil. Cela arriveroit sans doute, si les planetes n'etoient pas continuellement attirées par la gravité, vers le soleil. Mais quelle que soit la cause de la gravité, il suffit de savoir, que sa force soit la raison renversée des quarrés de la distance au soleil. Or, la derniere force de la lumiere, en cas qu'elle choquat sur les planetes, en les repoussant du soleil, devroit suivre la meme raison, vu que la force de la lumiere diminue aussi en raison du quarré des distances au soleil; elle ne feroit donc que diminuer, tant soit peu, la force de la gravité, et il en seroit de meme, que si les planetes etoient attirées vers le soleil, par une force un peu moindre, mais qui suivroit egalement la meme raison à l'egard de la distance. Or, si les planetes n'eprouvoient qu'une telle force, qui les poussat au soleil, elles devroient toujours decrire des ellipses autour du soleil, et leur grandeur devroit demeurer la meme eternellement: Ainsi, quoique la lumiere agisse par impulsion sur les planetes, en les repoussant au soleil, cela ne sauroit alterer la grandeur de l'orbite de chacune. Mais l'autre effet de cette matiere subtile, qui resulte de la resistence, doit necessairement, tant soit peu, retarder le mouvement. Ou bien, apres que la planete aura parcourru un certain arc, sa vitesse sera un peu moindre, que si elle n'avoit pas eprouvé cette resistance. Or, dés que le mouvement d'une planete est devenu plus lent que la conservation de son orbite l'exige, la force de la gravité l'emporte, et la planete doit s'approcher davantage du soleil. Cela est evident de là, que si le mouvement d'une planete etoit entierement detruit, elle tomberoit directement dans le soleil. Il est donc certain, qu'à cause d'un fluide, dont le ciel est rempli, de quelque nature qu'il soit, *les orbites des planetes doivent insensiblement etre retrecies*, quelque violente que puisse d'ailleurs etre la force repoussante de la lumière.

2. Pour la seconde question, comme l'Hypothese de l'emanation actuelle de la lumiere du soleil est insoutenable, et que la lumiere est plustot de la meme

maniere repandue par l'ether, que le son par l'air, aussi peu qu'un bruit passé augmente un son present, la lumiere du temps passé pourra augmenter celle d'aujourd'huy. Aussi voit on, que dés qu'on rend une chambre obscure, la lumiere qui y etoit entrée auparavant, s'eteint subitement.

3. A la troisieme question, il n'y a point de doute, que si la lumiere etoit une emanation reelle du soleil, son corps en devroit souffrir une perte tres considerable: Si nous considerons la rapidité que ce courrant continuel, qui remplit tout l'univers, devroit avoir, quelque subtile que nous concevions la matiere de la lumiere, je doute fort, que le soleil pourroit subsister seulement une heure; Mais dans l'autre systeme, ou la propagation de la lumiere est semblable à celle du son, le soleil ne perdra plus en nous eclairant, qu'une cloche en sonnant.

4. La quatrieme question n'a aucun doute: car comme il est certain, que nul homme et nulle creature semblable aux notres, ne sauroit subsister dans saturne, ainsi, si la terre avoit eté jamais aussi eloignée du soleil, que saturne l'est aujourd'huy, elle auroit eté absolument inhabitable aux creatures qui l'occupent à present. Donc quoique le monde ait existé depuis toute eternité (ce qui contredit deja au retrecissement successif des orbites) et que l'orbite de la terre soit enfin parvenue à l'etendue qu'elle occupe actuellement, il faudroit absolument, que dans un certain temps, toutes les creatures y fussent produites par un vray miracle.

5. Comme je crois l'hypothese de l'emanation actuelle de la lumiere insoutenâble, cette question[2]) n'a point de difficulté.

6. La derniere question est la plus importante. Il s'agit de savoir, si l'on puisse s'assurer, par les observations, d'un racourcissement des années, ce qui seroit une suite necessaire de la contraction de l'orbite de la terre. J'ay cru que cela pouvoit etre prouvé par les observations rapportées par PTOLOMÉE et la longueur de l'année, qu'il met de 365 jours 5 heures 55′, tandis qu'elle n'est aujourd'hui, que de 365 jours, 5 heures et 48′, sembloit favoriser mon sentiment. Mais apres un examen plus mur, on trouve que PTOLOMÉE s'est trompé dans son année, et qu'elle

2) PONTOPPIDAN a demandé: «En cinquieme lieu, ne pouroit on pas ecarter ces inconveniens par la [sup]position, que le corps du soleil; amoindri par l'emanation de la matiere lumineuse, avoit eté, du commencement, plus vaste et ne s'etoit diminué qu'à proportion de l'approche du globe et des autres planetes?» EJA.

n'a pas eté plus longue alors qu'aujourd'hui. Mais il est certain, que le mouvement de la lune est aujourd'hui un peu plus vite qu'il n'etoit autrefois, ce qui me paroit suffisant pour mon sentiment. Car puisque nous mesurons les années par le nombre des jours, qui nous assure que les jours sont aujourd'hui aussi long qu'autrefois? En cas que les jours soyent devenus un peu plus courts, les ans le seroient aussi, quoiqu'ils continssent le meme nombre de jours. J'ay l'honneur d'etre avec un tres profond respect

Monsieur etc.

Berlin, ce 11 de May 1754.

RECHERCHES
SUR
LA RÉSISTANCE
DU MILIEU
DANS LEQUEL
LES PLANETES SE MEUVENT,
PAR
J. A. EULER,
DE L'ACADÉMIE ROYALE DES SCIENCES ET BELLES-LETTRES
DE PRUSSE.

Hæc super imposuit liquidum, gravitate carentem,
Aethera, nec quicquam terrenæ fæcis habentem.

A' BERLIN,
Chez Chrêtien Fréderic Voſs.
MDCCLXII.

Commentatio A8 indicis ENESTROEMIANI

MÉMOIRE

DANS LEQUEL ON EXAMINE

Si les Planetes ſe meuvent dans un milieu dont la réſiſtance produiſe quelque effet ſenſible ſur leur mouvement ?

Compoſé à l'occaſion du Prix de M. DCC. LXII.

Par M. JEAN ALBERT EULER, de l'Académie Royale des Sciences & Belles-Lettres de Pruſſe.

Hæc ſuper impoſuit liquidum & gravitate carentem
Æthera nec quicquam terrenæ facis habentem.
OVID.

Prix de 1762. A

Commentatio A8[2] indicis ENESTROEMIANI
Recueil des pièces qui ont remporté les prix de l'académie royale des sciences 8, 1771
Septième pièce, 50 p.

MÉMOIRE DANS LEQUEL ON EXAMINE SI LES PLANETES SE MEUVENT DANS UN MILIEU DONT LA RÉSISTANCE PRODUISE QUELQUE EFFET SENSIBLE SUR LEUR MOUVEMENT?

Pour répondre à cette question, il y a trois choses à examiner.

Premierement: Quelle est la nature du milieu dans lequel les Planetes se meuvent?

En second lieu: Il faut rechercher si ce milieu est capable de causer quelque altération dans le mouvement des Planetes?

Et enfin: Il faut déterminer, par le calcul, tous les dérangemens qui sont en effet causés par cette résistance dans le mouvement des Planetes?

PREMIERE PARTIE

Sur la nature du milieu dans lequel les Planetes se meuvent

1. D'abord, on ne sauroit soutenir que l'espace dans lequel les planetes se meuvent soit un vuide parfait. Outre tant d'autres raisons, la seule lumiere prouve suffisamment, que tout l'espace du ciel est rempli de cette matiere subtile dont les rayons de lumiere sont formés. Si les rayons de lumiere étoient des émanations actuelles des corps luisans, lancées avec cette prodigieuse vîtesse qui leur fait parcourir l'espace immense du soleil jusqu'à nous en moins de huit minutes de tems, tout l'espace du ciel seroit rempli de ces émanations lumineuses qui le traverseroient presque en tout sens avec la même rapidité. Mais quoique le grand NEWTON ait soutenu ce sentiment, il est assujetti à tant d'inconvéniens, qu'il me sera permis de l'abandonner et d'embrasser l'autre sentiment, qui explique la propagation de la lumiere d'une maniere semblable à celle du son.

2. Sans parler de l'épuisement que les corps luisans devroient souffrir, suivant le sentiment de NEWTON, le seul phénomene, que plusieurs rayons de lumiere passent sans se troubler par le même point, le détruit tout-à-fait. Il est contre les principes les mieux établis de la Mécanique, que plusieurs particules, quelques subtiles qu'on les conçoive, passent à la fois le même point et en tout sens avec une vîtesse aussi prodigieuse que celle de la lumiere, sans se choquer et troubler le mouvement les unes les autres. Or dans l'autre sistême, nous savons non-seulement par l'expérience, que plusieurs sons traversent le même point sans se troubler; mais M. DE LA GRANGE[1]) a fait voir très-clairement, dans les Mémoires de la Société de Turin, que ce phénomene est parfaitement d'accord avec les principes de Mécanique, et qu'il en est même une suite nécessaire.

3. Il est donc certain, que la lumiere est produite par les corps luisans de la même maniere que le son est produit par les corps sonores, et que la propagation dans l'un et l'autre cas suit les mêmes loix. Il faut donc que tout l'espace des cieux soit rempli d'une matiere propre à transmettre les petites impulsions ou ébranlemens qui constituent la nature de la lumiere, tout comme nous savons à présent, par les heureuses recherches de M. DE LA GRANGE, que le son est transmis par l'air. De-là il s'ensuit que cette matiere céleste doit être fluide et semblable à l'air, en joignant à un certain degré de densité un certain degré d'élasticité, pour produire à la propagation de la lumiere la même vîtesse que l'expérience nous donne à connoître.

4. Or puisque la vîtesse de la lumiere est connue, étant environ six cent mille fois plus grande que celle du son, nous en pouvons conclure un beau rapport entre ce milieu du ciel et notre air. Sachant que la vîtesse des ébranlemens transmis par un milieu élastique est comme la racine quarrée de l'élasticité divisée par la densité, si nous posons l'élasticité du milieu des cieux m fois plus grande que celle de l'air, et sa densité n fois plus petite que celle de l'air, nous aurons $\sqrt{mn} = 600\,000$, ou bien $mn = 360\,000\,000\,000$. Et partant, si nous savions l'élasticité de ce milieu, nous en pourrions conclure sa densité, et réciproquement; par exemple, si son élasticité étoit 600 000 fois plus grande que celle de l'air, sa densité seroit précisément d'autant de fois plus petite.

1) JOSEPH LOUIS LAGRANGE (1736–1813), *Recherches sur la nature, et la propagation du son*, dans *Miscellanea Philosophico-Mathematica Societatis Privatae Taurinensis*, Tomus I (1759), pp. x et 112. EJA.

5. Qu'il me soit permis de donner à ce milieu le nom d'*Ether*, qui est donc une matiere fluide et élastique semblable à l'air, mais qui en differe tant par sa densité que par son élasticité; et quoique nous ne sachions ni l'une ni l'autre séparément, nous connoissons que, posant l'élasticité de l'éther m fois plus grande et sa densité n fois plus petite que celle de l'air, le produit de ces deux nombres mn doit être 360 000 000 000. Il n'y a donc aucun doute que l'un et l'autre de ces deux nombres ne soit très-grand. Puisque l'air, en montant, devient de plus en plus rare, et qu'il se perd enfin dans l'éther, il faut bien que l'éther soit incomparablement plus rare que l'air, et si l'élasticité de l'éther est la cause de la cohésion, dureté et du ressort des corps terrestres, comme il paroît très-vraisemblable, il faut bien que son élasticité soit pour le moins 1000 fois plus grande que celle de l'air. Or, supposant $m = 1000$, la densité de l'éther seroit 360 000 000 fois plus petite que celle de l'air; et si l'on ne supposoit l'élasticité que 100 fois plus grande que celle de l'air, la densité de l'éther deviendroit encore dix fois plus petite.

6. Puisque nous sommes assurés que les planetes et cometes ne souffrent presque aucune résistance dans leur mouvement par l'éther, il faut absolument que cette matiere ou l'éther soit plusieurs mille fois plus rare que l'air, ce qui s'accorde parfaitement bien avec ce que la vîtesse de la lumiere nous vient de fournir. Donc un pied cubique d'éther renfermeroit plusieurs mille fois moins de matiere qu'un pied cubique d'air; et puisque l'air est 800 fois plus léger que l'eau, et celle-ci 19 fois plus légere que l'or, si nous supposons l'éther 360 000 000 fois plus rare que l'air, un pied cubique d'éther contiendra $19 \times 800 \times 360\,000\,000$ moins de matiere qu'un pied cubique d'or; ou bien, un pied cubique d'or contiendroit autant de matiere que 5 472 000 000 000 pieds cubique d'éther; ou encore, qu'un cube d'éther dont le côté seroit de 17500 pieds, c'est-à-dire à-peu-près d'une lieue de France.

7. Ici il se présente d'abord une question très-importante: S'il seroit possible de diviser et subtiliser la matiere d'un pied cubique d'or, ensorte qu'elle remplisse tout l'espace d'une lieue cubique? Je sais bien que KEIL[2]) a prétendu en avoir démontré la possibilité, ayant prouvé que les intervalles entre les particules pourroient devenir moindres qu'une quantité donnée, quelque petite qu'elle fût; mais l'élasticité semble absolument exiger que les moindres particules se touchent tout-à-fait, et qu'elles se trouvent dans une continuité dont il s'en faut beaucoup que

2) JOHN KEILL (1671–1721), «Theoremata quaedam infinitam materiae divisibilitatem spectantia», *Philosophical Transactions* 29 (1714), 82–86. EJA.

Keil ait démontré la possibilité, à moins qu'on ne veuille donner aux particules une figure linéaire presque géométrique, qui ne se touchent que par les pointes; mais une telle structure seroit trop révoltante pour en parler sérieusement dans la Physique. Je crois plutôt qu'on peut hardiment nier qu'il soit possible de former une lieue cubique d'éther d'un pied cubique d'or, par la subtilisation, quoique la quantité de matiere soit de part et d'autre la même. Il y a ici un équivoque qui semble avoir trompé tous ceux qui ont écrit sur cette matiere.

8. Pour mettre cette matiere dans tout son jour, je commence par une remarque générale, que dans tous les corps il faut bien distinguer *leur véritable étendue* de la *masse* ou *quantité de matiere* dont ils sont composés. Or je nomme *la véritable étendue* d'un corps le volume ou la solidité géométrique, qui resteroit, si l'on retranchoit de son volume apparent tous les pores dont il est rempli. On sait que l'or même est tout rempli de pores: donc la véritable étendue d'une masse d'or sera toujours beaucoup plus petite que son volume apparent. Cette véritable étendue de chaque corps est une quantité géométrique, et partant bien différente de la quantité de matiere ou de la masse, qui est une quantité mécanique, en vertu de laquelle les corps s'opposent au changement de leur état; c'est donc *l'inertie*, et ces termes, *quantité de matiere*, *masse* et *inertie* signifient la même chose.

9. Les expériences sur la gravité prouvent suffisamment, que le poids de chaque corps est proportionnel à sa masse ou à son inertie. Donc puisque un pied cubique d'or est 19 fois plus pesant qu'un pied cubique d'eau, il est certain que celui-là contient 19 fois plus de matiere que celui-ci; mais il ne s'ensuit pas que la véritable étendue de l'or soit 19 fois plus grande que la véritable étendue de l'eau; ou bien qu'il seroit possible de réduire une masse d'eau en en ôtant tous les pores à un volume plus que 19 fois plus petit. Il n'est pas encore prouvé que deux corps, dont les masses sont égales, aient aussi la même véritable étendue; et je ne vois nulle nécessité pourquoi deux étendues égales de matiere devroient toujours avoir la même inertie; ou bien, pourquoi la quantité mécanique devroit toujours suivre la quantité géométrique.

10. Cependant, quand nous réfléchissons sur la cause de la gravité, quoiqu'elle nous soit inconnue, il semble qu'on ne la sauroit chercher que dans la pression d'un fluide extrêmement subtil, qui passe librement, même à travers les moindres pores des corps. Or une telle pression agit toujours en raison des volumes; et cela posé, le poids de chaque corps seroit toujours proportionnel à sa véritable

étendue. Donc puisque le poids est aussi proportionnel à l'inertie ou à la masse de chaque corps, il en suivroit que la véritable étendue fût toujours proportionnelle à l'inertie, comme presque tous les Philosophes ont cru jusqu'ici. Mais quelque fort que puisse paroître cet argument, il ne regarde que les corps terrestres sur lesquels la gravité agit, et par la même raison aussi sur tous les corps grossiers dont les planetes sont formées, parce qu'elles sont soumises à la même loi de gravitation; mais on n'en sauroit encore rien conclure de certain à l'égard des matieres subtiles étendues par tout le monde, qui apparemment ne sont pas assujetties à la gravitation, et qui en contiennent plutôt la cause.

11. Il est pourtant fort remarquable, que quoique nous ne voyons aucune liaison entre l'inertie et la vraie étendue d'un corps, tous les corps grossiers non-seulement de la terre mais aussi de tous les corps célestes aient cette propriété, que dans tous l'inertie soit proportionnelle à la vraie étendue, en vertu de laquelle une certaine étendue de matiere ne sauroit exister, sans qu'elle ait une certaine inertie ou masse. De-là on peut soutenir que tous les corps grossiers, quelques différens qu'ils soient en eux-mêmes, sont composés d'une matiere homogene; en prenant, par exemple, des morceaux de toutes sortes de matieres différentes, chacun d'une livre, si nous les concevons réduits à n'avoir plus de pores, tous auront la même étendue et aussi la même inertie. Donc, n'ayant plus de pores, il seroit difficile de dire en quoi tous ces morceaux de matieres seroient différens entr'eux.

12. Mais quelque essentielle que puisse être cette liaison dans les corps grossiers, rien n'empêche que les matieres subtiles ne soient d'une espece différente, et qu'une certaine étendue vraie de ces matieres subtiles ait beaucoup moins d'inertie qu'une égale étendue vraie des matieres grossieres: ce seroit alors une autre espece de matiere; et peut-être y a-t-il encore plusieurs especes dont chacune joint à la même étendue vraie une inertie plus petite que les précédentes. De sorte que le dernier degré, où à une étendue ne conviendroit aucune inertie, seroit une étendue purement géométrique et un vuide véritable. Mais sans admettre un tel vuide, pourvu qu'on accorde deux especes de matieres, dont l'une contienne sous la même étendue moins de masse ou d'inertie que l'autre, on est en état de lever toutes les difficultés qu'on fait ordinairement contre le *Systême du plein*.

13. Puisque dans les corps grossiers l'étendue vraie est le plus étroitement liée avec l'inertie, et que l'inertie d'un corps ne sauroit être changée par quelque cause que ce fût; il s'ensuit que la vraie étendue d'un corps grossier ne souffre aucun

changement dans sa quantité. Mais pour les matieres subtiles, peut être que leur nature est tout-à-fait différente à cet égard. Il semble être bien nécessaire que la même quantité conserve toujours la même inertie; mais ne seroit-il pas possible que la vraie étendue qui exclut tous les pores, devînt tantôt plus grande tantôt plus petite? Ne seroit-il pas encore possible qu'une telle matiere soit douée d'une force de s'étendre continuellement davantage dans sa propre substance, sans y recevoir des pores ou des espaces vuides? Cela ne seroit pas un aggrandissement réel; il est bien vrai, l'inertie qui semble constituer l'essence des matieres y demeurant la même, un tel aggrandissement ne sauroit être admis sans un miracle; mais dans ce cas, ne seroit-on pas moins embarrassé de la cause de l'élasticité de l'éther? Je n'ose entrer dans ces sublimes recherches; la pénétration m'y manque, et le sujet présent ne l'exige pas.

14. Je me contente d'avoir prouvé la possibilité que les espaces du ciel, par lesquels les planetes semblent se mouvoir librement, peuvent être remplis d'une matiere extrêmement subtile et très-élastique, sans les supposer presque vuides, comme on est obligé de faire quand on joint partout à la même inertie la même étendue vraie de matiere. Non seulement un tel vuide choque notre esprit, mais il paroît encore très-incompatible avec cette grande élasticité qu'on est obligé d'attribuer à l'éther. Donc quand on dit que *l'éther est tant de mille fois plus rare que l'air*, il ne faut pas entendre que *l'étendue propre d'un certain volume d'éther soit autant de fois plus petite que l'étendue propre d'un égal volume d'air*, mais que *cette proportion regarde uniquement l'inertie renfermée en des volumes égaux.*

SECONDE PARTIE

Sur la Résistance de l'Éther

15. Ici il se présente d'abord la question: S'il ne seroit pas possible que les planetes se meussent par l'éther sans souffrir la moindre résistance? Quoiqu'elles en soient poussées en arriere, ne seroit-il pas possible qu'elles en fussent poussées aussi en avant avec une force égale? car lorsqu'un corps se meut dans l'éther, il en déplace continuellement une partie; et en lui imprimant un mouvement, il en perd bien autant; et puisque l'éther derriere le corps est poussé par son élasticité dans les lieux que le corps quitte, il semble qu'il le pourroit accélérer autant qu'il aura été retardé en avant. Ce sentiment a été soutenu par de grands Géometres, qui l'ont cru conforme à la conservation des forces vives; ils tombent bien d'accord que dès le

premier instant le corps communique à l'éther une partie de sa force vive, pour y produire le mouvement dont l'éther, chassé en avant, va suivre le corps par un détour; mais dès que ce mouvement est une fois engendré, ils prétendent qu'il suffit pour accompagner le corps par tout son mouvement, sans qu'il ait besoin de souffrir une nouvelle perte. Ils regardent cette conservation comme l'effet de la parfaite élasticité de l'éther. Si les planètes, disent-ils, perdoient continuellement de leur mouvement, cette force vive ou périroit tout-à-fait, ou s'accumuleroit dans l'éther. Or l'un et l'autre leur paroît également absurde.

16. Mais quelque fondé que puisse paroître ce raisonnement, il est détruit par l'expérience. L'air étant un fluide assez parfaitement élastique, il devroit au moins participer de la même qualité, et causer une moindre résistance aux corps qui s'y meuvent, que s'il étoit destitué de toute élasticité. Or nous savons que tous les corps qui se meuvent dans l'air y souffrent une résistance très-considérable, et M. LULOFFS[3]) prétend même avoir prouvé, par la force que le vent exerce sur les aîles de moulins à vent, que la résistance de l'air est encore plus grande que celle qu'on trouve par les regles ordinaires de Mécanique. Il est aussi incontestable, qu'un boulet de canon éprouve une plus grande résistance que selon ces regles, parce qu'il laisse derriere lui un espace vuide, que l'air ne sauroit remplir assez rapidement. D'ou l'on peut conclure que quoique l'éther soit beaucoup plus élastique que l'air, cela n'empêche point qu'il n'oppose une résistance très-réelle au mouvement des planetes.

17. Puisque les planetes se meuvent incomparablement plus vîte qu'un boulet de canon, on en pourroit conclure de même, qu'en arriere l'éther dût être plus rare qu'ailleurs, et en avant plus dense et plus accumulé, tout comme cela arrive dans l'air. En appliquant cela à la terre, on verra que la plus grande rareté de l'éther répond aux lieux qui voient le soleil dans le couchant, et la plus grande densité aux lieux auxquels le soleil se lève. Donc partout au soir l'atmosphere sera le moins chargée d'éther, et au matin le plus; et cette variation ne sauroit manquer de produire des phénomenes bien singuliers. Si l'électricité est causée par un dérangement dans l'état d'équilibre de l'éther, et que l'électricité positive ait lieu où l'éther se trouve en trop grande abondance, et la négative où l'éther est trop rare, il

3) JOHAN LULOFS, «Wiskundige en Werktuigkundige Beschouwinge der Wind-Molens», *Verhandelingen uitgegeeven door de Hollandse Maatschappy der Weetenschappen te Haarlem* 2 (1755), 525–621, particulièrement pp. 590–600. EJA.

s'ensuivroit que partout vers le soir il regne dans l'atmosphere une électricité négative, et vers le matin une positive. Il s'agit donc de consulter l'expérience, si une telle variation a lieu ou non? mon but ne me permet pas d'entrer dans cette discussion.

18. Cependant lorsqu'un corps se meut dans l'éther, on n'en peut pas déterminer la résistance sur le même pied que dans l'air ou dans l'eau, où toute la surface antérieure du corps reçoit l'impulsion du fluide. Comme l'éther est une matiere extrêmement subtile, il pénetre presque librement tous les pores des corps; il en est presque de même que si par exemple un crible se mouvoit dans l'eau ou dans l'air, qui sans doute souffriroit une résistance beaucoup plus petite qu'une surface solide. Les planetes ne rencontrent donc de résistance dans l'éther qu'entant que leurs parties solides empêchent que l'éther ne passe tout-à-fait librement à travers de leur masse. D'où l'on voit que la résistance determinée par la regle ordinaire doit être diminuée de la partie qui répond au libre passage de l'éther; ou bien il ne faut considérer qu'une certaine partie de la surface de la planete qui est exposée à la résistance, et qui, selon toutes les apparences, sera très-petite à l'égard de toute la surface. D'ailleurs, l'obliquité dont les particules solides sont choquées par l'éther, peut encore considérablement diminuer la résistance.

19. Concevons donc un corps sphérique, dont la masse soit A, le rayon $= a$, et qui se meut avec une vîtesse égale à celle qu'un corps pesant sur la terre acquerreroit en tombant de la hauteur $= v$. Soit la densité du corps n fois plus grande que celle de l'éther, et selon la regle commune la résistance qu'éprouve le grand cercle est exprimé par le poids d'un cylindre d'éther, dont le rayon de la base $= a$ et la hauteur $= v$, et partant la solidité $= \pi aav$. Or la solidité du globe etant $\frac{4}{3}\pi a^3$ et la masse $= A$, la masse de ce cylindre, s'il étoit de la même matiere, seroit $= \frac{3Av}{4a}$, donc la masse de ce cylindre d'éther $= \frac{3Av}{4na}$, qui exprimeroit la résistance d'un grand cercle du globe, réduisant la masse au poids que ce même corps auroit sur la terre. Mais la résistance qu'éprouve le globe même est d'abord deux fois plus petite[4]), ou $\frac{3Av}{8na}$, et il la faudra encore diminuer, à cause de la pénétration de l'éther, et peut-être aussi à cause d'une plus grande obliquité d'impulsion.

4) Voir Isaac Newton, *Philosophiae Naturalis Principia Mathematica* (1726), troisième édition, Livre II, proposition 35, scholie. EJA.

20. Comme nous devons nous contenter de savoir cette diminution en gros, et qu'il est impossible de la déterminer *à priori*, posons la véritable resistance $= \frac{3Av}{8\lambda na}$ où, selon toutes les apparences, le nombre λ doit être assez considérable; or le nombre n est prodigieusement grand; car si nous supposons la densité de l'éther 360 000 000 fois plus petite que celle de l'air, et que la densité du corps soit égale à celle de l'eau, le nombre n sera $= 800 \times 360\,000\,000 = 288\,000\,000\,000$. Ensuite la vraie étendue du corps étant pour le moins 19 fois plus petite que l'apparente, à cause de sa nature criblée à l'égard de l'éther, le nombre λ pourroit bien surpasser 10; donc posant pour abréger $\frac{3}{8\lambda n} = \mu$, la valeur de cette fraction seroit environ $\mu = 1:10\,000\,000\,000\,000$, laquelle demeureroit à-peu-près la même, si le corps étoit plus ou moins dense. Or si nous supposons l'éther encore dix fois moins dense, nous aurions pour μ une fraction encore dix fois plus petite; et comme il est possible que la valeur de λ soit considérablement plus grande que dix, la fraction μ pourroit être encore plus petite que $1:100\,000\,000\,000\,000$; et on verra par la suite que la résistance qui en résulte, pourra très-bien subsister avec les observations.

21. Appliquons cela au mouvement d'une planete qui se meuve autour du soleil dans un cercle parfait, puisque nous savons que l'effet de cette résistance est extrêmement petit, et que le mouvement continuera de se faire dans un cercle pendant très-longtems, cherchons la diminution du mouvement pendant un tems quelconque. Soit la distance de la planete au soleil $= c$, sa vîtesse dûe à la hauteur $= v$, et l'attraction du soleil à cette distance $c = \frac{ff}{cc}$; prenant pour l'unité la gravité sur la terre, et parce que le mouvement se fait dans un cercle, il faut que la force centrifuge soit $= \frac{2v}{c} = \frac{ff}{cc}$, et partant $v = \frac{ff}{2c}$; donc la résistance de l'éther $\frac{\mu v}{a} = \frac{\mu ff}{2ac}$, que nous pouvons regarder comme constante pour un très-long tems. De-là, pendant que la planete parcourre l'espace ds, nous aurons $dv = -\frac{\mu ff}{2ac}ds$, et par conséquent[5]) $v = \frac{ff}{2c} - \frac{\mu ffs}{2ac} = \frac{ff}{2c}\left(1 - \frac{\mu s}{a}\right)$.

5) Edition originale: $= \frac{ff}{2c}$ manque. Corrigé par EJA.

22. Pour mieux connoître les élémens de cette expression, soit le tems périodique de la planète $= \Theta$ secondes, pendant lequel elle parcourre toute la circonférence du cercle $= 2\pi c$; de sorte que dans une seconde elle parcourre l'espace $= \frac{2\pi c}{\Theta}$, sa vîtesse étant dûe à la hauteur $= v$. Or posant g pour la hauteur par laquelle un corps grave tombe dans une seconde, cet espace parcouru par la planete dans une seconde, est aussi[6]) $= 2\surd gv$; donc $\surd gv = \frac{\pi c}{\Theta} = \sqrt{\frac{gff}{2c}}$: par conséquent $ff = \frac{2\pi\pi c^3}{g\Theta\Theta}$; d'où l'on connoîtra le rapport de la force accélératrice du soleil à chaque distance à la force accélératrice de la gravité naturelle sur la terre.

23. Quoique le retardement du mouvement dérange d'abord le mouvement circulaire; avant que d'entreprendre les recherches réquises pour développer cette question, je considererai ici la chose comme si la planete se mouvoit dans un canal circulaire, qui l'empêchât d'en sortir. Ce cas, quoique imaginaire, nous fera connoître en gros après combien de tems l'effet de la résistance peut devenir sensible; parce que pendant une révolution la planete parcourt l'espace $= 2\pi c$, et pendant ν révolutions ou dans le tems de $\nu\Theta$ secondes l'espace $= 2\nu\pi c$; posant cette valeur pour s, nous aurons pour la vitesse de la planete après ce tems

$$v = \frac{ff}{2c}\left(1 - \frac{2\mu\nu\pi c}{a}\right),$$

et la vitesse même

$$\surd v = \left(1 - \frac{\mu\nu\pi c}{a}\right)\sqrt{\frac{ff}{2c}};$$

d'où nous voyons que la diminution de la vitesse vaut la partie $\frac{\mu\nu\pi c}{a}$ de la vitesse initiale.

24. Faisons l'application de cette formule au mouvement de la terre, où l'on a environ $c = 18\,000\,a$ et $\Theta = 31\,556\,930''$, et cherchons après combien de tems la diminution de la vitesse pourroit devenir

$$\frac{\mu\nu\pi c}{a} = 18\,000\,\mu\nu\pi = \frac{1}{31\,556\,930},$$

6) $\surd gv$ signifie $\surd(gv)$, etc. EJA.

puisqu'un changement d'une seconde dans le tems périodique de la terre seroit déja remarquable. Supposons que cela arrive après v ans, et nous aurons

$$v = \frac{1}{18\,000 \times 31\,556\,930\,\mu\pi}.$$

Donnons à μ la valeur marquée ci-dessus (§ 20), et nous obtiendrons

$$v = \frac{10\,000\,000\,000\,000}{18\,000 \times 31\,556\,930\,\pi},$$

ou bien

$$v = \frac{100}{18} \text{ à-peu-près}.$$

De-là il s'ensuivroit qu'un tel effet pourroit être produit en six ans. Or quand même la valeur de μ seroit encore plus petite, l'effet de la résistance de l'éther sur le mouvement des planetes, pourroit toutefois devenir très-sensible après un assez grand nombre d'années.

25. Mais il ne suit pas de la diminution de la vitesse, que le tems périodique doive devenir plus long; il en doit même résulter un effet entierement contraire. La planete étant rallentie dans son mouvement, s'approchera plus du soleil, et décrira une orbite plus petite, à laquelle répond nécessairement un tems périodique plus court. Par cette raison, il s'en faut beaucoup que la détermination précédente soit juste, quand même le nombre μ seroit assez exact. Le véritable dérangement, que la résistance de l'éther peut causer dans le mouvement des planetes, demande des recherches beaucoup plus profondes, et qui feront le sujet de ma Troisieme Partie.

TROISIEME PARTIE

Sur le dérangement du mouvement des Planetes causé par la résistance de l'Éther

26. Considérons donc une planete quelconque [fig. 1], dont la masse soit A, son rayon $= a$, et laquelle étant attirée vers le soleil en $\odot$ et assujettie à la résistance de l'éther, décrive la ligne courbe AM; que cette planete ait parcouru après un tems de t secondes l'arc AM, et soit l'angle au soleil $A \odot M = \varphi$. Soit

ensuite la distance de la planete au soleil $\odot M = z$, et la force accélératrice du soleil étant posée $= \frac{ff}{zz}$, nous avons vu que si nous exprimons le tems périodique de la

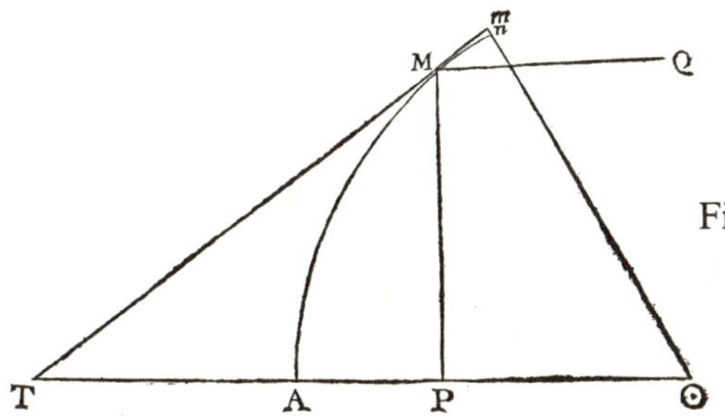

Fig. 1

terre par Θ secondes, sa distance moyenne au soleil par c et la hauteur par laquelle un corps grave tombe sur la terre dans une seconde par g, il sera $ff = \frac{2\pi\pi c^3}{g\Theta^2}$ où π marque la circonférence d'un cercle dont le diametre est $= 1$. Or la force accélératrice du soleil $\frac{ff}{zz}$ agit sur la planete selon la direction $M\odot$, et en la décomposant selon les directions fixes des coordonnées $\odot P = x$; $PM = y$; à cause de $x = z\cos\varphi$ et $y = z\sin\varphi$, nous aurons les forces selon $MP = \frac{ff}{zz}\sin\varphi$ et selon $MQ = \frac{ff}{zz}\cos\varphi$.

27. Maintenant pour connoître la vîtesse de la planete de laquelle dépend la résistance, soit Mm l'élément d'espace parcouru dans le tems dt et à cause de $Mn = z\,d\varphi$ et $mn = dz$, nous aurons $Mm = \sqrt{(dz^2 + zz\,d\varphi^2)}$ que je nommerai pour abréger $= ds$ et faisant la proportion $dt : ds = 1 : \frac{ds}{dt}$. Or prenant v pour la hauteur dûe à la vitesse de la planete, ce même espace parcouru par la planete dans une seconde est aussi $= 2\sqrt{gv}$, d'où nous tirons $v = \frac{ds^2}{4gdt^2}$, et partant la force accélératrice de la résistance, contraire au mouvement et agissant selon la tangente MT, sera suivant les principes établis ci-dessus $= \frac{\mu v}{a} = \frac{\mu\,ds^2}{4ga\,dt^2}$, où μ est une fraction extrêmement petite, que j'ai estimée ci-dessus (§20). Ensuite pour la dé-

composition de cette force qui est selon MT, ayant $MT:PT:MP = ds:-dx:dy$, nous trouvons la force selon $PT = -\frac{\mu\, dx\, ds}{4ga\, dt^2}$ et selon $MP = \frac{\mu\, dy\, ds}{4ga\, dt^2}$.

28. Donc conjointement la planete en M est sollicitée par ces deux forces accélératrices:

$$\text{L'une selon } MQ = \frac{ff\cos\varphi}{zz} + \frac{\mu\, dx\, ds}{4ga\, dt^2},$$

$$\text{Et l'autre selon } MP = \frac{ff\sin\varphi}{zz} + \frac{\mu\, dy\, ds}{4ga\, dt^2}.$$

De-là les principes de la Mécanique nous fournissent ces deux équations:

$$ddx = -\frac{2gff\, dt^2\cos\varphi}{zz} - \frac{\mu\, dx\, ds}{2a},$$

et

$$ddy = -\frac{2gff\, dt^2\sin\varphi}{zz} - \frac{\mu\, dy\, ds}{2a};$$

où il faut remarquer que l'on a $x = z\cos\varphi$; $y = z\sin\varphi$; donc

$$dx = dz\cos\varphi - z\, d\varphi\sin\varphi;$$

$$dy = dz\sin\varphi + z\, d\varphi\cos\varphi;$$

et

$$ddx = ddz\cos\varphi - 2dz\, d\varphi\sin\varphi - z\, d\varphi^2\cos\varphi - z\, dd\varphi\sin\varphi;$$

$$ddy = ddz\sin\varphi + 2dz\, d\varphi\cos\varphi - z\, d\varphi^2\sin\varphi + z\, dd\varphi\cos\varphi;$$

ensuite $ds = \sqrt{(dx^2 + dy^2)} = \sqrt{(dz^2 + zz\, d\varphi^2)}$; d'où nous tirons aisément ces deux équations:

I. $$ddx\sin\varphi - ddy\cos\varphi = -2dz\, d\varphi - z\, dd\varphi = \frac{\mu z\, d\varphi\, ds}{2a};$$

II. $$ddx\cos\varphi + ddy\sin\varphi = ddz - z\, d\varphi^2 = -\frac{2gff\, dt^2}{zz} - \frac{\mu\, dz\, ds}{2a};$$

où l'élément du tems dt est supposé constant, et $2gff = \frac{4\pi\pi c^3}{\Theta\Theta}$, de sorte que la quantité g sort du calcul.

29. La premiere de ces deux équations étant divisée par $zd\varphi$, donne

$$\frac{2dz}{z} + \frac{dd\varphi}{d\varphi} + \frac{\mu ds}{2a} = 0,$$

dont l'intégrale est

$$zz\,d\varphi \cdot e^{\mu s/2a} = G\,dt,$$

e étant le nombre dont le logarithme hyperbolique est l'unité, et G une quantité constante. Ensuite puisque $ds\,dds = dz\,ddz + z\,dz\,d\varphi^2 + zz\,d\varphi\,dd\varphi$, si nous multiplions la premiere par $-z\,d\varphi$, et la seconde par dz, la somme en sera

1. $$ds\,dds = -\frac{2gffdt^2\,dz}{zz} - \frac{\mu\,ds^3}{2a};$$

or l'équation $zz\,d\varphi \cdot e^{\mu s/2a} = G\,dt$, à cause de $zz\,d\varphi^2 = ds^2 - dz^2$, en éliminant $d\varphi$, donne

2. $$e^{\mu s/a}\,zz\,(ds^2 - dz^2) = GG\,dt^2.$$

De sorte que nous avons deux équations entre les trois variables z, s, t; si nous multiplions celle-là (1.) par $\frac{2}{dt^2}$, l'intégration fournira

$$\frac{ds^2}{dt^2} = 4gff\left(\frac{1}{z} + \frac{1}{b}\right) - \frac{\mu}{a}\int\frac{ds^3}{dt^2},$$

qui à cause de μ presque évanouissant se réduit à

$$\frac{ds^2}{dt^2} = 4gff\left(\frac{1}{z} + \frac{1}{b}\right) - \frac{4\mu gff}{a}\int ds\left(\frac{1}{z} + \frac{1}{b}\right).$$

30. Puisque nous savons que l'effet de la résistance est extrêmement petit, négligeons d'abord les termes affectés par μ; et ayant ces deux équations

$$\frac{ds^2}{dt^2} = 4gff\left(\frac{1}{z} + \frac{1}{b}\right) \quad \text{et} \quad GG\,dt^2 = zz\,(ds^2 - dz^2),$$

en éliminant dt^2, nous trouvons

$$ds^2 = \frac{4gff}{GG}\,zz\,(ds^2 - dz^2)\left(\frac{1}{z} + \frac{1}{b}\right);$$

donc posant, pour abréger,

$$\frac{4gff}{GG} = \frac{8\pi\pi c^3}{GG\Theta\Theta} = \frac{1}{h},$$

de sorte que

$$G = \frac{2\pi c}{\Theta}\sqrt{2ch} \quad (\S 26)$$

nous obtiendrons

$$ds = \frac{z\,dz\sqrt{\left(\frac{1}{z} + \frac{1}{b}\right)}}{\sqrt{\left(zz\left(\frac{1}{z} + \frac{1}{b}\right) - h\right)}} \quad \text{ou} \quad ds = \frac{dz}{\sqrt{\left(1 - \frac{bh}{z(z+b)}\right)}}.$$

31. Cette valeur de ds suffit pour être introduite dans les petits termes de nos équations affectés par μ. Soit donc

$$\frac{dz}{\sqrt{\left(1 - \frac{bh}{z(z+b)}\right)}} = d\sigma,$$

ou bien qu'on écrive dans tous les termes affectés par μ, au lieu de s, la lettre σ, et nous aurons les équations suivantes:

I. $$e^{\mu\sigma/a}zz(ds^2 - dz^2) = G\,dt^2 = \frac{8\pi\pi c^3 h}{\Theta\Theta}\,dt^2;$$

II. $$\frac{ds^2}{dt^2} = \frac{8\pi\pi c^3}{\Theta\Theta}\left(\frac{1}{z} + \frac{1}{b} - \frac{\mu}{a}\int d\sigma\left(\frac{1}{z} + \frac{1}{b}\right)\right);$$

III. $$e^{\mu\sigma/2a}zz\,d\varphi = \frac{2\pi c\sqrt{2ch}}{\Theta}\,dt;$$

où σ est une fonction de z. Donc éliminant dt^2 des deux premieres équations, nous obtiendrons:

$$e^{\mu\sigma/a} zz\,(ds^2 - dz^2)\left(\frac{1}{z}+\frac{1}{b}-\frac{\mu}{a}\int d\sigma\left(\frac{1}{z}+\frac{1}{b}\right)\right) = h\,ds^2$$

et de-là

1. $$ds = \frac{e^{\mu\sigma/2a}\,z\,dz\sqrt{\left(\frac{1}{z}+\frac{1}{b}-\frac{\mu}{a}\int d\sigma\left(\frac{1}{z}+\frac{1}{b}\right)\right)}}{\sqrt{\left(e^{\mu\sigma/a} zz\left(\frac{1}{z}+\frac{1}{b}-\frac{\mu}{a}\int d\sigma\left(\frac{1}{z}+\frac{1}{b}\right)\right)-h\right)}}.$$

D'où l'on trouve plus exactement s par z. Ensuite on aura,

2. $$dt = \frac{\Theta}{2\pi c\sqrt{2c}}\cdot\frac{e^{\mu\sigma/2a}\,z\,dz}{\sqrt{\left(e^{\mu\sigma/a} zz\left(\frac{1}{z}+\frac{1}{b}-\frac{\mu}{a}\int d\sigma\left(\frac{1}{z}+\frac{1}{b}\right)\right)-h\right)}}$$

et

3. $$d\varphi = \frac{dz\sqrt{h}}{z\sqrt{\left(e^{\mu\sigma/a} zz\left(\frac{1}{z}+\frac{1}{b}-\frac{\mu}{a}\int d\sigma\left(\frac{1}{z}+\frac{1}{b}\right)\right)-h\right)}}.$$

32. Pour profiter de ces équations, il faut tâcher d'en appliquer la solution à l'usage de l'Astronomie. Pour cet effet, prenant $h = \frac{k}{2}$ et $b = -\frac{2k}{1-nn}$, on a pour le cas où la résistance évanouit, $z = \frac{k}{1+n\cos\omega}$ et $d\varphi = d\omega$; de sorte que k exprime le demi-parametre, n l'excentricité, et ω l'anomalie vraie de l'orbite comptée depuis le périhélie. Cette expression de z, comme très-approchante, étant introduite pour avoir la valeur de σ, on trouve,

$$d\sigma = \frac{k\,d\omega\sqrt{(1+2n\cos\omega+nn)}}{(1+n\cos\omega)^2};$$

laquelle expression est assez exacte, pourvu qu'on ne suppose pas un tems assez long, pour que l'effet de la résistance devienne trop considérable. Or en n'étendant le calcul qu'à quelques siecles, pendant lesquels l'effet de la résistance demeure encore presque insensible, on peut ensuite répéter le même calcul pour les siecles suivans de la même maniere.

33. Puisque dans le cas de nulle résistance on a

$$z = \frac{k}{1 + n\cos\omega},$$

supposons, en tenant compte de la résistance

$$z = \frac{k}{1 + n\cos\omega + \mu u},$$

et tâchons de déterminer u en sorte qu'il demeure $d\varphi = d\omega$. Pour cet effet, ayant

$$\frac{1}{z} = \frac{1 + n\cos\omega + \mu u}{k},$$

la différentiation nous fournit

$$\frac{dz}{zz} = \frac{nd\omega\sin\omega - \mu du}{k};$$

laquelle valeur doit être introduite dans la troisieme équation (§ 31), qui à cause de $d\varphi = d\omega$ et $e^{\mu\sigma/a} = 1 + \dfrac{\mu\sigma}{a}$ se réduit à cette forme:

$$d\omega = \frac{dz\sqrt{h}}{zz\sqrt{\left(\frac{1}{z} + \frac{1}{b} - \frac{\mu}{a}\int d\sigma\left(\frac{1}{z} + \frac{1}{b}\right) + \frac{\mu\sigma}{a}\left(\frac{1}{z} + \frac{1}{b}\right) - \frac{h}{zz}\right)}}$$

ou bien à cause de

$$-\int d\sigma\left(\frac{1}{z} + \frac{1}{b}\right) + \sigma\left(\frac{1}{z} + \frac{1}{b}\right) = \int \sigma d\left(\frac{1}{z} + \frac{1}{b}\right) = -\int\frac{\sigma\, dz}{zz} = -\frac{n}{k}\int d\omega\sin\omega$$

$$= -n\int d\omega\sin\omega\int\frac{d\omega\sqrt{(1 + 2n\cos\omega + nn)}}{(1 + n\cos\omega)^2}$$

à celle-ci:

$$d\omega = \frac{dz\sqrt{h}}{zz\sqrt{\left(\frac{1}{z}+\frac{1}{b}-\frac{h}{zz}-\frac{\mu n}{a}\int d\omega \sin\omega \int \frac{d\omega\sqrt{(1+2n\cos\omega+nn)}}{(1+n\cos\omega)^2}\right)}}.$$

34. Or $\frac{1}{b}$ étant $=\frac{-1+nn}{2k}$, nous aurons

$$\frac{1}{z}+\frac{1}{b}=\frac{1+2n\cos\omega+nn+2\mu u}{2k};$$

et puisque $h=\frac{k}{2}$ nous obtiendrons[7])

$$\frac{1}{z}+\frac{1}{b}-\frac{h}{zz}=\frac{nn\sin\omega^2-2\mu u\, n\cos\omega}{2k}.$$

Par conséquent

$$d\omega = \frac{n\,d\omega\sin\omega-\mu\,du}{\sqrt{\left(nn\sin\omega^2-2\mu nu\cos\omega-\frac{2\mu n}{a}\int\sigma\,d\omega\sin\omega\right)}}$$

d'où prenant les quarrés, on trouve, en négligeant les termes qui renfermeroient $\mu\mu$,

$$nn\,d\omega\sin\omega^2-2\mu nu\,d\omega\cos\omega-\frac{2\mu n}{a}d\omega\int\sigma\,d\omega\sin\omega=nn\,d\omega\sin\omega^2-2n\mu\,du\sin\omega;$$

et partant,

$$du\sin\omega-u\,d\omega\cos\omega=\frac{d\omega}{a}\int\sigma\,d\omega\sin\omega,$$

qui étant divisée par $\sin\omega^2$ et intégrée, donne

$$\frac{u}{\sin\omega}=\frac{1}{a}\int\frac{d\omega}{\sin\omega^2}\int\sigma\,d\omega\sin\omega=-\frac{\cos\omega}{a\sin\omega}\int\sigma\,d\omega\sin\omega+\frac{1}{a}\int\sigma\,d\omega\cos\omega;$$

7) $\sin\omega^2$ signifie $\sin^2\omega$. EJA.

ou bien

$$u = \frac{1}{a}(\sin\omega \int \sigma\, d\omega \cos\omega - \cos\omega \int \sigma d\omega \sin\omega),$$

ayant

$$\sigma = k \int \frac{d\omega \sqrt{(1 + 2n\cos\omega + nn)}}{(1 + n\cos\omega)^2}.$$

35. Pourvu que l'excentricité soit moindre que l'unité ou $n < 1$, comme il arrive toujours dans les ellipses, l'arc σ se réduit à une telle expression:

$$\sigma = k(A\omega + Bn\sin\omega + Cn^2\sin 2\omega + Dn^3\sin 3\omega + En^4\sin 4\omega + \text{etc.}).$$

Or pour en trouver d'autant plus facilement la valeur de u, prenons la différentielle de l'équation trouvée ci-dessus,

$$du\sin\omega - u\, d\omega\cos\omega = \frac{d\omega}{a}\int \sigma\, d\omega \sin\omega,$$

qui est:

$$ddu\sin\omega + u\, d\omega^2 \sin\omega = \frac{d\omega^2}{a}\sigma\sin\omega,$$

ou bien

$$\frac{ddu}{d\omega^2} + u = \frac{\sigma}{a} = \frac{k}{a}(A\omega + Bn\sin\omega + Cn^2\sin 2\omega + Dn^3\sin 3\omega + En^4\sin 4\omega + \text{ etc.}).$$

Posons maintenant

$$u = \alpha\omega + \beta\sin\omega + \varDelta\omega\cos\omega + \gamma\sin 2\omega + \delta\sin 3\omega + \sigma\sin 4\omega + \text{ etc.}$$

et puisque

$$\frac{du}{d\omega} = \alpha + \beta\cos\omega + \varDelta\cos\omega - \varDelta\omega\sin\omega + 2\gamma\cos 2\omega + 3\delta\cos 3\omega + 4\varepsilon\cos 4\omega + \text{ etc.}$$

ensuite

$$\frac{ddu}{d\omega^2} = -\beta\sin\omega - 2\varDelta\sin\omega - \varDelta\omega\cos\omega - 4\gamma\sin 2\omega - 9\delta\sin 3\omega - 16\varepsilon\sin 4\omega - \text{ etc.}$$

donc

$$\frac{ddu}{d\omega^2} + u = \alpha\omega - 2\varDelta \sin\omega - 3\gamma \sin 2\omega - 8\delta \sin 3\omega - 15\varepsilon \sin 4\omega - \text{ etc.}$$

d'où nous tirons $\alpha = \frac{Ak}{a}$, $\varDelta = -\frac{Bnk}{2a}$, $\gamma = -\frac{Cn^2k}{3a}$, $\delta = -\frac{Dn^3k}{8a}$, $\varepsilon = -\frac{En^4k}{15a}$ etc. et par conséquent

$$u = \frac{k}{a}\Big(A\omega + \beta \sin\omega - \frac{Bn}{2}\,\omega\cos\omega - \frac{1}{3}\,Cn^2 \sin 2\omega - \frac{1}{8}\,Dn^3 \sin 3\omega - \frac{1}{15}\,En^4 \sin 4\omega - \text{ etc.}\Big)$$

où le coëfficient β demeure indéterminé; on le pourra donc prendre $=0$, car quelque valeur qu'on donneroit à β, on ne feroit que changer l'excentricité et le lieu de l'aphélie.

36. Donc posant

$$\int \frac{d\omega\sqrt{(1+2n\cos\omega+nn)}}{(1+n\cos\omega)^2} = A\omega + Bn\sin\omega + Cn^2\sin 2\omega + Dn^3\sin 3\omega + En^4\sin 4\omega + \text{ etc.}$$

nous aurons, après que la planete aura achevé l'angle $\varphi = \omega$ pour sa distance au soleil:

$$z = \frac{k}{1 + n\cos\omega + \frac{\mu k}{a}(A\omega - \frac{1}{2}Bn\omega\cos\omega - \frac{1}{3}Cn^2\sin 2\omega - \frac{1}{8}Dn^3\sin 3\omega - \text{etc.})}.$$

Ensuite le tems t se trouvera déterminé par cette équation:

$$\frac{2\pi c\sqrt{ck}}{\Theta}\,t = \int\Big(1 + \frac{\mu\sigma}{2a}\Big)\,zz\,d\omega\,.$$

Or

$$zz = \frac{kk}{(1+n\cos\omega)^2} - \frac{2\mu k^3}{a(1+n\cos\omega)^3}\,(A\omega - \tfrac{1}{2}Bn\omega\cos\omega - \tfrac{1}{3}Cn^2\sin 2\omega$$
$$- \tfrac{1}{8}Dn^3\sin 3\omega - \text{ etc.})$$

et

$$1 + \frac{\mu\sigma}{2a} = 1 + \frac{\mu k}{2a}(A\omega + Bn\sin\omega + Cn^2\sin 2\omega + Dn^3\sin 3\omega + \text{ etc.})$$

donc

$$\frac{2\pi c\sqrt{c}}{\Theta k\sqrt{k}}\,t = \int\frac{d\omega}{(1+n\cos\omega)^2}$$
$$-\frac{2\mu k}{a}\int\frac{d\omega\,(A\omega - \tfrac{1}{2}Bn\omega\cos\omega - \tfrac{1}{3}Cn^2\sin 2\omega - \tfrac{1}{8}Dn^3\sin 3\omega - \text{ etc.})}{(1+n\cos\omega)^3}$$
$$+\frac{\mu k}{2a}\int\frac{d\omega\,(A\omega + Bn\sin\omega + Cn^2\sin 2\omega + Dn^3\sin 3\omega + \text{ etc.})}{(1+n\cos\omega)^2}.$$

37. Ces formules suffisent pour déterminer tous les dérangemens que la résistance de l'éther peut causer dans le mouvement de toutes les planetes; ce qui étant la question proposée par l'*Illustre Académie Royale des Sciences*, je m'en vais en developper tous les phénomenes qui doivent se trouver dans le mouvement des Planetes. On ne sera pas surpris que je fasse ici abstraction des dérangemens que les planetes se causent par leur action mutuelle, il est évident qu'elle n'influe point sur l'effet de la résistance. Je considererai donc trois cas, le premier où l'orbite de la planète n'a aucune excentricité, le second où l'excentricité est très-petite, et le troisieme où l'excentricité est médiocre.

I. L'Orbite de la Planete n'ayant aucune excentricité

38. Dans ce cas $n = 0$ on aura $\sigma = k\omega$ (§32), $A = 1$, $B = 0$, $C = 0$, etc. (§35) et partant nos formules seront:

$$z = \frac{k}{1 + \dfrac{\mu k\omega}{a}} \quad \text{et} \quad \frac{2\pi c\sqrt{c}}{\Theta k\sqrt{k}}\,t = \omega - \frac{3\mu k}{4a}\,\omega\omega \quad (\S 36)$$

où k est le rayon de l'orbite au tems $t = 0$, a le rayon du corps de la planete, et ω l'angle décrit autour du soleil dans un tems quelconque t. Là-dessus je fais les réflexions suivantes:

1° Après une révolution où $\omega = 2\pi$, la distance de la planete au soleil sera

$$z = \frac{k}{1 + \frac{2\mu k}{a}\pi},$$

et partant tant soit peu plus petite qu'au commencement.

2° Après m révolutions où $\omega = 2\pi m$, la distance de la planete au soleil sera

$$z = \frac{k}{1 + \frac{2\mu\pi mk}{a}}.$$

La planete s'approchera donc de plus en plus du soleil, mais si insensiblement, que pendant chaque révolution l'orbite ne differe point d'un cercle.

3° Le tems pendant lequel la planete acheve m révolutions est

$$t = \frac{k\sqrt{k}}{c\sqrt{c}}\,\Theta m\left(1 - \frac{3\mu k}{2a}\pi m\right);$$

or le tems pendant lequel elle acheve $m+1$ révolutions, est

$$t' = \frac{k\sqrt{k}}{c\sqrt{c}}\,\Theta\,(m+1)\left(1 - \frac{3\mu k}{2a}\pi\,(m+1)\right).$$

4° Donc après m périodes, le tems d'une révolution sera

$$t' - t = \frac{k\sqrt{k}}{c\sqrt{c}}\,\Theta\left(1 - \frac{3\mu k}{2a}\pi\,(2m+1)\right),$$

d'où l'on voit que le tems périodique va en diminuant, le premier étant $\frac{k\sqrt{k}}{c\sqrt{c}}\,\Theta$.

5° La résistance se réduira donc à ces deux effets: le *premier*, que la distance au soleil diminue; et le *second*, que le tems périodique devient plus court. Or le mouvement demeure circulaire, sans qu'il en résulte une excentricité.

39. Pour le mouvement de la terre, si nous en négligeons l'excentricité, nous avons $k = c$ et $\Theta = 365^{\mathrm{j}}\, 6^{\mathrm{h}}\, 9'\, 36'' = 31\,558\,176''$, qui est le tems périodique par rapport aux étoiles fixes. Ensuite posant la parallaxe horizontale du soleil $= 12''$, nous avons $\frac{k}{a} = 17\,000$ et[8]) $\frac{3\pi k}{2a} = 77\,000$. Donc après m révolutions, la diminution du tems périodique sera de $77\,000 \times 31\,558\,176 \times \mu\,(2m+1)$ secondes, ce qui fait, en mettant pour μ le nombre donné ci-dessus par estime

$$\mu = \frac{1}{10\,000\,000\,000\,000}$$

environ $0{,}2429979552\,(2m+1)''$, ou bien $\frac{243}{1000}\,(2m+1)$ secondes. Après un siecle, la diminution de l'année solaire seroit donc $= \frac{201 \times 243}{1000} = 49''$, laquelle étant très-sûrement trop grande, puisque la diminution séculaire est moindre que $5''$, il est clair que la fraction μ est au moins 10 fois plus petite que je ne l'ai supposée ici. Il semble donc que le nombre λ employé ci-dessus (§20) doit surpasser 100, à moins qu'on ne veuille diminuer la densité de l'éther aux dépens de son élasticité; l'un et l'autre paroît également probable[9]).

8) Il doit être 80 000 environ. Les petites erreurs de calcul n'ont pas été corrigées. EJA.

9) L'édition originale (A8) a l'apostille suivante:

«Si l'on supposoit la parallaxe horizontale du Soleil de 9 secondes, comme on vient de la conclurre par le dernier passage de Venus sur le disque du Soleil: on auroit $\frac{k}{a} = 22\,000$ et $\frac{3\pi k}{2a} = 100\,000$ environ, car comme on n'est pas encore trop sur de la parallaxe du Soleil, et qu'elle peut aussi bien être de 8 que de 9 secondes, il nous sera toujours permis de prendre pour $\frac{k}{a}$ et $\frac{3\pi k}{2a}$ des nombres ronds.

La diminution du tems périodique sera donc après m révolutions de $100\,000 \times 31\,558\,176 \times \mu(2m+1)$ secondes, ou posant $\mu = 1 : 10\,000\,000\,000\,000$ de $0{,}31558176\,(2m+1)$ secondes, ou encore de $\frac{315}{1000}(2m+1)$ secondes, et partant après un siecle de 63 secondes.

Comme cette diminution de l'année solaire est encore plus grande que celle qu'on a trouvé en supposant la parallaxe du Soleil de $12''$, la fraction μ doit à plus forte raison être moindre que $1 : 10$ billions, et même moindre que $1 : 100$ billions.» EJA.

II. L'Orbite de la Planete ayant une très-petite excentricité

40. Soit donc la fraction n, qui exprime l'excentricité si petite, qu'on puisse négliger dans le calcul son quarré nn, et nous aurons, pour l'arc σ, cette expression:

$$\sigma = k\int d\omega\,(1 + n\cos\omega)\,(1 - 2n\cos\omega)$$
$$= k\int d\omega\,(1 - n\cos\omega) = k\,(\omega - n\sin\omega) \quad (\S\,32)\,,$$

et partant $A = 1$ et $B = -1$; de-là nous concluons la distance

$$z = \frac{k}{1 + n\cos\omega + \dfrac{\mu k}{a}\,(\omega + \frac{1}{2}n\omega\cos\omega)}\,.$$

Lorsque la planete aura parcouru dans son orbite l'angle ω, posant le tems qu'elle y a employé $= t$, nous aurons

$$\frac{2\pi c\sqrt{c}}{\Theta k\sqrt{k}}\,t = \int d\omega\,(1 - 2n\cos\omega) - \frac{2\mu k}{a}\int d\omega\,(1 - 3n\cos\omega)\,(\omega + \tfrac{1}{2}n\omega\cos\omega)$$
$$+ \frac{\mu k}{2a}\int d\omega\,(1 - 2n\,\cos\omega)\,(\omega - n\,\sin\omega)\,;$$

ou bien

$$\frac{2\pi c\sqrt{c}}{\Theta k\sqrt{k}}\,t = \omega - 2n\sin\omega - \frac{\mu k}{2a}\int d\omega\,(3\omega - 8n\omega\cos\omega + n\sin\omega)\,;$$

et prenant toutes les intégrales:

$$\frac{2\pi c\sqrt{c}}{\Theta k\sqrt{k}}\,t = \omega - 2n\sin\omega - \frac{\mu k}{2a}\,(\tfrac{3}{2}\omega^2 - 8n\omega\sin\omega - 9n\cos\omega)\,,$$

où k marque le demi-parametre, ou plutôt la distance moyenne de la planete au soleil.

41. Examinons d'abord l'orbite de la planete après qu'elle aura achevé m révolutions; posons pour cette fin $\omega = 2\pi m + \varphi$, de sorte que φ marque l'angle

parcouru dans la révolution suivante, et la distance de la planete au soleil sera exprimée de cette façon:

$$z = \frac{k}{1 + n\cos\varphi + \frac{\mu k}{a}(2\pi m + \varphi)(1 + \frac{1}{2}n\cos\varphi)}.$$

Ici je remarque que la plus petite distance n'est pas celle où $\varphi = 0$, comme il arriveroit s'il n'y avoit point de résistance, mais celle où

$$\varphi = \frac{\mu k(1 + \frac{1}{2}n)}{n(a + \mu\pi mk)},$$

de sorte que pendant ce tems la ligne des absides est avancée de l'angle

$$\varphi = \frac{\mu k(1 + \frac{1}{2}n)}{n(a + \mu\pi mk)}.$$

Mais posant $m = 0$, dès le premier mouvement la résistance fait que le périhélie réponde à l'angle

$$\frac{\mu k(1 + \frac{1}{2}n)}{na},$$

donc pendant m révolutions, il recule par

$$\frac{\mu\mu\pi mkk(1 + \frac{1}{2}n)}{na(a + \mu\pi mk)} = \frac{\mu\mu\pi m}{n} \cdot \frac{kk}{aa},$$

ce qui est absolument insensible; car si n est extrêmement petit, on ne peut plus distinguer le périhélie.

42. Posons donc $\varphi = \frac{\mu k}{na}$ pour trouver la plus petite distance de la planete au soleil, qui sera

$$z = \frac{k}{1 + n + \frac{\mu k}{a} 2\pi m(1 + \frac{1}{2}n)}.$$

Or posant $\varphi = \pi + \frac{\mu k}{na}$, la plus grande résulte

$$z = \frac{k}{1 - n + \frac{\mu k}{a}\pi(2m+1)(1-\tfrac{1}{2}n)}.$$

Mais puisque l'excentricité est extrêmement petite, et que l'effet ne devient sensible que lorsque le nombre m est assez grand, nous aurons la plus petite distance

$$z = \frac{k}{1 + \frac{2\pi\mu mk}{a} + n\left(1 + \frac{\pi\mu mk}{a}\right)},$$

et la plus grande distance

$$z = \frac{k}{1 + \frac{2\pi\mu mk}{a} - n\left(1 + \frac{\pi\mu mk}{a}\right)}.$$

Donc après m révolutions de la planete, nous trouvons le demi-parametre de son orbite

$$= \frac{ak}{a + 2\pi\mu mk} = k\left(1 - \frac{2\pi\mu mk}{a}\right)$$

et l'excentricité

$$= n\frac{a + \pi\mu mk}{a + 2\pi\mu mk} = n\left(1 - \frac{\pi\mu mk}{a}\right),$$

d'où nous voyons que la résistance de l'éther diminue tant le parametre que l'excentricité.

43. Pour trouver le tems pendant lequel la planete acheve m révolutions autour du soleil, posons $\omega = 2\pi m$, ce tems sera

$$t = \frac{\Theta k\sqrt{k}}{c\sqrt{c}}\left(m - \frac{3\pi\mu m^2 k}{2a} + \frac{9\mu nk}{4\pi a}\right);$$

Donc le tems de $m+1$ révolutions étant

$$t = \frac{\Theta k\sqrt{k}}{c\sqrt{c}}\left(m+1-\frac{3\pi\mu(m+1)^2k}{2a}+\frac{9\mu nk}{4\pi a}\right)$$

après que la planete aura achevé m révolutions, son tems périodique sera

$$= \frac{k\sqrt{k}}{c\sqrt{c}}\,\Theta\left(1-\frac{3\pi\mu k}{2a}(2m+1)\right)$$

tout comme nous l'avons trouvé dans le cas précédent. Ainsi une excentricité fort petite ne change rien dans la diminution du tems périodique, et le changement de l'excentricité même est si petit, qu'il sera presque impossible de s'en appercevoir, même après le plus grand nombre de révolutions. Car si pour la terre la valeur de $\frac{3\pi\mu k}{2a}\Theta$ est $\frac{243}{10\,000}$ prenant μ dix fois plus petit que je l'avois d'abord estimé (§ 39), nous aurons à-peu-près $\frac{3\pi\mu k}{2a} = \frac{1}{41\times 31\,558\,176}$, et partant $\frac{\pi\mu k}{a} = \frac{1}{2\,000\,000\,000}$, donc l'excentricité après m révolutions $=\left(1-\frac{m}{2\,000\,000\,000}\right)$ d'où l'on voit que même après 100 000 révolutions, cette diminution ne deviendroit pas encore sensible. Or le demi-parametre de l'orbite de la terre, après m révolutions, sera encore $= c\left(1-\frac{m}{1\,000\,000\,000}\right)$.

III. L'Orbite de la Planete étant considérablement excentrique

44. Je supposerai ici l'excentricité n plus grande qu'auparavant, mais pourtant plus petite que l'unité, puisque $n=1$ donneroit déja une parabole pour l'orbite de la planete. Les puissances de n deviendront donc de plus en plus petites, de sorte qu'on pourra négliger les puissances plus hautes que le cube n^3 ou le quarré-quarré n^4. Développons en premier lieu l'intégrale

$$\int \frac{d\omega\sqrt{(1+2n\cos\omega+nn)}}{(1+n\cos\omega)^2} \quad (§\,34),$$

et puisque la résolution en séries nous donne

$$\sqrt{(1+2n\cos\omega+nn)} = 1+\tfrac{1}{2}nn-\tfrac{1}{8}n^4+n(1-\tfrac{1}{2}nn)\cos\omega-nn(\tfrac{1}{2}-\tfrac{3}{4}nn)\cos\omega^2$$
$$+\tfrac{1}{2}n^3\cos\omega^3-\tfrac{5}{8}n^4\cos\omega^4,$$

et

$$(1+n\cos\omega)^{-2} = 1-2n\cos\omega+3n^2\cos\omega^2-4n^3\cos\omega+5n^4\cos\omega^4;$$

nous en obtiendrons

$$\frac{\sqrt{(1+2n\cos\omega+nn)}}{(1+n\cos\omega)^2} = 1+\tfrac{1}{2}nn-\tfrac{1}{8}n^4-n(1+\tfrac{3}{2}nn)\cos\omega+nn(\tfrac{1}{2}+\tfrac{13}{4}nn)\cos\omega^2$$
$$+\tfrac{1}{2}n^3\cos\omega^3-\tfrac{17}{8}n^4\cos\omega^4;$$

et à cause de

$$\cos\omega^2 = \tfrac{1}{2}+\tfrac{1}{2}\cos 2\omega;$$
$$\cos\omega^3 = \tfrac{3}{4}\cos\omega+\tfrac{1}{4}\cos 3\omega;$$
$$\cos\omega^4 = \tfrac{3}{8}+\tfrac{1}{2}\cos 2\omega+\tfrac{1}{8}\cos 4\omega;$$

cette expression se réduit à

$$1+\tfrac{3}{4}nn+\tfrac{45}{64}n^4-n(1+\tfrac{9}{8}nn)\cos\omega+nn(\tfrac{1}{4}+\tfrac{9}{16}nn)\cos 2\omega$$
$$+\tfrac{1}{8}n^3\cos 3\omega-\tfrac{17}{64}n^4\cos 4\omega;$$

d'où nous tirons l'intégrale

$$\int\frac{d\omega\sqrt{(1+2n\cos\omega+nn)}}{(1+n\cos\omega)^2} = (1+\tfrac{3}{4}nn+\tfrac{45}{64}n^4)\,\omega-n(1+\tfrac{9}{8}nn)\sin\omega$$
$$+\tfrac{1}{2}nn(\tfrac{1}{4}+\tfrac{9}{16}nn)\sin 2\omega+\tfrac{1}{24}n^3\sin 3\omega-\tfrac{17}{256}n^4\sin 4\omega;$$

et partant, nous aurons

$$A = 1+\tfrac{3}{4}nn+\tfrac{45}{64}n^4;\quad B=-1-\tfrac{9}{8}nn;\quad C=\tfrac{1}{8}+\tfrac{9}{32}nn;\quad D=\tfrac{1}{24};\quad E=-\tfrac{17}{256}.$$

45. De-là nous trouvons la distance de la planete au soleil, comme elle est représentée au §36; c'est-à-dire, z égal à k divisé par

$$1+n\cos\omega+\frac{\mu k}{a}\left(A\omega-\tfrac{1}{2}Bn\omega\cos\omega-\tfrac{1}{3}Cn^2\sin 2\omega-\tfrac{1}{8}Dn^3\sin 3\omega-\tfrac{1}{15}\,En^4\sin 4\omega\right).$$

Ensuite, pour l'expression du tems, nous avons d'abord

$$\int\frac{d\omega}{(1+n\cos\omega)^2}=(1+\tfrac{3}{2}nn+\tfrac{15}{8}n^4)\,\omega-n(2+3nn)\sin\omega+nn(\tfrac{3}{4}+\tfrac{5}{4}nn)\sin 2\omega$$
$$-\tfrac{1}{3}n^3\sin 3\omega+\tfrac{5}{32}n^4\sin 4\omega\,;$$

et les deux autres formules se réunissent dans celle-ci[10]:

$$\frac{\mu k}{2a}\int d\omega\Big(-3(1+\tfrac{13}{4}nn+\tfrac{421}{64}n^4)\,\omega+(8n+\tfrac{93}{4}\,n^3)\,\omega\cos\omega-(\tfrac{15}{2}\,n^2+22n^4)\,\omega\cos 2\omega$$
$$+6n^3\omega\cos 3\omega-\tfrac{35}{8}n^4\cos 4\omega-(n+\tfrac{9}{4}n^3)\sin\omega+(\tfrac{31}{24}nn+\tfrac{163}{48}n^4)\sin 2\omega$$
$$-\tfrac{17}{16}n^3\sin 3\omega+\tfrac{879}{1280}n^4\sin 4\omega\Big)$$

d'où, pour le tems t, on trouve[11]:

$$\frac{2\pi c\sqrt{c}}{\Theta k\sqrt{k}}\,t=(1+\tfrac{3}{2}n^2+\tfrac{15}{8}n^4)\,\omega-n(2+3n^2)\sin\omega+nn(\tfrac{3}{4}+\tfrac{5}{4}n^2)\sin 2\omega$$
$$-\tfrac{1}{3}n^3\sin 3\omega+\tfrac{5}{32}n^4\sin 4\omega+\frac{\mu k}{2a}\{-\tfrac{3}{2}(1+\tfrac{13}{4}n^2+\tfrac{421}{64}n^4)\,\omega\omega$$
$$+(8n+\tfrac{93}{4}n^3)\,\omega\sin\omega-(\tfrac{15}{4}n^2+11n^4)\,\omega\sin 2\omega+2n^3\omega\sin 3\omega$$
$$-\tfrac{35}{32}n^4\omega\sin 4\omega+(9n+\tfrac{51}{2}n^3)\cos\omega-(\tfrac{121}{48}n^2+\tfrac{691}{96}n^4)\cos 2\omega$$
$$+\tfrac{47}{48}n^3\cos 3\omega-\tfrac{2279}{5120}n^4\cos 4\omega\}\,.$$

Et si nous nommons le demi-grand-axe $=h$, à cause de $h=\dfrac{k}{1-nn}$, nous aurons[12]:

10) Edition originale (A8²): $-\frac{5}{16}$ au lieu de $-\frac{17}{16}$. Le terme $\frac{3}{2}Bn^3\sin\omega\cos 2\omega$ se réduit à $\frac{3}{4}n^3\sin\omega-\frac{3}{4}n^3\sin 3\omega$. J. A. EULER a laissé échapper la partie $-\frac{3}{4}n^3\sin 3\omega$. Corrigé par EJA.

11) Edition originale (A8²): $-\frac{3}{2}n^2$ au lieu de $+\frac{3}{2}n^2$, $+\frac{35}{32}$ au lieu de $-\frac{35}{32}$ et $\frac{37}{48}$ au lieu de $\frac{47}{48}$. Corrigé par EJA.

12) Edition originale (A8²): $-\frac{37}{96}$ au lieu de $-\frac{47}{96}$. Corrigé par EJA.

$$\frac{2\pi c\sqrt{c}}{\Theta h\sqrt{h}}\,t = \omega - 2n\sin\omega + nn(\tfrac{3}{4}+\tfrac{1}{8}n^2)\sin 2\omega - \tfrac{1}{3}n^3\sin 3\omega + \tfrac{5}{32}n^4\sin 4\omega$$
$$-\frac{\mu k}{a}\,\{\tfrac{3}{4}(1+\tfrac{7}{4}n^2+\tfrac{133}{64}n^4)\,\omega^2 - n(4+\tfrac{45}{8}n^2)\,\omega\sin\omega$$
$$+n^2(\tfrac{15}{8}+\tfrac{43}{16}n^2)\,\omega\sin 2\omega - n^3\omega\sin 3\omega + \tfrac{35}{64}n^4\omega\sin 4\omega$$
$$-n(\tfrac{9}{2}+6n^2)\cos\omega + n^2(\tfrac{121}{96}+\tfrac{41}{24}n^2)\cos 2\omega - \tfrac{47}{96}n^3\cos 3\omega$$
$$+\tfrac{2279}{10240}n^4\cos 4\omega\}\,.$$

46. Supposons que la planete ait déja achevé m révolutions, et pour celle qui suit posons $\omega = 2\pi m + \varphi$, alors sa distance au soleil sera z égal à k divisé par

$$1 + n\cos\varphi + \frac{\mu k}{a}\,(2\pi Am + A\varphi - \tfrac{1}{2}Bn(2\pi m+\varphi)\cos\varphi - \tfrac{1}{3}Cn^2\sin 2\varphi$$
$$-\tfrac{1}{8}Dn^3\sin 3\varphi - \tfrac{1}{15}En^4\sin 4\varphi)$$

et il est clair que l'orbite aura souffert le même changement que dans le cas précédent, parce que les termes qui ne renferment point le nombre m sont évanouissans. L'orbite même se rétrécira un tant soit peu, et l'excentricité deviendra plus petite, mais si peu, qu'il est impossible de s'en appercevoir; de même le lieu de l'aphélie ou du périhélie n'en souffrira aussi aucun changement sensible. L'unique effet sensible que la résistance de l'éther est capable de produire, consiste dans la diminution du tems périodique, ce qu'il convient d'examiner plus soigneusement.

47. Pendant que la planete acheve m révolutions, il s'écoule un tems t, et en posant dans l'expression ci-dessus (§45) $\omega = 2\pi m$, ce tems est[13])

$$t = \frac{h\sqrt{h}}{c\sqrt{c}}\,\Theta\left(m - \frac{\mu k}{a}\left(\tfrac{3}{2}(1+\tfrac{7}{4}n^2+\tfrac{133}{64}n^4)\,\pi mm - \frac{9n}{4\pi} + \frac{121n^2}{192\pi}\right.\right.$$
$$\left.\left. - \frac{623n^3}{192\pi} + \frac{59317n^4}{61440\pi}\right)\right).$$

13) Edition originale (A 8[2]): $k\sqrt{h}$ au lieu de $h\sqrt{h}$, $\frac{3}{4}n^2$ au lieu de $\frac{7}{4}n^2$ et 613 au lieu de 623. Corrigé par EJA.

Le tems de $m+1$ révolutions se trouvant de la même maniere, après que la planete aura achevé m révolutions, le tems de la révolution suivante est[14])

$$\frac{h\sqrt{h}}{c\sqrt{c}}\,\Theta\left(1-\frac{3\pi\mu k}{2a}\,(1+\tfrac{7}{4}nn+\tfrac{133}{64}n^4)\,(2m+1)\right),$$

ou bien[15])

$$\frac{h\sqrt{h}}{c\sqrt{c}}\,\Theta\left(1-\frac{3\pi\mu h}{2a}\,(1+\tfrac{3}{4}nn+\tfrac{21}{64}n^4)\,(2m+1)\right).$$

Et alors le demi-parametre qui fut au commencement k, sera

$$=k\left(1-\frac{2\pi\mu mk}{a}\right),$$

l'excentricité

$$=n\left(1-\frac{\pi\mu mk}{a}\right)$$

et le demi-grand-axe

$$=\frac{k\left(1-\dfrac{2\pi\mu mk}{a}\right)}{1-nn\left(1-\dfrac{2\pi\mu mk}{a}\right)}=h\left(1-\frac{2\pi\mu mh}{a}\right).$$

Que le tems périodique de la planete ait été $=T$, s'il n'y avoit point de résistance; et à cause de $T=\dfrac{h\sqrt{h}}{c\sqrt{c}}\,\Theta$, si l'on néglige l'excentricité, le tems périodique après m révolutions sera encore

$$=T\left(1-\frac{3\pi\mu h}{2a}\,(2m+1)\right).$$

Or en tenant compte de la résistance, le premier tems périodique étoit

14) Edition originale (A8[2]): $\frac{3}{4}nn$ au lieu de $\frac{7}{4}nn$. Corrigé par EJA.

15) Edition originale (A8[2]): $3\pi\mu k$ au lieu de $3\pi\mu h$, $-\frac{1}{4}$ au lieu de $+\frac{3}{4}$ et $\frac{85}{64}$ au lieu de $\frac{21}{64}$. Corrigé par EJA.

$= T\left(1 - \frac{3\pi\mu h}{2a}\right)$, donc après m révolutions la diminution du tems périodique sera $= \frac{3\pi\mu h}{a} mT$.

48. Quelque petite que paroisse cette diminution, l'effet devient néanmoins très-sensible pendant plusieurs siecles. Supposons que le tems périodique d'une planete soit à présent $= T$, le tems périodique de la Terre autour du Soleil étant aussi à présent $= \Theta$, et que pendant m révolutions le tems périodique de la planete diminue de la particule de tems $m\theta$, nous aurons $\theta = \frac{3\pi\mu h}{a} T$, et partant $\frac{3\pi\mu h}{a} = \frac{\theta}{T}$. Maintenant qu'on calcule des tables du moyen mouvement de cette planete selon la maniere ordinaire, comme si le tems périodique T demeuroit toujours le même; et parce qu'au tems T répond le moyen mouvement 2π, on aura pour un an ou pour le tems Θ le moyen mouvement $\frac{2\pi\Theta}{T}$, et pour le tems de N années le moyen mouvement sera $= \frac{2\pi N\Theta}{T}$. Mais à cause de la résistance, le mouvement moyen de la planete pour le même tems de N années sera différent. Pour le trouver, nous n'avons qu'à poser $t = N\Theta$, et l'angle ω, en omettant les termes qui renferment le sinus ou cosinus, donnera son mouvement moyen actuel. Or puisque $\frac{c\sqrt{c}}{h\sqrt{h}} = \frac{\Theta}{T}$, nous avons cette équation[16]):

$$\frac{2\pi N\Theta}{T} = \omega - \frac{3\mu h}{4a}\omega\omega = \omega - \frac{\theta\omega\omega}{4\pi T},$$

d'où nous trouvons le moyen mouvement actuel:

$$\omega = \frac{2\pi N\Theta}{T} + \frac{\theta\omega\omega}{4\pi T} = \frac{2\pi N\Theta}{T} + \frac{\pi NN\Theta\Theta\theta}{T^3}.$$

Il faut donc ajoûter au lieu moyen trouvé par les tables ordinaires du moyen mouvement la particule

$$\frac{\Theta\Theta\theta}{T^3}\pi NN = 180^\circ NN\frac{\Theta\Theta\theta}{T^3}.$$

16) Edition originale (A 8^2): $3\mu k$ au lieu de $3\mu h$. Corrigé par EJA.

49. Cette même correction a lieu, soit que N marque un nombre positif ou negatif. Donc si les tables des moyens mouvemens ont été dressées sur le tems périodique T d'une certaine époque, qui après m révolutions diminue de la particule $m\theta$, alors pour N ans tant avant qu'après cette époque, il faudra ajoûter la particule $180^\circ \cdot N^2 \cdot \frac{\Theta^2}{T^2} \cdot \frac{\theta}{T}$ à la longitude moyenne qu'on aura tiré de ces moyens mouvemens. Si l'on fait $T:\theta = 360^\circ:\delta$, ce petit arc δ exprime le moyen mouvement de la planete qui répond au tems θ, et maintenant notre correction de la longitude moyenne sera $= \frac{1}{2}\delta \cdot N^2 \cdot \frac{\Theta^2}{T^2}$, où Θ marque le tems d'une année. Donc si nous posons la distance moyenne du soleil à la terre $= c$ et à la planete $= h$, à cause de $\frac{\Theta^2}{T^2} = \frac{c^3}{h^3}$ notre correction sera $= \frac{1}{2}\delta \cdot N^2 \cdot \frac{c^3}{h^3}$; or cette fraction $\frac{c^3}{h^3}$ se tire aisément des tables. Je remarque encore que cette correction est proportionnelle au quarré de l'intervalle du tems, et partant dans un intervalle de quelque milliers d'années elle doit devenir très-considérable, quelque petite que soit la diminution d'une révolution à la suivante, que je suppose ici $= \delta$.

50. Donc tout l'effet de la résistance de l'éther sur le mouvement des planetes se réduit à la diminution de leurs tems périodiques, puisque nous avons vu que ni l'excentricité, ni leurs aphélies n'en souffrent aucun changement sensible. Or pour l'usage de l'Astronomie cet effet ne sauroit être représenté plus commodément que par une correction du mouvement moyen qui doit se faire ainsi. Ayant établi le mouvement moyen pour une certain époque et en ayant déduit une table des longitudes moyennes pour tous les ans tant avant qu'après cette époque, la longitude moyenne ne sera juste que pour cette époque même; pour tout autre tems qui en differe de N ans tant avant qu'après cette époque, il faudra ajoûter à la longitude moyenne tabulaire la particule $\frac{1}{2}\delta \cdot N^2 \cdot \frac{c^3}{h^3} = \frac{1}{2}\delta \cdot N^2 \cdot \frac{\Theta^2}{T^2}$. La correction d'un an sera donc $= \frac{1}{2}\delta \cdot \frac{\Theta^2}{T^2}$, soit cette correction $= \alpha$, et puisque $\delta = 2\alpha \frac{T^2}{\Theta^2}$, nous aurons $\theta = \frac{\alpha}{\pi} \frac{T^3}{\Theta^2}$, et partant, $\frac{3\pi\mu h}{a} = \frac{\alpha}{\pi} \frac{T^2}{\Theta^2}$, ou bien $\alpha = \frac{3\pi^2\mu h}{a} \frac{\Theta^2}{T^2}$ $= 3\pi^2\mu \frac{c^3}{ahh}$; de sorte que la correction pour un an soit réciproquement proportionnelle au quarré de la distance moyenne de la planete au soleil multiplié par le demi-diametre du corps de la planete.

51. Il suffit donc d'avoir déterminé cette correction pour une planete, et il sera aisé d'en déduire celles de toutes les autres planetes. Il n'y a aucun doute que les observations du soleil ne soient les plus propres pour nous fournir les éclaircissemens nécessaires sur cet important article, et si l'on peut parvenir à déterminer la correction d'un an α pour le soleil, on en déduira aisément la valeur de la fraction μ résultante de la résistance de l'éther, car à cause de $h = c$, on aura $\mu = \frac{\alpha a}{3\pi\pi c}$. Mais quoique la correction d'un an α soit si petite, qu'elle échapperoit aux plus grands soins des Astronomes, elle devient très-sensible pour un grand nombre d'années, parce que pour un intervalle de N années elle sera $= NN\alpha$, et partant pour un siecle $= 10\,000\,\alpha$, et pour dix siecles $= 1\,000\,000\,\alpha$. En consultant donc les plus anciennes observations, il ne sera plus difficile d'en conclure la véritable valeur de la correction annuelle α.

52. Pour cet effet, il faudroit conclure la quantité d'une année solaire moyenne par les plus modernes observations faites tout au plus dans l'espace d'un demi-siecle; de-là on devroit dresser des tables du moyen mouvement et de la longitude moyenne du soleil pour des époques les plus reculées, où l'on trouve de bonnes observations. Alors pour une telle époque, qui précede la présente de N ans, on compareroit la longitude moyenne tirée de ces tables avec celle qui conviendroit aux observations anciennes, et si l'on trouvoit que celle-ci surpassoit celle-là de la quantité Q, on auroit $NN\alpha = Q$, et partant $\alpha = \frac{NN}{Q}$. Mais il faudroit être bien assuré que la terre n'eût pendant ce tems souffert aucune altération dans son mouvement par l'action de quelque comete. A l'égard de l'action des autres planetes, on en connoît assez l'effet, pour en tenir compte dans cette recherche; mais il est fort à craindre que les cometes ne la rendent tout-à-fait inutiles.

53. La méthode dont les Astronomes se servent pour déterminer la quantité de l'année solaire, nous fournit encore un autre moyen de nous assurer de la correction annuelle α. Supposons qu'en comparant les observations présentes avec des observations faites avant L ans, on conclue la quantité de l'année $= A$, et qu'en les comparant avec des observations faites avant M ans, on la trouve $= B$. Maintenant soit la quantité de l'année solaire $= \Theta$, et la diminution annuelle $= \theta$; et le tems écoulé depuis L ans jusqu'à présent sera $= L\Theta + LL\theta$, et le tems écoulé depuis M ans jusqu'à présent $= M\Theta + MM\theta$. De-là on aura $A = \Theta + L\theta$, et

$B = \Theta + M\theta$, et partant $\theta = \dfrac{B-A}{M-L}$ et $\Theta = A - L\theta$; d'où l'on connoîtra la véritable année solaire Θ, qui convient à l'époque d'à-présent, et la diminution annuelle θ, et de-là on tirera $\alpha = \dfrac{\pi\theta}{\Theta}$.

54. Mais outre les cometes qui peuvent troubler cette détermination, on rencontre encore d'autres obstacles presque insurmontables du côté des observations même, qui ne sont pas si exactes, au moins dans les siecles passés, que l'erreur dans la quantité de l'année ne puisse être de plusieurs secondes. Or puisque $B - A$ est un nombre de secondes fort peu considérable, l'erreur commise dans les quantités de l'année A et B le rend tout-à-fait incertain. Ensuite les plus anciennes observations ont non seulement le défaut qu'elles sont peu exactes, mais il semble aussi qu'on n'est pas trop sûr du tems dans lequel elles ont été faites. Surtout la réduction de l'Almanac Egyptien au Romain, dont PTOLOMÉE s'est servi, ne paroît pas assez constatée, à cause des désordres auxquels l'Almanac Romain fut assujetti dans ce tems; et les corrections que RICCIOLI et d'autres Chronologues ont tâché d'y apporter, sont ouvertement fondées sur la quantité de l'année qu'ils ont supposée; ce qui est précisément la question. Il se pourroit donc bien faire que les momens des observations marqués par PTOLOMÉE dussent être reculés d'un ou de deux jours de plus qu'on les suppose dans le calcul; et alors les anciennes observations comparées aux modernes, donneroient incontestablement l'année plus longue que les Astronomes ne la supposent dans leurs tables.

55. En effet, feu M. CASSINI[17]), après les plus scrupuleuses recherches sur la quantité de l'année solaire moyenne, en prenant un milieu entre tous ses résultats, conclut l'année solaire moyenne de $365^{\mathrm{j}}\,5^{\mathrm{h}}\,48'\,47''$, qui convient avec celle qu'il a déduite de ses propres observations. Cependant, dans ses tables astronomiques[18]), il la suppose de $365^{\mathrm{j}}\,5^{\mathrm{h}}\,48'\,53''\,24'''$ et partant de $6''\,24'''$ plus grande; sans doute qu'il n'a fait ce changement que pour satisfaire aux anciennes observations; d'où il faut conclure que les anciennes observations demandent une année plus longue; ce qui prouve l'effet de la résistance de l'éther. Aussi ne sauroit-on douter que les observations de HIPPARQUE, comparées avec celles de PTOLOMÉE, ne donnassent l'année au moins de quelques minutes plus longue; d'où il suit encore, que l'année solaire moyenne est à présent plus courte qu'autrefois.

17) JACQUES CASSINI, *Éléments d'astronomie* (Paris, 1740), p. 232. EJA.
18) JACQUES CASSINI, *Tables astronomiques* (Paris, 1740). EJA.

56. Si M. CASSINI, dans ses Tables, avoit supposé l'année de $365^j\ 5^h\ 48'\ 47''$, le mouvement moyen pour cent années Juliennes seroit devenu $0^s\ 0°\ 46'\ 3''$, qui est dans ses Tables de $0^s\ 0°\ 45'\ 42''$, la différence étant de $21''$. Donc pour 200 ans la différence seroit $42''$, et partant, si nous supposons que les Tables de CASSINI représentent bien la longitude du soleil jusqu'à 200 ans en arriere, des tables construites sur la quantité de l'année de $365^j\ 5^h\ 48'\ 47''$ demanderont, pour l'intervalle de deux siecles, la correction de $42''$, et pour l'intervalle d'un siecle $10\frac{1}{2}''$. Si nous supposions la véritable quantité de l'année de $365^j\ 5^h\ 48'\ 46''$, on trouveroit, pour l'intervalle d'un siecle, la correction $=12''$, et partant $10\,000\alpha = 12''$, et $\alpha = \frac{1}{833}'' = \frac{\theta}{\Theta}\cdot 180° = \frac{\theta\cdot 60\times 60\times 180}{31\,558\,176}$; d'où $\theta = \frac{31\,558\,176}{833\times 60\times 60\times 180}'' = 0{,}058''$, ce qui est la diminution annuelle de l'année, et partant la diminution séculaire $= 5''\ 48'''$.

57. M. l'Abbé DE LA CAILLE[19]) ayant supposé l'année, dans ses tables, de $365^j\ 5^h\ 48'\ 49''$, qu'il a uniquement conclu des observations modernes, de maniere que ce doit être la juste quantité de l'année pour le milieu de ce siecle, le mouvement moyen pour 100 années Juliennes est $0^s\ 0°\ 45'\ 55\frac{6}{10}''$; donc si, pour satisfaire aux observations faites avant deux siecles, il falloit supposer le mouvement moyen séculaire de $0^s\ 0°\ 45'\ 42''$, il faudroit ajoûter à la longitude trouvée par les Tables de M. DE LA CAILLE $27\frac{2}{10}'' = 40\,000\alpha$, et partant α seroit $= \frac{1}{1470}''$, d'où l'on trouveroit $\theta = 0{,}033''$, et la diminution séculaire de l'année $= 3''\ 18'''$, d'où il semble qu'on devroit supposer la fraction $\mu = \dfrac{1}{150\,000\,000\,000\,000}$; ce qu'on pourroit encore assez bien accorder avec les principes établis ci-dessus sur la densité de l'éther, et sur la maniere dont il résiste au mouvement des corps.

58. Comme on ne sauroit encore rien décider de précis là-dessus, posons la diminution séculaire de l'année $= v''$ et à cause de $\theta = \dfrac{v}{100}$, nous aurons:

$$\alpha = 180° \frac{v}{100\times 31\,558\,176} = 0{,}000\,205\,v'',$$

19) NICOLAS-LOUIS DE LACAILLE (1713–1762), *Tabulae solares ad meridianum Parisinum* (Vindobonae, 1763), p. 5. EJA.

donc l'augmentation de la longitude moyenne pour l'intervalle d'un siecle sera $= 2\frac{1}{20}v$ secondes, et pour l'intervalle de 10 siecles ou de 1000 ans, elle sera $= 205v$ secondes, et de-là $\mu = \frac{v}{536\,000\,000\,000\,000}$. De-là, en tenant compte du demi-diametre des autres planetes, on aura l'augmentation de leur longitude moyenne pour l'intervalle de 1000 ans,

Pour Saturne	$= 135v''$;	Pour la Terre	$= 205v''$;
Pour Jupiter	$= 62v''$;	Pour Venus	$= 103v''$;
Pour Mars	$= 254v''$;	Pour Mercure	$= 135v''$;

après avoir établi des tables du moyen mouvement sur le tems périodique qui convient à chaque planete pour l'époque proposée.

59. Ce que je viens d'exposer, prouve suffisamment, que les planetes éprouvent dans leur mouvement quelque résistance de l'Ether, quoiqu'il soit impossible d'en déterminer la vraie quantité par la seule Théorie, moyennant laquelle nous venons cependant d'apprendre que cette résistance est très-petite, et rien n'empêche qu'elle ne soit encore considérablement plus petite que je ne l'ai estimée. Non seulement la densité de l'Ether pourroit être diminuée au-delà, en diminuant en même raison son élasticité; mais l'obliquité sous laquelle les particules vraiment solides des planetes reçoivent l'impulsion de l'Ether, pourroit causer encore une plus grande diminution. Cependant je doute fort qu'on puisse par ce moyen réduire la diminution de l'année solaire pendant un siecle au-dessous d'une seconde; il me semble plutôt qu'on la pourroit bien admettre de quelques secondes, sans porter la moindre atteinte aux observations anciennes. Il n'est pas encore suffisamment prouvé, que les observations rapportées par PTOLOMÉE, ne doivent être reculées d'un jour au-delà de ce qu'on les suppose communément; RICCIOLI lui-même en marque le jour differement dans son Almageste[20]) et dans son Astronomie réformee[21]), à cause de l'incertitude de l'Almanac Romain, et ceux qui ont voulu mieux regler ces tems, se sont ouvertement fondés sur l'hypothèse de la quantité inaltérable de l'année.

20) GIAMBATTISTA RICCIOLI (1598–1671), *Almagestum novum* (Bononiae, 1651), tomus I, pars prior, p. 362. EJA.

21) GIAMBATTISTA RICCIOLI, *Astronomiae reformatae* (Bononiae, 1665), tomus I, pars prior, p. 115. EJA.

60. Mais ici je rencontre un Antagoniste bien redoutable, le savant M. MEYER[22]), Professeur de Gottingue, qui soutient positivement que l'année solaire n'a souffert aucune altération, quoiqu'il admette une petite altération dans le mouvement moyen de la lune, ce qui paroît déja fort paradoxe. Pour maintenir ce sentiment, il rejette les observations des équinoxes de PTOLOMÉE, qui le renverseroient tout-à-fait; et ensuite il regarde le soupçon, qu'on auroit pu commettre une erreur d'un jour dans le calcul chronologique, comme très-absurde. Il dit que la différence d'un jour produiroit dans le mouvement moyen de la lune une différence de 13 degrés, qu'on ne sauroit admettre en aucune façon. Mais cette différence suppose déja un mouvement moyen bien établi et constant; ce qui est précisément de quoi on a raison de douter. Ensuite, si en reculant d'un jour les époques de PTOLOMÉE, on en change conformément le moyen mouvement du soleil et de la lune. On conservera précisément les mêmes éclipses et toujours le même rapport entre le soleil et la lune. Outre cela, il n'est pas à craindre que cette augmentation du moyen mouvement ne soit plus d'accord avec les observations plus modernes; le mouvement moyen de ces derniers siecles n'en est point altéré; mais en remontant aux siecles passés, il y faut apporter une correction proportionnelle au quarré du tems écoulé; d'où l'on voit, que quand même cette correction pour les tems de PTOLOMÉE et HIPPARQUE seroit très-considérable, elle pourroit bien subsister avec les observations faites depuis quelques siecles, même en remontant jusqu'aux tems d'ALBATEGNIUS[23]).

22) JOHANN TOBIAS MAYER, *Commentarii Societatis Regiae Scientiarum Gottingensis* 1 (1751) 1752, p. 373. EJA.

23) AL-BATTĀNĪ (av. 858–929). EJA.

CONIECTURA CIRCA NATURAM AERIS, PRO EXPLICANDIS PHAENOMENIS IN ATMOSPHAERA OBSERVATIS

Commentatio 527 indicis ENESTROEMIANI
Acta academiae scientiarum Petropolitanae [3 : I] (1779 : I), 1782, p. 162–187

1. Quanquam nobis in intima naturae mysteria penetrare, indeque veras caussas Phaenomenorum agnoscere neutiquam est concessum: tamen evenire potest, ut hypothesis quaedam ficta pluribus phaenomenis explicandis aeque satisfaciat, ac si vera caussa nobis esset perspecta, quemadmodum felicissimo successu omnes fere motus coelestes ex hypothesi attractionis universalis determinari solent, etiamsi haec ipsa hypothesis ex Physica prorsus sit profliganda.

2. Quam ob rem fortasse similimodo quaepiam hypothesis excogitare poterit, quae omnibus Phaenomenis aëris et atmosphaerae explicandis sufficiat. Talem ideam iam ante quinquaginta annos in Tomo II veterum Commentariorum[1]) proposui, quae ad plura aëris phaenomena explicanda satis apta videbatur, etiamsi facile agnovissem, talem aëris structuram revera admitti non posse. Imprimis huius Phaenomeni explicatio: quod, dum aër vaporibus est inquinatus, eius elasticitas diminuatur, mihi omni attentione digna est visa, cum eius caussa a nemine adhuc dilucide sit exposita. Illo vero tempore Theoria fluidorum nondum satis erat exculta, ut istam ideam penitus evolvere valuissem; quamobrem operae pretium fore existimo, illam aëris structuram, quam finxeram, accuratius examinare, et quaenam Phaenomena inde oriri debeant, maiori cura inquirere.

3. Naturam aëris autem ita animo conceperam, quasi ex innumerabilibus minimis bullulis seu sphaerulis esset compositus, quae singulae cuticula tenuissima

---

1) E7, p. 1 huius voluminis. EJA.

aquosa circumdarentur, intra quas propria aëris materia motu rapidissimo in gyrum circumagatur, in cuius vi centrifuga elasticitas aëris produci erat visa. Totum negotium igitur huc redit, ut ista hypothesis accuratius perpendatur, et ad cuiusmodi phaenomena producenda sit accommodata, inquiratur. Nisi enim manifestas contradictiones involvat, satis probabile videtur, aëris phaenomena, actu observata, non multum discrepare posse.

4. Quod igitur primo ad pelliculas illas aqueas sive vaporosas attinet, earum realitas in aqua spumosa atque bullis saponaceis manifesto deprehenditur, unde recte concludere poterimus, vapores in aëre adscendentes ita dissipari, ut elementa aërea instar cuticulae involvant, quae si per omnia elementa aequaliter dispergatur, atmosphaera nihil de pelluciditate amittet, sin autem intra sphaerulas aëreas confuse hospitentur, refractio radiorum lucis eorumque transitus non mediocriter perturbabitur, unde in aëre inferiore nebulae, in superiore autem nubes oriri videntur. Praeterea, quo plures vapores aëri fuerint admixti, cuticulae illae, sphaerulas aëreas ambientes, evadent densiores, quoad scilicet constitutio harum sphaerularum sufferre valet.

5. Deinde etiam non desunt rationes, quae suadent, propriam aëris materiam in his sphaerulis motu rapidissimo circumagi, cum aliunde caussa elasticitatis repeti nequeat. Praeterea cum iam satis evictum sit, calorem in certa agitatione aetheris consistere, hinc utique materia illa aëris in sphaerulis motum quendam recipere debet, qui cum in tali angusto spatio sit inclusus, non aliter nisi per motum vorticosum continuari potest, id quod eo magis sit probabile, quod aucto calore, indeque propterea motu isto verticoso, elasticitas aëris quoque augeatur; unde manifestum est, motum gyratorium in sphaerulis aëreis cum caussa caloris arctissime esse connexum.

6. Deinde vero etiam assumsi, singula aëris elementa in memoratis sphaerulis per circulos maximos circumagi, ut in iis undequaque aequalis vis centrifuga, a centro cuiusque sphaerulae recedens, oriatur, haecque hypothesis nuncquidem cum principiis mechanicis neutiquam consistere posse videtur, cum demonstratum sit, nullum alium motum circa punctum quodpiam fixum in corporibus dari posse, nisi qui peragatur circa axem quempiam fixum. A tali autem motu alia vis centrifuga generari non potest, nisi quae ab axe gyrationis recedat, ideoque in eadem sphaerula maxime esset irregularis, cum certum sit, elasticitatem aëris in omnes plagas aequaliter tendere.

7. Obiectio autem hinc ab illo motu oriunda tantum locum habet in corporibus solidis: in fluidis enim res longe aliter se habere potest; atque adeo clare hic demonstrabo, motum illum intestinum in singulis sphaerulis revera ita comparatum esse debere, ac si singula elementa in circulis maximis circa centrum revolverentur.

8. Primo autem, cum materia aërea sit homogenea, omnes eius particulas inter se aequales concipere licet, quibus adeo initio celeritates aequales sint impressae, quibus ergo singulae, si essent solitariae, in plano uniformiter in directum ferrentur, in superficie autem sphaerica in circulis maximis essent progressurae; unde si cuiusque celeritas fuerit $=c$ et r radius sphaerae, cuiusque vis centrifuga foret $=\frac{cc}{2gr}$, qua scilicet a centro sphaerae recedere conaretur; ubi g exprimit altitudinem lapsus gravium uno minuto secundo, siquidem celeritas per spatium uno minuto secundo percursum definiatur.

9. Consideremus iam duas huiusmodi particulas A et B [fig. 1], secundum directiones AC et BC ita motas, ut in C convenirent et collisionem paterentur, quippe sine qua corpusculum A descripturum esset spatium $Ca=c$, alterum vero BC secundum directionem Cb celeritate eadem c. Iam ut videamus quamnam mutationem collisio sit productura, toti systemati, mente saltem, imprimamus celeritatem $=c$, secundum directionem bC, quo pacto corpusculum B in quiete sistetur in puncto C, corpusculum vero A, sumto $CB=Cb=c$, motum habebit secundum diagonalem CD parallelogrammi $CBDa$, qua retro producta in d, collisio perinde peragetur, ac si corpusculum A motu dC in alterum corpusculum B, in C quiescens, impingeret. Constat autem, tum corpusculum A in C esse quieturum, alterum vero B celeritatem esse acquisiturum CD. Iam restituatur motus mente impressus celeritate Cb. Hoc modo corpus prius A nunc motum

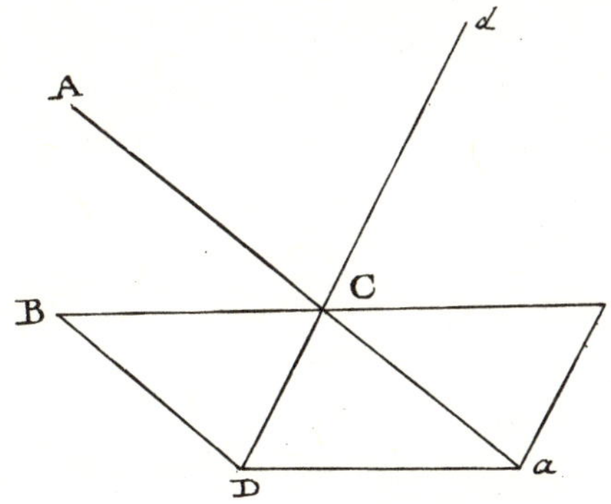

Fig. 1

acquiret Cb, alterius vero B motus componetur ex motu CD et Cb, unde si compleatur parallelogrammam $CDab$, istud corpus iam motum habebit Ca. Unde patet, motus utriusque corporis per collisionem inter se permutari, ita ut utriusque motus ab altero continuetur. Hinc cum inter bina corpuscula nullum discrimen intercedat, totus motus se habebit ac si utrumque corpusculum motum insitum sine ulla mutatione prosequeretur, unde etiam utriusque vis centrifuga nullam mutationem ob collisionem subibit.

10. Quod si iam tales collisiones in infinitum multiplicentur, omnes motus nihilominus in circulis sphaerae maximis, at vero successive ab aliis aliisque corpusculis, peragentur; quamobrem omnes vires centrifugae directe a centro recedent, et quidem eadem quantitate $\frac{cc}{2gr}$.

11. In motu ergo vorticoso, quem materiae aëris propriae, in singulis sphaerulis memoratis, tribuimus, tuto assumere possumus, omnes plane motus in circulis maximis peragi, atque singulas particulas pari vi centrifuga a centro sphaerulae recedere conari; quamobrem omnes obiectiones, quae olim contra vortices CARTESIANOS sunt motae, omnem vim amittunt. Neque tamen idcirco arbitror, illos CARTESII vortices admitti posse; at vero illi vortices, per quos Vir Celeb. DANIEL BERNOULLI olim attractionem universalem in mundo explicare est annisus, hinc summam vim acquirunt, ita ut omnes obiectiones contra factae quasi sponte evanescant.

12. Contemplemur nunc sphaerulam quamcunque, naturam aëris constituentem, cuius extrema crusta aqua, seu materia vaporosa constet, intra quam materia aëris propria motu ante descripto in gyrum agatur. Sumto igitur centro

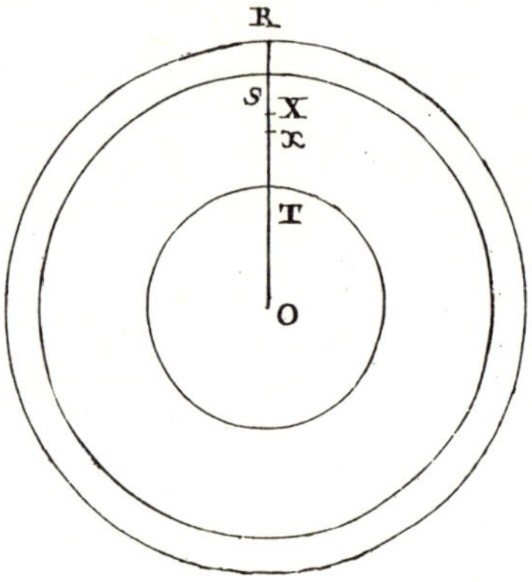

Fig. 2

sphaerulae in O [fig. 2], sit $OR = r$ radius totius sphaerulae, seu extimae eius superficiei, sitque RS crassities crustae aqueae, ponaturque $OS = s$, ita ut crassities crustae aqueae sit $RS = r - s$; tum vero sit ST crassities crustae aëreae gyrantis, quae ergo, posito $OT = t$, erit $= s - t$. Intimum autem huius sphaerulae spatium, a centro O usque ad T, repletum sit aethere puro, gravitate destituto, a cuius scilicet agitatione materia aëris propria perpetuo ad motum cieatur, modo magis modo minus, pro gradu caloris seu frigoris.

13. Hinc ergo, denotante π peripheriam circuli, cuius diameter $= 1$, erit volumen totius sphaerulae $= \frac{4}{3}\pi r^3$; unde patet, volumen crustae aqueae fore $\frac{4}{3}\pi(r^3 - s^3)$, et volumen crustae aëreae $\frac{4}{3}\pi(s^3 - t^3)$; volumen denique aethereum erit $\frac{4}{3}\pi t^3$. Quod si iam densitatem aquae unitate designemus, erit massa crustae aqueae $\frac{4}{3}\pi(r^3 - s^3)$. At si densitatem materiae aëris propriae vocemus δ, erit eius massa $\frac{4}{3}\pi\delta(s^3 - t^3)$. Quare cum ipse aether densitate carere sit censendus, erit tota massa in sphaerula contenta $\frac{4}{3}\pi(r^3 - s^3 + \delta s^3 - \delta t^3)$, quae scilicet simul exprimet pondus totius sphaerulae, cuius radius $= r$; unde manifestum est, aetherem in medio contentum recte negligi posse.

14. Hic autem statim liquet, densitatem δ non ex statu aëris ordinarii, qui iam ob elasticitatem plurimum est expansus, aestimari debere. Cum enim aër, in spatio ST inclusus, omni elasticitate destituatur, quandoquidem demum eius agitatione elasticitas aëris naturalis producitur, iste aër in nostra sphaerula contentus in eo statu reperiri censendus est, ac si ad summum densitatis gradum iam esset compressus. Phaenomena autem consulentes deprehendimus, aërem naturalem in spatium quasi octingenties minus comprimi posse, qui numerus respondet rationi inter densitatem aëris et aquae; neque ergo errabimus, si densitatem aëris, ad summum compressionis gradum reductum, densitati aquae aequalem statuamus, ita ut sit $\delta = 1$. Probabile autem admodum videtur, aërem ad talem statum reductum, simulque omni elasticitate carentem, vix a natura aquae esse discrepaturum. Hinc igitur, posito $\delta = 1$, massa atque etiam pondus nostrae sphaerulae erit $\frac{4}{3}\pi(r^3 - t^3)$ sive aequabitur ponderi massae aqueae idem volumen habentis.

15. Investigemus nunc totam vim centrifugam, quae ex motu gyratorio crustae aëreae ST oriri debet; quem in finem consideremus punctum quoddam medium X, posita eius distantia a centro $OX = x$, eiusque celeritate gyratoria $= c$, erit vis centrifuga in puncto $X = \frac{cc}{2gx}$. Hac scilicet vi elementum materiae in X

conatur a centro O recedere, unde in ista crusta aërea orietur status pressionis a termino T ad S continuo crescens.

16. Constat autem, in fluidis statum pressionis commodissime exprimi posse per certam altitudinem, quam hic vocemus $=p$, qua designatur, pressionem fluidi aequalem esse ponderi cylindri, ex eadem materia constantis, et cuius altitudo sit $=p$. Pro nostro ergo casu designet p altitudinem columnae aqueae, cuius pondus aequetur pressioni in puncto X, dum scilicet in eandem basin premit. Hinc ergo, sumto elemento $Xx = dx$, ita ut pressio in x sit $p + dp$, evidens est, incrementum pressionis dp aequari debere vi centrifugae, qua elementum Xx a centro repellitur; unde patet fore $dp = \frac{cc}{2gx}dx$, sicque integrando nanciscimur[2]) $p = \frac{cc}{2g}lx$, quod integrale ita determinari debet, ut sumto $x = t$ evanescat, ita ut pro puncto X pressio sit $p = \frac{cc}{2g}l\frac{x}{t}$. Quare promoto puncto X usque ad S, pressio hoc loco erit $p = \frac{cc}{2g}l\frac{s}{t}$. Tanta scilicet pressione ista crusta aërea, simulque tota prorsus sphaerula, conabitur se expandere, ac revera se expanderet, nisi undequaque paribusque viribus comprimeretur.

17. Quantumvis autem talis bullula sive expandatur sive comprimatur, in ea semper eadem quantitas materiae manet inclusa, quia neque materiae contentae exitus, neque novae ingressus conceditur. Ponamus ergo quantitatem materiae inclusae aequari globulo aqueo, cuius radius $=a$, quandoquidem omnem calculum ad densitatem aquae reducimus. Hinc ergo quantitas materiae in hac bullula contentae erit $\frac{4}{3}\pi a^3$, quae cum partim ex crusta aquea partim ex aërea eiusdem cum aqua densitatis constet, ponamus massam aqueam esse $\frac{4}{3}\pi\lambda a^3$, ita ut quantitas materiae aëreae propriae sit $\frac{4}{3}\pi(1-\lambda)a^3$: quantitas ergo aquae per aërem dispersae erit ad quantitatem aëris propriam ut $\lambda : 1-\lambda$.

18. Supra autem invenimus, volumen crustae aqueae esse $\frac{4}{3}\pi(r^3 - s^3)$, unde erit $r^3 - s^3 = \lambda a^3$. Deinde cum volumen materiae aëreae inventum sit $\frac{4}{3}\pi(s^3 - t^3)$, erit $s^3 - t^3 = (1-\lambda)a^3$. Hinc igitur tam s^3 quam t^3 per a^3 et r^3 definire poterimus: erit scilicet $s^3 = r^3 - \lambda a^3$ et $t^3 = r^3 - a^3$. Quare cum altitudinem pressioni debitam invenerimus

2) l denotat logarithmum. EJA.

$$p = \frac{cc}{2g} l \frac{s}{t} = \frac{cc}{6g} l \frac{s^3}{t^3},$$

erit nunc

$$p = \frac{cc}{6g} l \frac{r^3 - \lambda a^3}{r^3 - a^3}.$$

19. Cum porro densitas reperiatur, si massa per volumen dividatur, quia nostro casu massa est $\frac{4}{3}\pi a^3$, volumen autem $\frac{4}{3}\pi r^3$, erit densitas hoc loco $\frac{a^3}{r^3}$. Quod si ergo densitatem hanc designemus littera q, erit $q = \frac{a^3}{r^3}$, ideoque $r^3 = \frac{a^3}{q}$, qui valor in nostra formula substitutus praebet $p = \frac{cc}{6g} l \frac{1 - \lambda q}{1 - q}$; ubi, ut notavimus, q exprimit densitatem aëris, dum aquae densitas unitate designatur, ideoque q semper erit fractio quam minima. In superficie scilicet terrae erit quasi $q = \frac{1}{700}$, vel $q = \frac{1}{800}$.

20. Cum igitur q sit fractio tam exigua, erit satis exacte

$$l(1 - \lambda q) = -\lambda q - \tfrac{1}{2}\lambda^2 q^2 - \tfrac{1}{3}\lambda^3 q^3 \quad \text{et}$$
$$l(1 - q) = -q - \tfrac{1}{2}q^2 - \tfrac{1}{3}q^3,$$

qui posterior logarithmus a priore subtractus relinquit

$$l \frac{1 - \lambda q}{1 - q} = (1 - \lambda) q + \tfrac{1}{2}(1 - \lambda^2) q^2 + \tfrac{1}{3}(1 - \lambda^3) q^3,$$

quocirca habebimus pro pressione p hanc formulam:

$$p = \frac{cc}{6g} \left((1 - \lambda) q + \tfrac{1}{2}(1 - \lambda^2) q^2 + \tfrac{1}{3}(1 - \lambda^3) q^3\right)$$

cuius seriei plerumque sufficiet primum terminum, vel ad summum bina priora accepisse.

21. Iam igitur insignem relationem inter densitatem aëris q et altitudinem pressioni debitam p sumus adepti, cum sit

$$p = \frac{cc}{6g}\left((1-\lambda)q + \tfrac{1}{2}(1-\lambda\lambda)qq + \tfrac{1}{3}(1-\lambda^3)q^3\right),$$

ubi tam litterae p et q quam λ determinatos sortiuntur valores. Erit scilicet p altitudo Barometri aquei, pressionem Atmosphaerae aequilibrantem, cuius igitur pars circiter decima quarta dabit altitudinem Barometri mercurialis; tum vero erit q ad 1 ut densitas aëris ad densitatem aquae. Denique λ exprimit portionem vaporum aqueorum per Atmosphaeram dispersorum. His observatis solus primus terminus seriei pro p inventae, $\frac{cc}{6g}(1-\lambda)q$, insigne phaenomenon iam nobis egregie explicat. Inde enim patet, quo plures vapores cum aëre sint permixti, quorum quantitas est ut λ, pressionem p esse debere tanto minorem, pro eadem scilicet aëris densitate q, atque haec explicatio, quam iam olim loco supra citato inveneram, notatu maxime digna est visa.

22. Neque vero solus primus seriei terminus istam elasticitatis aëris diminutionem declarat, sed etiam omnes sequentes termini minores sunt quam si esset $\lambda = 0$, nullique vapores in aëre versarentur. Ceterum hic quoque notari oportet, etiam litteram c insignem variationem subire posse, pro rapiditate motus gyratorii in nostris sphaerulis vel bullulis, quae cum gradui caloris proportionalis esse videatur, aucto vel diminuto calore etiam quantitas cc vel increscet vel diminuetur.

23. Quin etiam hinc ipsa celeritas c, qua materia aërea in bullulis gyratur, absolute determinari potest, pro data scilicet altitudine p et densitate q cum humiditate λ. Sumto enim primo tantum seriei termino, qui ad hoc institutum prorsus sufficit, erit $cc = \frac{6g}{1-\lambda}\frac{p}{q}$; unde patet, hanc celeritatem directe proportionalem esse altitudini Barometri p, reciproce vero densitati aëris q; tum vero, auctum ob humiditatem λ, celeritatem c etiam augeri. Constat autem in pedibus Rhenanis esse $6g = 93\frac{3}{4}$. Istius igitur formulae radix quadrata dabit celeritatem gyrationis in paribus pedibus expressam: Indicabit enim numerum pedum, qui hac celeritate uno minuto secundo percurrerentur.

24. Cum calor a celeritate procul dubio pendeat, indagemus istam celeritatem tam pro maximo calore, qui in aëre aperto observari solet, et qui in Thermometro Delisliano[3]) respondet quasi centum gradibus, quam pro summo frigore, quod respondet 200 gradibus in eodem Thermometro. Pro priore ergo casu, quia aër est rarissimus, sumamus $q = \frac{1}{1000}$, simulque ipsi p summam altitudinem tribuamus, quae est quasi 34 pedum in Barometro aqueo. Ipsam humiditatem vero hic penitus negligamus, ut sit $\lambda = 0$. Ex his iam valoribus colligitur $cc = 93{,}75 \times 34 \times 1000 = 3\,187\,500$, ideoque ipsa celeritas $c = 1790$ ped. quae ergo respondet 100 gradibus Thermometri Delisliani.

25. Simili modo pro summo frigore 200 gradibus respondente, sumamus densitatem $q = \frac{1}{700}$, altitudinem vero Barometri etiam minimam accipiamus, scilicet $p = 31$ pedum, hincque colligitur $cc = 93{,}75 \times 31 \times 700 = 2\,034\,375$, ideoque $c = 1430$[4]), quae ergo celeritas ducentis gradibus Thermometri Delisliani respondet, ita ut differentia inter has duas celeritates sit 360 pedum. Hinc intelligitur, si plura huiusmodi experimenta instituantur, haud difficile fore pro singulis gradibus huius Thermometri respondentes celeritates assignare. Quo pacto istud Thermometrum ad multo maiorem perfectionis gradum evehetur.

26. Eodem modo etiam reliqua instrumenta, quibus aëris indoles explorare solet, beneficio nostrae formulae magis perfici poterunt. Quod quidem ad Barometrum attinet, id vix ulla emendatione indiget, si modo pro quovis caloris gradu ratio densitatis mercurii habeatur; quo enim mercurius minorem habuerit densitatem, quod fit in magno calore, tum altitudo barometrica secundum eandem rationem imminui debet, ut ad certam densitatem fixam revocetur. Summo autem frigore, quo Mercurius in spatium aliquanto minus contrahitur, ideoque eius densitas augetur, ob hanc rationem altitudo Barometri observata aliquantillum augeri debebit.

27. Praecipuum autem instrumentum, quod ad Theoriam nostram explorandam requiritur, est Manometrum, nunc quidem fere prorsus obsoletum, quo

3) Joseph Nicolas de L'Isle, *Mémoires pour servir à l'histoire et aux progrès de l'astronomie, de la géographie et de la physique* (St. Pétersbourg, 1738), pp. 267–284. EJA.

4) Editio princeps: 1330 loco 1430. Correxit EJA.

densitas aëris indicatur, et cuius descriptio exstat in WOLFII *Elementis Matheseos Tomo II*[5]), ut et in *Mémoires de l'Académie Royale de Paris* 1705[6]). Pro usu autem nostro optandum esset, ut gradibus arbitrariis in tali instrumento signatis adscriberentur numeri, indicantes, quoties densitas aëris minor sit quam densitas aquae, quam tanquam fixam spectare possumus, quippe cuius exiguae variationis, a maiore vel minore calore oriundae, ratio facile haberi poterit. Pluribus autem experimentis erit opus, antequam hoc instrumentum ad summum perfectionis gradum perducatur.

28. Superest autem adhuc instrumentum, humiditati aëris dimetiendae aptum, quod Hygrometrum appellare solet. Plura huiusmodi instrumenta passim exstant descripta; verum valde dubitandum videtur, num veram aquae quantitatem, per aërem dispersam, declarent. Interim tamen plurimis etiam novis experimentis opus erit, haec instrumenta ita instruere, ut quovis tempore verum valorem nostrae litterae λ, hoc est eam fractionem, quae indicet, quotam totius voluminis partem aqua seu humiditas in aëre occupet, doceat, a quo perfectionis gradu etiamnunc plurimum sumus remoti.

29. Cum igitur in Thermometro DELISLIANO gradui 200, quo insigne frigus indicatur, respondeat celeritas 1430 pedum in minuto secundo, quia congelatio Mercurii adhuc maiorem gradum frigoris postulat, ei circiter respondebit celeritas 1200 ped. ita ut, nisi celeritas ista fuerit maior, Mercurius fluiditatem penitus amittat. Deinde cum gradui 100 respondeat celeritas 1790 pedum, quia terminus congelationis in gradum 150 cadit, cui ergo respondebit celeritas 1610 ped., celeritate maiore opus erit, ad aquam in statu fluiditatis conservandam.

30. Quia porro gradus ebullitionis aquae in hoc Thermometro est 0, ei propemodum conveniet celeritas 2150 ped. ubi ergo aqua ebullire incipiet. Et quia in nostra formula altitudo Barometri potissimum ingreditur, hinc intelligere licet, cur, aucta aëris elasticitate, maior gradus ad ebullitionem aquae requiratur, et cur, minuta elasticitate, aqua etiam in minore gradu caloris ebullire queat, quemadmodum experimentis est comprobatum, cuius phaenomeni ergo caussa in nostra formula sine dubio erit quaerenda.

---

5) CHRISTIAN WOLFF, *Elementa matheseos universae*, editio nova (Halae Magdeburgicae, 1733), tomus 2, pp. 398–400. EJA.

6) PIERRE VARIGNON, «Manometre ou machine pour trouver le raport des raretes ou raréfactions de l'air naturel», *Mémoires de l'Académie Royale des Sciences* (Paris), (1705), 1706, pp. 300–331. EJA.

31. Neque vero celeritas, ad quemvis caloris gradum aëri inducendum requisita, tantum ad aërem spectare est censenda, cuius scilicet minimae particulae tanta celeritate commoveri debent, sed etiam in omnibus plane corporibus perinde locum habere videtur. Omnes quoque naturae scrutatores in hoc conveniunt, quod caussa caloris in motu quodam pernicissimo minimarum particularum consistat. Quae ergo sententia non solum nostrae Theoriae maxime est conformis, sed etiam ipsam celeritatem, cuilibet gradui caloris convenientem, assignare valemus. Quanquam autem haec celeritas maxime enormis videatur, tamen perpendendum est, in natura dari celeritates adhuc incomparabiliter maiores, cuiusmodi est celeritas radiorum lucis, in quibus cum caussa omnis caloris sit quaerenda, mirum non est, hinc tam insignem celeritatis gradum generari posse.

32. Sed revertamur ad nostram formulam supra inventam et ad solum aërem proprie accommodatam, quae haec quatuor elementa: 1°) altitudinem pressioni debitam p; 2°) densitatem aëris q; 3°) gradum caloris, formula $\frac{cc}{6g}$ contentum et 4°) gradum humiditatis λ complectitur. Ex datis horum elementorum ternis quibusque quartum assignari potest; ita, si dentur densitas aëris q, gradus caloris $\frac{cc}{6g}$, cuius loco br[evitatis] gr[atia] scribamus h et gradus humiditatis λ, hinc altitudo, pressioni debita, p in genere ita exprimitur, ut sit $p = hl\frac{1-\lambda q}{1-q}$, quam pro faciliori calculo in hanc seriem resolvimus:

$$p = h\big((1-\lambda)q + \tfrac{1}{2}(1-\lambda\lambda)qq + \tfrac{1}{3}(1-\lambda^3)q^3 + \text{etc.}\big)$$

cuius applicatio iam satis est exposita.

33. Ponamus nunc dari primo altitudinem pressioni debitam p, secundo densitatem aëris q, et tertio humiditatem λ, hinc gradus caloris $h = \frac{cc}{6g}$ ita definitur, ut sit

$$h = \frac{p}{l\dfrac{1-\lambda q}{1-q}},$$

hincque, logarithmis per seriem expressis, erit

$$h = p : \left((1-\lambda)q + \tfrac{1}{2}(1-\lambda\lambda)qq + \tfrac{1}{3}(1-\lambda^3)q^3 + \text{etc.}\right)$$

cuius seriei plerumque sufficit solum primum terminum cum secundo sumsisse.

34. Sin autem detur altitudo pressioni debita p, mensura caloris h, ac tertio humiditas λ, inde ad densitatem q inveniendam recurrendum est ad exponentialia, cum sit

$$e^{p/h} = \frac{1-\lambda q}{1-q},$$

unde posito br[evitatis] gr[atia] $e^{p/h} = k$, erit $q = \dfrac{k-1}{k-\lambda}$, et quia h plerumque plurimum excedit p, cum per seriem sit

$$k = 1 + \frac{p}{h} + \frac{pp}{2hh} + \frac{p^3}{6h^3} + \text{etc.}$$

erit

$$q = \left(\frac{p}{h} + \frac{pp}{2hh} + \frac{p^3}{6h^3} + \text{etc.}\right) : \left(1 - \lambda + \frac{p}{h} + \frac{pp}{2hh} + \frac{p^3}{6h^3}\ \text{etc.}\right).$$

35. Denique si detur altitudo pressioni debita p, mensura caloris h, et densitas q, per formulam exponentialem

$$e^{p/h} = k = \frac{1-\lambda q}{1-q}$$

humiditas λ ita determinetur, ut sit

$$\lambda = \frac{1-k(1-q)}{q},$$

quae expressio hoc laborat defectu, ut minimus error, in elementis datis p, k et q commissus, enormes errores in valore λ producat.

36. Imprimis autem nostra formula commodissime adhiberi potest ad Problema aërometricum maximi momenti resolvendum, quo quaeri solet, quanta vi opus sit ad aërem in spatium quantumvis minus coarctandum, cuius ergo solutionem hic subiungimus.

Problema

Investigare, quanta vis requiratur, ad datum aëris volumen in spatium quantumvis minus comprimendum.

Solutio

37. Ponamus aërem comprimendum in tubo cylindrico contineri, cuius amplitudo sit $=ff$ et compressionem per emboli intrusionem produci; tum si altitudo, pressioni debita, fuerit $=p$, vis requisita aequabitur ponderi columnae aqueae, cuius basis $=ff$ et altitudo $=p$, ita ut ista vis per massam aqueam expressa sit $=ffp$. Ponamus in statu naturali, unde compressio inchoat, altitudinem pressioni debitam esse $=a$, densitatem vero aëris naturalem $=b$; at vero mensurae caloris cum humiditate, quoniam durante compressione nullam mutationem patiuntur, sint ut ante h et λ, ita ut sit $a = hl\dfrac{1-\lambda b}{1-b}$, sive proxime $a = (1-\lambda)bh$.

38. Ponamus nunc intrusione emboli istud aëris volumen iam in spatium n vicibus minus esse compressum, ita ut iam eius densitas sit $q = nb$, unde quaeri debet altitudo pressioni debita p, huic densitati respondens, quae ergo, loco q scribendo nb, erit $hl\dfrac{1-\lambda nb}{1-nb}$; unde nisi compressio iam satis sit notabilis, satis exacte erit $p = h(1-\lambda)nb$. Hinc patet, pressionem p exacte proportionalem esse numero n, sive pressionem semper densitati esse proportionalem, nisi compressio iam sit vehementer magna.

39. In maioribus autem compressionibus adhiberi etiam conveniet secundum seriei terminum, ita ut sit

$$p = h\big((1-\lambda)nb + \tfrac{1}{2}(1-\lambda\lambda)nnbb\big),$$

cum initio fuisset $p = a$, quae ergo altitudo quanto iam sit maior, definiri debet ex hac formula:

$$\frac{p}{a} = n + \tfrac{1}{2}(1-\lambda)nnb;$$

unde patet, vim comprimentem plusquam n vicibus esse maiorem, prorsus uti per experimenta est observatum.

40. Sin autem compressio longe ulterius continuari concipiatur, recurrendum erit ad formulam logarithmicam $p = hl\frac{1-\lambda nb}{1-nb}$, quae, comparata cum pressione initiali $a = hl\frac{1-\lambda b}{1-b}$, dabit

$$\frac{p}{a} = l\frac{1-\lambda nb}{1-nb} : l\frac{1-\lambda b}{1-b}.$$

Quia autem posterior logarithmus est satis exacte $(1-\lambda)b$, erit

$$\frac{p}{a} = \frac{1}{(1-\lambda)b} \cdot l\frac{1-\lambda nb}{1-nb};$$

unde patet, casu $nb = 1$ sive $n = \frac{1}{b}$, vim requisitam fieri infinitam.

41. Quo haec clarius perspiciantur, ponamus pro statu initiali esse $b = \frac{1}{800}$, sive densitatem aëris ad aquae densitatem esse ut 1 ad 800, humiditatem autem λ penitus seponamus, eritque ergo a pressio Atmosphaerae naturalis, et quaeramus nunc, quotuplo maior pressio requiratur ad aëris volumen in spatium n vicibus minus coarctandum; tum igitur formula ante data nobis praebebit, $\frac{p}{a} = 800\,l\frac{800}{800-n}$, ubi logarithmis utendum erit hyperbolicis. Ita si aër in spatium quadringenties minus comprimatur, hoc est, si $n = 400$, fiet $\frac{p}{a} = 554$; scilicet vis, quae tantum esset quadringenties maior, non sufficit, sed requiritur vis 554 vicibus maior. Quodsi autem condensatio in spatium 700^{es} minus postulatur, vi opus erit 1663 vicibus maiore.

42. Operae igitur pretium videtur, pro hac hypothesi $b = \frac{1}{800}$ et $\lambda = 0$ tabulam construere, indicantem, quotuplo maior fiat vis comprimens requisita ad aërem in spatium n cuplo minus redigendum. Sequens tabula igitur ostendet pro singulis numeris n valorem formulae

$$\frac{p}{a} = 800\,l\frac{800}{800-n},$$

ubi pro condensationibus minoribus erit

$$\frac{p}{a} = n + \tfrac{1}{2}\frac{nn}{800} + \tfrac{1}{3}\frac{n^3}{800^2},$$

ita ut tantum logarithmis hyperbolicis indigeamus, quando numerus n ultra 100 assurgit[7]).

n	$p:a$	n	$p:a$
1	1,0006	125	135,92
2	2,0025	150	166,11
3	3,0056	175	197,49
4	4,0100	200	230,14
5	5,0157	225	264,19
6	6,0226	250	299,75
7	7,0308	275	336,97
8	8,0403	300	376,00
9	9,0510	350	460,28
10	10,063	400	554,48
20	20,254	450	661,34
30	30,577	500	784,64
40	41,035	550	930,52
50	51,630	600	1109,0
60	62,369	650	1339,2
70	73,254	700	1663,5
80	84,288	750	2218,1
90	95,477	800	infinit.
100	106,82		

43. Manifestum est vires, ad aërem comprimendum requisitas, in hac tabula exhibitas, iam in se complecti pressionem Atmosphaerae; unde si tantum vires actu adhibendae quaerantur, subtrahi debet inde illa pressio, seu, quod eodem redit, ab omnibus his viribus subtrahi debet vis casui $n = 1$ respondens. Praeterea ob humiditatem, quam hic neglеximus, omnes vires hic designatae aliquod incrementum nanciscuntur; scilicet, si pro compressione n cupla vis fuerit $= n + \pi$, ob humiditatem λ ista vis erit $n + (1 + \lambda)\pi$.

7) Ab $n = 40$ usque ad $n = 100$ terminus $\frac{1}{4}\frac{n^4}{800^3}$ inclusus est. EJA.

De variatione status aëris per universam Atmosphaeram

Fig. 3

44. Quae hactenus sunt tradita ad aërem in certo loco existentem restringuntur. Nunc autem videamus, qua lege status aëris per Atmosphaeram sive ascendendo sive descendendo immutetur. Hic igitur in certo Terrae loco A [fig. 3] statum aëris tanquam cognitum assumemus, hincque verticaliter ascendendo investigabimus, quomodo pro quavis altitudine $AZ = z$, status aëris se sit habiturus; ubi evidens est, si ad interiora Terrae descendere velimus, altitudinem z tantum negativam esse accipiendam. Hic igitur ante omnia pro loco fixo A statum aëris definiri oportet, siquidem pro quavis altitudine $AZ = z$ elementa calculi cum hoc loco comparari conveniat.

45. Sit igitur primo altitudo Barometri aquei pro loco $A = a$, et pro loco $Z = p$; secundo densitas aëris in A sit $= b$, in Z vero $= q$; tertio sit humiditas in $A = \lambda$, in Z vero $= \varphi$; quarto denique ponatur celeritas motus gyratorii in subsidium vocati pro loco $A = c$, pro Z vero $= u$. Tum vero pro loco A sit brevitatis gratia $\frac{cc}{6g} = h$, pro loco Z autem sit $\frac{uu}{6g} = v$; ubi notetur, quantitates h et v longitudinem plurium millium pedum designare, propterea quod celeritas c plerumque inter terminos 1400 et 1800 pedum continetur. Quia igitur valores litterae c iam ad gradus Thermometri relatos assumimus, pro praecipuis valoribus celeritatis c valores quantitatis h in sequenti tabula adiungamus, incipiendo a $c = 1200$ usque ad $c = 2000$ per 50 ascendendo:

c	h	c	h
1200	15360	1650	29040
1250	16667	1700	30827
1300	18026	1750	32667
1350	19440	1800	34560
1400	20907	1850	36507
1450	22427	1900	38507
1500	24000	1950	40560
1550	25627	2000	42667
1600	27307		

46.[8]) Quia densitas aëris ascendendo decrescit, evidens est, solum primum terminum seriei supra datae abunde sufficere; unde statuere licebit $p = (1 - \varphi)vq$, quae formula etiam tuto adhiberi poterit, si in viscera terrae descendamus. Hinc igitur densitas in loco Z erit $q = \frac{p}{(1-\varphi)v}$, unde si per elementum $Zz = dz$ ulterius ascendamus, pressio in z erit $p + dp$, quae autem minor erit quam in Z ponduscolo aëris in intervallo Zz contenti, quod reperiemus, si elementum dz per densitatem q multiplicemus, unde fit $dp = -q\,dz$. Ergo, loco q substituto valore modo dato erit $dp = -\frac{p\,dz}{(1-\varphi)v}$, hincque $dz = -\frac{(1-\varphi)v\,dp}{p}$, quae aequatio, quia tam v quam φ tanquam functiones altitudinis z sunt spectandae, ita repraesentari debet: $\frac{dz}{(1-\varphi)v} = -\frac{dp}{p}$, cuius integrale est

$$\int \frac{dz}{(1-\varphi)v} = C - lp,$$

quam constantem C ita definiri oportet, ut casu $z = 0$, quo simul integrale

$$\int \frac{dz}{(1-\varphi)v}$$

evanescere debet, evadat $p = a$, unde erit $C = la$, ideoque

$$\int \frac{dz}{(1-\varphi)v} = l\frac{a}{p}.$$

47. Sinistrum igitur membrum huius aequationis erit certa functio ipsius z, pendens a ratione, secundum quam tam calor quam humiditas ascendendo sive crescit sive decrescit; membrum vero dextrum tantum altitudinem barometricam tam in A quam in Z involvit, cuius logarithmus hyperbolicus sumi debet. Atque hic perinde est, sive altitudo Barometri aquei sive mercurialis in calculum introducatur, quia fractio $\frac{a}{p}$ inde non mutatur. Cum igitur sit $l\frac{a}{p} = la - lp$, valores

8) Editio princeps: Numeri paragraphon usque ad finem huius commentationis differunt. EJA.

horum logarithmorum hyperbolicorum pro singulis altitudinibus Barometri mercurialis, quae per digitos indicari solent, in sequenti tabula ab altitudine 36 pollicum, ad quam certe infra Terram descendendo nunquam pervenietur, diminuendo per semipollices, exhibeamus:

Alt. Bar.	Logar. hyperb.	Alt. Bar.	Logar. hyperb.	Alt. Bar.	Logar. hyperb.
36,0	3,583519	26,0	3,258097	16,0	2,772589
35,5	3,569533	25,5	3,238679	15,5	2,740840
35,0	3,555348	25,0	3,218876	15,0	2,708050
34,5	3,540960	24,5	3,198673	14,5	2,674149
34,0	3,526361	24,0	3,178054	14,0	2,639057
33,5	3,511545	23,5	3,157001	13,5	2,602690
33,0	3,496508	23,0	3,135494	13,0	2,564949
32,5	3,481240	22,5	3,113515	12,5	2,525729
32,0	3,465736	22,0	3,091042	12,0	2,484907
31,5	3,449988	21,5	3,068053	11,5	2,442347
31,0	3,433987	21,0	3,044522	11,0	2,397895
30,5	3,417727	20,5	3,020425	10,5	2,351375
30,0	3,401197	20,0	2,995732	10,0	2,302585
29,5	3,384390	19,5	2,970415	9,5	2,251292
29,0	3,367296	19,0	2,944439	9,0	2,197225
28,5	3,349904	18,5	2,917771	8,5	2,140066
28,0	3,332205	18,0	2,890372	8,0	2,079442
27,5	3,314186	17,5	2,862201	7,5	2,014903
27,0	3,295837	17,0	2,833213	7,0	1,945910
26,5	3,277145	16,5	2,803361	6,5	1,871802

Hic perinde est quanam digitorum mensura uti velimus, quia tantum ratio in computum ingreditur.

48. Quod si tam calor quam humiditas per totam altitudinem z constans assumatur, ut sit $v = h$ et $\varphi = \lambda$, aequatio nostra integralis erit

$$\frac{z}{(1-\lambda)h} = l\frac{a}{p},$$

unde pro quavis altitudine barometrica altitudo $AZ = z$ innotescit, cum sit $z = (1-\lambda)\, h\, l\frac{a}{p}$. Vicissim vero pro altitudine AZ data, posito

$$e^{\frac{z}{(1-\lambda)h}} = y,$$

erit $y = \frac{a}{p}$, ideoque $p = \frac{a}{y}$. Quia autem haec fractio $\frac{z}{(1-\lambda)h}$ est plerumque quam minima, erit proxime

$$\frac{1}{y} = 1 - \frac{z}{(1-\lambda)h} + \frac{zz}{2(1-\lambda)^2 hh} - \text{etc.}$$

ideoque

$$p = a\left(1 - \frac{z}{(1-\lambda)h} + \frac{zz}{2(1-\lambda)^2 hh} - \frac{z^3}{6(1-\lambda)^3 h^3} + \text{etc.}\right).$$

49. Hic autem casus vix usquam locum habebit: certum enim est, per Atmosphaeram ascendendo, gradum caloris continuo diminui. Etsi autem ratio diminutionis maxime latet, tamen non multum a veritate aberrabimus, si valorem ipsius v hoc modo repraesentemus:

$$v = \frac{h}{1 + \frac{z}{f}},$$

ita ut in altitudine $z = f$ valor ipsius v ad dimidium redigatur; tum enim, cognita hac altitudine f, ista formula vix a veritate aberrare poterit. Hanc ob rem sumamus $v = \frac{hf}{f+z}$, humiditatem vero per totam altitudinem invariatam spectemus, ut sit $\varphi = \lambda$, quia plane non constat, quomodo variatio humiditatis in calculum introduci posset, cuius etiam effectus vix sensibilis esse potest, quia maximus valor ipsius λ nunquam $\frac{1}{30}$ superare posse videtur.

50. His igitur valoribus substitutis aequatio nostra erit

$$\frac{1}{h(1-\lambda)}\int\frac{(f+z)\,dz}{f} = l\,\frac{a}{p},$$

unde integratione instituta erit

$$\frac{1}{(1-\lambda)h}\left(z+\frac{zz}{2f}\right)=l\frac{a}{p},$$

ex qua aequatione, si modo constet altitudo f, ubi mensura caloris v ad dimidium redigitur, ad quamvis altitudinem z assignari poterit altitudo barometrica p.

51. Hic autem occurrit quaestio maximi momenti, quomodo pro quavis altitudine Barometri, supra Terram ascendendo observata, inde ipsa elevatio loci, sive altitudo $AZ = z$ definiri possit. Evidens autem est, talem conclusionem ex sola altitudine Barometri neutiquam derivari posse, nisi simul innotescat quantitas illa f. Quod si autem insuper in Z observetur altitudo Thermometri, ex ea valor ipsius v concludi poterit, quandoquidem hic assumere licet, pro quolibet gradu Thermometri innotescere celeritatem motus gyratorii c, sive nostro casu u, unde fit $v=\frac{uu}{6g}$. Invento igitur vero valore ipsius v, ob $v=\frac{fh}{f+z}$, erit vicissim $f=\frac{vz}{h-v}$, qui valor in nostra postrema aequatione substitutus dabit hanc:

$$\frac{(h+v)z}{2(1-\lambda)hv}=l\frac{a}{p},$$

consequenter habebimus

$$z=\frac{2(1-\lambda)hv}{h+v}\,l\frac{a}{p}.$$

52. Ope igitur huius formulae par solas observationes barometricas et thermometricas elevatio cuiusque loci super horizontem assignari poterit, quae, cum sit $h=\frac{cc}{6g}$ et $v=\frac{uu}{6g}$, ita se habebit:

$$z=\frac{2(1-\lambda)huu}{cc+uu}\,l\frac{a}{p};$$

unde si Thermometrum iam sit instructum, uti supra monuimus, quaevis altitudo satis facile assignari poterit ope observationum barometricarum et thermometricarum. Ac si pro z prodeat valor negativus, quod evenit quando $p > a$, inde profunditas infra superficiem Terrae patescet.

53. Quia autem celeritatum c et u valores adhuc quandam incertitudinem involvunt, propterea quod observationes multo accuratiores requirunt, quam quidem instituere licet, optandum esset, ut nostra formula primum ad casus tales, ubi elevatio loci aliunde iam est cognita, applicetur; hinc enim facile accuratiores determinationes litterarum c et u concludi poterunt; hocque pacto simul constructio Thermometrorum ad maiorem perfectionis gradum perducetur. Qui autem indolem nostrarum formularum omni studio perpendere dignabitur, longe plura ad scientiam naturalem promovendam inde derivare poterit.

DE STATU AEQUILIBRII MARIS A VIRIBUS SOLIS ET LUNAE SOLLICITATI

Commentatio 546 indicis ENESTROEMIANI
Acta academiae scientiarum Petropolitanae [4 : I] (1780 : I), 1783, p. 132–153

1. Cum olim quaestio de fluxu et refluxu maris summo studio sit pertractata, atque adeo tres[1]) dissertationes ab Academia Regia Scientiarum Parisina praemio sint condecoratae, hoc argumentum iam penitus exhaustum videbitur, ita ut nihil amplius desiderari queat. Quoniam vero illo tempore theoria de aequilibrio et motu fluidorum adhuc parum erat exculta, plurima ibi occurrunt, quae vel nimis operose et per plures ambages ex primis principiis sunt deducta, vel etiam hypothesibus precario assumtis et a veritate alienis sunt superstructa. Variae opiniones a veritate abhorrentes invaluerunt, cuiusmodi sunt, quod quantitas aestus marini potissimum ab interna Terrae structura, quae nobis penitus est incognita, atque adeo etiam a gravitate specifica aquae pendeat, ita ut, si ea esset vel maior vel minor, etiam aestus marini inde prodirent vel minores vel maiores, cum tamen secundum vera principia huius Theoriae postmodum stabilita hae circumstantiae nihil ad scopum conferant.

2. Quanquam autem hic praecipua quaestio circa motum maris versabatur, tamen omnes, qui eius enodationem illo tempore susceperunt, unanimi consensu agnoverunt, nihil in hac re determinari posse, nisi status aequilibrii, in quo oceanus, a viribus tam Solis quam Lunae agitatus, aequiescere queat, accuratae definiatur; unde etiam istum aequilibrii statum ex principiis tum temporis cognitis determinare

1) Quattuor dissertationes condecoratae sunt: DANIEL BERNOULLI, «Traité sur le flux et reflux de la mer»; LEONHARD EULER, «Inquisitio physica in causam fluxus ac refluxus maris»; COLIN MACLAURIN, «De causa physica fluxus et refluxus»; ANTOINE CAVALLERI, «Dissertation sur la cause physique du flux et du reflux de la mer»; *Recueil des pièces qui ont remporté les prix de l'Académie Royale des Sciences*, tome 4 (Paris, 1752). EJA.

sunt conati, dum ex viribus, quibus singula puncta in superficie maris sollicitantur, eam curvam investigarunt, quae ad omnes earum virium directiones medias esset normalis, quod principium utique in natura rei est fundatum, sed difficillimas evolutiones analyticas postulat, quas vix, ac ne vix quidem, expedire licuit. Tum vero etiam ad canales usque ad centrum Terrae excavatos confugerunt et in eam figuram maris inquisiverunt, ut omnes iste columnae fierent aequiponderantes, quae investigatio utique exactam cognitionem structurae internae globi terraquei requirebat, quam quisque pro arbitrio ita finxerat, prouti scopo proposito maxime convenire videbatur, cum tamen nuncquidem actum sit, inde vix quicquam ad hunc scopum conferri.

3. Cum igitur post illud tempus theoriam de aequilibrio fluidorum ita mihi excolere et in lucem constituere contigerit, ut nihil insuper desiderari queat, his veris principiis insistens istud argumentum denuo sum suscepturus, quoniam hinc ad accuratam cognitionem status aequilibrii maris a viribus quibuscunque sollicitati perveniemus, qua omnes illae opiniones perversae penitus removeantur. Praeterea vero per haec nova principia tota ista investigatio a maximis illis difficultatibus, quibus ante premebatur, penitus liberabitur; atque hic imprimis summus usus principii minimae actionis, quod illustrissimo Praesid[ent]i DE MAUPERTUIS acceptum referre debemus, clarissime elucebit, cum eo utentes plurimas easque difficillimas integrationes evitare queamus. Ab isto igitur principio inchoëmus.

Principium Universale Aequilibrii Fluidorum[2])

4. Si singulae fluidi particulae Z [fig. 1] ad quotcunque virium centra fixa C, C′, C″ urgeantur, viribus, quae sint cuicunque functioni distantiarum propor-

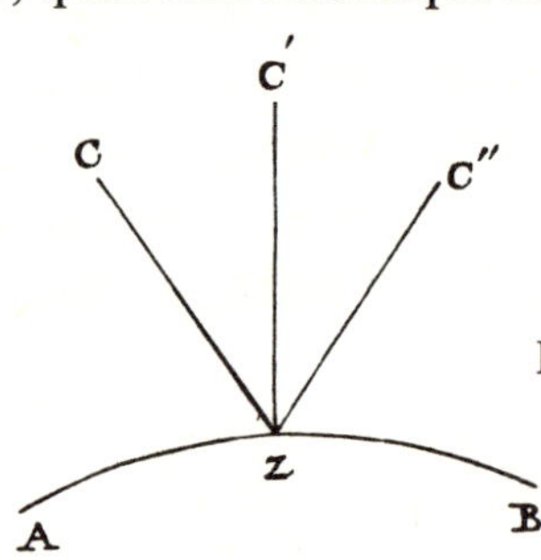

Fig. 1

2) E 225, «Principes généraux de l'état d'équilibre des fluides», *Mémoires de l'Académie Royale des Sciences de Berlin* [11] (1755), 1757, 217–273 = *Opera omnia* II 12, pp. 2–53. Vide p. 15. EJA.

tionales, tum superficies fluidi AZB erit in aequilibrio, si summa actionum, quas vires in singula puncta Z exerunt, ubique fuerit eadem. Hinc ergo si vocentur distantiae $CZ = z$, $C'Z = z'$, $C''Z = z''$ etc., ipsae autem vires ad haec centra tendentes sint Z, Z', Z'', etc., tum quia actiones harum virium per istas formulas exprimuntur: $\int Z\,dz$, $\int Z'\,dz'$, $\int Z''\,dz''$, pro statu aequilibrii statim habemus istam aequationem:

$$C = \int Z\,dz + \int Z'\,dz' + \int Z''\,dz'',$$

ubi C denotat quantitatem constantem ex circumstantiis definiendam.

Status quaestionis

5. Consideremus igitur totum globum terraqueum, sive circumquaque aqua circumfusum, sive ex aliqua tantum parte, cuius figura foret perfecte sphaerica, si nullae vires externae adessent; eius centrum sit in C [fig. 2], radius vero CA

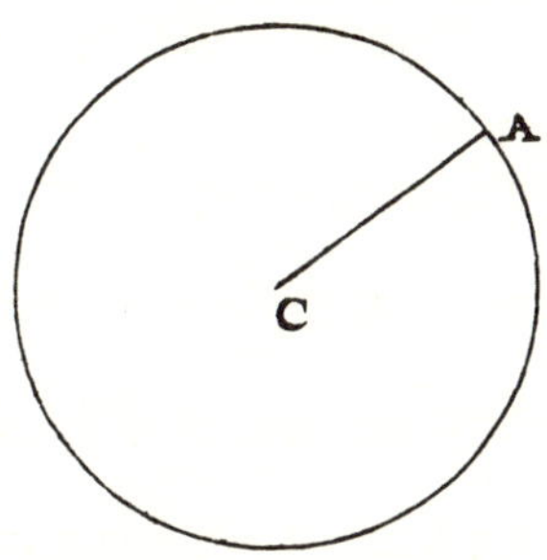

Fig. 2

unitate designetur. Hic enim a motu vertiginis Terrae animum abstrahimus, ita ut totus globus in perfecta quiete versetur, et singulae eius particulae a sola gravitate versus centrum C urgeantur, viribus utcunque a distantia pendentibus. In ipsa autem superficie vis gravitatis unitate exprimatur. Quamcunque enim legem gravitas ad centrum accedendo tenuerit, si mutationes, quae eius figurae a viribus externis induci possunt, respectu totius magnitudinis Terrae fuerint quam minimae, prouti eas fore novimus prope superficiem, vis gravitatis nunquam ab unitate discrepare censeri potest. Tum vero facile intelligitur, quamcunque mutationem vires externae superficiei aquae induxerint in hypothesi perfectae quietis, tum etiam admisso motu vertiginis eandem mutationem verae figurae Terrae, quae tum erit sphaeroidica, accidere debere, quandoquidem etiam hoc casu differentia a figura sphaerica est quam minima. Tota enim investigatio in eo versatur, ut, quantum superficies aquae a viribus externis ultra statum naturalem sive attollatur sive deprimatur, inquiri debeat, ac tum perinde erit, sive figura naturalis fuerit sphaerica, sive paulisper ab ea recedat. His igitur constitutis quaestio eo redit; ut dum vel Sol vel Luna in quocunque coeli loco versentur, ibique tanquam fixi concipiantur, pro singulis punctis in superficie maris assumtis quantitas actionis a viribus horum corporum nata rite definiatur. Vires autem tam Solis quam Lunae quadrato distantiarum reciproce proportionales assumemus.

De quantitate actionis a viribus Solis et Lunae orta

6. Quoniam pro statu aequilibrii maris indagando ad quodvis tempus tam Sol quam Luna in eodem coeli loco permanere concipi debent, siquidem pro singulis momentis status aequilibrii variatur, ponamus distantiam Solis a centro Terrae perpetuo $= a$, distantiam vero Lunae ab eodem centro $= b$, quae igitur pro diverso luminarium situ diversos valores recipere possunt, inter quos medium sumendo hae distantiae per radium Terrae unitate designatum ita definientur, ut sit $a = 24000$ et $b = 60\frac{1}{4}$ circiter. Praeterea vero vis Solis absoluta tanta sit, ut in distantia a centro Solis $= \alpha$ aequetur vi gravitatis $= 1$; pro Luna autem sit ista distantia, in qua eius vis gravitati aequaretur $= \beta$, unde, si distantia a Sole fuerit $= z$, vis Solis erit $\frac{\alpha\alpha}{zz}$; sin autem z sit distantia a Luna, vis ibi ad Lunam tendens erit $= \frac{\beta\beta}{zz}$, ita ut hoc loco quantitates $\alpha\alpha$ et $\beta\beta$ idem denotent, quod vulgo per massam Solis vel Lunae indicari solet. Colligitur autem ex novissimis parallaxis solaris determinationibus, fore $\alpha\alpha = 355426$, unde fit $\alpha = 596$, ita ut in distantia Solis tot semidiametris Terrae aequali vis Solis aequetur ipsi vi gravitatis in superficie Terrae. Pro Luna autem quantitas $\beta\beta$ a NEWTONO[3]) aestimata est $\frac{1}{40}$, Celeb. autem BERNOULLI[4]) ostendit, eam esse notabiliter minorem, ita ut propemodum aestimari possit $\beta = \frac{1}{8}$, unde fieret $\beta\beta = \frac{1}{64}$.

3) NEWTON, *Principia*, Liber III, propositio 37, corollarium 4. EJA.

4) D. BERNOULLI, «Traité sur le flux et reflux de la mer», *Recueil des pièces qui ont remporté les prix de l'Académie Royale des Sciences*, tome 4, (Paris, 1752), pp. 53–191, vide pp. 113–117 = *Die Werke von Daniel Bernoulli*, Band 3, ed. D. SPEISER et al. (Basel, 1987), pp. 327–438, vide pp. 379–382. EJA.

7. His igitur positis, si distantia sive Solis sive Lunae a puncto quocunque Z fuerit $=z$, erit vis hoc punctum sollicitans, scilicet $Z=\frac{\alpha\alpha}{zz}$, pro Sole, et $Z=\frac{\beta\beta}{zz}$, pro Luna. Hinc ergo quantitas actionis pro Sole erit $\int Z\,dz=-\frac{\alpha\alpha}{z}$ et pro Luna $=-\frac{\beta\beta}{z}$. Sit igitur C [fig. 3] centrum Terrae et S centrum Solis, ac

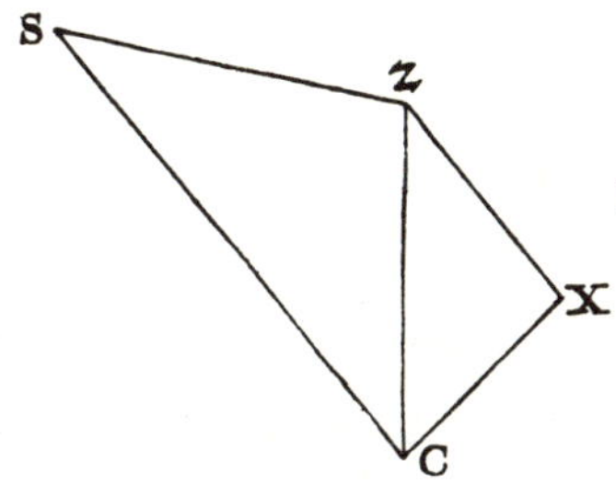

Fig. 3

ponatur distantia CS $=a$; tum vero consideretur in superficie maris, postquam se iam ad aequilibrium composuerit, punctum quodcunque Z, cuius distantia a centro Terrae, quae iam erit aliquantillum sive maior sive minor unitate, sit CZ $=r$, angulus vero SCZ vocetur $=\varphi$, eritque distantia ZS $=z=\sqrt{(aa-2ar\cos\varphi+rr)}$, sicque quantitas actionis erit

$$=\frac{-\alpha\alpha}{\sqrt{(aa-2ar\cos\varphi+rr)}}.$$

Simili modo pro Luna, cuius distantiam a centro Terrae ponimus $=b$, si eius elongationem ab eodem puncto Z statuamus $=\psi$, erit actio Lunae in punctum Z

$$=\frac{-\beta\beta}{\sqrt{(bb-2br\cos\psi+rr)}}.$$

8. Hic autem probe perpendendum est, Solem et Lunam eatenus tantum in punctum Z agere, quatenus actio discrepat ab ea, quae in centrum Terrae exeritur. Cum igitur hoc centrum in S in directione CS sollicitetur vi $=\frac{\alpha\alpha}{zz}$, punctum Z a pari vi in directione contraria ZX urgeri censendum est, existente ZX parallela ipsi CS; ad eam igitur ex C ducatur perpendicularis CX, voceturque intervallum ZX $=x$. Haec igitur vis ad punctum fixum, in recta ZX infinite remotum, trahere censenda est, ita ut distantia, quam in genere posuimus $=z$, hic sit $x+\infty$. Quia

igitur ipsa vis, quam posuimus Z, hic est constans $= \frac{\alpha\alpha}{aa}$, erit eius actio in punctum Z

$$\int Z\,dz = \int \frac{\alpha\alpha}{aa}\,dx = \frac{\alpha\alpha}{aa}\,x.$$

Cum igitur ob angulum $SCZ = \varphi$ sit $ZX = x = r\cos\varphi$, ista actio erit

$$\frac{\alpha\alpha r\cos\varphi}{aa},$$

unde actio tota a Sole oriunda in punctum Z erit:

$$-\frac{\alpha\alpha}{\sqrt{(aa - 2ar\cos\varphi + rr)}} + \frac{\alpha\alpha r\cos\varphi}{aa},$$

similique modo tota actio a Luna in punctum Z exerta:

$$= -\frac{\beta\beta}{\sqrt{(bb - 2br\cos\psi + rr)}} + \frac{\beta\beta r\cos\psi}{bb}.$$

Aequatio pro statu aequilibrii maris a viribus Solis et Lunae sollicitati

9. Hic igitur ante omnia vis gravitatis, qua punctum maris quodcunque Z [fig. 4] ad ipsum centrum Terrae C urgetur, considerari debet, quam, quia nullam mutationem sensibilem subire potest, ob variatam distantiam $CZ = r$, unitate designamus, ita ut hoc casu sit $z = r$ et $Z = 1$. Hinc ergo actio gravitatis in punctum maris Z erit $= r$. Quam ob rem si curva AZB denotet superficiem maris iam ad aequilibrium reductam; tum vero hoc tempore Sol et Luna versentur in

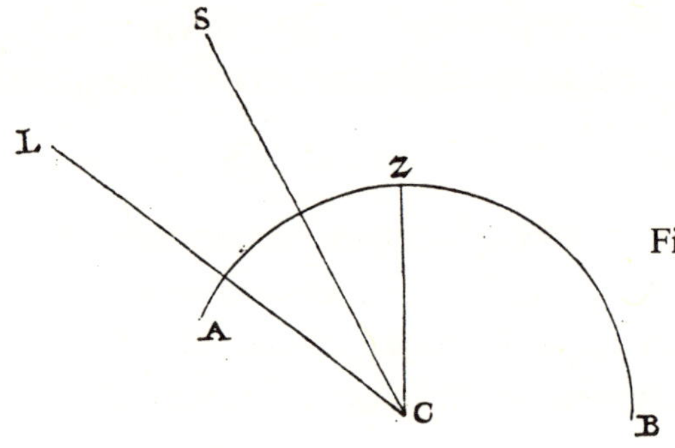

Fig. 4

punctis S et L, ita ut distantiae sint CS $= a$ et CL $= b$, anguli vero SCZ $= \varphi$ et LCZ $= \psi$, summa omnium actionum, quibus punctum Z urgetur, erit:

$$= r - \frac{\alpha\alpha}{\sqrt{(aa - 2ar\cos\varphi + rr)}} + \frac{\alpha\alpha r\cos\varphi}{aa} - \frac{\beta\beta}{\sqrt{(bb - 2br\cos\psi + rr)}} + \frac{\beta\beta r\cos\psi}{bb}$$

quae ergo pro statu aequilibrii debet aequari quantitati constanti, quam designemus littera C, atque hoc modo pro figura, quam mare accipiet, statim nanciscimur istam aequationem finitam atque adeo algebraicam:

$$C = r - \frac{\alpha\alpha}{\sqrt{(aa - 2ar\cos\varphi + rr)}} + \frac{\alpha\alpha r\cos\varphi}{aa} - \frac{\beta\beta}{\sqrt{(bb - 2br\cos\psi + rr)}} + \frac{\beta\beta r\cos\psi}{bb}.$$

Hinc igitur praestantia huius methodi, principio minimae actionis innixae, clarissime elucet, cum nobis statim aequationem algebraicam pro statu aequilibrii maris sit largita, dum olim ista determinatio per calculos molestissimos, ex hypothesibus precario assumtis, derivari debuerit. Hi enim calculi plerumque abstrusissimas sectionum conicarum proprietates requirebant, simulque ita erant comparati, ut nullo modo pateret, quomodo ad aequationem algebraicam perveniri posset, nisi adhibitis approximationibus. Hic autem nulla adhuc approximatione sumus usi.

10. Nunc autem facillime eas approximationes, quae ad scopum nostrum erant accommodatae, expedire poterimus. Cum enim distantiae a et b sint praegrandes ratione radii CZ $= r$, evolvamus in genere hanc formulam irrationalem: $(aa - \Omega)^{-1/2}$, quae praebet sequentem seriem:

$$\frac{1}{a} + \frac{1\cdot\Omega}{2\cdot a^3} + \frac{1\cdot 3\cdot\Omega^2}{2\cdot 4\cdot a^5} + \frac{1\cdot 3\cdot 5\cdot\Omega^3}{2\cdot 4\cdot 6\cdot a^7} + \text{etc.}$$

Nunc igitur pro Sole habemus $\Omega = 2ar\cos\varphi - rr$, unde elicimus[5])

$$\frac{1}{\sqrt{(aa - 2ar\cos\varphi + rr)}} = \frac{1}{a} + \frac{r\cos\varphi}{aa} - \frac{rr}{2a^3}(1 - 3\cos\varphi^2) - \frac{r^3\cos\varphi}{2a^4}(3 - 5\cos\varphi^2),$$

5) Intellegimus $\cos^2\varphi$ loco $\cos\varphi^2$. EJA.

simili autem modo erit pro Luna:

$$\frac{1}{\sqrt{(bb-2br\cos\psi+rr)}}=\frac{1}{b}+\frac{r\cos\psi}{bb}-\frac{rr}{2b^3}(1-3\cos\psi^2)-\frac{r^3\cos\psi}{2b^4}(3-5\cos\psi^2).$$

Quoniam igitur hic prima membra constantia sub constante C complecti licet, tota aequatio in sequentem formam contrahetur:

$$0=C+r+\frac{\alpha\alpha rr}{2a^3}(1-3\cos\varphi^2)+\frac{\alpha\alpha r^3\cos\varphi}{2a^4}(3-5\cos\varphi^2)$$
$$+\frac{\beta\beta rr}{2b^3}(1-3\cos\psi^2)+\frac{\beta\beta r^3\cos\psi}{2b^4}(3-5\cos\psi^2).$$

Quod si ergo brevitatis gratia statuamus $\frac{\alpha\alpha}{2a^3}=A$ et $\frac{\alpha\alpha}{2a^4}=\mathfrak{A}$, tum vero $\frac{\beta\beta}{2b^3}=B$ et $\frac{\beta\beta}{2b^4}=\mathfrak{B}$, aequatio nostra erit:

$$0=C+r+Arr(1-3\cos\varphi^2)+\mathfrak{A}r^3\cos\varphi(3-5\cos\varphi^2)$$
$$+Brr(1-3\cos\psi^2)+\mathfrak{B}r^3\cos\psi(3-5\cos\psi^2)$$

mox autem patebit litteras A, B, $\mathfrak{A}$, $\mathfrak{B}$, fractiones esse quam minimas.

11. Hactenus radium Terrae in statu naturali unitate expressimus. Quoniam vero excessum vel defectum quantitatis r supra vel infra hanc unitatem in mensuris absolutis, scilicet pedibus, desideramus, ponamus radium Terrae naturalem $=k$, ita ut sit uti supra ostendimus $\alpha=596k$ et $\beta=\frac{k}{8}$; tum vero pro distantiis mediis Solis et Lunae $a=24000k$ et $b=60k$, tantum opus est, ut hac littera k omnes termini aequationis inventae ad eundem dimensionum numerum reducantur; unde cum in termino principali, qui est r, unica sit dimensio, etiam in reliquis terminis unica dimensio inesse debet; quare quia litterae α, β, a et b iam unam dimensionem continent, aequatio nostra sequenti modo exhiberi debet:

$$0=C+r+\frac{\alpha\alpha rr}{2a^3}(1-3\cos\varphi^2)+\frac{\alpha\alpha r^3\cos\varphi}{2a^4}(3-5\cos\varphi^2)$$
$$+\frac{\beta\beta rr}{2b^3}(1-3\cos\psi^2)+\frac{\beta\beta r^3\cos\psi}{2b^4}(3-5\cos\psi^2)$$

quae ergo a praecedente non differt, et hoc nobis praestat commodum, ut nunc radium Terrae k tuto per quamcunque mensuram absolutam exprimere possimus, veluti in pedibus. Ex mensuris autem actu institutis iste radius terrae k deprehensus est continere 19 601 352 pedes parisinos.

12. Quoniam igitur iam certi sumus distantiam r non ultra aliquot pedes a radio naturali k differre posse, statuamus $r = k + v$, ita ut v sit quantitas quam minima prae k, ideoque $rr = kk + 2kv$ et $r^3 = k^3 + 3kkv$, quibus valoribus introductis aequatio nostra pro quantitate v determinanda ita se habebit[6]):

$$0 = C + k + \frac{\alpha\alpha kk}{2a^3}(1 - 3\cos\varphi^2) + \frac{\alpha\alpha k^3 \cos\varphi}{2a^4}(3 - 5\cos\varphi^2)$$

$$+ \frac{\beta\beta kk}{2b^3}(1 - 3\cos\psi^2) + \frac{\beta\beta k^3\cos\psi}{2b^4}(3 - 5\cos\psi^2)$$

$$+ v\left(1 + \frac{\alpha\alpha k}{a^3}(1 - 3\cos\varphi^2) + \frac{3\alpha\alpha kk\cos\varphi}{2a^4}(3 - 5\cos\varphi^2)\right)$$

$$+ v\left(\frac{\beta\beta k}{b^3}(1 - 3\cos\psi^2) + \frac{3\beta\beta kk\cos\psi}{2b^4}(3 - 5\cos\psi^2)\right),$$

ubi ergo $C + k$ quantitatem minimam exprimere debebit. Quod si ergo loco $C + k$ scribamus $-c$ et priores terminos ad alteram partem aequationis transferamus, aequatio nostra hanc induet formam:

$$v\left(1 + \frac{\alpha\alpha k}{a^3}(1 - 3\cos\varphi^2) + \frac{3\alpha\alpha kk\cos\varphi}{2a^4}(3 - 5\cos\varphi^2)\right.$$

$$\left. + \frac{\beta\beta k}{b^3}(1 - 3\cos\psi^2) + \frac{3\beta\beta kk\cos\psi}{2b^4}(3 - 5\cos\psi^2)\right)$$

$$= c - \frac{\alpha\alpha kk}{2a^3}(1 - 3\cos\varphi^2) - \frac{\alpha\alpha k^3\cos\varphi}{2a^4}(3 - 5\cos\varphi^2)$$

$$- \frac{\beta\beta kk}{2b^3}(1 - 3\cos\psi^2) - \frac{\beta\beta k^3\cos\psi}{2b^4}(3 - 5\cos\psi^2),$$

6) Editio princeps: $9 - \cos\varphi^2$ loco $3 - 5\cos\varphi^2$ passim. Atque expressiones mathematicae usque ad finem paragraphes errata parva continent. Correxit EJA.

ubi c est constans per integrationem ingressa, quam quovis casu ita determinari oportet, ut totum mare in statu mutato adhuc eandem aquae quantitatem contineat atque in statu naturali.

Evolutio numerica Aequationis inventae per Pedes Parisinos

13. Primo igitur littera c certum numerum pedum designabit, ex quantitate omnis aquae, quam mare continet, definiendum; tum vero, ut supra vidimus, est $\alpha = 596k$, $\beta = \frac{k}{8}$, pro distantiis autem mediis Solis et Lunae sumamus $a = 24000k$ et $b = 60k$. Hinc ergo pro aequationis nostrae parte priore habebimus ut sequitur:

$$\frac{\alpha\alpha k}{a^3} = \frac{596^2}{24000^3} = 0{,}0000000257$$

$$\frac{\alpha\alpha kk}{2a^4} = \frac{596^2}{2 \cdot 24000^4} = 0{,}0000000000$$

$$\frac{\beta\beta k}{b^3} = \frac{1}{64 \cdot 60^3} = 0{,}0000000723$$

$$\frac{\beta\beta kk}{2b^4} = \frac{1}{128 \cdot 60^4} = 0{,}0000000006$$

unde patet, pro prima parte nostrae aequationis sine ullo errore scribi posse v. Pro altera autem aequationis parte habebimus[7]):

$$\frac{\alpha\alpha kk}{2a^3} = \frac{596^2 \cdot k}{2 \cdot 24000^3} = 0{,}25183$$

$$\frac{\alpha\alpha k^3}{2a^4} = \frac{596^2 \cdot k}{2 \cdot 24000^4} = 0{,}00001$$

$$\frac{\beta\beta kk}{2b^3} = \frac{k}{2 \cdot 64 \cdot 60^3} = 0{,}70896$$

$$\frac{\beta\beta k^3}{2b^4} = \frac{k}{2 \cdot 64 \cdot 60^4} = 0{,}01182.$$

7) Editio princeps: Expressiones mathematicae errata parva continent. Correxit EJA.

Hinc igitur aequatio nostra hanc induet formam[8]):

$$v = c - 0{,}25183\,(1 - 3\cos\varphi^2) - 0{,}70896\,(1 - 3\cos\psi^2)$$
$$- 0{,}00001\cos\varphi\,(3 - 5\cos\varphi^2) - 0{,}01182\cos\psi\,(3 - 5\cos\psi^2)$$

unde valor ipsius v in pedibus Parisinis elicietur.

14. Ponamus brevitatis gratia istos valores numericos modo inventos, pro Sole: $0{,}25183 = m$, pro Luna autem: $0{,}70896 = n$ et $\nu = 0{,}01182$, ut sit:

$$v = c + m(3\cos\varphi^2 - 1) + n(3\cos\psi^2 - 1) + \nu\cos\psi\,(5\cos\psi^2 - 3).$$

Ubi notetur, valorem ipsius m, ad Solem pertinentem, satis esse iustum, dummodo ad distantiam mediam Solis et Terrae referatur, at vero valores n et ν quodammodo adhuc esse incertos, quoniam a massa Lunae pendent, pro qua assumsimus litteram $\beta = \frac{1}{8}k$, quemadmodum praecessio aequinoxiorum postulare videtur; fieri igitur posset ut aliquanto maior vel minor accipi deberet. Praeterea vero etiam hi valores ad distantiam mediam Lunae a Terra sunt accommodati, unde eos, si Luminaria fuerint in Apogeo, paulisper diminui, in Perigeo autem augeri oportet. Quare cum excentricitas Solis sit quasi $\frac{1}{60}$, si Sol fuerit in Apogeo, littera m diminui debet in ratione triplicata distantiae, sive ut $1:(1+\frac{1}{60})^3$, quae ratio proxime est ut $1:1\frac{1}{20}$, unde pro Apogeo Solis fit $m = 0{,}23924$, pro Perigeo autem tanto maior, id est $m = 0{,}26442$. Deinde quia Lunae excentricitas est quasi $\frac{1}{18}$, pro eius Apogeo valor litterae n diminui debet sua parte sexta, ita ut sit $n = 0{,}59080$, pro Perigeo vero tantundem augeri, fietque $n = 0{,}82712$. Littera denique[9]) ν, quae biquadrato distantiae reciproce est proportionalis, pro Apogeo fiet $\nu = 0{,}00919$, pro Perigeo vero $\nu = 0{,}01445$. Hoc igitur modo harum litterarum valores minimi, medii et maximi ita se habebunt:

Valor	min.	med.	max.
m	0,23924	0,25183	0,26442
n	0,59080	0,70896	0,82712
ν	0,00919	0,01182	0,01445

8) Editio princeps: 0,01197 loco 0,01182. Correxit EJA.
9) Editio princeps: 0,00931 loco 0,00919 et 0,01463 loco 0,01445. Correxit EJA.

Sufficiet autem hos valores nosse, quoniam ridiculum foret, in hoc negotio minutiis inhaerere.

Applicatio Formulae inventae ad superficiem Sphaericam

15. Repraesentet iam circulus ABCD [fig. 5] globum terraqueum, in cuius superficie notentur puncta S et L, quibus hoc tempore Sol et Luna verticaliter immineant, ita ut Sol versetur in zenith loci S, Luna vero in zenith loci L, et

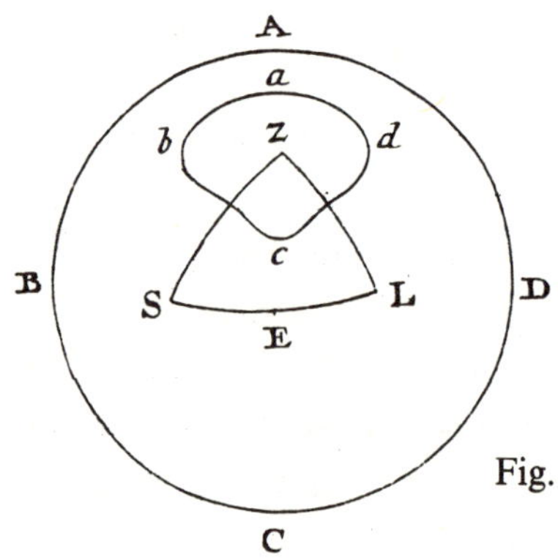

Fig. 5

quaeramus statum maris, quem pro hoc luminarium situ esset accepturum. Consideremus igitur in maris superficie quodcunque punctum Z, quod supra libellam naturalem ab actione luminarium attollatur per intervallum $= v$, in pedibus Parisinis determinandum. Hinc ergo ad puncta S et L ducantur arcus circulorum maximorum ZS et ZL, qui vocentur $ZS = \varphi$ et $ZL = \psi$, ac manifestum est angulum φ exprimere distantiam Solis a puncto zenith loci Z, similique modo angulus ψ distantiam Lunae ab eodem zenith, quibus positis spatium elevationis v supra libellam definietur hac aequatione:

$$v = c + m(3\cos\varphi^2 - 1) + n(3\cos\psi^2 - 1) + \nu\cos\psi\,(5\cos\psi^2 - 3)\,,$$

ubi quantitas constans c quovis casu ita debet determinari, ut facta mutatione status etiamnunc eadem copia aquae in mari proposito reperiatur.

16. Ad hanc igitur constantem c inveniendam, tota figura eius maris, cuius status quaeritur, probe est perpendenda, nisi quaestio instituatur de universo oceano totam Terram ambiente. Sit igitur *abcd* [fig. 5] mare illud undequaque clausum et terminatum, cuius status aequilibrii pro descripto luminarium situ desidere-

tur; eius quodvis elementum circa punctum Z situm designetur per dS, cui ergo insistit columna aquea altitudinis $= v$, ita ut eius volumen sit $v\,dS$. Haec igitur formula differentialis integretur, eiusque integrale per totum spatium *abcd* extendatur, ut obtineatur tota massa aquea huic spatio insistens, quae cum esse debeat nihilo aequalis, ex aequatione $\int v\,dS = 0$ colligetur valor constantis c. Perspicuum autem est, hoc fieri non posse, nisi per totum spatium *abcd* valores negativi ipsius v valoribus positivis aequivaleant. Ex quo facile intelligitur, valorem istius constantis c non solum a figura maris propositi *abcd*, sed etiam a positione luminarium S et L potissimum pendere, ita ut pro quovis alio situ ista investigatio de novo institui debeat. Deinceps autem videbimus, quomodo hanc rationem expedire conveniat.

17. Hic autem assumamus, istius litterae c valorem iam rite esse assignatum, et spatium *abcd* per totam Terrae superficiem esse extensum, ut, quantum aqua in singulis Terrae locis sive supra libellam attolli sive infra eam deprimi debeat, definire queamus. Ac primo quidem facile patet, aquam maxime elevari debere in loco quodam inter puncta S et L sito, qui sit in E, quandoquidem novimus, ab utriusque luminaris actione seorsim considerata aquam maxime attolli in eo loco, cui luminare verticaliter imminet. Ponamus igitur distantiam luminarium $\mathrm{SL} = \zeta$, ac translato puncto indefinito Z in E habebimus arcum $\mathrm{SE} = \varphi$, arcum vero $\mathrm{LE} = \zeta - \varphi = \psi$, ita ut sit $d\psi = -d\varphi$. Quare pro maximo valore ipsius v inveniendo, eius differentiale nihilo aequemus, at ob $d\psi = -d\varphi$ prodibit ista aequatio:

$$-6m\sin\varphi\cos\varphi + 6n\sin\psi\cos\psi + 15\nu\sin\psi\cos\psi^2 - 3\nu\sin\psi = 0,$$

ubi cum ν sit valde parvum prae m et n, erit proxime:

$$-m\sin\varphi\cos\varphi + n\sin\psi\cos\psi = 0$$

sive

$$m\sin 2\varphi = n\sin 2\psi = n\sin(2\zeta - 2\varphi),$$

sicque punctum E ibi reperietur, ubi erit $m\sin 2\mathrm{SE} = n\sin 2\mathrm{LE}$.

18. Cum igitur sit $\sin(2\zeta - 2\varphi) = \sin 2\zeta\cos 2\varphi - \cos 2\zeta\sin 2\varphi$, hinc colligitur fore

$$\operatorname{tang} 2\varphi = \frac{n\sin 2\zeta}{m + n\cos 2\zeta}.$$

Quaeratur ergo angulus θ, cuius tangens sit

$$= \frac{n \sin 2\zeta}{m + n \cos 2\zeta},$$

et cum eadem quoque sit tangens anguli $\theta + 180°$, hinc nanciscemur duos valores pro angulo φ, quorum alter erit $\varphi = \frac{1}{2}\theta$, alter vero $\varphi = \frac{1}{2}\theta + 90°$. Prior igitur manifesto dat locum E, ubi aqua maxime elevabitur, posterior vero eum locum, ubi maxime deprimetur, qui ergo a loco E intervallo quadrantis erit remotus in arcu SL producto. Simul vero etiam patet, in locis diametraliter oppositis aquam fore vel maxime elevatam vel depressam.

19. Haec autem, quoniam ad distantiam quamcunque inter luminaria spectant, nimis sunt generalia, quam ut inde conclusiones concinnas derivare queamus. Infra autem has determinationes ad casus particulares accommodabimus. Nunc autem rem in genere considerantes videamus statum aquae tam in ipso loco S quam in loco L. Pro priore igitur, translato puncto Z in S, erit $\varphi = 0$ et $\psi = \zeta$, unde fit

$$v = c + 2m + n(3\cos\zeta^2 - 1) + \nu\cos\zeta\,(5\cos\zeta^2 - 3);$$

at vero translato Z in L erit $\varphi = \zeta$ et $\psi = 0$, hincque

$$v = c + m(3\cos\zeta^2 - 1) + 2n + 2\nu.$$

Sin autem punctum Z ita capiatur, ut ambo arcus SZ et LZ fiant quadrantes, tum ob $\varphi = 90°$ et $\psi = 90°$ erit $v = c - m - n$, ubi ergo aqua semper erit maxime depressa.

Applicatio ad hypothesin, qua tota Terra aqua circumdata ponitur

20. Contemplemur nunc casum, quo tota Terra esset fluida, vel saltem continuo maris tractu cincta, et quaeramus valorem quantitatis constantis c pro quovis situ amborum luminarium, ut inde vera elevatio maris super libellam naturalem in omnibus locis assignari queat. Ad hoc igitur necesse est, ut formula integralis $\int v\,dS$ per totam Terrae superficiem extendatur et valor resultans nihilo aequetur; sic enim pervenietur ad aequationem, ex qua quantitas c facile determinari poterit. Ubi notetur, differentiale dS exprimere elementum quodcunque superficiei sphaericae.

21. Cum igitur valor quantitatis v ex quatuor partibus sit compositus, quarum prima est ipsa constans c, quae quaeritur; secunda, actio a Sole orta $m(3\cos\varphi^2 - 1)$; tertia, actio principalis a Luna orta $n(3\cos\psi^2 - 1)$; quarta denique, correctio istius actionis $= \nu\cos\psi(5\cos\psi^2 - 3)$: singulas istas partes in elementum superficiei dS ducamus et integralia per totam superficiem Terrae extendamus, quo facto aggregatum omnium nihilo aequari debebit. Prima igitur pars, seu constans c statim praebet $\int c\,dS = cS$, ubi loco S totam superficiem Terrae assumi oportet. Hic iterum radium Terrae unitate designemus, quandoquidem non quantitas absoluta spectatur, ac posita ratione diametri ad peripheriam ut $1:\pi$, notum est, totam superficiem Sphaerae esse $= 4\pi$, unde integrale ex prima parte ortum, erit $= 4\pi c$.

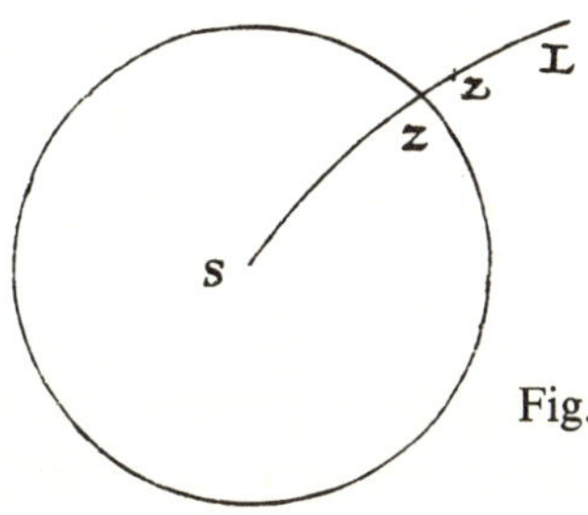

Fig. 6

22. Pro secunda parte $m(3\cos\varphi^2 - 1)$ circa punctum S [fig. 6] in superficie sphaerica describamus circulum, intervallo SZ $= \varphi$, eritque $\sin\varphi$ radius huius circuli in plano considerati, eiusque propterea peripheria $= 2\pi\sin\varphi$. Iam tribuamus arcui SZ $= \varphi$ incrementum infinite parvum $Zz = d\varphi$, quod in peripheriam $2\pi\sin\varphi$ ductum dabit incrementum areae istius areae circularis in superficie sphaerica, quod ergo erit $= 2\pi\,d\varphi\sin\varphi$; quamobrem istud incrementum per secundam partem $m(3\cos\varphi^2 - 1)$ multiplicetur, et integrari debebit ista formula: $2m\pi\,d\varphi\sin\varphi\,(3\cos\varphi^2 - 1)$. Hunc in finem ponamus $\cos\varphi = x$, eritque $d\varphi\sin\varphi = -dx$, et formula integranda evadet $= -2m\pi\,dx(3xx - 1)$, cuius ergo integrale fit

$$2m\pi(x - x^3) + C = 2m\pi(\cos\varphi - \cos\varphi^3) + C,$$

quae constans C autem evanescit, quia integrale inventum sponte fit $= 0$, sumto arcu $\varphi = 0$, ita ut integrale ex hac parte ortum sit $2m\pi\cos\varphi\,\sin\varphi^2$, cuius valor ut per totam superficiem sphaericam extendatur, arcus indefinitus φ sumi debet semicirculo aequalis; unde, cum fiat $\sin\varphi = 0$, evidens est, integrale ex secunda parte ipsius v natum sponte fieri $= 0$, atque hinc simul manifestum est, etiam integrale ex tertia parte oriundum ad nihilum reduci.

23. Pro parte quarta ipsius v in figura loco S sumatur punctum L, voceturque arcus $LZ = SZ = \psi$, et quia elementum spatii circularis nunc erit $2\pi\, d\psi \sin\psi$, formula differentialis ex quarta parte nata erit $2\nu\pi\, d\psi \sin\psi \cos\psi\, (5\cos\psi^2 - 3)$. Faciamus hic iterum $\cos\psi = x$, ut habeamus hanc formulam: $-2\nu\pi x\, dx\, (5xx - 3)$, cuius integrale est $-2\nu\pi\, (\frac{5}{4}x^4 - \frac{3}{2}xx)$ $= -2\nu\pi\, (\frac{5}{4}\cos\psi^4 - \frac{3}{2}\cos\psi^2)$[10] $+ C$. Erit autem[11] $C = -\frac{1}{2}\nu\pi$. Extendatur nunc hoc integrale per totam Sphaeram, ponendo $\psi = 180°$, atque etiam haec quarta pars evadet $= 0$.

24. Hinc igitur patet, si tota terra continuo oceano esset obducta, tum semper fore constantem $C = 0$, in quocunque loco coeli ambo luminaria versentur, ita ut pro hac hypothesi semper valeat ista aequatio:

$$v = m\,(3\cos\varphi^2 - 1) + n\,(3\cos\psi^2 - 1) + \nu\cos\psi\,(5\cos\psi^2 - 3),$$

unde pro singulis Terrae locis elevatio aquae supra libellam in pedibus parisinis definiri poterit. Scilicet si valor ipsius v prodeat positivus, elevatio supra libellam naturalem indicabitur; ubi autem obtinuerit valorem negativum, ibi depressio infra libellam indicabitur. Haec autem omnia intelligenda sunt de statu aequilibrii, quem actio luminarium Terrae esset inductura, si tam Terra quam ipsa luminaria perpetuo in eodem loco quiescerent. In hac igitur hypothesi sequens Problema resolvi poterit.

Problema

25. In hypothesi Terrae quiescentis et oceano continuo circumdatae, si dentur in coelo loca Solis et Lunae, ubi per aliquot tempus commorari concipi debent, ut totus Oceanus se ad statum aequilibrii componere possit, in omnibus Terrae locis sive elevationem sive depressionem aquae supra vel infra libellam naturalem determinare.

10) Editio princeps: $= 2\nu\pi$ loco $= -2\nu\pi$. Correxit EJA.
11) Editio princeps: $C = \frac{1}{2}\nu\pi$ loco $C = -\frac{1}{2}\nu\pi$. Correxit EJA.

SOLUTIO

Referat circulus AEBF [fig. 7] superficiem Terrae, ubi A et B sint Poli borealis et australis, semicirculus vero AEB repraesentet primum meridianum, tum vero puncta S et L ea sint loca, quibus Sol et Luna verticaliter immineant,

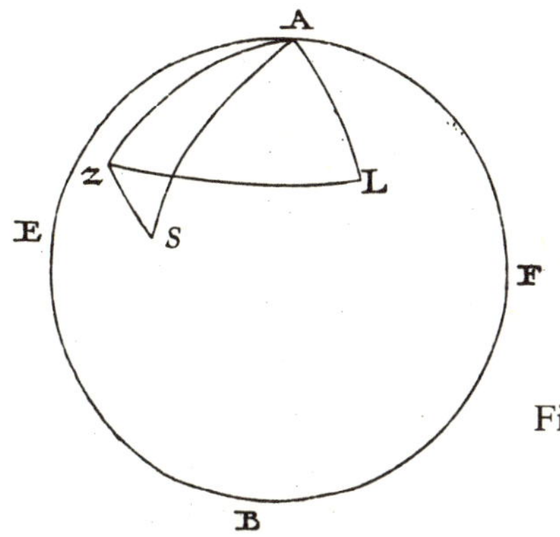

Fig. 7

pro quibus ductis meridianis AS et AL, sint longitudines EAS $= \zeta$ et EAL $= \eta$, distantiae vero a polo boreali A vocentur AS $= f$ et AL $= g$. Iam proponatur Terrae locus quicunque Z, cuius longitudo sit EAZ $= \omega$, et distantia a Polo AZ $= z$. Quo igitur pro hoc loco quantitas v possit definiri, eius distantiae a punctis S et L, scilicet arcus ZS $= \varphi$ et ZL $= \psi$ quaeri debent. Pro priore ZS $= \varphi$ ex triangulo sphaerico ZAS, in quo habentur latera AZ $= z$ et AS $= f$, cum angulo intercepto ZAS $= \zeta - \omega$, deducitur:

$$\cos SZ = \cos\varphi = \cos f \cos z + \sin f \sin z \cos(\zeta - \omega).$$

Simili modo ex triangulo ZAL, in quo habentur latera ZA $= z$ et AL $= g$, cum angulo intercepto ZAL $= \eta - \omega$, colligitur

$$\cos LZ = \cos\psi = \cos g \, \cos z + \sin g \, \sin z \cos(\eta - \omega).$$

His igitur duobus angulis φ et ψ inventis, erit elevatio aquae in loco Z supra libellam naturalem:

$$v = m(3\cos\varphi^2 - 1) + n(3\cos\psi^2 - 1) + \nu\cos\psi(5\cos\psi^2 - 3),$$

ubi litterarum m, n et ν valores pro ratione distantiae tam Solis quam Lunae ex §14 sunt desumendi.

Corollarium 1

26. Quod si ergo locus Z in ipso polo boreali A accipiatur, ut sit $z = 0$, statim habebitur $\varphi = f$ et $\psi = g$, unde pro hoc loco erit

$$v = m(3\cos f^2 - 1) + n(3\cos g^2 - 1) + \nu \cos g(5\cos g^2 - 3).$$

Sin autem punctum Z in polo australi B accipiatur, fiet $\varphi = 180^\circ - f$ et $\psi = 180^\circ - g$, et pro hoc loco erit

$$v = m(3\cos f^2 - 1) + n(3\cos g^2 - 1) - \nu \cos g(5\cos g^2 - 3),$$

ubi ultimum membrum contrario signo est affectum.

Corollarium 2

27. Si punctum Z in aequatore accipiatur, ut sit $z = 90^\circ$, manente longitudine indefinita $EAZ = \omega$, reperietur

$$\cos\varphi = \sin f \cos(\zeta - \omega) \quad \text{et} \quad \cos\psi = \sin g \cos(\eta - \omega),$$

unde elevatio aquae erit,

$$v = m\big(3\sin f^2 \cos(\zeta - \omega)^2 - 1\big) + n\big(3\sin g^2 \cos(\eta - \omega)^2 - 1\big) + \nu \sin g \cos(\eta - \omega)\big(5\sin g^2 \cos(\eta - \omega)^2 - 3\big).$$

Corollarium 3

28. Si Sol et Luna fuerint in coniunctione, ut sit $\eta = \zeta$ et $g = f$, erit[12])

$$\cos\varphi = \cos\psi = \cos f \cos z + \sin f \sin z \cos(\zeta - \omega),$$

12) Editio princeps: z loco ζ. Correxit EJA.

quibus inventis elevatio in loco Z reperietur:

$$v = (m+n)(3\cos\varphi^2 - 1) + \nu\cos\varphi\,(5\cos\varphi^2 - 3)\,.$$

Sin autem Sol et Luna fuerint in oppositione, erit $\eta - \zeta = 180°$ et $g = 180° - f$, unde cum sit $\sin\eta = -\sin\zeta$, $\cos\eta = -\cos\zeta$, $\sin g = \sin f$, et $\cos g = -\cos f$, sequitur fore

$$\cos\varphi = \cos f\cos z + \sin f\sin z\cos(\zeta - \omega)$$

et

$$\cos\psi = -\cos f\cos z - \sin f\sin z\cos(\zeta - \omega)\,.$$

Sicque patet fore $\cos\psi = -\cos\varphi$, ex quibus elevatio aquae in Z colligitur:

$$v = (m+n)(3\cos\varphi^2 - 1) - \nu\cos\varphi\,(5\cos\varphi^2 - 3)\,.$$

Scholion

29. In hac hypothesi commode accidit, ut pro omnibus locis Solis et Lunae quantitas constans c eundem valorem nanciscatur. Sin autem mare undique clausum proponatur, quod tantum modicam portionem superficiei Terrae occuparet, tum pro quovis situ luminarium valor istius quantitatis constantis c seorsim computari deberet, quod utique calculum non parum molestum postularet. Plerumque autem sufficere poterit differentiam tantum inter elevationem aquae in variis locis talis maris determinasse, ita ut tum non opus sit veram quantitatem litterae c definire. Ceterum quoniam hoc loco mihi tantum fuit propositum, pro quovis situ Solis ac Lunae, statum aequilibrii, in quo maria acquiescere queant, ex veris principiis determinare, hoc argumentum ulterius non prosequor, neque quicquam de Phaenomenis fluxus et refluxus maris hic attingam, quippe quae iam dudum satis dilucide sunt tractata.

ENODATIO DIFFICULTATIS SUPER FIGURA TERRAE A VI CENTRIFUGA ORIUNDA

Commentatio 619 indicis ENESTROEMIANI
Nova acta academiae scientiarum Petropolitanae 2 (1784), 1788, p. 121–130
Summarium ibidem, Histoire p. 74–76

SUMMARIUM

L'Auteur commence par la remarque de la grande différence, qu'on trouve dans le rapport du diamètre de l'Equateur à l'axe de la Terre, suivant que la figure de cette planète est déterminée par la combinaison de la force de la gravité avec la force centrifuge, ou par la mesure de différens dégrés du Méridien. Ce rapport varie de celui de 578 : 577 à celui de 201 : 200. Il ajoute que quoique Mrs. HUGHENS et NEWTON, qui les premiers sont parvenus par leurs calculs au rapport de 578 : 577, ayent regardé la terre comme uniformément épaisse, on trouveroit cependant le même résultat pour le rapport dont il s'agit, quelque différente qu'on supposat la structure des parties intérieures de la Terre, aussi longtems qu'on regarderoit l'action de la gravité comme dirigée vers le centre de la Terre. C'est ce que l'Auteur démontre par la Théorie de l'équilibre des fluides; pourvu cependant qu'à distances égales du centre la force de la gravité soit égale, et qu'on puisse regarder comme extrêmement petite la différence entre le Diamètre de l'Equateur et l'axe de la Terre. Il faut donc nécessairement, continue M. EULER, que chaque particule de la Terre soit non seulement attirée vers le centre, mais qu'il y ait encore d'autres forces latérales, dont la direction soit perpendiculaire à celle vers le centre: et en effet l'hypothèse de la gravité universelle, par laquelle chaque particule est attirée vers l'autre, démontre, que cette seconde espèce de force existe actuellement, et doit être considérée dans le calcul. Mais il n'est pas possible de déterminer ces forces, sans connoitre auparavant la figure de la Terre et toute sa structure. Aussi cette recherche est-elle si couverte d'obscurité, qu'on ne doit pas, dit l'Auteur, s'attendre à une explication parfaite, et tout ce que les géomètres ont proposé jusqu'à présent sur ce sujet, ne repose que sur des hypothèses précaires, et déstituées de toute probabilité. Pour traiter donc, autant que possible, ce sujet dans toute sa généralité, M. EULER considère les deux forces perpendiculaires entre elles, dont nous avons parlé, et d'après ce qui est probable dans la nature, il fait celle, qui agit vers le centre, proportionelle à une fonction quelconque Z de la distance z du centre, et l'autre, latérale, qu'il nomme

S, comme dépendante non seulement de la distance z, mais aussi de l'angle φ, que forme le diamètre de l'Equateur avec la ligne tirée de la particule vers le centre; il ajoute, que cette force doit s'évanouïr pour les particules situées dans l'axe et dans l'Equateur. D'après cela l'Auteur recherche la figure de la Terre selon les principes qu'il a expliqués dans le 13e Tome des Nouveaux Commentaires pour l'équilibre des fluides. Il parvient donc à une équation différentielle, et trouve que pour que l'intégration puisse avoir lieu, (ce qui est une condition absolument nécessaire de l'équilibre,) il faut que la force S soit une fonction homogène de -1 dimension des coordonnées perpendiculaires x et y, auxquelles chaque particule est réduite. Ensuite après avoir substitué à ces coordonnées leurs valeurs en z et φ, il remarque que la condition de l'équilibre demande encore, qu'à distances égales du centre la chaleur et la densité soyent les mêmes. En faisant maintenant la pression des particules $=0$, il trouve l'équation pour la surface de la Terre, où cependant l'ignorance, dans laquelle nous sommes sur la structure intérieure de ce globe, permet de faire encore plusieurs suppositions arbitraires. Pour expliquer donc la chose par un exemple, l'Auteur fait la force centrale égale à une puissance n de z, et la latérale égale à $\alpha \sin\varphi \cos\varphi$, et après avoir substitué ces valeurs dans l'équation, il détermine la constante introduite par l'intégration, et le surplus du diamètre de l'Equateur sur l'axe de la Terre; ce qui se fait en transportant d'abord la particule quelconque sous les poles et ensuite sous l'Equateur. Il ne reste plus alors qu'à substituer pour α une fraction qui convienne au rapport trouvé de 201 : 200, et ainsi α devient $=\frac{1}{154}$. Le *Maximum* de la force latérale a lieu pour l'angle $\varphi = 45°$, et devient, à la surface de la mer, presque égale à la force centrifuge.

D'un autre coté il est manifeste, que la loi trouvée pour les forces latérales S ne peut pas avoir lieu, parcequ'autrement elles deviendroient infiniment grandes à des distances infiniment petites du centre: d'où il s'en suit, que si la Terre étoit toute fluide, sa surface ne pourroit jamais être tranquille ou en équilibre. Mais comme selon toute probabilité l'Ocean n'est nulle part assés profond, pour que la différence entre la formule trouvée pour S et la vraye loi d'attraction, quelle qu'elle puisse être, devienne sensible, il sera incontestablement possible, que l'Ocean se tienne en équilibre, si nous faisons abstraction de plusieurs autres causes physiques, qui peuvent y exciter des agitations.

---

1. Notum est, si Terrae figura ex sola vi gravitatis cum vi centrifuga coniuncta definiatur, rationem Diametri aequatoris ad axem Terrae non maiorem reperiri quam 578 : 577, cum tamen haec ratio post mensuras diversorum graduum institutas multo maior deprehendatur 201 : 200 propemodum. HUGENIUS quidem et NEWTONUS, qui primi hanc rationem investigarunt, totam Terram tanquam ex materia uniformi compositam sunt contemplati, interim tamen quaecunque diversa structura in partibus Terrae interioribus statuatur, eadem semper ratio diametri aequatoris ad axem resultat, quamdiu scilicet gravitatis directio ad centrum Terrae tendens assumitur.

2. Quod si enim intra Terram gravitatio potestati cuicunque distantiae a centro, quae sit $= z$, proportionalis statuatur, ut ea sit $= z^{n-1}$, dum semi-axis Terrae per unitatem exprimitur, ex Theoria aequilibrii fluidorum deducitur ista aequatio:

$$\frac{z^n}{n} = C + \frac{xx}{2f},$$

ubi si CA [fig. 1] pro radio aequatoris et CB $= 1$ pro semi-axe terrae accipiatur, littera z denotat distantiam cuiusque particulae Z a centro terrae C, at x intervallum CX demisso ex Z ad CA perpendiculo CX, littera vero f denotat

Fig. 1

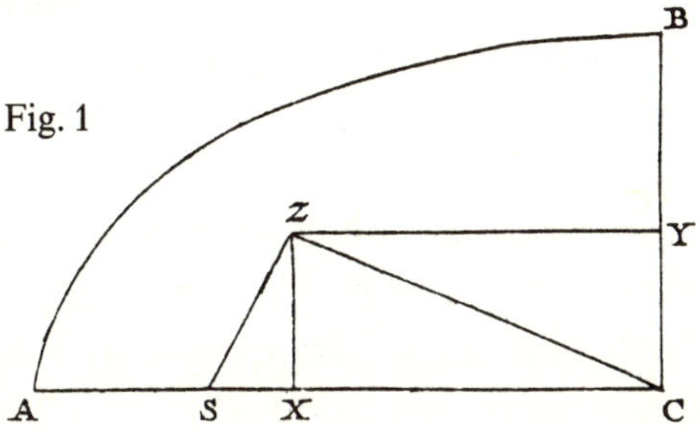

numerum 289 ex motu vertiginis Terrae ortum; tum vero littera C est quantitas constans ex ipso statu Terrae determinanda. Primo igitur punctum indefinitum Z capiatur in ipso polo B, fietque $x = 0$; at prodire debet $z = 1$, unde colligitur constans $C = \frac{1}{n}$. Nunc punctum Z transferatur in aequatorem A, ut fiat $x = z$, atque habebitur ista aequatio:

$$\frac{z^n}{n} = \frac{1}{n} + \frac{zz}{2f}.$$

Hinc autem, quia novimus valorem ipsius z quam minime unitatem esse superaturum, ponamus $z = 1 + \omega$, eritque satis exacte $z^n = 1 + n\omega$, et ob $2f = 578$ loco $\frac{zz}{2f}$ scribi sufficiet $\frac{1 + 2\omega}{578}$, hincque aequatio nostra praebebit

$$\frac{1 + n\omega}{n} = \frac{1}{n} + \frac{1 + 2\omega}{578},$$

unde colligitur $\omega = \frac{1}{576}$, ita ut hinc fiat radius aequatoris $C = 1 + \frac{1}{576}$, ideoque diameter aequatoris ad axem Terrae ut $577:576$, unde patet hanc rationem ab exponente indefinito n prorsus non pendere.

3. Ex his etiam manifestum est, quaecunque alia functio ipsius z pro gravitate accipiatur, perpetuo eandem conclusionem inde sequi debere. Quamobrem cum vera proportio inter axem Terrae et diametrum aequatoris tantopere ab ista inventa ratione dissideat, necessario statui oportet, singulas terrae particulas Z non solum ad centrum Terrae C urgeri, sed insuper alias vires adesse debere, quibus particula in Z secundum directionem ZS ad CZ normalem sollicitetur; tales etiam vires hypothesis gravitatis universalis qua singulae particulae ad omnes alias attrahi supponuntur revera ostendit, ita ut in figura Terrae determinanda etiam istae vires laterales in computum duci debeant. Verum has ipsas vires ex theoria gravitationis ne determinare quidem licet, nisi iam ante figura Terrae cum universa eius structura fuerit cognita, quandoquidem harum virium determinatio non solum a densitate materiae per totam Terram dispositae sed etiam ab ipsa figura externa totius Terrae pendet; unde satis intelligitur, hanc investigationem tantopere esse absconditam, ut eius perfecta explicatio nullo modo sperari possit. Quicquid enim a Geometris super hoc argumento in medium est allatum, meris hypothesibus iisque precario assumtis innititur, quae plerumque adeo omni probabilitate destituuntur.

4. Quod si vero hanc inquisitionem generalissime suscipere velimus, binas illas vires, quibus singulae Terrae particulae Z secundum directiones ZC et ZS ob gravitatem universalem sollicitantur, generaliter in computum introduci conveniet, unde autem ob summam generalitatem vix quicquam concludere licebit; cum non constet, a quibusnam elementis istae vires pendere sint censendae; vis quidem prior ad centrum C urgens probabili ratione functioni cuipiam ipsius distantiae $\mathrm{CZ} = z$ proportionalis statui posse videtur, quam designemus littera Z, altera vero vis lateralis secundum ZS quae sit $= S$, manifesto non solum a distantia $\mathrm{CZ} = z$ pendere potest, sed insuper angulum ACZ ita involvere debet, ut ea evanescat tam casu quo iste angulus evanescit, quam ubi fit rectus; quoniam ex rei natura evidens est, istam vim lateralem tam in aequatore CA quam in axe CB evanescere debere, siquidem nullum est dubium quin tam sub polis quam in aequatore omnia corpora directe versus centrum C sollicitentur, quamobrem vim illam alteram S tanquam functionem binarum variabilium $\mathrm{CZ} = z$ et $\mathrm{CX} = x$ spectari oportebit.

5. His igitur praenotatis figuram Terrae secundum principia aequilibrii fluidorum, quemadmodum ea in Tom. XIII novorum Commentariorum[1]) exposui investigemus, ac primo quidem statum pressionis in puncto quocunque Z definiamus, quae altitudine p definiatur. Hunc in finem pro puncto Z vocemus binas coordinatas $\mathrm{CX} = x$ et $\mathrm{XZ} = \mathrm{CY} = y$: hic enim tertia coordinata, quae ibi vocata erat $= z$, carere possumus, quandoquidem certum est, in omnibus sectionibus per axem factis terram eandem figuram habere debere. Nunc igitur ambae vires Z et S secundum binas directiones ZX et ZY resolvantur, ac prior quidem Z secundum ZC pro directionibus ZY et ZX, praebet vires $\frac{Zx}{z}$ et $\frac{Zy}{z}$. Altera vero vis $\mathrm{ZS} = S$ pro iisdem directionibus dat has vires: $-\frac{Sy}{z}$ et $\frac{Sx}{z}$. Praeterea vero vis centrifuga a motu diurno Terrae orta praebet vim secundum YZ $= \frac{x}{f}$, unde ex ternis viribus quas in genere designavi per litteras[2]) P, Q, R, primo erit $R = 0$, secundo $P = \frac{x}{f} - \frac{Zx}{z} + \frac{Sy}{z}$ et tertio $Q = -\frac{Zy}{z} - \frac{Sx}{z}$. Ex his autem viribus, sumta littera q pro densitate in puncto Z, principia aequilibrii hanc dederunt aequationem[3]):

$$\frac{\partial p}{q} = P\,\partial x + Q\,\partial y,$$

quae ergo nostro casu induet hanc formam,

$$\frac{\partial p}{q} = \frac{x\,\partial x}{f} - \frac{Zx\cdot\partial x}{z} + \frac{Sy\cdot\partial x}{z} - \frac{Zy\cdot\partial y}{z} - \frac{Sx\cdot\partial y}{z},$$

sive

$$\frac{\partial p}{q} = \frac{x\,\partial x}{f} - \frac{Z}{z}\,(x\,\partial x + y\,\partial y) + \frac{S}{z}\,(y\,\partial x - x\,\partial y).$$

Haec autem aequatio porro, ob $x\,\partial x + y\,\partial y = z\,\partial z$, contrahitur in hanc

$$\frac{\partial p}{q} = \frac{x\,\partial x}{f} - Z\,\partial z + \frac{S}{z}\,(y\,\partial x - x\,\partial y).$$

1) E 375, *Novi Commentarii Academiae Imperialis Scientiarum Petropolitanae* 13 (1768), 1769, 305–416 = *Opera omnia* II 13, pp. 1–72. EJA.

2) Id est, vires in spatio secundum directiones trium axium coordinatarum. EJA.

3) Intellegimus d loco ∂. EJA.

6. Nunc autem ante omnia tenendum est, nisi haec formula integrationem admittat, statum aequilibrii nullo modo locum invenire posse; quamobrem, cum tuto assumere queamus, in Terra dari statum aequilibrii, quia alioquin quaestio de figura ne suscipi quidem posset, necesse est ut formula haec inventa integrationem admittat, quod quidem in primo termino $\frac{x\,\partial x}{f}$ sponte evenit; tum vero etiam integratio in secundo termino semper succedit, dummodo Z fuerit functio ipsius z uti assumsimus; quamobrem superest ut postremum membrum $\frac{S}{z}(y\,\partial x - x\,\partial y)$ integrationem admittat, quod cum hoc modo repraesentari possit $\frac{Sxy}{z}\left(\frac{\partial x}{x} - \frac{\partial y}{y}\right)$, integratio locum habere nequit, nisi $\frac{Sxy}{z}$ sit functio ipsius $\frac{x}{y}$, ideoque functio nullius dimensionis ipsarum x et y; quare cum $z = \sqrt{(xx+yy)}$ unam habeat dimensionem, necesse est ut S sit functio homogenea ipsarum x et y, cuius dimensionum numerus sit -1, unde iam satis clare cognoscimus indolem functionis S, siquidem pro certo assumamus, figuram terrae aequilibrio esse praeditam.

7. Quo hoc clarius appareat, loco coordinatarum x et y in postremo membro introducamus angulum $\mathrm{ACZ} = \varphi$, eritque $x = z\cos\varphi$ et $y = z\sin\varphi$, unde fit $y\,\partial x - x\,\partial y = -zz\,\partial\varphi$, ita ut iam nostra aequatio hanc induat formam:

$$\frac{\partial p}{q} = \frac{x\,\partial x}{f} - Z\,\partial z - Sz\,\partial\varphi,$$

quae manifesto integrationem non admittit, nisi fuerit Sz functio anguli φ. Sit igitur Φ ista functio, eritque vis lateralis $S = \frac{\Phi}{z}$; ubi patet, istam functionem Φ ita debere esse comparatam, ut evanescat tam posito $\varphi = 0$, quam $\varphi = 90°$. Hinc igitur integrando adipiscemur

$$\int\frac{\partial p}{q} = C + \frac{xx}{2f} - \int Z\,\partial z - \int\Phi\,\partial\varphi.$$

Atque hic porro observandum est aequilibrium subsistere non posse, nisi etiam formula $\int\frac{\partial p}{q}$ sit integrabilis. Quoniam autem hic densitatem aquae q ubique constantem assumere licet, erit utique

$$\frac{p}{q} = C + \frac{xx}{2f} - \int Z\,\partial z - \int \Phi\,\partial\varphi\,.$$

Si enim ob diversos caloris gradus densitas q esset variabilis, iam satis evictum est, aequilibrium locum habere non posse, nisi q sit functio ipsius p tantum, hoc est nisi per singula strata ubi eadem est pressio p etiam densitas sit eadem.

8. His de aequilibrio per totam fluidi massam praemissis, nil aliud superest nisi ut aequatio inventa ad supremam aquae superficiem accommodetur, ubi cum pressio p sit evanescens, posito $p = 0$ aequatio haec

$$0 = C + \frac{xx}{2f} - \int Z\,\partial z - \int \Phi\,\partial\varphi\,,$$

exprimet figuram quam suprema aquae superficies in statu aequilibrii accipiet. Ex iam allatis autem patet, parum referre, cuiusmodi functio ipsius z pro Z assumatur, quoniam discrimen inter diametrum aequatoris et axem Terrae nimis est parvum, quam ut ex natura functionis Z sensibilis diversitas oriri possit.

9. Designemus igitur uti incepimus unitate semi-axem Terrae CB, sitque sub ipso polo in B vis gravitatis acceleratrix etiam unitate expressa, ita ut posito $z = 1$, fieri quoque debeat $Z = 1$; quamobrem statuamus aliquanto generalius $Z = z^{n-1}$, ut fiat $\int Z\,\partial z = \frac{z^n}{n}$. Deinde quia functio Φ evanescere debet casibus $\varphi = 0$ et $\varphi = 90°$, pro Φ capiamus functionem simplicissimam huic conditioni satisfacientem, ponendo $\Phi = \alpha \sin\varphi \cos\varphi$, unde fit[4]) $\int \Phi\,\partial\varphi = \frac{1}{2}\alpha \sin\varphi^2$. His igitur valoribus substitutis aequatio pro superficie aquae erit

$$0 = C + \frac{xx}{2f} - \frac{z^n}{n} - \tfrac{1}{2}\alpha \sin\varphi^2,$$

quae ergo simul exprimit figuram Terrae, quam ob vim centrifugam recipere debet, ubi ut ante est $f = 289$.

10. Ante omnia hic constantem C definire oportet, id quod commodissime fiet transferendo punctum Z in ipsum B, ubi fieri necesse est $z = 1$, $x = 0$ et

4) Intellegimus $\sin^2\varphi$ loco $\sin\varphi^2$. EJA.

$\varphi = 90°$, unde colligitur constans $C = \frac{1}{n} + \frac{1}{2}\alpha$. Sicque aequatio pro figura Terrae erit

$$0 = \frac{1}{n} + \tfrac{1}{2}\alpha + \frac{xx}{2f} - \frac{z^n}{n} - \tfrac{1}{2}\alpha \sin\varphi^2,$$

sive

$$0 = \frac{1 - z^n}{n} + \frac{xx}{2f} + \tfrac{1}{2}\alpha \cos\varphi^2,$$

seu quia $\cos\varphi = \frac{x}{z}$, erit

$$z^n = 1 + \frac{nxx}{2f} + \frac{\alpha nxx}{2zz},$$

unde ob $z = \sqrt{(xx + yy)}$ facile deducitur aequatio inter binas coordinatas x et y.

11.[5]) Hinc igitur quaeramus semi-diametrum aequatoris transferendo punctum Z in A, ubi ergo fiet $x = z =$ CA, cuius propterea valor ex hac aequatione elici debet

$$z^n = 1 + \frac{nzz}{2f} + \tfrac{1}{2}\alpha n.$$

Quia vero novimus, valorem ipsius z parum ab unitate discrepare, ponamus $z = 1 + \omega$, ut fiat $z^n = 1 + n\omega$ et $zz = 1 + 2\omega$, quibus valoribus inductis fiet

$$\omega = \frac{1 + \alpha f}{2(f - 1)}.$$

Sin autem hunc valorem accuratius desideremus, loco z^n scribamus $1 + n\omega + \frac{1}{2}n(n-1)\omega\omega$ et $1 + 2\omega + \omega\omega$ loco zz, et nostra aequatio fiet

$$\omega + \tfrac{1}{2}(n-1)\omega\omega = \frac{1 + 2\omega + \omega\omega}{2f} + \tfrac{1}{2}\alpha,$$

ubi cum sit $\big(2(f-1) + (f(n-1) - 1)\omega\big)\omega = 1 + \alpha f$, inde fit

5) Editio princeps: Numeri paragraphes usque ad finem huius commentationis non accurati sunt. Correxit eosdem EJA.

$$\omega = \frac{1+\alpha f}{2(f-1)+\big(f(n-1)-1\big)\omega},$$

ubi si loco f scribatur 289, fiet

$$\omega = \frac{1+289\alpha}{576+(289n-290)\omega},$$

in quo denominatore loco ω sufficit scripsisse valorem vero proximum, qui est

$$\frac{1+289\alpha}{576},$$

unde obtinebitur valor correctus

$$\frac{(1+289\alpha)\,576}{576^2+(289n-290)(1+289\alpha)}.$$

12. Quoniam igitur ex mensuris variorum graduum meridiani ratio diametri aequatoris ad axem Terrae conclusa est ut 201 ad 200, erit pro prima approximatione $\omega = \frac{1}{200}$, atque hinc valor coefficientis α definiri poterit, cum esse debeat

$$\frac{1}{200} = \frac{1+289\,\alpha}{576},$$

unde igitur fiet $\alpha = \frac{376}{200\cdot 289} = \frac{47}{25\cdot 289} = \frac{1}{154}$, proxime. Superfluum autem foret determinationem magis exactam desiderare, cum ratio assumta 201 : 200 satis notabiliter a veritate recedere possit; quamobrem hinc plus concludere non licet, quam esse propemodum $\alpha = \frac{1}{150}$, unde simul patet, exponentem n prorsus non in computum ingredi.

13. Hinc igitur discimus, ut superficies oceani in statu quo Terra actu reperitur in aequilibrio subsistere possit, necessario requiri, ut in visceribus Terrae singulae particulae non solum ad centrum Terrae in directione ZC sollicitentur, sed praeterea accedat vis lateralis secundum directionem ad ZC normalem agens, cuius quantitas propemodum erit

$$\frac{\Phi}{z} = \frac{\sin\varphi\cos\varphi}{150z},$$

dum scilicet gravitas sub ipso Polo unitate exprimitur. Ita igitur vis lateralis maxima evadet ubi angulus ACZ $= \varphi$ fit semirectus, quippe cui respondebit vis lateralis $= \frac{1}{300z}$, quae ergo in superficie maris ubi fit proxime $z = 1$ evadit $\frac{1}{300}$, ideoque vi centrifugae fere aequalis.

14. Nisi igitur istae vires laterales ita fuerint comparatae, maria in superficie Terrae in aequilibrio subsistere nequeunt, sed in perpetua agitatione versarentur, id quod imprimis intelligendum est, si totus Terrae globus ex materia fluida constaret; tum enim, etiamsi suprema superficies talem figuram accepisset, ut vires totales, quibus singula puncta ibi sollicitantur, essent ad ipsam superficiem normales, tamen quia in maioribus profunditatibus aqua non foret in quiete, mox ille situs perturbaretur, neque igitur tota Terra ad certam figuram se componere posset.

15. Manifestum autem est, talem legem circa vires laterales in ipsa natura locum habere nullo modo posse, quandoquidem pro minimis a centro Terrae distantiis hae vires laterales in infinitum excrescerent, cuiusmodi effectus ab attractione mutua neutiquam oriri potest; unde pro certo affirmare possumus: Si tota Terra esset fluida, eius superficiem nunquam ad ullum statum aequilibrii pervenire posse.

16. Cum autem maxime verisimile sit, maria nusquam tantam profunditatem occupare, ut discrimen nostrae formulae pro S inventae a vera lege attractionis, quaecunque ea fuerit, unquam sentiri queat, ideoque perinde, utrum vires laterales nostram legem sequantur an vero quamcunque aliam, utique fieri poterit, ut universus oceanus in aequilibrio subsistat, siquidem hic ab exiguis agitationibus, quae a plurimis caussis oriri possunt, mentem abstrahamus, cuiusmodi sunt venti, imprimis autem varietas caloris. Cum enim sub aequatore calor perpetuo multo maior sit quam versus polos, quoniam ibi densitas aquae aliquanto minor evadit, aequilibrium etiam ob hanc caussam locum habere nequit, sed per ea, quae in Theoria motus fluidorum demonstravi, aqua suprema perpetuo fluxu ab aequatore versus polos deferri debet, quae autem iactura ab aqua prope fundum a polis ad aequatorem affluente iterum resarcietur. Talis igitur motus oceano aeque est naturalis atque ille, quo perpetuo ab oriente occidentem versus profertur.

DE ATMOSPHAERA LUNAE EX ECLIPSI SOLIS ANNULARI EVICTA

Commentatio 839 indicis ENESTROEMIANI[1])
Opera postuma 2, 1862, p. 391–401

1) Vide introductionem huius voluminis, p. LXVIII. EJA.

RECHERCHES SUR LA DÉCOUVERTE DES COURANTS DE LA MER

Commentatio 850 indicis ENESTROEMIANI
Opera postuma 2, 1862, p. 790–792

1. Pour développer cette question, proposée par l'Académie Royale des Sciences de Paris pour l'année prochaine[1]) 1749, je remarque d'abord, que sans avoir égard ni au ciel, ni au continent ou autres parties de la terre, qui n'ont d'autre mouvement, que celui qui leur est commun avec la terre même, il est absolument impossible de s'apercevoir d'aucun mouvement, que puisse avoir la mer. Car quelque adresse que nous puissions employer pour connaître le mouvement du vaisseau où nous nous trouvons, nous n'en saurions déterminer que son mouvement relatif par rapport à l'eau; mais du mouvement de l'eau même, rien ne saurait être conclû de là.

2. Il n'y a donc que deux voies pour s'apercevoir si la mer a quelque mouvement ou courant, ou non. La première nous est fournie par le ciel, car si nous sommes en état de déterminer par les observations célestes tant la latitude que la longitude du lieu où nous nous trouvons, et que nous puissions aussi exactement assigner la situation du même lieu, par l'estime du sillage que le vaisseau a fait, supposant que la mer soit en repos. Tant que ces deux conclusions ne seront pas d'accord entr'elles, la différence sera attribuée au mouvement de la mer, par lequel le vaisseau aura été transporté dans un autre endroit, que l'estime du chemin parcouru ne paraîtra indiquer. C'est de là que je tirerai le premier moyen de découvrir le mouvement ou le courant de la mer, quoique les conclusions qu'on en peut tirer, soient pour la plupart fort mal assurées.

---

1) Le mémoire présent est resté inachevé et par là inédit; le mémoire de DANIEL BERNOULLI, «Sur la nature et la cause des courans», *Recueil des pièces qui ont remporté les prix de l'Académie Royale des Sciences*, tome 7 (Paris, 1769), fut couronné en 1751. EJA.

3. Le continent ou les terres, qu'on rencontre sur la route qu'on poursuit, fournissent le second moyen de reconnaître les courants de la mer; soit que la situation de ces terres soit connue ou non. Quand le courant de la mer s'étend jusqu'aux côtes de ces terres, ce sera le plus sûr et le plus aisé moyen de s'en apercevoir; mais si le courant ne se trouve qu'en pleine mer, la rencontre des terres tant connues qu'inconnues ne manquera pas de nous éclaircir sur ce sujet, ce que j'examinerai plus au long dans la seconde partie de ce discours.

4. Dans les endroits où il est permis de pénétrer jusqu'au fond de la mer, on concevra aisément qu'il doit être possible d'en tirer sans beaucoup de peine la connaissance des courants s'il y en a, et que ce moyen sera encore le moins sujet à des inconvénients, qui se rencontrent dans les deux autres. Quoique cette méthode semble appartenir à la précédente, à cause que c'est à quelque partie immobile de la terre qu'on rapporte l'état de la mer, néanmoins puisque cette méthode demande une exécution tout à fait différente, j'y destinerai la troisième partie de mes recherches, et qui me paraît renfermer les plus sûrs moyens de découvrir les courants.

5. La raison nous assure, et l'expérience le confirme, que les courants pénètrent rarement jusque'au fond de la mer. Le mouvement est pour la plupart le plus vite à la surface, de laquelle en descendant il décroit insensiblement, jusqu'à ce qu'il s'évanouisse tout à fait au fond même. Peut-être qu'il y a aussi des cas, où les courants ne s'étendent qu'à une profondeur assez médiocre, et alors il y aura des moyens de s'assurer des courants par des expériences faites dans l'eau dormante au-dessous[2]) du courant. Pour cet effet j'ajouterai la quatrième partie, qui, quand même on ne serait jamais assuré que ce cas ait lieu, servira toujours à connaître la différence du mouvement de la mer à la surface, et à quelque profondeur; et quoique cela ne regarde pas immédiatement la question proposée, il contribuera toujours considérablement à la connaissance des mouvements de la mer, à quoi le but de la question paraît s'étendre.

6. Je conviens que plusieurs des méthodes, que je proposerai, seront très imparfaites, de sorte qu'on ne s'en pourra promettre beaucoup de succès. Mais comme la cause de cette imperfection se trouve dans l'imperfection des méthodes et des instruments, propres à faire les observations tant du ciel que du sillage, je ne crois pas que je les aurais dû omettre pour cette raison. Je suis plutôt assuré, que

2) Edition originale et manuscrit: au-dessus *au lieu de* au-dessous. Corrigé par EJA.

l'Académie Royale demande un parfait dénombrement de toutes les manières possibles pour arriver à ce but, soit qu'elles soient praticables dans l'état présent de la marine, ou non. En tout cas l'utilité de ces méthodes sera que, bien qu'on ne saurait rien conclure d'une seule, deux ou plusieurs ensemble, qui conduiront à la même conclusion, ne manqueront pas d'en augmenter la certitude.

PREMIÈRE PARTIE

Des moyens tirés des observations célestes pour découvrir les courants de la mer

7. Je commencerai par considérer un vaisseau, qui n'ait aucun mouvement par rapport à l'eau, ou qui dans un calme paraisse être dans un parfait repos, de sorte qu'il demeure constamment environné des mêmes parties de l'eau. Il est vrai que ce cas est très rare et qu'il ne dure guère longtemps, néanmoins il est à propos de le considérer avant que d'envisager l'autre cas, où le vaisseau a un mouvement par rapport à la mer. Que le vaisseau donc tienne constamment le même lieu sur la mer; et il sera effectivement dans un repos parfait, si la mer sera destituée de tout courant. Dans ce cas il est clair, que les observations célestes marqueront toujours tant la même élévation du pôle, que la même longitude.

8. Donc si nous supposons, qu'on soit en état de déterminer exactement par les observations du ciel, tant la latitude que la longitude du lieu où on se trouve, il s'ensuit réciproquement, que dans le cas proposé, où le vaisseau paraît demeurer toujours au même endroit, si l'on observe toujours la même élévation du pôle et la même longitude, que la mer n'aura aucun courant, mais qu'elle persévère dans un parfait repos aussi bien que le vaisseau, par rapport à la terre. Il est vrai qu'il s'en faut beaucoup qu'on ne soit en état de s'assurer par cette méthode de la tranquillité de la mer; mais je crois qu'il me sera permis de ne pas regarder ici à l'imperfection des manières d'observer, que j'envisagerai comme étant portées au plus haut degré de perfection. Cependant je ne laisserai pas de remarquer à chaque méthode tout ce qui la pourrait rendre impraticable.

9. Dans la même supposition, que le vaisseau demeure en repos par rapport à la mer, si après quelque temps on remarque que la latitude, ou la longitude ou toutes les deux aient changé: on en concluera, qu'il y a quelque courant sur la mer,

et il sera aisé d'en déterminer tant la vitesse que la direction. Car puisque le vaisseau se trouve en repos par rapport à l'eau, ce sera l'eau, qui aura été transportée pendant ce temps d'un endroit à l'autre par son courant. Pour cet effet on n'aura qu'à marquer sur la carte depuis le lieu où le vaisseau était au commencement, la différence de latitude et de longitude qui auront été trouvées par les observations, et on aura l'espace que la surface de la mer aura parcouru pendant le temps qui s'est passé entre les observations, et de là on determinera aisément le courant.

10. Cette détermination aura lieu, quand même le vaisseau sera porté d'un mouvement quelconque par rapport à la mer. Car après avoir déterminé le vrai lieu où le vaisseau s'est trouvé au commencement de cette recherche, on observera soigneusement le sillage pour en connaître, combien on s'est avancé tant en longitude qu'en latitude, et on marquera ce lieu sur la carte. Ensuite on cherchera aussi par les observations du ciel combien on aura fait de chemin tant en longitude qu'en latitude, pour marquer cette place aussi sur la carte. Alors la distance entre ces deux points marqués par ces deux manières différentes donnera d'abord l'espace que la surface de la mer aura parcouru pendant le temps écoulé entre les observations.

11. Cette manière est si claire d'elle même, que je ne trouve pas nécessaire de m'y arrêter davantage; j'ajouterai seulement quelques remarques, qui pourraient servir à en faciliter la pratique, si elle était d'ailleurs applicable. La connaissance du sillage me paraît assez exacte pour en pouvoir déduire le mouvement relatif du vaisseau par rapport à l'eau pendant quelques jours de suite sans se tromper sensiblement, à moins que l'estime ne soit troublée par une forte tempête. Cependant il est à remarquer que quelque petite que soit l'erreur à laquelle on est assujetti, elle devient plus grande, plus longtemps qu'on continue cette estime sans la corriger par quelque observation du ciel. Mais dans le sujet présent les observations ne peuvent pas être employées à ce dessein, puisqu'elles doivent servir à connaître le courant. Par cette raison donc l'intervalle du temps qu'on met entre les observations, ne doit pas être trop grand, ou il ne doit pas excéder quelques jours.

RECENSIO DISSERTATIONIS DE VENTIS

Quae ab Auctore (Cel. D'Alembert) sequente symbolo est insignita:

Haec ego de ventis: dum ventorum ocior alis
Palantes pellit populos Fridericus, et Orbi
Insignis lauro, ramum praetendit Olivae.

Commentatio 851 indicis ENESTROEMIANI
Opera postuma 2, 1862, p. 793–797

Auctor dissertationis[1]), quae prae reliquis praemio digna est visa, quaestionem ab Academia propositam de causa et ordine ventorum tam accurate tantoque studio pertractavit, ut si instituto non penitus satisfecerit, plurimos tamen nodos, quibus ista quaestio est complicata, feliciter resolvisse, atque viam ad veram ventorum theoriam complanavisse sit censendus. Quam ob causam etiam reliquis, qui suas de hoc argumento meditationes cum Academia communicaverunt, palmam praeripuisse est visus, cum horum alii ne ipsam quidem quaestionem recte intellexerint, alii nimis negligenter tractaverint, alii denique intra limites nimis arctos sese continuerint, neque singulas causas, unde venti oriuntur, debita diligentia evolverint, neque effectus ad omnes globi terrestris regiones transtulerint. Ita autem haec eximia de ventis commentatio abstrusissimis calculis ac profundissimis sublimioris analyseos et mechanicae problematibus est referta, ut lectores non solum attentissimos, sed etiam in subtiliori Analysi probe versatos requirat; hancque ob rem praecipua capita huius dissertationis modumque, quo phaenomena ventorum explicare et determinare conatur, breviter et quantum potero dilucide enarrabo.

---

1) *Réflexions sur la cause générale des vents.* Pièce qui a remporté le prix proposé par l'Académie R[oyale] des sc[iences] et belles-lettres de Prusse pour l'an 1746; par M. D'ALEMBERT. Berlin 1747 petit in-4to.

Primum igitur, antequam ullum atmosphaerae motum examini subiicit, eius statum aequilibrii definit, figuramque determinat, ad quam sese, tam ob gravitatem quam ob vim centrifugam, componere deberet: quam invenit sphaeroidicam ad polos aliquantum compressam. Hanc ergo figuram atmosphaera perpetuo esset conservatura, nisi aliae causae accederent, quae istum aequilibrii statum turbare valeant. In hoc autem primo problemate copiosius versatur, et dum rem generaliter prosequitur inquirit, quid evenire deberet, si forte copia materiae atmosphaerae nimis esset parva, quam ut exuberantiae illi sub aequatore formandae sufficeret, quo loco ad considerationes non parum curiosas digreditur. Deinde statu aequilibrii definito, perpendit quomodo atmosphaera primum in eum pervenerit, seu quo motu etiamnum, si a causa quacunque figura immutaretur, ea in statum aequilibrii reverti debeat. Ostendit autem hoc casu totam atmosphaeram motum quendam reciprocum seu oscillatorium instar penduli esse recepturam, quo demum amisso ad quietis statum sit perventura. Hunc igitur motum oscillatorium, quo in mechanica sublimiori vix quicquam solutu difficilius excogitari potest, admirabili sollertia determinat. Ad hoc methodo plane nova utitur, quae in aliis quaestionibus cum mechanicis tum hydrodynamicis maximum usum habere potest. Hactenus autem vim gravitatis ubique ad centrum terrae directam, ac certam distantiarum rationem tenentem assumserat; nunc igitur idem argumentum accuratius aggreditur, atque gravitatem tanquam vim ex singulis attractionum viribus, quibus partes materiae in se invicem agere concipiuntur, resultantem contemplatur, ita ut nunc tam directio quam quantitas ab ipsa figura terrae plurimum pendeat. Non solum autem particulis terrae huiusmodi vim attractricem tribuit, sed etiam particulis, ex quibus ipsa atmosphaera est composita. Nemo autem non videt hoc modo tam quaestionem de statu aequilibrii, quam de motu illo oscillatorio, quo atmosphaera statum aequilibrii sit assecutura, multo fieri difficiliorem: nihilo tamen minus Auctor noster et has difficultates superavit. Calculum hunc ubique generaliter instituit, ut ex eo etiam figura cuiusvis corporis, puta planetae definiri possit, si materia liquida nucleum solidum ambiat; observat autem fieri posse, ut etiamsi nucleus sphaericus assumeretur, huiusmodi corpori tamen a motu vertiginis figura sphaeroidica allongata induceretur: hoc scilicet pendet a certa tam densitatum quam quantitatum materiae fluidae et solidae ratione. Huiusmodi autem meditationes hoc praesertim tempore, quo tam eximia opera de figura terrae extant edita, maxima attentione utique dignae sunt habendae, etiamsi ad praesens institutum minus pertinerent, verum Auctor eas cum tota tractatione tam arcte colligavit, ut superfluae videri nequeant. Si enim peregrina causa accedat, quae in aërem agendo statum aequilibrii turbet, tum statim motus ille oscillatorius concitabitur, qui dum a causa continuo urgente compescitur, in motum perennem, cuiusmodi est ventus, abibit. Inducit autem Auctor primum solem ac lunam respectu centri terrae quiescentes eidemque regioni constantes imminentes, ex eorumque vi attractiva, quam in aërem exercent, motum aëris hinc oriundum determinat. Praeterea vero globum terrae solidum ac sphaericum, seu saltem proxime talem assumit, aëremque elateris adhuc expertem[2]) contemplatur, ut ab his viribus motum fere similem aestui maris recipere debeat. Has scilicet hypotheses a veritate abhorrentes ideo fingit, ut ventorum determinatio ex praemissis fiat facilior, indeque transitus ad ipsam veritatem commodius suscipi queat. Primum igitur, quod ad motum terrae vertiginis attinet, quo atmosphaera sub aequatore altior redditur quam sub polis, si motus huius initio figura atmosphaerae fuisset sphaerica, statim aër a polis ad aequatorem fluere inceperit necesse est, et quia ob motum conceptum maior copia qua par est, ad aequatorem deferetur, iterum inde ad polos defluet, sicque motu reciproco venti boreales et australes se alternatim excipere debebunt. Tempus huiusmodi reciprocationis seu oscillationis aëris fere unius diei ope calculi ante traditi determinat. Sol deinde vel luna si accedat, quia in loco, cui imminet, huicque opposito aërem intumescere facit, similem motum reciprocum aëri inducet, cuius directio semper erit in plano verticali, quod in quovis terrae loco per locum sive solis sive lunae ducitur, huiusque motus tam tempus unius reciprocationis, quam celeritatem quovis momento determinat. Et si autem hos motus, qui a motu terrae diurno et a viribus attractricibus solis et lunae seorsim pendent, definivit, tamen exinde non difficulter motum aëris ab his causis coniunctim agentibus oriundum colligit; ei autem, quia a veritate adhuc ob nimis liberam hypothesium fictionem vehementer abludit, fusius non inhaeret; sed hoc loco momenta quaedam in physica maxime notatu digna attingit.

Ostendit nimirum hanc solis et lunae actionem in altitudine barometri nullam sensibilem mutationem efficere valere, etiamsi aër elasticitate careret. In quo a Celeb. Daniele Bernoulli dissentit, qui in tractatu de aestu maris satis ingentem mutationem in pressione aëris ab actione solis et lunae oriri debere statuit, si quidem elater abesset: ob aëris autem elasticitatem fieri demum existimat, ut talis mutatio imperceptibilis reddatur. Argumenta igitur Bernoullii Auctor noster omni studio evolvit, atque ostendit theoriam suam inde nullum detrimentum pati. In his autem, quae hactenus sunt proposita, Auctor primam tractationis suae partem constituit, quae fundamenti loco duabus sequentibus partibus inserviat.

2) Secundum Manuscriptum. Editio princeps: expertam loco expertem. EJA.

Hic scilicet aërem elateris expertem atque luminaria eidem constanter terrae loco imminentia assumsit, qua hypothesi tantum primo initio atmosphaera motum oscillatorium induere debuit, qui autem mox cessare et in statum aequilibrii desinere debuisset.

Nunc igitur in parte secunda tam soli quam lunae verum motum tribuit, quo fit, ut aër nunquam ad statum aequilibrii pervenire possit. Motum vero terrae vertiginis non amplius inter causas ventorum numerat, quoniam est uniformis, aërque iamdudum ab eo in statum quemdam permanentem redigi debuit, quo saltem in infima regione nullus ventus sentiri posset. Neque etiam aberrationis figurae terrae a sphaerica rationem habet, quoniam est valde parva, nullumque sensibile discrimen tam in directione quam celeritate venti producere valet. Primum igitur animadvertit, quoniam vires, quibus singulae atmosphaerae particulae urgentur, sunt inaequales, motum iis induci non posse, quin altitudo atmosphaerae fiat minor, ubi celeritas maior deprehendatur, maior autem, ubi aëris motus sit tardior. Ob hanc autem causam determinatio motus ex actione vis sive solis sive lunae oriunda fit longe difficillima, unde quo facilius scopum attingere possit, singulas primo aëris particulas tanquam a se invicem solutas, ut aliae aliarum motus turbare nequeant, considerat atque earum motus investigat. Ostendit vero hoc casu aërem sub aequatore constanter ab ortu in occasum moveri debere; in locis autem ab aequatore remotioribus motum fore compositum ex motu orientali et alio, cuius directio in planum meridiei incidat. Celeritatem autem hinc adhuc longe minorem invenit, quam revera sub aequatore deprehenditur. Observat autem multum interesse, sive superficies ipsius terrae sit dura, sive fluida: in hoc enim casu, quo terra mari cirumdata ponitur, aqua quoque viribus sollicitantibus cedit, sicque spatium, quod ab atmosphaera occupatur, modo augendo, modo diminuendo ipse aëris motus hinc non exiguam mutationem subibit. Ad hoc discrimen accuratius definiendum, totamque tractationem propius ad veritatem revocandam, nunc globum terrae solidum, aërum autem unum continuum fluidum terram ambiens assumit, eiusque motum ab actione solis vel lunae oriundum indagat, quem iterum sub aequatore ab ortu in obitum directum invenit. Notat autem si altitudo atmosphaerae eiusque densitas alia esset, atque revera est, fieri potuisse, ut ventus contrarius oriretur, atque ut atmosphaera non sub astro sed in regionibus 90 ab eo gradibus remotis intumesceret, quod insigne paradoxum uberius exponit causamque in inertia aëris tum sitam fore declarat. In statu autem aëris naturali atmosphaera semper figuram sphaeroidicam allongatam habebit, cuius maior axis per solem vel lunam transeat, et cum differentia valde esset exigua, si sol vel luna quiesceret, ita ut hinc actio attractiva solis vel lunae ulli vento producendo impar sit visa: Auctor docet ob motum solis vel lunae elongationem atmosphaerae notabiliter augeri, eiusque tumorem sub luminaribus longe fieri maiorem, ita ut hinc modicus ventus, qualem experientia exhibet, oriri queat. Hactenus vero, quoniam Auctor potissimum motum aëris sub aequatore spectaverat, ad caclulum facilius expediendum, assumserat motum ubique fieri secundum directionem in plano verticali per astrum ducto positam: nunc igitur problema difficillimum aggreditur, ista relicta hypothesi veram motus aëris directionem pro singulis regionibus investigaturus. Quanquam autem pluribus adhibitis artificiis totum negotium ad resolutionem aequationum reduxit, tamen ad formulas tantopere implicatas pervenit, ut vera venti directio tantum proxime ex iis colligi possit: quantum autem inde concludere licet, experientiae satis est consentaneum, ut de veritate theoriae non dubitandum videatur. Animadvertit tamen circumstantiam notatu dignissimam, problema scilicet hoc non penitus esse determinatum, quia motus aëris non solum a viribus iugiter sollicitantibus pendeat, sed etiam a motu ipsi primo impresso: qui cum sit incognitus, calculus non aliter nisi per experientiam ad veritatem accomodari potest. Cum autem hactenus superficiem terrae duram et immobilem considerasset, nunc terram undequaque oceano opertam contemplatur; densitatem autem aëris pro variis altitudinibus utcunque variabilem assumit, unde ad resolutionem plurium aequationum valde intricatarum deducitur. Hic occasionem arripit de integratione aequationum differentialium altiorum ordinum agendi, ubi nonnulla elegantia et utilia artificia patefacit, quae ad analysin promovendam non parum conferre videntur: tum vero etiam praeclaras quasdam notationes circa quantitates imaginarias, earumque, quando formulas integrales afficiunt, reductione ad arcus circulares tradit. In hoc autem problemate, cuius solutionem in aliquot aequationibus complexus est Auctor, contineri putat universam de ventis quaestionem, ideoque varias evolvit hypotheses densitatis aëris, quae ad statum naturalem proxime accedere videantur, indeque complicatissimarum aequationum integrationes perficit, et quantum rei difficultas permittit ad veram venti directionem et celeritatem ubique locorum definiendam accomodat. Progreditur deinceps ad motum aëris non liberum determinantium, casumque primo examinat, quo sub aequatore duae series altissimorum montium sibi parallelae existant, intra quos aër alium motum praeter orientalem vel occidentalem recipere nequeat. Deinde has series montium ab aequatore ad quemvis parallelum transfert, ita ut semper ab ortu in occasum extendantur, et ad quodvis tempus venti celeritatem determinat, cum directio per se sit cognita. Tum has montium series secundum duos meridianos sibi proximos extensas contemplatur, et difficillimo calculo celeritatem venti assignat, quae omnia ita sunt comparata, ut si debito studio evolvantur, in omni terrae statu, quo motus aëris a montibus coërcetur, inde ventus definiri posse videatur. Multo magis autem arduum fit problema, quando binae illae montium series inter se non sunt parallelae, interim tamen tanta sagacitate Auctor pollet, ut etiam has novas difficultates maximam partem superare potuerit. Concludit denique hanc satis amplam tractationem problemate generali, quo methodum exponit in hypothesi, quod terra profundo oceano undique tecta sit, pro quovis loco et tempore verum aëris motum, hoc est directionem ac celeritatem venti determinandi, quod ope praecedentium problematum fieri posse luculenter ostendit.

Quoniam vero hactenus causas ventorum in solis viribus attrahentibus solis et lunae quaesivit, nunc demum in effectum caloris solis, cui sine dubio primariae partes in productione ventorum sunt tribuendae, inquirit. Causam autem et effectum caloris ita comparatum iudicat, ut ad calculum accurate revocari nequeat: proxime autem scopum attingere arbitratur, si rarefactionem aëris perpetuo a quadrato sinus distantiae solis a zenith pendere, seu esse proportionalem quadrato sinus elevationis solis super horizonte, quemadmodum ergo hic effectus in formulas ante exhibitas introduci debeat, ostendit. Quia vero calor ac proinde aëris rarefactio non solum ab altitudine solis, sed etiam a tempore, quo idem locus actioni solis iam erat expositus, pendet, ita ut pro eadem solis altitudine calor post meridiem maior sit quam ante meridiem; ab hac diversitate alium effectum non proficisci statuit, nisi constantem atmosphaerae motum ab oriente in occasum, qui cum omnibus aliis ventis a causis ante perpensis oriundis coniungi debeat, hocque modo non solum ventos orientales intra tropicos spirantes, eorumque a vero oriente deflexiones statis temporibus observatas satis dilucide explicat, sed etiam quales venti in reliquis terrae zonis regnare deberent, nisi perturbatio a terris orta accederet, non obscure indicat. Quas ob causas hic Auctor quaestioni propositae multo accuratius satisfecisse est visus, quam ab aliis factum esse deprehendimus. Postquam autem haec dissertatio huc esset allata, eius Auctor misit additamentum, in quo nonnulla loca clarius explicat, et levem quemdam errorem, qui in calculum § 89 ingresserat, emendat: imprimis autem in hoc additamento tam celeritatem quam directionem venti sub quovis parallelo commodius exprimit, dum sol vel luna non in aequatore sed circulo quocunque parallelo circa terram revolvitur. Motum venti ubique resolvit secundum directiones paralleli et meridiani; quarum illa praebet plagam orientis, haec vero continuum affluxum aëris a polis ad aequatorem arguit: qui tamen effectus, cum tandem omnis aëris massa ad aequatorem accumularetur, admitti non potest. Interim tamen hoc incommodum ita conciliat, ut motum affluxus aequatorem versus continuo imminui dicat; calculus enim declarabat celeritatem secundum directionem meridiani non solum sub aequatore evanescere, sed etiam sub polis, et in zonis tantum temperatis esse notabilem. Fortasse etiam in regionibus atmosphaerae supremis aëris portio quaedam ab aequatore versus polos defluit, sicque aequabilem aëris statum conservat. Verum utcunque fiat haec restitutio, quae proprie ad quaestionem propositam non pertinet, consensus experientiae ad theoriam confirmandam sufficit; ea enim constat, prope tropicos, ubi venti adhuc constantes deprehenduntur, ventos orientales in hemisphaerio quidem boreali boream versus, in altero autem austrum versus aliquantum declinare: ita ut dubium sit nullum, quin haec declinatio in locis ab aequatore magis remotis adhuc maior esset futura, nisi totus ventorum ordo a vicinitate continentis interrumperetur. Notatu autem maxime dignum est, quod Auctor invenit, istam declinationem ab aequatore tantum usque ad certum latitudinis gradum crescere, ultraque eum iterum decrescere, atque sub ipsis polis aeque ac sub aequatore denuo evanescere. Secundum hanc ergo Auctoris theoriam ventus pro quovis loco ac tempore proposito ita definiri debebit. Quaeratur primo ventus ex calore solis oriundus, habita simul ratione eius venti constantis, qui ob gradum caloris pomeridianum maiorem antemeridiano locum habet. Deinde seorsim determinetur ventus, quem vis attractiva cum solis tum lunae producit, cognitisque his singulis ventis tam ratione directionis quam ratione celeritatis non erit difficile secundum regulas compositionis motus ventum ex iis ortum proxime saltem assignare.

DE CAUSA GRAVITATIS

Commentatio ab indice ENESTROEMIANI abest
Miscellanea Berolinensia 7, 1743, p. 360–370

Gravitatio corporum in se mutua tam latissime patet, ut plurimi philosophi non dubitaverint eam ad primas materiae proprietates, cuiusmodi sunt extensio et inertia, referre, mutuamque corporum sese attrahentium actionem a viribus insitis deducere. Cum enim corporibus extensio et inertia ideo tribuantur, quod omnibus constanter et perpetuo inesse deprehendantur, similis fere ratio gravitatis videbatur, propterea quod vix ullum adhuc sit observatum corpus, quod gravitatione penitus careret. Non solum enim omnia corpora circa terram posita deorsum nituntur, sed etiam terram ipsam cum reliquis planetis solem versus urgeri, observationes astronomicae ad leges Mechanicae examinatae luculenter docuerunt; quin in minimis etiam corporibus, quae quidem experimentis subiici possunt, similis quaedam vis sive magnetica sive electrica quotidie magis magisque evincitur. Sane in Astronomia attractionis hypothesis utiliter adhibetur in explicandis gravitatis phaenomenis. Planetae enim et Cometae perinde moveri observantur, acsi cum a sole tum a se mutuo attraherentur, iisdemque legibus progrediuntur, quas eis ex hac hypothesi Vir summus NEWTONUS praescripsit. Mathematico ad determinandam quantitatem effectuum et virium minime opus est, ut qualitates et causas earum rerum perspectas habeat, adeoque omni iure utitur termino attractionis ad indigitandam causam, etsi hanc prorsus ignoret. Cumque in natura effectus unus sit causa alterius effectus, Physicus rationem effectus ab alio, cuius causa ipsum fugit, redditurus, hunc eius causam allegare, adeoque ex attractione tanquam ex indubitabili Phaenomeno aliorum Phaenomenorum rationem reddere tenetur. Imo etiamsi vel maxime vera attractionis causa detecta fuerit, non tamen in explicandis singulis eius effectibus ab ista causa remota ratio petenda esset, quae ab effectu eius observabili tanquam a proxima merito derivatur. Ast battologiam a CARTESIO feliciter profligatam in Philosophiam iure quasi postliminii reducunt, sibique et aliis viam ad solidiorem rerum naturalium cognitionem praecludunt, quotquot de investiganda gravitatis causa desperantes nodum gordium secare conantur, dum attractionem vim primitivam corporibus inhaerentem esse, acriter contendunt. Nimirum ex notione effectuum seu Phaenomenorum attractionis, quae sola observatione innotescunt, per praecipantiam fingunt eandem causae notionem, idemque per idem explicantes committunt circulum maxime vitiosum. Attractionem enim concipiunt per vim, qua corpora ad se invicem urgentur, et gravitas ipsis est vis, qua corpus fertur versus centrum terrae; rationem autem reddituri cur id fiat, affirmant, id fieri propter attractionem seu gravitatem, quod idem est, ac si affirmarent, corpora ad se accedere mutuo, quia mutuo ad se accedunt, corpora tendere ad centrum terrae quia ad centrum terrae tendunt. An luculentius circuli vitiosi afferri possit exemplum, merito dubitandum. Ex primis Philosophiae principiis, seu ex notionibus communibus, quibus axiomata omnia constant, demonstratum est, in causae notione contineri notionem principii, quod nec admitti nec fingi debet, nisi ex eius determinationibus, per quas concipitur, effectus inde dependens sit explicabilis; alias Philosopho quidlibet pro lubitu fingere liceret. Causa itaque et principium differunt ab effectu et principiato, uti determinans a determinato. Admodum diversae sunt istae notiones, resque per illas repraesentatae; trianguli latera, quibus anguli determinantur, non sunt ipsi anguli. Ergo nec attractio et gravitas seu phaenomena, quae hoc nomine compellantur, sunt eorum causa. Non solum vero admissa attractione seu gravitate instar vis corporibus insitae mirum in modum confunduntur notiones causae et effectus, sed et praeterea pullulant innumerae difficultates inextricabiles. In hac enim hypothesi attractio producit omnes suos effectus absque omni contractu et impulsu et sine ullo concursu sive materiae propriae ex corporibus emanantis, sive alterius materiae intermediae: Stabilitur itaque eo ipso realitas adeoque multo magis possibilitas actionis in distans, quae fieri concipitur absque ullo contactu. Non ergo repugnat Sonum ad aures meas deferri sine aeris interventu; conspici a me objecta, quae ob infinitam locorum intercapedinem organa mea sensoria tangere nequeunt; herbas in planeta Saturno crescentes vi attractrice in se incomprehensibili in corpore meo vomitus aliosque vehementes intestinorum motus excitare, etiamsi neque nares neque os quicquam inde hauserint. Admissa enim corporum actione in distans sine impulsu, ex maiori vel minori eorum distantia ratio diversitatis effectuum reddi nequit. Concedunt attractionis patroni, corpora eo quod sint entia composita extensa et mobilia, in conflictu adeoque per contactum iuxta regulas motus infinitis mutationibus obnoxia esse, eamque ob causam ea tanquam machinas considerari posse ac debere. Nec minori evidentia demonstratum est, nullam in composito adeoque corpore accidere posse mutationem, nisi quoad figuram, magnitudinem partium, et locum totius, omnesque inde proficiscentes modificationes ex motu originem trahere. Sane scientia Physices ex eo demum tempore incrementa cepit, quo, explosis qualitatibus occultis, Philosophia Mechanica invaluit, quippe quae, facem praeferente Geometria, modificationes corporum ad causas suas revocans in eo versatur, ut ex eorum structuris texturis et mixtionibus, id est, ex modo compositionis mutationes inde pendentes secundum leges motus intelligibili modo explicet. Harum legum prima est, quod corpus unumquodque perseveret in statu suo quiescendi vel movendi uniformiter in directum, nisi quatenus illud a viribus impressis cogitur statum suum mutare. Quoniam veritas istius propositionis per exempla patet, sumunt eam Mathematici instar axiomatis, illique tanquam firmissimae basi arduam motus Scientiam superstruunt. Negari tamen nequit, evidentiam huius axiomatis pendere ab alio antiquiore, quod ex nihilo nihil fiat, seu ex principio rationis sufficientis; hoc enim sublato nihil amplius obstat, quo minus mobile ipsummet suam mutet celeritatem atque directionem. Evanescit itidem evidentia utriusque axiomatis in conceptu attractionis instar vis primitivae, corpori insitae, quippe per supra notata attractio seu gravitatio Phaenomenon tantum nihilque praeterea denotat, causam vero mentitur termino inani, cui nulla notio, id est nihilum respondet, nisi fingendi licentiam tibi arrogans in gratiam attractionis corporibus animam quandam inque ea appetitum, si non rationalem, saltem sensitivum, amorem vel instinctum, consequenter in corpore duplicem Mechanismum concipere velis, quorum unus ex mechanicis explicabilis legibus contiguitatem et contactum necessario requirit, alter vero absque contactu motum adeoque mechanismum immaterialem generat: In hac igitur hypothesi gravitatio aeque ac extensio omnibus corporibus tam sunt propriae, ut etiamsi corpus quodcunque per se a reliquis corporibus remotum spectetur, gravitate tamen tanquam essentiali attributo carere nequeat, unde et praeterea prono alveo fluit, gravitationem pariter ac motum concipi per se in uno corpore absque ulla relatione ad alia corpora, hoc est extra omnem nexum et Systema mundi materialis. Etsi autem ex hoc labyrintho exitus nullus pateat, attractionis tamen quidam defensores in eo ultimum praesidium quaerunt, quod nemo adhuc causam physicam gravitationis, quae omnibus phaenomenis satisfaciat, detegere potuerit. Fatentur nihilominus iniquissimam atque Philosopho indignam sententiam esse, cuius effectus causam assignare nequeunt, eiusdem omnino causam dari negare. Videamus igitur an gravitatis causa mechanica indicari possit? Quid tentasse nocet?

In Mechanica ex prima motus lege demonstratur, a corpore A non posse moveri corpus B nisi ipsum A in motu sit constitutum, consequenter in nullo corpore status mutationem concipiendam nisi per conflictum, quo fit ut unius status conservari nequeat salvo alterius impingentis statu. Inde ostendi posset, in omnibus, quibus mundus materialis componitur, partibus, statu quidem discrepantibus, inter se tamen arctissimo nexu cohaerentibus, continuam mutationem consequi, ita ut nullum corpus vel puncto temporis in statu suo conservetur. Dum igitur corpus, quod etsi circa terrae superficiem quiescit, tamen deorsum nititur, a vi quadam externa deorsum urgeatur, sanciunt leges motus. Haec pressio, cum a nulla materia visibili efficiatur, ab invisibili ortum trahat necesse est, quae in corpus vel directe impingendo vel ob vim centrifugam premendo totum nisum producat: Tertium causae mechanicae modum dari non suspicantur attractionis Patroni. Verum gravitatem ab allisione materiae subtilis deorsum incurrentis oriri recte negant, neque fere minori iure explicationes a vorticibus petitas, in quibus corpora per vim centrifugam deorsum pellantur, reprehendunt, quoniam hac ratione vel phaenomenis non fit satis, vel ipsi vortices tam complicati excogitari debent, ut simplicitati, quam natura constanter affectat, nimium adversentur. Hinc igitur, cum praeter binos hos explicandi modos tertium non dari autument, merito sibi concludere videntur, gravitatis causam mechanicam seu explicabilem omnino non dari. Quamvis autem ista, quam faciunt, causarum enumeratio perfecta videatur, tamen in eo peccat iudicium, quod gravitas immediate ex uno vortice oriri assumatur, cum esse queat effectus mediate ex pluribus vorticibus oriundus. Ex minimis autem vorticulis primum elasticitatem provenire posse dubitare vix licet, atque hinc summa aetheris elasticitas ortum habere merito videtur. Sub aetheris nomine intelligo fluidum subtilissimum universum mundi spatium ita implens, ut vix dentur pori ab eo vacui: cuius fluidi existentiam praeter alias rationes radiorum lucis sive emissio sive propagatio evidentissme comprobant. Simul vero ex his phaenomenis perspicitur, universum aetherem in statu maximopere compresso versari, cuius compressionis causa, si cui minus recte vorticulis tribui videatur, is saltem aetherem tanquam aquam undique maxima vi comprimi concedet; quicquid autem sit, nihil contra leges philosophandi saniores mihi quidem assumere videor, si aetherem esse fluidum undique circumfusum ac summopere compressum statuam; haec enim hypothesis tum experimentis et observationibus maxime est consentanea, tum a suspicione qualitatum occultarum, qualis est attractio, alienissima. Cum igitur aether vehementer sit compressus, omnia corpora, quae in eo versantur, undique ingenti vi comprimantur necesse est, a quo effectu duritiem et cohaesionem corporum oriri verisimillimum videtur; corpus autem sic undique compressum ad motum non ciebitur, nisi pressiones, quas quaquaversus sentit, fuerint inaequales. Quamobrem si aether ubique aequali vi esset compressus, in corporibus nullus prorsus motus generaretur; sin autem compressiones aetheris fuerint inaequales, tum utique corpora sollicitari deberent eam regionem versus, ubi compressio aetheris fuerit minor. Quodsi ergo causa investigari posset; ob quam compressio aetheris minor esset futura prope terram, quam alibi, causa gravitatis eo ipso foret assignata. Cum autem aether sit fluidum subtilissimum compressiones per eum ita aequaliter distribui debere videntur, ut eiusmodi inaequalitas, qualis ad gravitatem producendam requiritur, locum habere nequeat; statim enim ac compressio uno in loco fuerit minor quam in adiacentibus, in his compressio relaxatur, atque quasi puncto temporis in aequilibrium reducetur. Verumtamen ex hoc ipso fonte explicari poterit, quemadmodum inaequalitas in aetheris compressione locum habere queat; cum enim si eiusmodi inaequalitas affuerit, motus in aethere subsequi debeat, ita vicissim, si motus in aethere adsit, ibidem compressio minor sit necesse est. Quoniam namque motus generari non potest sine dispendio virium, si aetheri motus prope terram tribuatur, iste motus neque produci neque conservari poterit, nisi compressio aetheris his in locis diminuatur. Quod etiamsi ex natura motus fluidorum ostendi posset, tamen facilius per observationes in aere et aqua agnoscitur, ubi compressiones, quamprimum motus accesserit, fiunt inaequales. Cum igitur valde sit probabile, aetherem circa terram in continuo motu esse positum, quod cum plurima alia phaenomena suadent, tum vero maxime virtus terrae magnetica evincit, nihil, quod legibus motus non sit conforme, assumere mihi equidem videor, si aetheris compressionem prope terram minorem statuam, quam in regionibus a terra remotis, ubi iste aetheris motus minus viget. Hac autem inaequalitate compressionis in aethere admissa, omnia corpora, quae circa terram sunt posita, maiori vi deorsum prementur, quam sursum, propterea quod supra corpora aetheris compressio maior est quam infra ea, et hanc ob rem ad descensum sollicitabuntur. Simili modo si aether prope solem in motu continuo fuerit constitutus, gravitatio planetarum versus solem hinc modo satis plano explicabitur. Effectus autem magnetici perpetuum aetheris motum vorticosum circa tellurem declarant, cuius generationem et conservationem alia occasione ex legibus motus explanabo, unde manifestum est, quae causa virtutem magneticam producat, eandem gravitatis principium in se continere; quae cum ob simplicitatem et conformitatem naturae maxime conveniant, ob hoc ipsum nullum dubium relinquunt, quin causa gravitatis in diminutione vis, qua aether prope terram comprimitur, collocari debeat. Quo autem gravitas ad terram appropinquando crescat in ratione duplicata viciniae a centro, quam legem phaenomena indicant, necesse est, ut diminutiones compressionis aetheris in ratione distantiarum simplici decrescant, quae ratio cum sit omnium simplicissima, simul veritatem huius explicationis non mediocriter confirmat. Sit compressio aetheris absoluta seu non diminuta $= c$, erit ea in distantia a centro terrae x aequalis c minus quantitate ipsi x reciproce proportionali; ponatur ergo compressio aetheris in distantia x a centro terrae C $= c - \frac{cg}{x}$. Hinc si corpus AABB circa terram versetur, eius superficies superior AA deorsum premetur vi $= c - \frac{cg}{\mathrm{CA}}$, inferior autem superficies BB sursum premetur vi $= c - \frac{cg}{\mathrm{CB}}$, quae vis cum minor sit quam prior, corpus deorsum premetur

$$\mathrm{vi} = cg\left(\frac{1}{\mathrm{CA}} - \frac{1}{\mathrm{CB}}\right) = \frac{cg \cdot \mathrm{AB}}{\mathrm{AC} \cdot \mathrm{BC}}.$$

Cum igitur magnitudo corporis AB sit incomparabiliter minor quam distantia CB, erit AC = BC, hinc eiusdem corporis gravitas in distantia a centro quacunque erit ut $\frac{1}{\mathrm{AC}^2}$ hoc est reciproce ut quadratum distantiae a centro. Similis ergo erit ratio diminutionis compressionis aetheris circa solem atque planetas, propterea quod vis, qua corpora eo urgentur, pariter reciproce est proportionalis quadratis distantiarum. Aetheris vero constitutionem prope quaevis corpora alterari, inflexio radiorum quae observatur, dum radii praetereunt, non obscure indicat. Cum enim radii in motu aetheris undulatorio constent, iste autem motus a compressione aetheris pendeat, evidens est, ubi radiorum inflexio eveniat, ibi quoque compressionem aetheris diversam esse oportere. Ex hoc vero fonte vis electrica corporum pariter atque adhaesio aquae ad vitrum in tubulis capillaribus, aliaque similia phaenomena explicationem facilem nanciscuntur; quae omnia attentius perpendenti sponte fient clariora.